Molecular and Cellular Biology of the Allergic Response

CLINICAL ALLERGY AND IMMUNOLOGY

Series Editor

MICHAEL A. KALINER, M.D.

Medical Director
Institute for Asthma and Allergy
Washington, D.C.

1. Sinusitis: Pathophysiology and Treatment, *edited by Howard M. Druce*
2. Eosinophils in Allergy and Inflammation, *edited by Gerald J. Gleich and A. Barry Kay*
3. Molecular and Cellular Biology of the Allergic Response, *edited by Arnold I. Levinson and Yvonne Paterson*
4. Neuropeptides in Respiratory Medicine, *edited by Michael A. Kaliner, Peter J. Barnes, Gert H. H. Kunkel, and James N. Baraniuk*

ADDITIONAL VOLUMES IN PREPARATION

Provocation Testing in Clinical Practice, *edited by Sheldon Spector*

Molecular and Cellular Biology of the Allergic Response

edited by
Arnold I. Levinson
Yvonne Paterson
University of Pennsylvania School of Medicine
Philadelphia, Pennsylvania

Marcel Dekker, Inc. New York • Basel • Hong Kong

Library of Congress Cataloging-in-Publication Data

Molecular and Cellular biology of the allergic response / edited by Arnold I. Levinson, Yvonne Paterson.
p. cm. — (Clinical allergy and immunology; 3)
Includes bibliographical references and index.
ISBN 0-8247-8876-1 (alk. paper)
1. Allergy—Pathophysiology. 2. Molecular immunology. 3. Cellular immunity. I. Levinson, Arnold I. II. Paterson, Yvonne. III. Series.
[DNLM: 1. Immunity, Cellular. 2. Hypersensitivity—immunology. 3. IgE—immunology. 4. T-Lymphocytes—immunology. 5. Antigens—immunology. QW 568 M7175 1994]
QR188.M64 1994
616.07'9—dc20
DNLM/DLC
for Library of Congress 94-4449
CIP

The publisher offers discounts on this book when ordered in bulk quantities. For more information, write to Special Sales/Professional Marketing at the address below.

This book is printed on acid-free paper.

Marcel Dekker, Inc.
270 Madison Avenue, New York, New York 10016

Current printing (last digit):
10 9 8 7 6 5 4 3 2 1

PRINTED IN THE UNITED STATES OF AMERICA

Series Introduction

The decision to initiate a series of books on clinical allergy and immunology was based upon the need to create a library of texts useful for both clinicians and scientists in these rapidly enlarging fields. There already are excellent textbooks providing overviews of the fields of allergy and immunology, and the scientific journals attempt to provide concise reviews of selected topics of interest. However, there is no library of books that takes relevant topics and expands them into texts, for the express purpose of making them of interest to both clincians and scientists. Thus, this new series.

Clinical Allergy and Immunology will develop into the premier series of texts for our field. The initial book, edited by Howard Druce, was directed at sinusitis and represents the type of amalgamation of pathophysiology with treatment that will be reflected in most of the books yet to be published. The second book was an extremely timely discussion of the eosinophil by Gleich and Kay. It, too, will be a useful addition to many clinicians' or scientists' collections. The fourth book, to be released in early 1994, is a scientific review of neuropeptides and their relationship to the upper and lower airways. This book combines the efforts of 30 scientists and clinicians, and should prove to be a standard reference in this area for some years to come.

The series will subsequently include books focused upon areas of allergy, immunology, specific diseases and important clinical entities, developing areas of research relevant to clinicians, major changes in therapeutic approaches, and research areas that warrant a text. The benefactor of this series should be the patient, because physicians will now have a series to seek when searching for concise but authoritative summaries of a field. The other major benefactor will be the clinician. Despite the fact that allergy is the single most prevalent chronic disease suffered by humans, only a small number of medical schools incorporate allergy in their curricula, and medical students are notably deficient in the knowledge of these diseases. As allergy is largely an outpatient specialty, few house officers see allergic patients other than asthmatics or those experiencing anaphylaxis. To clinicians, this series offers the chance to develop an extensive knowledge about selected topics, each chosen for its clinical relevance.

Immunology is the field of the future. Our understanding of immune processes and their relevance to health and disease is advancing at a breakneck pace. As therapeutic approaches become defined, the relevant immunological advances will become incorporated into texts in this series.

Molecular and Cellular Biology of the Allergic Response, edited by Levinson and Paterson, addresses the basic biology of the allergic response and puts the most timely scientific advances into perspective. This book aims to provide scientific explanations for the underlying mechanisms in allergy, and to do so in a way that both scientists and clinicians will appreciate. It discusses each of the basic components of the allergic response and provides well-written presentations from experts on each of the topics. The focus on the molecular biology of each of these presentations is unique and allows the reader to enter into the next era of science as it applies to allergic biology. This reference should find its way into the libraries of anyone who wishes to get into the underlying immunological and molecular bases for allergy. It is a valuable and appropriate addition to this series.

Michael A. Kaliner

Preface

Allergic disease afflicts approximately 28% of the population at some time during their lives. It accounts for a substantial amount of morbidity and, in some cases, mortality. We, as editors, are using the words "allergic disease" in a literal sense and, in so doing, refer only to those disorders caused by the action of IgE antibodies. Our understanding of the immunopathogenesis of these disorders and our ability to develop more rational therapeutic interventions are linked closely to advances made in our understanding of the cellular and molecular biology of the immune response in general. This book, therefore, attempts to examine IgE-mediated immune responses within the context of our current knowledge of immune mechanisms.

In assembling the book, we have been guided by a desire to present an exposition of the relevant current knowledge in a manner that is both focused and concise, and that captures the excitement of the rapid advances made in understanding the immune response in general, and IgE responses in particular. Accordingly, the book is divided into two parts. Part I deals with the essential ingredients of immune responsiveness and covers the molecular biology of T- and B-cell differentiation and activation, the molecular genetics of the major histocompatibility complex, and antigen processing and presentation. In Part II, we take an expanded look at how these processes give rise to an IgE-mediated immune response and how the unique properties of IgE, its cellular receptor, and the cells that bear this receptor render this extraordinary molecule a potent inflammatory activator. We also deal with how this information can be harnessed to develop more rational treatment strategies for allergic disease. Each chapter makes every attempt to explore these processes in molecular as well as cellular terms. We have assembled an international team of contributors who are themselves actively engaged in research on topics they address in the text. We are hopeful that this book will be a useful resource for all "students" of immunology who have an interest in the allergic response, whether their primary focus is treating patients with IgE-mediated inflammation or investigating the scientific basis of allergic disease.

We wish to thank the authors for their patient collaboration and the staff of Marcel Dekker, Inc., for their help in producing this volume.

Arnold I. Levinson
Yvonne Paterson

Contents

Contributors

Mark Ballow, M.D. Professor, Department of Pediatrics, Division of Allergy/Clinical Immunology and Pediatric Rheumatology, Children's Hospital of Buffalo, State University of New York, Buffalo, New York

Jean-Yves Bonnefoy, Ph.D. Head, Immunology Section, Glaxo Institute for Molecular Biology, Geneva, Switzerland

Vivian Lam Braciale, Ph.D. Associate Professor, Department of Microbiology, University of Virginia Health Sciences Center, Charlottesville, Virginia

I. Nicholas Crispe, M.B.B.S., Ph.D. Assistant Professor, Immunobiology Section, Yale University School of Medicine, New Haven, Connecticut

Stephen Desiderio, M.D., Ph.D. Associate Professor, Molecular Biology and Genetics, The Johns Hopkins University School of Medicine, and Associate Investigator, Howard Hughes Medical Institute, Baltimore, Maryland

Susan M. Dymecki, M.D. The Johns Hopkins University School of Medicine and Howard Hughes Medical Institute, Baltimore, Maryland

Glen N. Gaulton, Ph.D. Associate Professor, Department of Pathology and Laboratory Medicine, University of Pennsylvania, Philadelphia, Pennsylvania

Raif S. Geha, M.D. Chief, Division of Immunology, Children's Hospital, Harvard Medical School, Boston, Massachusetts

Gerard F. Hoyne, Ph.D. Department of Immunology, Imperial College of Science, Technology, and Medicine, St. Mary's Hospital Medical School, London, England

Carole L. Jelsema, Ph.D. Health Science Administrator, National Institute of Arthritis, Musculoskeletal and Skin Diseases, National Institutes of Health, Bethesda, Maryland

A. Barry Kay, Ph.D., D.Sc. Department of Allergy and Immunology, Royal Brompton National Heart and Lung Institute, London, England

Jean-Pierre Kinet, M.D. Head, Molecular Allergy and Immunology Section, National Institute of Allergy and Infectious Diseases, National Institutes of Health, Bethesda, Maryland

Jonathan R. Lamb, Ph.D. Department of Immunology, Imperial College of Science, Technology, and Medicine, St. Mary's Hospital Medical School, London, England

Elwyn Loh, M.D., Ph.D. Assistant Professor, Hematology-Oncology Section, Department of Medicine, University of Pennsylvania, Philadelphia, Pennsylvania

Gerald T. Nepom, M.D., Ph.D. Scientific Director, Diabetes and Immunology Programs, Virginia Mason Research Center, Seattle, Washington

Robyn E. O'Hehir, M.B.B.S., Ph.D. Department of Immunology, Imperial College of Science, Technology, and Medicine, St. Mary's Hospital Medical School, London, England

Joanne C. Pratt, Ph.D. Department of Pathology and Laboratory Medicine, University of Pennsylvania, Philadelphia, Pennsylvania

David M. Robinson, M.D. Member, Allergy Immunology Section, Department of Medicine, Virginia Mason Medical Center, Seattle, Washington

Douglas S. Robinson, M.D. Department of Thoracic Medicine, Royal Free Hospital, and Royal Brompton National Heart and Lung Institute, London, England

Frederick M. Schaffer, M.D. Professor, Department of Pediatrics, Division of Allergy/Clinical Immunology and Pediatric Rheumatology, Children's Hospital of Buffalo, State University of New York, Buffalo, New York

Lawrence B. Schwartz, M.D., Ph.D. Charles and Evelyn Thomas Professor of Medicine, Department of Internal Medicine, Virginia Commonwealth University, Richmond, Virginia

Donata Vercelli, M.D. Assistant Professor, Department of Pediatrics, Children's Hospital, Harvard Medical School, Boston, Massachusetts

Peter F. Weller, M.D. Associate Professor of Medicine, Division of Infectious Diseases, Department of Medicine, Harvard Medical School and Beth Israel Hospital, Boston, Massachusetts

I

BASIC FEATURES OF THE IMMUNE RESPONSE

1

Molecular Genetics of Immunoglobulin Genes and the Generation of Antibody Diversity

Mark Ballow and Frederick M. Schaffer
Children's Hospital of Buffalo, State University of New York, Buffalo, New York

I. INTRODUCTION

It has been over 40 years since the discovery that plasma cells are the immunoglobulin- or antibody-producing cells of the body (1). With the discovery of the thymus, and the bursa of Fabricius in birds, it became clear that the plasma cell and its precursors followed a different pathway of differentiation from their counterparts, the T cells, which mediate cellular immunity. The precursor of the plasma cell, the B cell, originates from the same hematopoietic stem cell that gives rise to T lymphocytes, but are generated by distinct anatomical microenvironments.

The primary function of B cells is to produce antibodies against a large variety of antigenic substances that are potentially harmful to the host. This concept is clearly seen in patients with immune-deficiency disorders in whom B-cell function is defective. This ability to produce a broad spectrum of antibodies by the generation of clonally diverse B cells, each of which has a unique antigen-binding specificity, has evolved during phylogeny to reach a remarkable state of sophistication in higher vertebrates.

II. THE IMMUNOGLOBULINS

Immunoglobulins (Ig) or *antibodies* (Ab) are specialized glycoproteins, found in the serum and tissue fluids, that play an important role in the adaptive immune response (2). Each antibody binds to a particular part of a putative "antigenic determinant" or epitope (3). Because of their specialized structures, antibodies are important in the detoxification process of harmful proteins. They also act by enhancement of the phagocytic process as an opsonin, and as an accessory effector molecule in the activation of the complement cascade. Five distinct classes of immunoglobulin molecules exist: IgG, IgA, IgM, IgD, and IgE. They each are distinct and differ by size, charge, amino acid composition, and carbohydrate content (Table 1). In addition to the differences between immunoglobulin classes, the immunoglobulins within each class are also very heterogeneous.

The basic structure of all immunoglobulin molecules consists of two identical polypeptide light chains and two identical polypeptide heavy chains linked together by disulfide bridges (4,5) (Fig. 1). The heavy chain type determines the class and subclass of the immunoglobulin molecule. Immunoglobulin G, IgD, and IgE occur only as monomers, IgA can occur both as a monomer or in a polymeric form (6). Immunoglobulin M occurs as a pentamer with five four-chain subunits linked together. The light chain consists of two distinct types called kappa (κ) and lambda (λ). Any of the light-chain types can combine with any of the heavy-chain types, but in any one immunoglobulin molecule both light chains are of the same type. The NH_2-terminal end of the immunoglobulin molecule forms the antigen-binding site, which contains the amino acid sequence variability (see later discussion). The remaining parts of the immunoglobulin molecule consist of constant region domains. The light chain contains one constant region domain, whereas the heavy chain can have three or four constant region domains, which are linked by intrachain disulfide bonds. These constant region domains have several important biological functions and control the metabolism of the immunoglobulin molecule (7). For example, the first constant region domain binds complement, whereas the second and third constant region domains of the IgG molecule contain the Fc determinant, which is important in binding to the Fc receptor on mononuclear cells, neutrophils, and other phagocytic cells.

Immunoglobulin G is the major immunoglobulin class in normal serum, accounting for 70–75% of the total immunoglobulin pool. It is distributed between the intravascular and extravascular plasma pools; it is the only antibody class that crosses the placenta and is the major antibody of the secondary immune response. The IgG immunoglobulin class is composed of four IgG subclasses, the characteristics of which are shown in Table 2. Immunoglobulin M accounts for approximately 10% of the immunoglobulin pool. Because of its size, it is largely confined to the intravascular plasma pool and is the preponderant early or primary antibody response (8). Similar to certain of the IgG subclasses, IgM is very efficient in activating complement. Immunoglobulin A represents 15–20% of the human serum immunoglobulin pool, and is the preponderant im-

Table 1 The Physical and Biological Properties of Human Immunoglobulins

Property	IgG	IgA	IgM	IgD	IgE
Molecular form	Monomer	Monomer, polymer	Pentamer	Monomer	Monomer
Subclasses	IgG1,2,3, and 4	IgA1 and 2	None established	None	None
H-chain allotypes	Gm	Am	Mm	none	none
Molecular weight	150,000	160,000	950,000	175,000	190,000
Carbohydrate (%)	3	7	12	9	13
Serum level (mg/dl)[a] (95% range)	1250 (639–1349)	210 (70–312)	125 (56–352)	4 (0–61)	0.03 (0–0.09)
Antibody valence	2	2	5 or 10	2	2
Half-life (days)	21	6	10	3	2
% distribution (% intravascular)[b]	45	42	80	75	50
Complement fixation (classic pathway)[b]	+(IgG1,2,3)	–	+	–	–
Complement activation (alternative pathway)	IgG4	+(IgA1,2)	–	+	–
Cells bound	Macrophages, neutrophils	Lymphocytes	Lymphocytes	None	Mast cells, basophils
Other properties	Secondary response, placental transfer	Mucous secretions	Primary response, rheumatoid factor	Class switching	Immediate hypersensitivity

[a]Adult normal values.

[b]The plus and minus signs indicate whether the pathway exists (+) or not (–).

Source: Adapted from D. J. Jerke and J. D. Capra. Immunoglobulins: Structure and Function. *In Fundamental Immunology*, edited by W. E. Paul, New York: Raven, 1984; D. W. Talmage and F. G. LaRosa. Specificity, Structure and Genetics of Antibodies. *In Immunological Diseases*, edited by M. Samter, D. Talmage, M. Frank, K. Austen, and H. Clamen, Boston, MA: Little, Brown and Co., 1988; and I. Roitt, J. Brostoff, and D. Male (eds). *Immunology*. St. Louis, MO: CV Mosby Co., 1989.

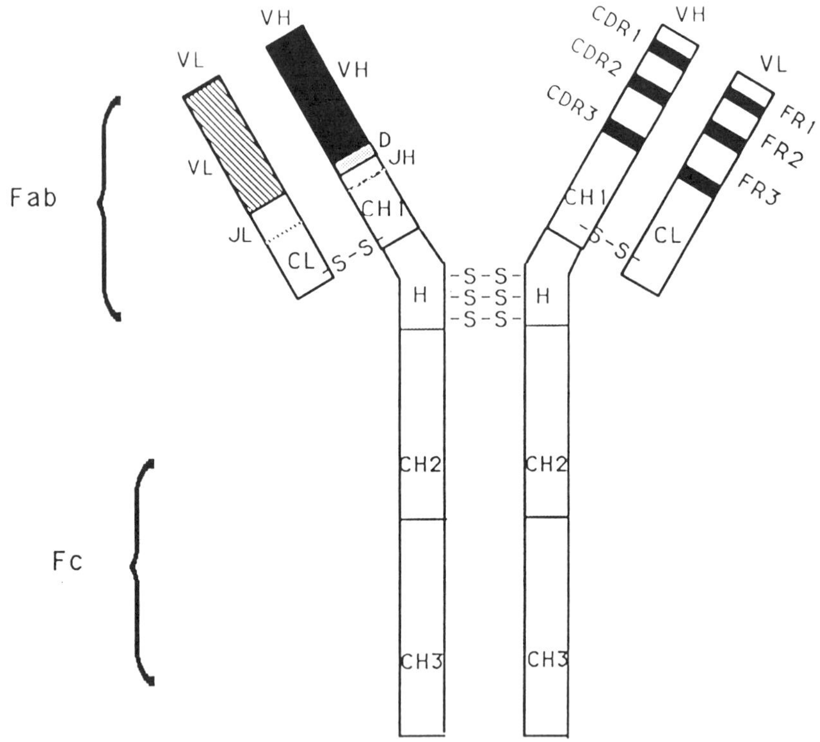

Figure 1 The IgG molecule is composed of two heavy chain and two light chain polypeptides joined together by disulfide bonds (-S-S-) at the hinge (H) region. Both the heavy (H) and the light (L) chains are composed of variable (V) regions at the NH_2-terminus end (the $F(ab)_2$ fragment), and constant (C) region domains at the COOH-terminus end (one C region for the L chain, and three C region domains for the IgG H chain). The V_H, C_H, V_L, and C_L region domains are depicted on one side of the $F(ab)_2$ fragment, and the complementary determining regions (CDR) and framework (FR) areas on the other side of the $F(ab)_2$ fragment. The FR regions are relatively conserved amino acid sequences of the variable regions that are interspersed among three highly variable regions called CDR. These hypervariable regions of the H and L chains interact to form a three-dimensional structure, which constitutes the antigen-binding region. (Adapted from Ref. 19.)

munoglobulin in mucous secretions (9). In secretions IgA (secretory) is complexed with specialized peptide structures called the J chain and secretory component. Secretory component facilitates the transport of secretory IgA into the secretions and protects IgA from proteolytic digestion. Immunoglobulin A1 is the preponderant subclass in serum, whereas IgA2 is found preponderantly in secretions. Immunoglobulin D accounts for less than 1% of the total plasma

Table 2 Characteristics of IgG Subclasses

Characteristic	IgG1	IgG2	IgG3	IgG4
%Normal serum IgG	60–70	20–30	5–8	1–3
Range in adult serum (mg/dl)	500–1200	110–600	50–100	4–290
T ½ (days)	21	21	7	21
Allotypes	4–10	1	14	0
Complement fixation	++	+	+++	–
Fc receptor bindings	+++	–	+++	+
Cryoglobulin	++	+	+++	+
Staph protein A binding	+	+	–	+
Susceptibility to digestion	++	+	+++	++
Transport across placenta	++	+	++	++
Antipolysaccharide antibodies	+	+++	–	++
Antiprotein antibodies	+++	+	+++	+
Anti-DNA	++++	+	+++	+

Source: Adapted from A. F. Kavanaugh and D. P. Huston. Insights, Allergy *4*: 4 (1989).

immunoglobulin pool, but is present on the cell membrane of many circulating B lymphocytes. Its exact biological function is unknown, but it may play a role in the regulation of B-cell development. Immunoglobulin E is present in the serum in the smallest amounts of all the immunoglobulin classes, and it is usually found bound to basophils and mast cells (10). This immunoglobulin class plays an important role in immunity to parasites and in immediate hypersensitivity diseases.

III. THE MOLECULAR BASIS OF ANTIBODY DIVERSITY

The heavy chain genes are encoded on chromosome 14 at band q32. The genes encoding for the immunoglobulin heavy chain are located as discontinuous gene segments on the chromosome (11). The variable regions of the immunoglobulin heavy chain consist of a framework of relatively conserved amino acids interspersed by three highly variable areas, also called hypervariable regions (see Figs. 1 and 2). These hypervariable regions on the heavy and light chains interact in a three-dimensional structure to form the antigen-binding structure of the immunoglobulin molecule (12). The hypervariable regions are also referred to as complementarity-determining regions or CDR (13). For the heavy chain, the variable region is assembled from three gene segments, a variable (V_H), a diverse (D), and a joining (J) region gene element, which are located in close proximity on the same region of the chromosome. These D and J region genes, together

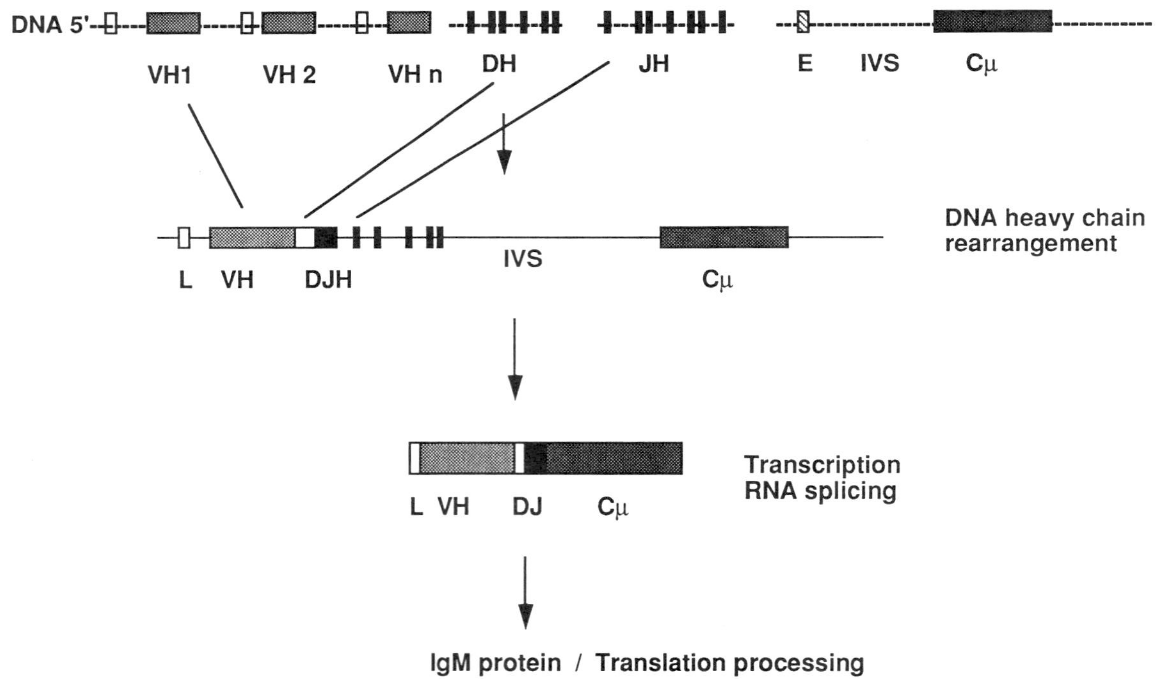

Figure 2 The immunoglobulin heavy (*H*) chain gene locus on chromosome 14 (q32). The mechanism for the assembly of germline genes into an IgM molecule is shown. The variable (V_H) region genes of the H chain are composed of seven multimember families stretched over 2500 kb of chromosome comprising ~ 200 V_H gene segments. Downstream (3′) from the V_H region genes are a family of diversity (*D*) gene segments and six functional joining (J_H) gene segments. The variable region is assembled from germline DNA by recombinatorial events in which segments of a single V_H, *D*, and J_H gene are brought together to form a V_HDJ_H segment. The intervening DNA (*IVS*) or introns are deleted by mRNA splicing in which the *C* region gene elements are juxtaposed with the *V* region segment to form an mRNA that can be translated into μ chain protein.

with the latter part of the V_H genes form the third hypervariable region (see Fig. 2). Any one V gene element can join to any D or J gene element to form the heavy-chain variable-region domain of the immunoglobulin molecule.

The number of human V_H gene segments have been estimated by band counting on Southern blots (14). These analyses have led to estimates that the human immunoglobulin heavy-chain gene complex contains approximately 100–200 V_H gene segments divided into seven families. Initially these V_H gene elements were divided into three subgroups, based on the protein sequences of myeloma proteins (3). Subsequently, additional V_H families, based on nucleotide homologies, were described by cross-hybridization with murine probes (14,15). Recently, Schroder and associates (16) described a seventh V_H gene family in humans. The human V_H3 family appears to have the greatest number of members (25), whereas the V_H6 family has only one member. In contrast with the mouse, in which V_H gene families are clustered on the chromosome, many of the V_H gene segments of the various human V_H gene families are interspersed over 2500 kb on chromosome 14. Of interest, the V_H6 family is only 80 kb from the J_H region (17).

A. Mechanism of the V_H Domain Assembly by DNA Recombinatorial Events

Closely linked to the V_H gene locus are the gene segments for the J_H locus (14). The human J_H cluster consists of six functional J_H genes with three pseudo-J_H gene segments (18). Downstream from the J segments are the constant region genes of the immunoglobulin heavy chain. All the different immunoglobulin isotypes use the same set of variable region genes in a unique series of events that occur during B-cell development and differentiation. This process of DNA rearrangement and recombination of the various immunoglobulin gene segments results in the final immunoglobulin molecule, with its uniqueness for a single antigen specificity. As shown in Figure 2 for the heavy chain immunoglobulin genes, one of the first events is the joining of a single D_H gene segment with a single J_H gene segment that then combines with a single V_H gene segment to form the complete variable region domain, $V_H DJ_H$ (11). The V gene segments contain highly specialized and conserved recombination–recognition sequences. These special regions consist of a palindromic heptamer sequence, and a less well-conserved A/T-rich nonamer sequence that lies adjacent to each of the coding elements. The recombinatorial process is directed by a pair of these signal sequences, one of which is associated with a 12-base sequence spacer between the heptamer and nonamer, and the other a 23-base sequence between the other signal elements (19) (Fig. 3). The asymmetry of these spacer sequences guides V gene segment assembly and is responsible for the juxtaposition of the D to J_H region and, subsequently, the joining of the V_H segment to the DJ_H complex.

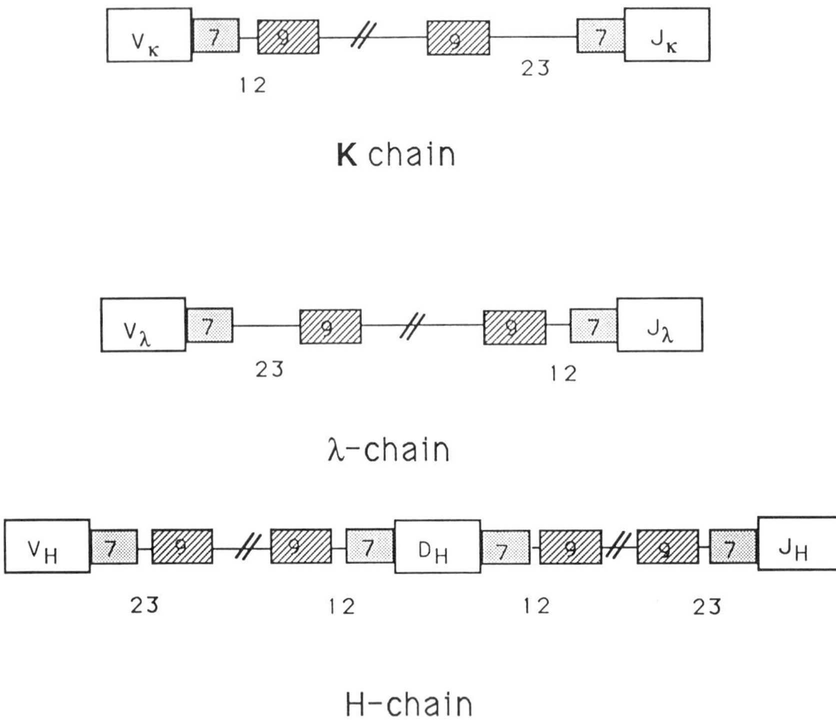

Figure 3 A map of the heptamer and nanomer sequences and base pair spacers within the heavy and light (κ and λ) chain gene loci. The stippled box represents the heptamer sequence, whereas the striped box represents the nanomer sequence. The Arabic numbers below a gene locus represent the size of the base pair spacers. (Adapted from Ref. 52.)

These conserved signal sequences flank the 3′ ends of the *V* segments, the 5′ ends of the *J* segments, and the 5′ and 3′ ends of the *D* segments (17). The recombination–recognition sequences are targets for specialized enzymes involved in the recombination process called recombinases. The recombinase cuts each signal sequence adjacent to the heptamer, with the joining of the two signal ends to form a "signal joint." The two coding ends (i.e., D and J_H or V_H and DJ_H) undergo nucleotide excision or addition, and then join to form the "coding joint." Nucleotides may be added to these cut DNA strands by a template-independent terminal deoxynucleotidyl (TdT) enzyme activity that creates *N* region sequences at the coding joints between D and J_H, or between V_H and DJ_H (20). The two coding strands (D and J_H, or V_H and DJ_H) and the two signal sequences are ligated. If the coding sequences of the two segments have the same polarity on the chromosome, the intervening sequences are joined in a circle and deleted.

Analysis of a variety of different B-cell lines and tumors have shown that the intervening DNA sequences between the recombined V_H gene elements is lost as a consequence of this rearrangement process. If the coding sequences have opposite polarity, the corresponding part of the locus is inverted. Although *N* region diversity at the junctions of D_H and J_H, and V_H to D_HJ_H are common in the adult, *N* region diversity is rare in prenatal and neonatal B cells (21,22), and absent in light chain recombinatorial events (11). The lack of *N* region diversity correlates with low levels of TdT enzyme (23) in neonatal B cells. Individual immunoglobulin gene segments can be transcribed before rearrangement. The transcripts from unrearranged immunoglobulin loci are expressed only in B cells, including pro-B cells and pre-B cells; they are termed *sterile transcripts* because they are not translated into protein (24). The transcripts of these DNA immunoglobulin segments may be a prerequisite for rearrangement, perhaps by opening up the chromatin for the recombinase enzyme (25).

Two types of regulatory elements are important in controlling immunoglobulin transcription: (1) promoters, which are sequence motifs located 5′ to each V_H gene segment, and (2) enhancers, which are located in the intron regions between the J_H and *C* region gene segments. Specific proteins called DNA-binding proteins bind to these specialized DNA sequences (e.g., promoters and enhancers) to regulate gene transcription (26). Several sequence motifs located upstream (5′) of the V_H elements serve as immunoglobulin promoters (Fig. 4). Two important immunoglobulin promoters are the octamer motif, ATTTGCAT, or its inverse, and a TATA box (27–30). This DNA region also contains several other sequence motifs that influence immunoglobulin gene transcription (31). The Oct-2-binding protein is most likely responsible for lymphoid-specificity of the immunoglobulin octamer promoter, since it is restricted to B cells and only a few other cell types, and is expressed throughout the B-cell lineage (32). Rearrangement of the V_H gene segment to the *D* segment brings the promoter-controlling sequence, which is present 5′ upstream from each V_H gene segment, in close proximity to a tissue-specific enhancer sequence between the J_H and the constant region genes. This is necessary to have enhanced transcription of mRNA for the production of the heavy-chain immunoglobulin molecule.

B. The Generation of Antibody Diversity

The delineation of the molecular biology of the immunoglobulin genes has provided a better understanding of the mechanisms of antibody diversity. The somatic assembly of immunoglobulin variable region genes by DNA recombinatorial events (V_HD and J_H) permits the generation of a nearly unlimited array of antibody specificities. These mechanisms give rise to a unique set of

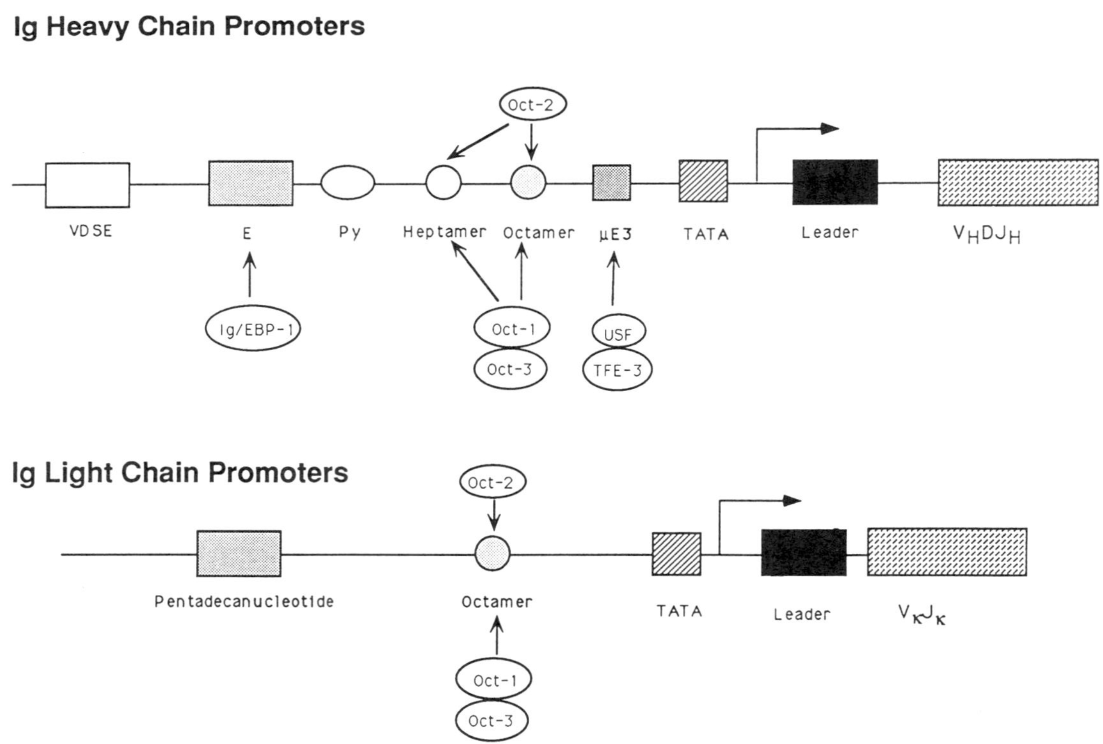

Figure 4 The Ig heavy and light chain promoter and enhancer motifs. The oval figures above the gene loci depicted represent B cell lineage specific transacting factors while those located below the gene loci are nonspecific transacting factors. The arrows point to the putative binding sites for these transacting factors. Adapted from Ref. 31.

B-lymphocyte clones that are capable of antigen-induced activation and expansion into plasma cells. Five molecular genetic mechanisms contribute to this process of antibody diversity. First, there are several germline variable V_H region genes, each encoding for a specific antibody combining site. Second, the combination of a variable region gene segment with random D and J_H gene segments contributes to the generation of variable region diversity in the third hypervariable region. Third, further diversity of antibody specificity occurs by the addition of random nucleotides at the recombinatorial sites during the DNA recombinatorial process of the V_H, D, and J_H segments (see foregoing). Fourth, antibody diversity is enhanced by the process of point mutations, resulting in amino acid changes in the hypervariable regions. This adds to antibody diversity and also changes in anti- body affinity. Fifth, pre-B cells can pass a functionally rearranged heavy chain gene onto multiple progeny that can then rearrange different light chain genes. This is one of several mechanisms for generating intraclonal antibody diversity. The cell surface expression of a single species of immunoglobulin by an immature B cell, that is, one heavy chain and one light chain variable region, results from allelic exclusion and is regulated by the immunoglobulin gene rearrangement process.

C. Assembly and Selection of the Constant Region Genes

The effector function of the immunoglobulin molecule varies between the different immunoglobulin classes or isotypes and is defined by the structure of the C_H region. Although antibodies may have the same antigen specificity, expression of different isotypes of immunoglobulin classes can lead to a variety of effector functions. Thus, the C_H region contributes to another aspect of immune diversity, other than those we described earlier, that generates antibody specificity. In addition, cytokines may regulate the expression of particular B-cell isotypes, either by directly influencing DNA rearrangements at particular switch regions or by selecting B cells precommitted to a particular isotype.

The human C_H gene complex is mapped at the q32 band of chromosome 14 and consists of nine functional genes and two pseudogenes (18,33,34) (Fig. 5). It is postulated that the whole immunoglobulin gene system has evolved from the T-cell receptor gene, since cell-mediated immunity preceded humoral immunity. In fact, the T-cell receptor and immunoglobulin genes might be derived from a common ancestral gene of the β_2-microglobulin and Thy-1 genes, which have only a single domain. The C_H gene evolved further by duplication of the C exons. From evolutionary studies of immunoglobulins, the C_μ gene is closest to the ancestral C_H gene, whereas the α and δ genes appear to be derived from the ancestral γ gene (35).

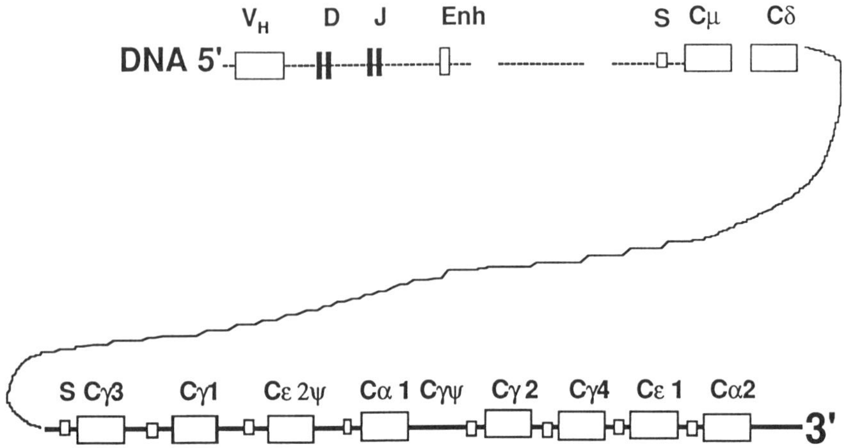

Figure 5 The heavy chain constant region-coding sequences shown in relation to the V_H, D, and J_H gene segments. A switch region is located 5′ to each coding gene (exon) except for the $C\delta$ gene segment. Each constant region segment is composed of a cluster of exons (three for γ, δ, and α and four for μ and ϵ).

The complete immunoglobulin molecule is formed by two recombinatorial events, the first recombinatorial mechanism, as outlined earlier, generates the V_HDJ_H recombination. In the second type of recombination called the S–S recombination, the regions on the 5′ side of each C_H gene, the switch sites or S region, pair together (Fig. 6). The initial S–S always involves the S_μ region as one of the pair, and the S region of another C_H isotype as the other. The S region is located about 1–4 kb upstream from each coding region except for the $C\delta$ gene. This second rearrangement leads to isotype switching from IgM to another one of the other immunoglobulin classes or isotypes. Both rearrangements result in deletion of the intervening DNA sequences from the chromosome (36,37). Studies of the S–S combination sites have shown that the structural basis of this S region is based on a tandem repetition of related nucleotide sequences (38). Although the mechanism of recombination in these S regions is not entirely clear, Honjo and colleagues (37) have proposed that homologous sequences in the S regions juxtapose and proceed through a recombinatorial event in which DNA material is deleted.

Two explanations have been given for the deletion of the intervening DNA material during the process of the class switch (35). The first model postulates that the DNA segment is looped out from the chromosome, and this loop is spliced out from the DNA sequence (see Fig. 6). The other model suggests that the deletion occurs by unequal crossover events between sister chromatids. In this model there is an asymmetric segregation in which one daughter cell with a deletion of C_H genes, and the other cell with a duplication results in a crossover event and an

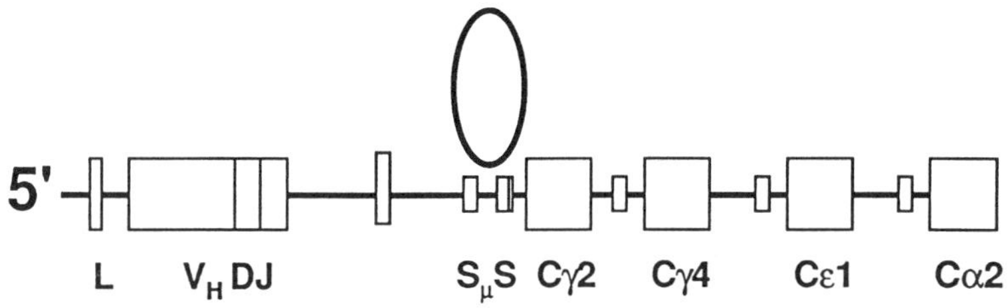

Figure 6 A molecular model for immunoglobulin isotype switching. Switch regions (*S*) adjacent to each *C* region segment, S_μ and $S_\gamma 2$ in this example, pair together, which results in a looping out of the intervening DNA. The deletion of this DNA results in the juxtaposition of the $V_H DJ_H$ segment to the *C*-region gene element.

exchange between sister chromatids (39). Studies of a pre-B-cell line transformed by the Abelson murine leukemia virus suggest the looping-out mechanism (40).

Many B cells in the peripheral blood express both IgM and IgD on their cell surface. The B lymphocytes in the neonate may, in fact, express three different cell surface isotypes (41). In the case of B cells expressing both IgM and IgD, the μ and δ exons are located close together on the germline DNA (see Fig. 5). There is no switch region upstream from the δ exon. The cell makes a long RNA transcript that includes both the μ and δ genes and the upstream V_H, DJ_H (42). Thus, there may be an intermediate step of RNA splicing, which is often associated with the expression of two constant region gene products on the cell surface of the B lymphocyte. These long mRNA transcripts have been isolated from nuclei. Hybridization studies of the DNA material and nuclear run-on transcription analysis have shown a germline configuration of the *C* region genes supporting the view that partial class switching may occur by differential mRNA splicing (43).

D. Membrane and Secreted Immunoglobulin

The μ heavy chain can be expressed as either a secreted molecule in the body fluids, or as a membrane-associated molecule. The *C*μ gene is composed of four exons, which encode for the C-region domains, and are common to both the membrane-associated and secreted forms of IgM. Directly 3′ to the *C*μ region exons is the exon associated with the secreted form of IgM—μ*s*, which encodes a 21-amino acid hydrophilic COOH-terminus: 3′ to the μ*s* exon (~1.8kb) are two membrane exons (μ*m*)—*M1* and *M2* that encode a 41-amino acid hydrophobic COOH-terminus. The switch from membrane μ to secreted IgM occurs by differential mRNA splicing (44–46). In the μ*m* molecule, the secreted μ*s* exon is spliced out of the mRNA while maintaining intact the two membrane exons. The regulation of μ mRNA production is very complex, involving several different mechanisms (reviewed in 47).

E. Immunoglobulin Heavy-Chain Repertoire Development

In the mouse, chromosomal position is a major determinant of rearrangement frequency. In fetal pre-B cells, there is a bias toward utilization of V_H gene segments that are proximal to the J_H region. In B cells obtained from the peripheral blood pool, the repertoire of V_H families is random and occurs in rough proportion to the complexity of the families, showing no correlation with J_H proximity. The position-dependent preference of V_H gene family use in mice may have important implications for B-cell ontogeny, differentiation, and development. Preferential expression of certain families of V_H genes could be linked to changes in the ability to synthesize antibodies to certain classes of antigens in postnatal life. Studies are in progress to determine if, in humans, there is a similar preferential use of certain V_H gene families during ontogeny.

IV. IMMUNOGLOBULIN LIGHT CHAIN GENES

A. Kappa Light Chain Genes

The light chain polypeptides of antibodies are classified as either κ or λ. When a B cell is committed to producing an antibody, all antibodies synthesized are of the same light chain class. There is as yet no strong evidence suggesting that either class of immunoglobulin light chain has a preferred or unique function. Genetic studies have demonstrated that the human κ locus of genes is located on the short arm of chromosome 2 (48), whereas the λ locus of genes is located on the long arm of chromosome 22 (49). The ratio of κ to λ in humans is 2:1, whereas murine studies have demonstrated a 20:1 ratio, suggesting a larger role for λ light chains in humans. The κ and λ light chain genes may reflect an evolutionary duplication process that has provided a greater diversity of light chain genes capable of being used in antibody synthesis.

The κ light chain gene locus consists of a single constant region gene segment (*C*κ), five functional joining segments (*J*κ), and approximately 50 functional variable region gene segments (*V*κ; Fig. 7). The *C*κ exon (coding region) is flanked by two enhancer elements. Each *V*κ gene segment contains an upstream (5′) promoter site, a leader sequence, and exons that encode the first three framework regions and the first two CDR regions and part of CDR 3. At the 5′ end, the *J*κ gene segments contain a recombination signal sequence that is complementary to the recombination signal site located at the 3′ end of the *V*κ gene segment similar to the V_H and J_H signal sequences (see Fig. 7).

The *V*κ gene segments are grouped into four families based upon homology of the DNA sequence. Each family is made up of *V*κ gene segments sharing greater than 80% DNA sequence homology. The sizes of these families range from a single member (*V*κ*IV*) to more than 30 members. The *V*κ genes are interspersed on chrom-

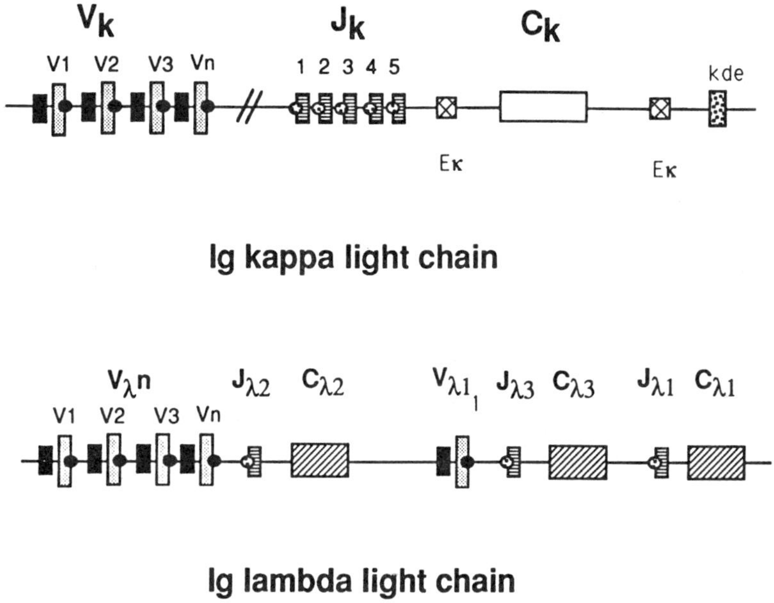

Figure 7 The κ and λ light chain gene loci: The black boxes represent leader sequences, whereas the boxes labeled Eκ represent the 5′ and 3′ Cκ enhancer motifs. The κ-deleting element (κde) is a recognition site involved in isotype switching to light chain production (see text). Note that unlike the κ gene locus, the λ gene locus has a *J* gene segment located just upstream of each *C* gene segment.

osome 2. In addition, a significant proportion of the identified *V*κ gene segments were found to be pseudogenes. *Pseudogenes* are genes that are not expressed and are characterized by changes in their nucleotide sequence that directly affect transcription or translation of the gene product. Several of the κ pseudogenes have been identified on chromosomes 1, 15, and 22. These pseudogenes may contribute to the κ gene diversity by the process of gene conversion, that is the correction of gene mutations (e.g., as found in pseudogenes) by unequal inheritance of alleles within a recombinant region. Through this process pseudogenes may become functional and, thereby, enlarge the *V*κ gene repertoire (50,51).

Recombinatorial events for light chain genes are somewhat distinctive. To form a functional κ light chain gene, a *V*κ gene segment joins to a *J*κ gene segment. Unlike heavy chain recombinatorial events, no diversity gene segment is involved in the formation of functional light chain genes. *V*κ gene segments have been identified as being in different transcriptional orientations. These findings suggest that both classic deletion and inversion rearrangement mechanisms are employed in the κ chain locus. In either event, the *V*κ gene segment is juxtaposed to the *J*κ gene segment. The complementary recombination signal sequences, located at the 3′ end of the *V*κ gene segment and at the 5′ end of the *J*κ gene segment, are joined and enzymatically separated as a distinct unit. These recombination signal sequences consist of well-conserved palindromic heptamers and nanomer sequences that are separated by 11- or 12-bp spacers at the 3′ end of the *V*κ gene segment and by a 23 ± 1-bp spacer at the 5′ end of the *J*κ gene segment (see Fig. 3). Since the length of a single DNA helical loop is approximately 11 bp, the length of the spacers in the *V*κ and *J*κ gene segments imply that the complementary joining (heptamer and nanomer) sequences are one or two loops apart on the same side of the DNA helix. During recombination, gene segments with an 11- or 12-bp spacer (*V*κ) usually join with gene segments containing a 23 ± 1-bp spacer (*J*κ), and this process diminishes the risk of joining inappropriate gene segments (see Fig. 3).

It is theorized that, during recombination, these conserved sequences are recognized by an immunoglobulin gene recombinase enzyme(s). The complementary sequences are joined, leading to the formation of a stem and loop configuration and, thereby, juxtaposing a *V*κ gene segment to a *J*κ gene segment (51,52). Eventually the stem and loop structures are enzymatically deleted. Thus, κ genes processed by deletional mechanisms result in loss of the deleted DNA (stem and loop structures) from the cell, whereas κ chain genes, generated by inversion mechanisms, result in conserving the inverted DNA (Fig. 8). After *V*κ-*J*κ joining occurs, a primary nuclear RNA transcript is made. This transcript undergoes splicing of the intervening introns (noncoding DNA sequences) and *J*κ exons. The spliced mRNA is then translated resulting in the formation of a κ light chain polypeptide. The leader peptide, which directs the κ light chain into the endoplasmic reticulum, is later removed from the κ chain polypeptide (see Fig. 8).

1. κ Light Chain Gene Diversity

Diversity of the κ light-chain repertoire is due to several mechanisms: the joining of 1 of 50 V_κ gene segments to any of 5 *J*κ gene segments, imprecise junctional recombination, and somatic mutations. During κ gene segment recombination, the joining of recombination signal sequences is precise, whereas the *V*κ–*J*κ juxtapositioning is somewhat imprecise, leading to the loss of a few nucleotides from both the *V*κ and *J*κ gene segments. This imprecise joining occurs within the CDR 3 region of the antibody molecule and may cause amino acid substitutions that increase the diversity of the κ light chain. Not all such joining events generate productive κ light chains. If joining results in a DNA sequence that is an out-of-phase reading frame (nonproductive rearrangement), either no protein, or a nonsense protein, may be formed. Since amino acids are encoded by a triplet set of nucleotides, two out of three random joining events could result in nonproductive rearrangements.

Immunoglobulin gene diversity is also increased by the process of somatic mutation. In particular, areas of hypermutation within the variable regions of immunoglobulin chain genes have been demonstrated to have point mutations within or near the V_LJ_L and $V_HD_HJ_H$ regions, which supports the concept of somatic mutation. These point mutations may result in changes in the amino acid sequence of a variable region peptide and, thereby, contribute to antibody diversity (53).

2. *Hierarchy of Immunoglobulin Gene Activation*

During pre-B-cell development, the μ heavy chain gene is rearranged. Usually this is followed by $V\kappa$–$J\kappa$ gene rearrangement and, if successful (productive), the formation of an IgM antibody ensues (51,52). Nonproductive rearrangements occur frequently (in as many as 33% of pre-B cells). Deletion of the κ genes can occur on both the maternal and paternal chromosomes, resulting in the failure of κ chain synthesis. Many λ-chain-producing B cells have deletions of the $C\kappa$ exon or $J\kappa$–$C\kappa$ region, and these deletions are thought to be possibly mediated by the same recombinase enzyme(s) involved in V_LJ_L recombination. Specifically, these deletions may occur by rearrangement between a κ-deleting element (kde) located 3′ to the $C\kappa$ exon and a recombination signal sequence positioned in the intron between the $C\kappa$ and $J\kappa$ exons or the $V\kappa$ and $J\kappa$ exons (see Fig. 7). The nonproductive rearrangement of κ light chain translation leads to the initiation of λ light chain gene rearrangement. If λ light chain gene rearrangement is successful, then a μ, λ IgM antibody is formed (54). Transgenic mouse experiments have demonstrated that once an exogenous rearranged light chain gene was introduced into murine zygotes, this resulted in feedback inhibition of endogenous light chain formation (55). This study supported the concept of allelic exclusion. Simply stated, only one of two alleles (e.g., κ and λ light chain genes) can be expressed by a B cell. In addition, this negative-feedback would usually suggest a mechanism that diminishes the chance for B cells to express both classes of light chain immunoglobulin. Thus, as a general rule, κ light chain gene rearrangement precedes λ light chain rearrangement, and only if there is aberrant rearrangement of, or deletion of, the κ light chain genes will λ light chain rearrangement occur (54,55).

B. λ Light Chain Genes

The λ light chains are found in approximately 40% of immunoglobulins secreted by human B cells. This is in sharp contrast with the less than 5% of murine immunoglobulins containing λ light chains. The human locus for λ light chain genes is on the long arm of chromosome 22, and its organization is different from other immunoglobulin gene loci. Early studies demonstrated that this locus contains six constant region genes of which two are pseudogenes ($C_{\lambda 4}$ and $C_{\lambda 5}$).

The six C_λ gene segments are regularly spaced at 5-kb intervals over a 40-kb region on chromosome 22 (50,56). The demonstration of duplications of $C_{\lambda 2}$–$C_{\lambda 3}$ region in some individuals plus the high degree of DNA sequence homology among the six C_λ gene segments strongly suggested that the C_λ gene segments originated from gene duplication (50,56). Recently, three additional C_λ gene segments, two of which may be functional genes, have been identified (57). These newly identified C_λ gene segments are not on chromosome 22, and their function is unknown. In contrast with the heavy chain and κ light chain gene loci, each of the functional C_λ gene segments has only one J_λ gene segment located upstream by approximately 1.3–1.5 kb. The exact number of V_λ gene segments is unknown; however, some estimates have suggested the existence of more than 70, which have been grouped into seven families (58). The V_λ gene segments are located upstream from the region containing the paired J_λ and C_λ gene segments. Like the $C\kappa$ gene segments, a recombination signal sequence is located at the 3′ end of the V_λ and at the 5′ end of the J_λ gene segments (see Figs. 3 and 7). Unlike the recombinatorial events for the κ light chain gene locus, there appears to be only a deletional rearrangement and no inversion associated with the rearrangement process of the λ light chain gene segments. A λ light chain gene rearrangement involves joining of a V_λ gene segment to a paired J_λ and C_λ gene segment, rather than to a J gene segment alone, as in the κ light chain gene locus. During λ light chain gene segment rearrangement, recombination signal sequence joining occurs; although, the base-pair spacers are arranged opposite that of the κ cluster of light chain genes (52). Specifically, the V_λ recombination signal sequence contains a 23-bp + 1 spacer at the 3′ end while at 5′ end of the J_λ gene segment is a recombination signal sequence containing an 11- or 12-bp spacer (see Fig. 3).

Diversity for the λ light chain genes is generated by similar processes as evident for the κ light chain genes. Diversity because of imprecise V_λ–J_λ joining has been associated with deletions identified at the 3′ end of the V_λ gene segment and at the 5′ end of the J_λ gene segment. A comparison of V_λ–J_λ junction sequences to germline gene sequences demonstrated differences that may have been consistent with somatic mutation. Sequence analysis demonstrated no evidence for N diversity for the λ and κ light chain gene loci, a process that results in the addition of amino acids by the terminal deoxynucleotide transferase activity, as is evident in the heavy chain and the T-cell receptor gene loci (58).

C. Control of Immunoglobulin Light Chain Gene Transcription

Transcription of immunoglobulin light chain genes is dependent on the presence of promoter sites and is affected by enhancer sequences (motifs). Enhancer motifs are DNA-binding sites that increase the level of specific mRNA transcription without regard to location or orientation. The V_L genes have promoter elements that are upstream from the leader sequence and contain a TATA box, a pentadecanucleotide sequence, and a highly conserved octamer sequence (see Fig. 4). The TATA box is a well-conserved sequence that is located about 30-bp upstream from the initiation site (cap site) of transcription by the RNA polymerase II enzyme. The octamer sequence (ATTTGCAT), or its inversion, is found in all immunoglobulin gene promoter sites. Since the TATA box is not restricted to lymphoid cells, recent work has implicated the octamer motif as a B-cell-specific "binding site" for the initiation of immunoglobulin gene transcription (31,59). Recent studies have demonstrated the existence of families of proteins (transacting factors) that bind to the octamer motif. There exist both lymphoid specific (oct-2) and nonspecific (oct-1 and oct-3) transacting factors that bind to the octamer motif and effect transcription of immunoglobulin light chain genes.

The C_κ gene segments are flanked by enhancer elements. The enhancer element at the 5′ end ($ENH_{i\kappa}$) demonstrates activity only in mature B cells and plasma cells, whereas its analogue at the 5′ end of the C_μ gene segment is expressed in all stages of B-cell differentiation. This stage-specific activity of the $ENH_{i\kappa}$ is due to the binding of a 10-bp sequence, within this enhancer motif, by the NF-κB-transacting factor. The NF-κB-transacting factor consists of two subunits. The 50-kDa protein subunit is the DNA-binding segment and the 65-kDa protein subunit binds to an inhibitory protein within the cytoplasm known as I-κB. When B cells are appropriately stimulated, the I-κB component of the cytoplasmic NF-κB–I-κB complex is phosphorylated. This leads to the release of the NF-κB-transacting factor, which can then translocate to the nucleus. Enhancement of the κ gene segments occurs once the $ENH_{i\kappa}$ is bound. The NF-κB-transacting factor is not restricted to B lymphocytes, and it has been associated with increased levels of transcription of several lymphokines and the major histocompatibility complex (MHC) class I and II genes. The specificity of the NF-κB-transacting factor for enhancing immunoglobulin gene transcription may, in part, be due to posttranslational proteolytic modification. In addition, the binding of other transacting factors to the octamer sequence or other enhancer motifs, associated with immunoglobulin light chain gene segments, may affect the specificity and level of light chain transcription.

The enhancer motif at the 3′ end of the *C*κ gene segment ($ENH_{3'K}$) has also been associated with activity only during the later stages of B-cell differentiation. The λ-expressing B cells have been identified with deletions of the $ENH_{3'K}$ motif (31). These findings may be related to the process of κ gene segment inactivation and allelic exclusion during λ light chain gene segment rearrangement.

Murine studies have demonstrated the presence of similar enhancer motifs for λ light chain gene segments; no functional binding of the NF-κB-transacting factor has been shown. Further research is required to fully assess the role of enhancer motifs and transacting factors in immunoglobulin gene transcription.

D. Molecular Aspects of B-Cell Development

The *B cells* are defined as lymphocytes that express cell surface immunoglobulin. The antibody receptor serves both to bind antigen as well as a signal-transducing molecule to initiate the production of secondary messengers that activate B cells to differentiate. The B cells are derived from the mesenchymal embryonic tissues starting in the embryonic liver at 8-weeks gestation. At approximately 18-weeks gestation the bone marrow becomes the principal site of production for the B-cell lineage and other blood cell types, except T cells, which are generated in the thymus.

The B cells are derived from pluripotent stem cells that can give rise to the erythroid and myeloid lineages and to T cells (60). The first derivative of the pluripotent stem cell that is committed to the B-cell lineage is the pro-B cell. This cell does not express any of the B-cell phenotypic markers, nor has it begun to rearrange its immunoglobulin genes (61,62). The first recognizable cell belonging to the B-cell lineage is the pre-B cell. These cells are characterized by the presence of intracytoplasmic μ-chains, but are surface immunoglobulin (slg)-negative, since they can only synthesize μ heavy chains, not κ or λ light chains. Light chain synthesis is required for immunoglobulin heavy chains to be transported to the surface (63). The intracytoplasmic μ chains are retained in the endoplasmic reticulum and cannot enter the Golgi apparatus to be exported into the extracellular space (64).

Recently, a new group of genes have been discovered in the mouse and in humans, which are expressed selectively in pre-B cells. These genes encode proteins that act as surrogate light chains to associate noncovalently with μ heavy chain proteins (65,66). These genes, $V_{pre\text{-}B}$ and $\lambda 5$, are expressed at the earliest stages of immunoglobulin gene rearrangement (i.e., DJ_H rearrangement) (67), and form a transition from pre-B cells to the immature B-cell stage (68,69). These genes are highly conserved in phylogeny (70). Alt and associates (71,72) have found that germline V_H transcripts are selectively expressed in pre-B cells. These germline V_H gene products together with the $D_HJ_HC_\mu$ protein are expressed together with the $V_{pre\text{-}B}$–$\lambda 5$ surrogate light chain on the pre-B cell surface (73).

The immature B-cell stage is reached when light chains are synthesized that can combine with the μ heavy chain to express IgM on the cell surface. The surface IgM-positive immature B cell undergoes further maturation to coexpress IgD. The mature IgM^+–IgD^+ B cell is ready to respond to factors produced by T cells that promote growth and differentiation to the plasma cell. A small percentage of IgM^+ B cells switch to express other immunoglobulin isotypes (e.g., IgG, IgA, or IgE). These immunoglobulins have the same antigen-binding specificity as the original IgM^+ B cell. Isotype diversity occurs normally in a germ-free environment, indicating that antigens are not required for this isotype switching. However, as discussed elsewhere, T cells appear to influence the process of isotype switching during antigen-induced B-cell activation.

As mature B cells encounter their specific antigen, they undergo activation and clonal expansion. The B cells develop into plasma cells, which are terminally differentiated and have an average life span of 2–3 days. A total of 10–12 days are required for stem cells to enter the B-cell pathway and progress to the plasma cell stage. Four to five days are required for the newly formed μ^+ pre-B cells to reach the mature IgM B-cell stage, and an additional 3–4 days is required for the B cell to differentiate into a plasma cell. Approximately 5×10^7 B cells are produced daily in the bone marrow. Recent date suggest that the stromal cells of the bone marrow play an important part in growth. Pre-B cells attach through the neural cell adhesion molecule (N-CAM) to the stromal cells of the bone marrow. These adhesion molecules may be important in retaining pre-B cells in the bone marrow until they have matured to a stage enabling them to successfully migrate to the peripheral lymphoid organs for further differentiation. Stromal cells may also produce factors necessary for the growth and differentiation of pre-B cells. Interleukin-7 (IL-7), recently identified as a product of stromal cells, is one important growth factor for pre-B cells. Other factors that promote the growth and differentiation of hematopoietic stem cells include IL-3, colony-stimulating factor (CSF)-1, and granulocyte–macrophage (GM)–CSF. When B cells migrate from the bone marrow into the spleen, many die within the first few days. Daily, 2×10^6–3×10^6 B cells enter the peripheral pool of mature B cells (74,75). Stimulation by antigen promotes growth and differentiation of selected B-cell clones, which leads to either an antibody-secreting plasma cell, or a long-lived memory B cell. Importantly, the surface immunoglobulin receptors of B cells are capable of binding specific antigens and can internalize and partially digest these antigenic molecules. Thus, B cells can function as antigen-presenting cells by recycling antigen on their surfaces in association with class II major histocompatibility molecules for presentation to antigen-binding T cells.

The B cells may express other cell surface glycoproteins that may be of functional importance or useful as markers for studying B-cell development and differentiation. There are a number of protein and glycoprotein molecules that are expressed on the surface of B-cell membranes that can be detected by monoclonal antibodies and are occasionally clustered because of their reactivity to a group of monoclonal antibodies. These cell surface glycoproteins undoubtedly are involved in several complex interactions between B cells and other components of the immune system. It must be kept in mind that the expression of these cell membrane determinants constitutes a dynamic process, changing continuously during cell activation and proliferation. Monoclonal antibodies provide a powerful probe to analyze these cell surface membrane constituents. Recently, new markers have been added to the analysis of B-cell differentiation and maturation, such as the *RAG-1* and *RAG-2* genes, which encode proteins that activate recombination of the variable, diversity, and joining gene segments (76), and certain oncogenes such as *abl*, c-*myb*, N-*myc*, and *bcl-2* (77).

V. CLINICAL B-LYMPHOCYTE DISORDERS

A. X-Linked Agammaglobulinemia

X-linked agammaglobulinemia (XLA) was first described by Bruton in 1952. It is associated with minimal concentrations of all serum immunoglobulins as a consequence of a B-cell differentiation arrest at the pre-B-cell stage. Most patients do not develop recurrent bacterial infections until 4–6 months of age because of the protection provided by transplacentally acquired maternal IgG. Examination of bone marrow aspirates from persons with XLA demonstrates a normal presence of pre-B cells and an absence of plasma cells. In the peripheral blood, there is an absence or paucity of immunoglobulin-bearing B cells ($<1\%$ of normal). The extent of the arrest in B-cell differentiation is somewhat variable. Limited B-cell differentiation has been exemplified by the identification of some IgM-bearing lymphocytes in the peripheral blood of individuals with XLA (78), and the ability to produce IgM-secreting B cells by the transformation of XLA lymphocytes with Epstein–Barr virus (EBV) (79). Molecular assessment of these EBV cell lines demonstrates normally rearranged V_HDJ_H gene segments, suggesting that the genetic mechanisms for recombination and expression of heavy and light chain genes are intact (80). Most persons with XLA have normal cellular (T-cell) immunity and since T and B cells are believed to share the same recombinase enzyme(s) for their respective antigen receptor genes (72), this would further confirm the integrity of the genetic components for rearrangement events. Studies by Schwaber and his colleagues infer that XLA may be associated with the inaccessibility of V_H gene segments to the immunoglobulin gene recombinase enzyme. They demonstrated that pre-B cells derived from the bone marrow of three XLA patients formed antibodies with a truncated μ chain which lacks a V_H gene segment (81). The inaccessibility of V_H gene segments is not evident in all cases, since some pre-B cells escape the B-cell differentiation arrest and undergo successful V_HDJ_H and V_LJ_L rearrangements and expression of surface IgM molecules (78). To determine if a defect in attaining complete immunoglobulin rearrangement is the underlying cause of the XLA differentiation arrest, Anker and his associates (82) evaluated the clonal diversity of 18 EBV B-cell lines established from persons with XLA. If the inaccessibility of immunoglobulin gene segments for rearrangement was evident, then one would expect to find a very limited repertoire of heavy and light chain gene utilization. The sequencing of B-cell lines for V_HDJ_H and V_LJ_L regions demonstrated normal rearrangement; B-cell clonal heterogeneity also existed. Clonal diversity, including B-cell lines that have undergone isotype switching, would support the conclusion that the defect in XLA is a regulatory one affecting B-cell differentiation and is not related to the accessibility of immunoglobulin gene segments for rearrangement.

A region of the midportion of the long arm of the X chromosome, between the DNA gene markers *DXS3* and *DXS94*, has been implicated by genetic linkage analysis to contain gene(s) associated with XLA (83). It is hypothesized that this gene affects pre-B-cell differentiation. The localization of the putative *XLA* gene(s) has made carrier detection and prenatal diagnosis possible by linkage with polymorphic X chromosomal markers. Genetic analysis has demonstrated the X-linked recessive inheritance of XLA. Furthermore, it has been demonstrated that immunologically normal females, who are (heterozygous) carriers of the *XLA* gene, demonstrate an absence of B cells with the mutant gene on the active X chromosome. During normal embryogenesis, there is a random X chromosomal inactivation that results in a mixture of X chromosomes used by all cell populations. However, in female carriers of *XLA*, the B lymphocyte precursors undergo a secondary selection in which cells with the mutant X chromosome are eliminated. Thus, a carrier's B cells will almost all have the normal X as the active X chromosome. Two methods have been developed to assess X inactivation patterns in B cells acquired from potential carriers of the *XLA* gene defect and are effective for identifying a carrier of the *XLA* gene loci. However, these assays are not effective for the families of sporadic male cases, which are often due to new mutations (84,85).

B. Immunoglobulin A Deficiency and Common Variable Immunodeficiency

Immunoglobulin A deficiency (IgA-D) is the most common primary immunodeficiency, which in symptomatic individuals is associated with a variety of infections, allergies, autoimmune disorders, gastrointestinal diseases, and genetic disorders (86). Selective IgA-D is characterized by serum IgA concentrations of less than 5 mg/dL, with normal serum concentrations of other immunoglobulin classes. The central phenotypic feature of this immunodeficiency is an arrest in B-cell differentiation, the extent of which may determine the clinical variability. Individuals with IgA-D tend to have normal numbers of IgA-bearing B cells, but these are relatively immature and fail to undergo terminal differentiation into IgA-secreting plasma cells (87).

Common variable immunodeficiency (CVID), also known as the late-onset form of hypogammaglobulinemia, is a diagnostic classification applied to individuals exhibiting decreased levels of multiple immunoglobulin classes. It is a heterogeneous disorder, typically manifested by recurrent bacterial infections and, similar to IgA-D, is often associated with systemic disorders (88). In most persons with CVID, B cells are produced in normal numbers, but they fail to undergo differentiation into antibody-secreting plasma cells, reflecting an arrest in B-cell differentiation.

The B cells in the circulation of persons with CVID or IgA-D are immature and resemble their counterparts evident in neonatal blood (89). In IgA-D, virtually all of these immature IgA-bearing B cells are also positive for surface IgM or IgD, whereas very few B cells (<10%) express only IgA (87). In CVID most B cells are immature and express IgM and another immunoglobulin class (Fig. 8). The presence of immature IgA-bearing B cells in IgA-D, plus the expression of IgG, IgA, and IgE by immature B cells in CVID, provide evidence for the integrity of the heavy and light chain gene segments. This concept is also supported by noting that offspring of IgA-D individuals demonstrate normal expression of the IgA genes (90). Furthermore, in a study of 66 IgA-D patients, all had intact *Cα1* and *Cα2* gene segments (91). This is not a surprising conclusion, since one would have to envision a deletion encompassing the *Cα1* and *Cα2* gene segments with the intervening three heavy chain gene segments (see Fig. 5) for IgA deficiency to be based on a structural defect of the constant region gene segments. A theoretic larger deletion would have to be envisioned to implicate a structural genetic defect in CVID. Homozygous deletions of C_H gene segments have been reported in a few persons (see the next section), but none of these have involved the deletions discussed in the foregoing. Because most B cells in these two disorders express all immunoglobulin isotypes, the basic genetic elements required for isotype switching are probably intact. Most patients demonstrate normal T-cell immunity, which further implies the integrity of the genetic elements for rearrangement events (since both T and B cells may share the same recombinase enzyme). These factors would support the hypothesis of an immunoregulatory gene defect as the underlying cause of IgA-D and CVID, rather than defects in the immunoglobulin structural genes.

We have suggested that both IgA-D and CVID have a common genetic basis (92). This hypothesis is based on several lines of evidence: both disorders frequently occur in members of the same family; an arrest in B-cell differentiation is a central phenotype of both disorders; IgA-D has been found in association with certain *MHC* haplotypes (A. N. Wilton, 1985) and we have shown the same *MHC* haplotypes are associated with CVID. Our studies suggest that the susceptibility to IgA-D and CVID is associated with an immunoregulatory gene, or genes, of the *MHC* class III region (92–94).

C. Immunoglobulin Gene Deletions

Evidence for structural immunoglobulin gene deletions has been identified in very few patients. In one study, 2 unrelated healthy persons of 11,000 studied had an absence of serum immunoglobulins owing to deletions within the heavy chain gene cluster on chromosome 14 (95). In a similar study (96), only 4

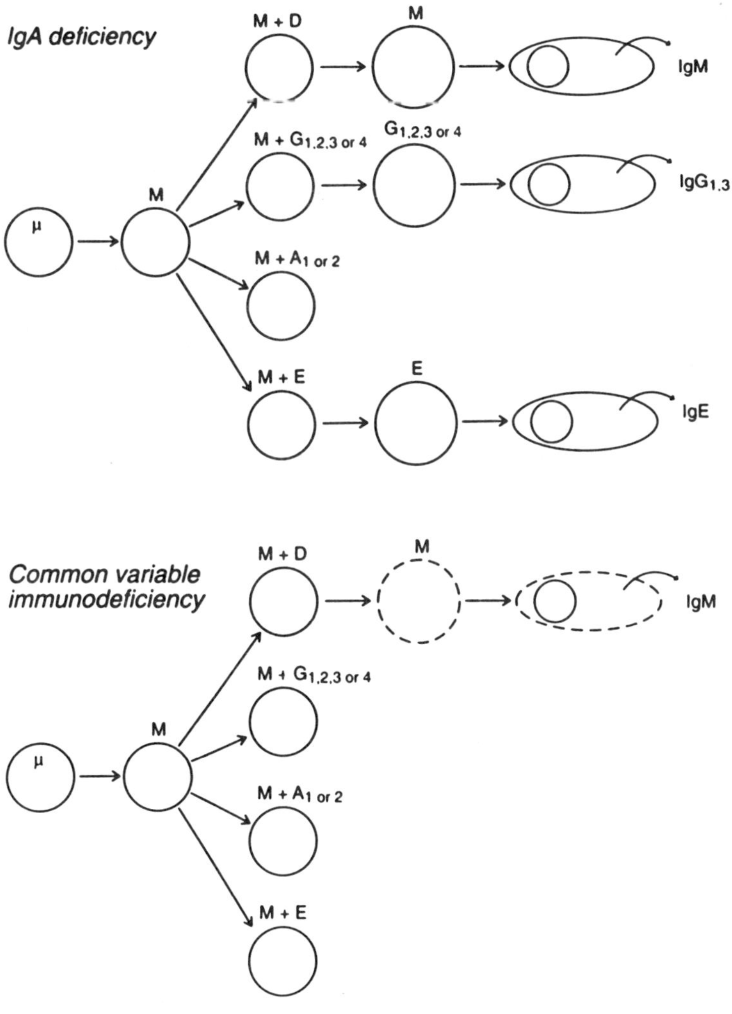

Figure 8 The B-cell differentiation arrest of IgA-D and CVID. Both disorders are associated with an arrest in B-cell differentiation, characterized by the presence of a normal number of immature B cells that express multiple immunoglobulin isotypes on the cell surface (capital letters, Arabic numbers represent IgG subclasses). In IgA-D, this results in a paucity of IgA and, in 20% of cases, IgG2 (or 4)-secreting (curved arrows) plasma cells (oval cells). In CVID, this results in a lack of all antibody-secreting plasma cells for most patients; however, a few of these individuals do have IgM-secreting plasma cells (illustrated as cells with a broken border). (From Ref. 86.)

unrelated persons of the 22,000 studied demonstrated absences of some serum immunoglobulins owing to immunoglobulin C_H gene deletions. The original studies of Lefranc and her colleagues (97,98) documented broad deletions on chromosome 14, one of which encompasses the $\gamma 1$–$\varphi\epsilon 1$–$\alpha 1$–$\varphi\gamma$–$\gamma 2$–$\gamma 4$ C_H genes, whereas another encompasses the $\varphi\epsilon 1$ and $\alpha 1$ C_H genes. Of note, the rare documentation of these latter studies was, in part, due to the possibility of consanguinity giving rise to homozygosity in the family studied (98). Other reported broad deletions consisted of the $\varphi\epsilon 1$–$\alpha 1$–$\varphi\gamma$–$\gamma 2$–$\gamma 4$, $\alpha 1$–$\varphi\gamma$–$\gamma 2$–$\gamma 4$–ϵ, $\varphi\gamma$–$\gamma 2$–$\gamma 4$,–ϵ–$\alpha 2$ C_H genes and a deletion possibly involving the $\gamma 2$ C_H gene alone (95,96,99). One possible mechanism responsible for these deletions is the crossing over of mispaired homologous immunoglobulin gene chromosomal regions (95,98). Alternatively, atypical intragenic recombination between repetitive sequences within the C_H gene switch region could lead to a germline deletion of the intervening C_H gene segments (98). Of interest, one patient with IgG2 and IgG4 deficiency had a deletion of the C_H $\gamma 2$ gene. The deletion may have included a switch-regulatory sequence that was located in the intron between the C_H $\gamma 2$ and the C_H $\gamma 4$ gene segments (96). Further work is needed to substantiate this possibility and to identify other potential switch-regulatory sequence deletions.

Light chain mutations have also been identified. In particular, one patient with IgA-D and cystic fibrosis was relatively asymptomatic; however, he failed to form κ light chains. Molecular studies demonstrated the presence of two distinct $C\kappa$ point mutations. These mutations resulted in amino acid substitutions that prevented the stable formation of intrachain disulfide bonds required for κ light chain stability (100).

Immunoglobulin gene deletions have been associated with a set of rare lymphoproliferative disorders known as heavy chain diseases (HCD). These disorders are characterized by the expansion of a clonal B-cell population that synthesizes immunoglobulins made of truncated heavy chains only. The sequencing of the immunoglobulin genes from an individual with HCD revealed that partial gene deletions and nonimmunoglobulin gene insertions caused the elimination of normal RNA-splicing sites. This, in turn, led to aberrant mRNA transcription and the formation of a truncated antibody associated with HCD (101).

D. Hyperimmunoglobulin M Syndrome

Hyper-IgM syndrome was initially identified in infant boys suggesting an X-linked mode of inheritance (102); however, other forms of this disorder have been reported. There is a late–onset–acquired form that has been diagnosed in both sexes. This syndrome has also been reported in some children as a complication of congenital rubella (103). Regardless of the form of presentation, the hallmark of this disorder is increased levels of serum IgM, with minimal serum levels of IgG and IgA. Phenotypic evaluation demonstrated the absence of IgG- and IgA-bearing B cells in the peripheral blood and lymph nodes of patients (104). The presence of only IgM-bearing B cells and the lack of production of specific IgG and IgA antibodies following immunizations, supported the hypothesis of an isotype-switch defect as the basis of hyper-IgM syndrome. Recent studies demonstrated the presence of IgM and IgG double-isotype-bearing cells in a cloned EBV B-cell line derived from a patient with hyper-IgM syndrome. Sequence analysis of the immunoglobulin genes demonstrated that no rearrangement of the switch regions occurred and that the same V_HDJ_H exon was used by both the μ and γ mRNA transcripts. These results imply that the IgM and IgG molecules were formed by alternate RNA splicing of the same V_HDJ_H primary transcript (105). Whether this double-isotype-bearing B cell is an intermediate step required before isotype switching remains to be determined. In any event, these results support the hypothesis of an isotype-switch defect.

E. Selective Immunoglobulin M Deficiency

Selective IgM deficiency is a rare disorder characterized by the absence of serum IgM and normal levels of other immunoglobulin classes. Persons with this disorder are susceptible to overwhelming bacterial infections, usually caused by pseudomonas, pneumococcus, or meningococcus (106). Most patients have normal T-cell immunity and some have normal responses to foreign antigens, as reflected by the synthesis of specific IgG and IgA antibodies. Phenotypic examination demonstrated the presence of a normal number of IgM-bearing B cells in some patients, implying the integrity of the $C\mu$ gene segments (64). The ability of some patients to form specific IgG and IgA antibody responses after immunization suggested that the genetic components required for isotype switching were intact. This disorder has not yet been associated with known structural gene defects of the IgM heavy chain locus. Coculture studies, with T cells from normal persons, demonstrated normal levels of in vitro synthesis of IgM (64,107,108). These results suggest a possible immunoregulatory defect as the basis of selective IgM deficiency.

F. Environmental Factors

Throughout this section we have discussed the known and implied immunogenetic basis for the B-cell clinical immunodeficiencies; however, environmental factors can clearly play a role in the pathogenesis of the B-cell maturation arrest associated with these disorders. It has been well established that congenital infections with rubella virus can result in IgA-D or, less frequently, CVID and hyper-IgM syndrome (109,110). Other congenital infections, specifically with cytomegalovirus and *Toxoplasma gondii* organisms have also been associated with the development of IgA-D. Late-onset forms of IgA-D and CVID have been diagnosed following Epstein-Barr viral infections.

The administration of some antisiezure medications and several medications used for the treatment of rheumatoid arthritis and inflammatory bowel disease have been associated with the development of IgA-D. Of interest, a possible genetic predisposition is associated with some forms of drug-induced IgA-D, as indicated by the high incidence of particular *MHC* haplotypes among affected individuals. In most cases the immunoglobulin deficiency state is reversible on discontinuation of the medication (reviewed in 86). The specific modes by which these drugs affect B-cell differentiation have yet to be determined.

G. Chromosomal Translocations Involving the Immunoglobulin Genes

In some B-cell cancers, such as Burkitt's lymphoma, chromosomal translocations involving the immunoglobulin gene loci occur. A typical translocation involves a reciprocal exchange of gene segments between chromosome 8 and 14 {t (8;14)}. Specifically, a region from band q24 of chromosome 8 (which contains the c-*myc* oncogene) is exchanged with a region from band q32 of chromosome 14 (which contains the heavy chain gene loci), as is illustrated in Figure 9. Less often, translocations involve exchanges of gene segments from chromosomes 2p11 and 22q11, which respectively contain the κ and λ light-chain gene loci, with chromosome 8q24. Identification of the c-*myc* oncogene at chromosome 8q24 implicated its possible role in Burkitt's lymphoma. Earlier studies identified similar translocations of the c-*myc* gene to the immunoglobulin gene loci in murine plasmacytomas. The actual insertion site and orientation of the translocated c-*myc* gene within the μ heavy chain gene locus on chromosome 14 is variable, but often occurs at the μ switch region or within the J_H gene segment. The translocation and reciprocal exchange of gene segments usually results in insertion of V_H gene segments into chromosome 8, whereas the nonaffected chromosome 14 maintains normal Cμ polypeptide synthesis. These tumors are characterized by deregulation of the translocated c-*myc* gene, which results in atypically high levels of c-*myc* RNA transcription. This results in overexpression of the c-*myc* protein, which is thought to play a role in malignant transformation (111).

Similar translocation events have been identified in chronic lymphocytic leukemia (CLL) and in follicular lymphoma. The translocation in some cases of CLL occurs between chromosomes 11q13 and 14q32, whereas in follicular lymphoma more than 80% of the patients studied demonstrated translocations between chromosomes 18q21 and 14q32 (Fig. 10). In both lymphoproliferative diseases, the insertion site on chromosome 14 was within a J_H gene segment

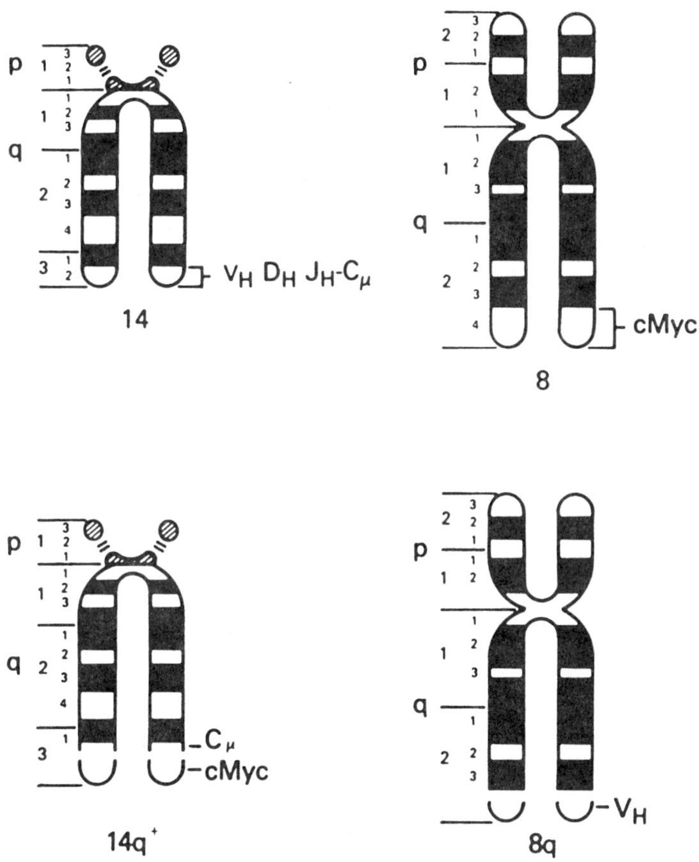

Figure 9 The reciprocal chromosomal t(8:14) translocation in Burkitt's lymphoma. The top two chromosomes illustrated depict normal configurations. The lower chromosome 14 contains a "deregulated" c-*myc* gene that was translocated from chromosome 8. In turn, the lower chromosome 8 contains V_H gene segments translocated from chromosome 14 (From Ref. 115.)

near the enhancer element. The chromosomal breakpoints have the 5′ end of a J_H gene segment on chromosome 14, whereas chromosome 18 contains the 3′ end of a D gene segment that, before translocation, was juxtaposed to the J_H within a V_HDJ_H complex. Extra nucleotides, that are not from chromosome 14 or 18, have been identified at the chromosomal breakpoints. The addition of these nucleotides to chromosome 14 is thought to occur in a process similar to N-diversity addition and is possibly mediated by TdT. The occurrence of

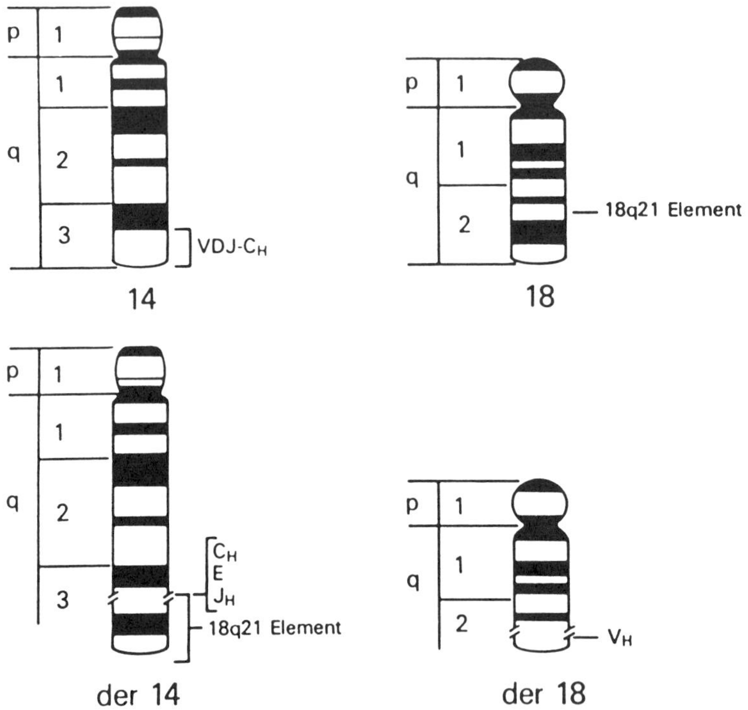

Figure 10 The reciprocal chromosomal t(14;18)(q32;q21) translocation in follicular lymphoma. The upper chromosomes represent the normal location of the *VDJ* and 18q21 elements, respectively, on chromosomes 14 and 18. The lower chromosomes illustrate the results of reciprocal translocation with the derivative (der) chromosome 14 having received the *Bcl-2* gene locus (18q21), near the J_H enhancer (*E*) element of the heavy chain gene locus, while chromosome 18 received the translocated V_H gene segment. (From Ref. 115.)

translocation during the time of V_HDJ_H rearrangement implies that this process happened during the early pre-B-cell stage of differentiation.

The breakpoint of chromosome 18q21 generated by the translocation process is within a small cluster of genes. The cloning and identification of genes within this cluster have been very informative. A new B-cell-associated gene at chromosome 18q21 has been identified and is known as *Bcl-2*. The breakpoint on chromosome 18 involves the *Bcl-2* gene and results in it being in the same transcriptional orientation as the immunoglobulin gene segments. Thus, a Bcl-2–Ig mRNA is transcribed, and translation of this mRNA leads to overexpression of the *Bcl-2* gene product. Although additional genetic events may be required, overexpression of the *Bcl-2* gene is thought to cause malignant transformation by prolonging survival and expansion of the B-cell population containing the translocation (112,113). Since the t(14;18)(q32;q21) translocation may occur during the pre-B-cell stage of development, further immunoglobulin gene rearrangement events are likely to occur. The translocation breakpoint remains unchanged and, thereby, remains a molecular marker that can be followed throughout the course of the tumor.

NOTE ADDED IN PROOF

Recent investigations, completed after this chapter was sent to the publisher, have shed light upon the B-cell differentiation abnormalities of X-linked agammaglobulinemia (XLA) and the x-linked hyper-IgM syndrome (HIM). Tsukada and his associates (116) demonstrated an absence or reduced expression of a cytoplasmic tyrosine kinase, termed B-cell progenitor tyrosine kinase (BPK) in pre-B and B-cell lines from patients with XLA. The decreased expression of the BPK gene may be linked to XLA since: (1) the BPK gene locus was mapped within 100 kb of the DXS 178 marker which defines a DNA polymorphism linked to XLA, and (2) BPK mRNA and protein expression was decreased or absent only in XLA pre-B and B-cell lines and not in cell lines derived from normal individuals or obligate XLA gene carriers. Further investigations focusing on the interplay between the BPK gene product and other molecules may lead to a better understanding of the role BPK plays in the pre-B cell arrest in XLA, and the importance of this kinase in B-cell differentiation.

Aruffo and colleagues (117) reported a defective interaction between the CD40 receptor on B-cells and its ligand on T-cells, gp39, in patients with X-linked hyper-IgM syndrome (HIM). B cells from HIM patients express functional CD40, but their T-cells cannot bind to the CD40 determinant. Ligation of the B-cell CD40 cell surface determinant has been demonstrated to play an important role in B-cell proliferation and, more importantly, in immunoglobulin isotype switching of B-cells. Although patients expressed normal levels of gp39 mRNA, the mRNA encoded for defective gp39 protein as evidenced by mutations in the extracellular domain of the gp39 molecule and defective binding of the gp39 ligand to the CD40 receptor on B-cells. Furthermore, the gene encoding the gp39 molecule was mapped to the region of the X-chromosome previously associated with HIM, the Xq26 region.

REFERENCES

1. A. Fagraeus, Antibody production in relation to the development of plasma cells, *Acta Med. Scand. Suppl., 204* (1948).
2. D. Davies, and H. Metzger, Structural basis of antibody function, *Annu. Rev. Immunol., 1*: 87 (1983).

3. E. Kabat, The antibody combing site, *Prog. Immunol.*, *5*: 67 (1983).
4. A. Nisonoff, and G. Thorbecke, Immunochemistry, *Annu. Rev. Biochem.*, *33*: 355 (1964).
5. B. Hanes, and D. Glover, eds., *Molecular Immunology*, IRL, Oxford, p. 1 (1988).
6. L. Glynn, and M. Steward, eds., *Structure and Function of Antibodies*, John Wiley & Sons, Chichester (1981).
7. D. Burton, Structure and function of antibodies, *Molecular Genetics of Immunoglobulins* (F. Calabi and M. Neuberger, eds.), Elsevier Science, Amsterdam, p. 1 (1987).
8. H. Metzger, Structure and functions of IgM immunoglobulins, *Adv. Immunol.*, *12*: 57 (1970).
9. T. Tomasi, *The Immune System in Secretions*, Prentice-Hall, Englewood Cliffs, NJ (1976).
10. K. Ishizaka, Structure and biological activity of immunoglobulin E, *The Biology of Immunologic Disease* (F. Dixon, and D. Fisher, eds.), Sinauer Associates, Sunderland, MA, p. 13. (1983).
11. S. Tonegawa, Somatic generation of antibody diversity, *Nature*, *302*: 575 (1983).
12. R. Poljak, L. Amzel, H. Avey, B. Chen, R. Phizackerley, and F. Saul, Three-dimensional structure of the Fab′ fragment of a human immunoglobulin at 2.8A resolution, *Proc. Natl. Acad. Sci. USA*, *70*: 3305 (1973).
13. T. Wu, and E. Kabat, An analysis of the sequences of the variable regions of Bence Jones proteins and myeloma light chains and their implications for antibody complementarity, *J. Exp. Med.*, *132*: 211 (1970).
14. J. Berman, S. Mellis, R. Pollock, et al., Content and organization of the human Ig V_H locus: Definition of three new V_H families and linkage to the Ig C_H locus, *EMBO J.*, *7*: 727 (1988).
15. G. Rechavi, B. Bienz, D. Ram, Y. Ben-Neriah, and J. Cohen, Organization and evolution of immunoglobulin VH gene subgroups, *Proc. Natl. Acad. Sci. USA*, *79*: 4405 (1982).
16. F. Mortari, J. Newton, J. Wang, and J. H. W. Schroeder, The human cord blood antibody repertoire. Frequent usage of the V_H7 gene family, *Eur. J. Immunol.*, *22*(1): 241–5 (1992).
17. G. Rathbun, J. Berman, C. Yancopoulos, and F. Alt, Organization and expression of the mammalian heavy-chain variable-region locus, *Immunoglobulin Genes* (T. Honjo, F. Alt, and T. Rabbitts, eds.), Academic Press, New York, p. 63 (1989).
18. J. Ravetch, U. Siebenlist, S. Korsmeyer, T. Waldmann, and P. Leder, The structure of the human immunoglobulin mu locus: Characterization of embryonic and rearranged *J* and *D* genes. *Cell*, *27*: 583 (1981).
19. M. Lieber, Site-specific recombination in the immune system, *FASEB J.*, *5*: 2934 (1991).
20. N. Landau, D. Shatz, M. Rosa, and D. Baltimore, Increased frequency of N-region insertion in a murine pre-B cell line infected with a terminal deoxynucleotidyl transferase retroviral expression vector, *Mol. Cell. Biol.*, *7*: 3237 (1987).
21. H. Gu, I. Forster, and K. Rajewsky, Sequence homologies, N sequence insertion and J_H gene utilization in V_HDJ_H joining: Implications for the joining mechanism

and the ontogenetic timing of Ly1B cell and B cell progenitor generation, *EMBO J.*, *6*: 3299 (1990).
22. K. Meek, Analysis of junctional diversity during B lymphocyte development, *Science*, *250*: 820 (1990).
23. K. Gregoire, I. Goldschneider, R. Barton, and F. Bollum, Ontogeny of terminal deoxynucleotidyl transferase-positive cells in lymphopoietic tissues of rat and mouse, *J. Immunol.*, *123*: 1347 (1979).
24. P. Rothman, S. Lutzker, B. Gorham, V. Stewart, R. Coffman, and F. Alt. Structure and expression of germline immunoglobulin λ3 heavy chain gene transcripts: Implications for mitogen and lymphokine directed class-switching, *Int. Immunol.*, *2*: 622 (1990).
25. F. Alt, P. Ferrier, B. Malynn, et al., Control of recombination events during lymphocyte differentiation. Heavy chain variable region gene assembly and heavy chain class switching, *Ann. N.Y. Acad. Sci.*, *546*: 9 (1988).
26. K. Calame, Immunoglobulin gene transcription: Molecular mechanisms, *Trends Genet.*, *5*: 395 (1989).
27. R. Grosschedl, and D. Baltimore, Cell-type specificity of immunoglobulin gene expression is regulated by at least three DNA sequence elements, *Cell*, *41*: 885 (1985).
28. J. Mason, G. Williams, and M. Neuberger, Transcription cell type specificity is conferred by an immunoglobulin V_H gene promoter that includes a functional consensus sequence, *Cell*, *41*: 479 (1985).
29. T. Parslow, D. Blair, W. Murphy, and D. Granner, Structure of the 5′ ends of the immunoglobulin genes: A novel conserved sequence, *Proc. Natl. Acad. Sci. USA*, *81*: 2650 (1984).
30. T. Wirth, L. Staudt, and D. Baltimore, An octamer oligonucleotide upstream of a TATA motif is sufficient for lymphoid-specific promoter activity, *Nature*, *329*: 174 (1987).
31. L. Staudt, and M. Lenardo, Immunoglobulin gene transcription, *Annu Rev. Immunol.*, *9*: 373 (1991).
32. L. Staudt, H. Singh, R. Sen, T. Wirth, P. Sharp, and D. Baltimore, A lymphoid-specific protein binding to the octamer motif of immunoglobulin genes, *Nature*, *323*: 640 (1986).
33. I. Kirsch, C. Morton, K. Nakahara, and P. Leder, Human immunoglobulin heavy chain genes map to a region of translocations in malignant B lymphocytes, *Science*, *216*: 301 (1982).
34. J. Flanagan, and T. Rabbitts, Arrangement of human immunoglobulin heavy chain constant region genes implies evolutionary duplication of a segment containing gamma, epsilon and alpha genes, *Nature*, *300*: 709 (1982).
35. T. Honjo, F. Alt, and T. Rabbitts, eds., *Immunoglobulin Genes*, Academic Press, London (1989).
36. A. Shimizu, N. Takahashi, Y. Yaoita, and T. Honjo, Organization of the constant-region gene family of the mouse immunoglobulin heavy chain, *Cell*, *28*: 499 (1982).
37. T. Honjo, A. Shimizu, and Y. Yaoita, Constant-region genes of the immunoglobulin heavy chain and the molecular mechanism of class switching,

Immunoglobulin Genes (T. Honjo, F. Alt, and T. Rabbitts, eds.), Harcourt Brace Jovanovich, London (1989).
38. T. Nikaido, S. Nakai, and T. Honjo, Switch region of immunoglobulin *C*mu gene is composed of simple tandem repetitive sequences, *Nature*, *292*: 845 (1981).
39. M. Obata, T. Kataoka, S. Nakai, et al., Structure of a rearranged gamma 1 chain gene and its implication to immunoglobulin class-switch mechanism, *Proc. Natl. Acad. Sci. USA*, *78*: 2437 (1981).
40. M. Wabl, J. Meyer, G. Beck-Engeser, M. Tenkhoff, and P. Burrows, Critical test of a sister chromatid exchange model for the immunoglobulin heavy-chain class switch, *Nature*, *313*: 687 (1985).
41. W. Gathings, A. Lawton, and M. Cooper, Immunofluorescent studies on the development of pre-B-cells, B lymphocytes, and immunoglobulin isotype diversity in humans, *Eur. J. Immunol.*, *7*: 804 (1977).
42. K. Moore, J. Rogers, I. Hunkapiller, et al., Expression of IgD may use both DNA rearrangement and RNA splicing mechanisms, *Proc. Natl. Acad. Sci. USA*, *78*: 1800 (1981).
43. A. Shimizu, T. Kinashi, M. Nussenzweig, T. Mizuta, P. Leder, and T. Honjo, Molecular mechanism for immunoglobulin double-isotype expression, *Cold Spring Harbor Symp. Quant. Biol.*, *54* (Part 1): 175 (1989).
44. J. Rogers, P. Early, C. Carter, et al., Two mRNAs with different 3′ ends encode membrane bound and secreted forms of immunoglobulin μ chain, *Cell*, *20*: 303 (1980).
45. P. Early, J. Rogers, M. Davis, et al., Two mRNAs can be produced from a single immuno-globulin mu gene by alternative RNA processing pathways, *Cell*, *20*: 313 (1980).
46. F. Alt, A. Bothwell, M. Knapp, et al., Synthesis of secreted and membrane-bound immunoglobulin mu heavy chains is directed by mRNAs that differ at their 3′ ends, *Cell*, *20*: 293 (1980).
47. J. Guise, G. Galli, J. Nevins, and P. Tucker, Developmental regulation of secreted and membrane forms of immunoglobulin μ chain, *Immunoglobulin Genes* (T. Honjo, F. Alt, and T. Rabbitts, eds.), Harcourt Brace Jovanovich, London, p. 275 (1989).
48. S. Malcolm, P. Barton, C. Murphy, M. Ferguson-Smith, D. Bentley, and T. Rabbitts, Localization of human immunoglobulin κ light chain variable region genes to the short arm of chromosome 2 by in situ hybridization, *Proc. Natl. Acad. Sci. USA*, *79*: 4957 (1982).
49. J. Erikson, J. Martinis, and C. Croce, Assignment of the genes for human λ immunoglobulin chains to chromosome 22, *Nature*, *294*: 173 (1981).
50. E. Lai, R. Wilson, and L. Hood, Physical maps of the mouse and human immunoglobulin-like loci, *Adv. Immunol.*, *46*: 1 (1989).
51. H. Zalchau, Immunoglobulin light-chain genes of the κ type in man and mouse, *Immunoglobulin Genes* (T. Honjo, F. Alt, and T. Rabbitts, eds.), Academic Press, New York, p. 92 (1989).
52. T. Waldmann, The arrangement of immunoglobulin and T cell receptor genes in human lymphoproliferative disorders, *Adv. Immunol.*, *40*: 247 (1987).

53. S. Tonegawa, Nobel lecture in physiology or medicine—1987. Somatic generation of immune diversity, *In Vitro Cell Dev. Biol.*, *24*: 253 (1988).
54. S. Korsmeyer, P. Hieter, J. Ravetch, D. Poplack, T. Waldmann, and P. Leder. Developmental hierarchy of immunoglobulin gene rearrangements in human leukemic pre-B-cells, *Proc. Natl. Acad. Sci. USA*, *78*: 7096 (1981).
55. T. Blackwell, and F. Alt, Mechanism and developmental program of immunoglobulin gene rearrangement in mammals, *Annu. Rev. Genet.*, *23*: 605 (1989).
56. E. Selsing, J. Durdik, M. Moore, and D. Persiani, Immunoglobulin λ genes, *Immunoglobulin Genes* (T. Honjo, F. Alt, and T. Rabbitts, eds.), Academic Press, London, p. 111 (1989).
57. H. Chang, E. Dmitrovsky, P. Hieter, et al., Identification of three new Ig λ-like genes in man, *J. Exp. Med.*, *163*: 425 (1986).
58. P. Chuchana, A. Blancher, F. Brockly, D. Alexandre, G. Lefranc, and M. Lefranc, Definition of the human immunoglobulin variable lambda (*IGLV*) gene subgroups, *Eur. J. Immunol.*, *20*: 1317 (1990).
59. K. Calame, Mechanisms that regulate immunoglobulin gene expression, *Annu. Rev. Immunol.*, *3*: 159 (1985).
60. L. Smith, I. Weissman, and S. Heimfeld, Clonal analysis of hematopoietic stem cell differentiation in vivo, *Proc. Natl. Acad. Sci. USA*, *88*: 2788 (1991).
61. W. Davidson, T. Fredrickson, E. Rudikoff, R. Coffman, J. Hartley, and H. Morse, A unique series of lymphomas related to the Ly-1^+ lineage of B-lymphocyte differentiation, *J. Immunol.*, *133*: 744 (1984).
62. C. Muller-Sieburg, C. Whitlock, and I. Weissman, Isolation of two early B lymphocyte progenitors from mouse marrow: A committed pre-pre-B cell and a clonogenic thy-1^{10} hematopoietic stem cell, *Cell*, *44*: 653 (1986).
63. B. Thorens, M. Schulz, and P. Vassalli, Bone marrow pre-B lymphocytes synthesize immunoglobulin μ chains of membrane-type with different properties and intracellular pathways, *EMBO J.*, *4*: 361 (1985).
64. D. Levitt, and M. Cooper, Mouse pre-B cells synthesize and secrete μ heavy chains but not light chains, *Cell*, *19*: 617 (1980).
65. S. Pillai, and D. Baltimore, Formation of disulphide-linked $\mu_2\omega_2$ tetramers in pre-B cells by the 18K ω-immunoglobulin light chain, *Nature*, *329*: 172 (1987).
66. S. Pillai, and D. Baltimore, The omega and iota surrogate immunoglobulin light chains, *Curr. Top. Microbiol. Immunol.*, *137*: 136 (1988).
67. A. Rolink, and F. Melchers, Molecular and cellular origins of B lymphocyte diversity, *Cell*, *66*: 1081 (1991).
68. B. Cherayil, and S. Pillai, The omega/lambda 5 surrogate immunoglobulin light chain is expressed on the surface of transitional B lymphocytes in murine bone marrow, *J. Exp. Med.*, *173*: 111 (1991).
69. B. Cherayil, and S. Pillai, The ω/λ5-surrogate immunoglobulin light chain is expressed on the surface of transitional B lymphocytes in murine bone marrow, *J. Exp. Med.*, *173*: 111 (1991).
70. S. Bauer, A. Kudo, and F. Melchers, Structure and pre B lymphocyte-restricted expression of the V_{preB} gene in humans and conservation of its structure in other mammalian species, *EMBO J.*, *7*: 111 (1988).

71. M. Reth, and F. Alt, Novel immunoglobulin heavy chains are produced from DJ_H gene segment rearrangements in lymphoid cells, *Nature*, *312*: 418 (1984).
72. G. Yancopoulos, and F. Alt, Developmentally controlled and tissue-specific expression of unrearranged V_H gene segments, *Cell*, *40*: 271 (1985).
73. T. Tsubata, R. Tsubata, and M. Reth, Cell surface expression of the short immunoglobulin μ chain (Dμ protein) in murine pre-B cells is differently regulated from that of the intact μ chain, *Eur. J. Immunol.*, *21*: 1359 (1991).
74. D. Osmond, B cell development in the bone marrow, *Semin. Immunol.*, *2*: 173 (1990).
75. D. Osmond, S. Priddle, and S. Rico-Vargas, Proliferation of B cell precursors in bone marrow of pristane-conditioned and malaria-infected mice: Implications for B cell oncogenesis, *Curr. Top. Microbiol. Immunol.*, *166*: 149 (1990).
76. M. Oettinger, D. Schatz, C. Gorka, and D. Baltimore, *RAG-1* and *RAG-2*, adjacent genes that synergistically activate *V(D)J* recombination, *Science*, *148*: 1517 (1990).
77. M. Potter, Neoplasic development in B lymphocytes, *Carcinogenesis*, *11*: 1 (1990).
78. M. Conley, B cells in patients with X-linked agammaglobulinemia, *J. Immunol.*, *134*: 3070 (1985).
79. S. Tsuchiya, T. Konno, K. Tada, and Y. Ono, Epstein–Barr virus-induced lymphoblastoid cell lines from patients with primary immunodeficiency diseases, *Scand. J. Immunol.*, *11*: 155 (1980).
80. E. Mensink, R. Schuurman, J. Schot, A. Thompson, and F. Alt, Immunoglobulin heavy chain gene rearrangements in X-linked agammaglobulinemia, *Eur. J. Immunol.*, *16*: 963 (1986).
81. J. Schwaber, H. Molgaard, S. Orkin, H. Gould, and F. Rosen, Early pre-B cells from normal and X-linked agammaglobulinemia produce Cμ without V_H region, *Nature*, *304*: 355 (1983).
82. R. Anker, M. Conley, and B. Pollok, Clonal diversity in the B cell repertoire of patients with X-linked agammaglobulinemia, *J. Exp. Med.*, *169*: 2109 (1989).
83. S. Kwan, J. Terwilliger, R. Parmley, et al., Identification of a closely linked DNA marker, DXS178, to further refine the X-linked agammaglobulinemia locus, *Genomics*, *6*: 238 (1990).
84. M. Conley, P. Brown, A. Pickard, et al., Expression of the gene defect in X-linked agammaglobulinemia, *N. Engl. J. Med.*, *79*: 564 (1986).
85. J. Puck, R. Nussbaum, and M. Conley, Carrier detection in X-linked severe combined immunodeficiency based on patterns of X chromosome inactivation, *J. Clin. Invest.*, *79*: 1395 (1987).
86. F. Schaffer, R. Monteiro, J. Volanakis, and M. Cooper, IgA deficiency, *Immunodef. Rev.*, *3*: 15 (1991).
87. M. Conley, and M. Cooper, Immature IgA B cells in IgA-deficient patients, *N. Engl. J. Med.*, *305*: 495 (1981).
88. R. Good, S. Zak, R. Condie, and R. Bridges, Clinical investigation of patients with agammaglobulinemia and hypogammaglobulinemia, *Pediatr Clin. North Am.*, *7*: 397 (1960).
89. W. Gathings, A. Lawton, and M. Cooper, Immunofluorescent studies of the development of pre-B cells, B lymphocytes and immunoglobulin isotype diversity in humans, *Eur. J. Immunol.*, *7*: 804 (1977).

90. L. Hammarstrom, G. de Lange, and C. Smith, *IgA2* allotypes determined by restriction fragment length polymorphism in IgA deficiency. Re-expression of the silent *A2m(2)* allotype in the children of IgA-deficient patients, *J. Immunogenet.*, *14*: 197 (1987).
91. L. Hammarstrom, B. Carlsson, C. Smith, J. Wallin, and L. Wieslander, Detection of IgA heavy chain constant region genes in IgA deficient donors: Evidence against gene deletions, *Clin. Exp. Immunol.*, *60*: 661 (1985).
92. F. Schaffer, J. Palermos, Z. Zhu, B. Barger, M. Cooper, and J. Volankis, Genotypic studies of *MHC* class III genes in individuals with IgA deficiency and common variable immunodeficiency, *Prog. Immunol.*, *111*: 544 (1989).
93. F. Schaffer, J. Palermos, Z. Zhu, B. Barger, M. Cooper, and J. Volanakis, Individuals with IgA deficiency and common variable immunodeficiency share polymorphisms of major histocompatibility complex class III genes, *Proc. Natl. Acad. Sci. USA*, *86*: 8015 (1989).
94. J. Volankis, Z. Zhu, F. Schaffer, et al., *MHC* class III gene(s) endow(s) susceptibility to both IgA deficiency and common variable immunodeficiency, *Progress in Immune Deficiency*, *III* (H. Chapel, R. Levinsky, and A. Webster, eds.), Royal Society of Medicine Services (1991).
95. N. Migone, S. Oliviero, G. de Lange, et al., Multiple gene deletions within the human immunoglobulin heavy-chain cluster, *Proc. Natl. Acad. Sci. USA*, *81*: 5811 (1984).
96. A. Bottaro, M. De Marchi, G. De Lange, et al., New types of multiple and single gene deletions in the human *IgCH* locus, *Immunogenetics*, *29*: 44 (1989).
97. G. Lefranc, H. Chaabani, E. Van Loghem, M. P. Lefranc, G. deLange, and A. N. Helal, Simultaneous absence of the human IgG1, IgG2, IgG4, and IgA1 subclasses: immunological and immunogenetical considerations, *Eur. J. Immunol.*, *13*(3): 240–4 (1983).
98. M. Lefranc, G. Lefranc, and T. Rabbitts, Inherited deletion of immunoglobulin heavy chain constant region genes in normal human individuals, *Nature*, *300*: 760 (1982).
99. H. Chaabani, N. Bech-Hansen, and D. Cox, A multigene deletion within the immunoglobulin heavy chain region, *Am. J. Hum. Genet.*, *37*: 1164 (1985).
100. J. Stavnezer, and B. Zegers, Aberrant kappa constant region genes in a patient lacking type kappa immunoglobulins, *Primary Immunodeficiency Diseases* (M. Eibl, and F. Rosen, eds.), Elsevier Publishing, Amsterdam, p. 55 (1986).
101. P. Guglielmi, A. Bakhshi, M. Cogne, M. Seligmann, and S. Korsmeyer, Multiple genomic defects result in an alternative RNA splice creating a human λ H chain disease protein, *J. Immunol.*, *141*: 1762 (1988).
102. F. Rosen, S. Kevy, E. Merler, C. Janeway, and D. Gitlin, Recurrent bacterial infections and dysgammaglobulinemia: Deficiency of 7S gamma-globulins in the presence of elevated 19S gamma-globulins, *Pediatrics*, *28*: 182 (1961).
103. R. Schimke, C. Bolano, and C. Kirkpatrick, Immunologic deficiency in the congenital rubella syndrome, *Am. J. Dis. Child.*, *118*: 626 (1969).
104. D. Levitt, P. Haber, K. Rich, and M. Cooper, Hyper IgM immunodeficiency. A pri- mary dysfunction of B lymphocyte isotype switching, *J. Clin. Invest.*, *72*: 1650 (1983).
105. Y. Akahori, Y. Kurosawa, Y. Kamachi, S. Torii, and H. Matsuoka, Presence of

immunoglobulin (Ig) M and IgG double isotype-bearing cells and defect of switch recombination in hyper IgM immunodeficiency, *J. Clin. Invest.*, *85*: 1722 (1990).
106. J. Hobbs, IgM deficiency, *Birth Defects*, *11*: 112 (1975).
107. D. Levitt, and M. Cooper, IgM production: Normal development and selective deficiency, *Primary Immunodeficiencies* (M. Seligmann, and W. Hitzig, eds.), Elsevier/North Holland Biomedical, New York, p. 3 (1980).
108. T. Inoue, Y. Okumura, M. Shirahama, H. Ishibashi, S. Kashiwagi, and H. Okubo, Selective partial IgM deficiency: Functional assessment of T and B lymphocytes in vitro, *J. Clin. Immunol.*, *6*: 130 (1986).
109. M. Hancock, C. Huntley, and J. Sever, Congenital rubella syndrome with immunoglobulin disorder, *J. Pediatr.*, *72*: 636 (1968).
110. J. Soothill, K. Hayes, and J. Dudgeon, The immunoglobulins in congenital rubella, *Lancet*, *1*: 1385 (1966).
111. S. Korsmeyer, B-lymphoid neoplasms: Immunoglobulin genes as molecular determinants of clonality, lineage, differentiation, and translocation, *Adv. Intern. Med.*, *33*: 1 (1988).
112. T. McDonnell, N. Deane, F. Platt, et al., *bcl-2*-immunoglobulin transgenic mice demonstrate extended B cell survival and follicular lymphoproliferation, *Cell*, *57*: 79 (1989).
113. T. McDonnell, G. Nunez, F. Platt, et al., Deregulated *Bcl-2*-immunoglobulin transgene expands a resting but responsive immunoglobulin M and D-expressing B-cell population, *Mol. Cell Biol.*, *10*: 1901 (1990).
114. J. Darnell, H. Lodish, and D. Baltimore, eds. *Molecular Cell Biology*, Scientific American Books, New York, p. 1104 (1986).
115. S. Korsmeyer, *Important Advances in Oncology*, J. B. Lippincott, Philadelphia (1987).
116. S. Tsukada, D. C. Saffran, D. J. Rawlings et al., Deficient expression of a B-cell cytoplasmic tyrosine kinase in human X-linked agammaglobulinemia, *Cell*, *72*: 279–90 (1993).
117. A. Aruffo, M. Farrington, D. Hollenbaugh, et al., The CD40 Ligand, gp39, is defective in activated T-cells from patients with X-linked hyper-IgM syndrome, *Cell*, *72*: 291–300 (1993).

2

Molecular Mechanisms of Transmembrane Signaling in B-Cell Differentiation and Activation

Stephen Desiderio and Susan M. Dymecki
The Johns Hopkins University School of Medicine and Howard Hughes Medical Institute, Baltimore, Maryland

I. INTRODUCTION AND OVERVIEW

A. Definitions

The term *activation*, as applied to B lymphocytes, is used variously in the immunological literature. In a broad sense, activation includes three distinct phases in the responses of resting B cells to external stimuli: progression from the resting or G_0 state to G_1; proliferation; and differentiation to become antibody-secreting cells. The ligands that, alone or in combination, can induce these responses are of five general types: antigens, lymphokines, cell surface adhesion molecules, complement components, and antibodies themselves. The term activation is sometimes used in a more restricted way to describe phenotypic changes in B cells that precede proliferation and differentiation. In this chapter we will employ the term in its more inclusive sense.

B. Principal Modes of Antigenic Activation

Activation of B cells is regulated by diverse stimuli, and proceeds by multiple pathways. The requirements for antigenic stimulation of B cells, for example, vary depending on the form of the antigen itself.

Some antigens, classically termed T-independent (TI), are so named because of their ability to induce antibody responses in congenitally athymic (*nu*/*nu* or nude) mice. It is now recognized, however, that functional T cells are present in *nu/nu* mice at about 5–10% of normal, a result of extrathymic T-cell maturation. Therefore, the ability to elicit antibody in *nu/nu* mice cannot be taken as proof that the response is independent of T cells or of their products (see, for example, type II TI antigens, following).

The TI antigens are typically polymers—bacterial capsular polysaccharides, for example—that present multiple copies of a given epitope to a B cell. A subset of these antigens are polyclonal B-cell activators: they can activate B cells regardless of the antigenic specificity of the cells. Depending on the number and density of antigenic units per polymer chain, the binding of a polymeric antigen to multiple antibody molecules on the B cells can stimulate proliferation and antibody secretion, but not differentiation. As a consequence, TI antigens do not produce high-affinity antibodies or yield immunological memory.

The TI antigens have been subdivided, based on their ability to induce antibodies in mice expressing the *xid* (X-linked immunodeficiency) mutation, which is associated with impaired B-cell function (1). Antigens with polyclonal B-cell activator properties elicit responses in *xid* mice and are designated type I TI antigens; polymeric antigens that fail to elicit responses in *xid* mice are designated type II TI antigens (2,3). From experiments with cultured cells, it is apparent that type II TI antigens are, in fact, not strictly T-independent, but require soluble T-cell products to elicit an antibody response (4).

Classic T-dependent (TD) antigens fail to elicit antibody responses in *nu/nu* mice. This category includes most soluble proteins, as well as red blood cells. The TD antigens can be divided into two types: those for which the T-cell requirement can be satisfied by supernatants from activated T cells (type I TD), and those that require the presence of T cell (type II TD) (5). Type I TD antigens include red blood cells; the type II TD class includes most proteins.

II. SYNOPSIS OF B-CELL DEVELOPMENT

Development of B cells can be divided into two phases: one occurring in primary lymphoid tissues independently of antigen, the other occurring in secondary lymphoid tissues, such as spleen, lymph nodes, and Peyer patches, and dependent on exposure to antigen. In developing mammals, hematopoietic stem cells, including those that eventually give rise to the lymphoid lineages, arise in the yolk sac. In the mouse (gestational period 19 days), stem cells enter the fetal liver by embryonic day 9 (E9). The earliest B-cell precursors arise in murine fetal liver at about E11–E12 (6). The murine fetal liver continues to produce B cells for about 3 weeks, but by E17, B-cell production begins in the bone marrow, which is the major site of B-cell production in the adult animal. In humans, the earliest B-cell production begins in the fetal liver during the eighth week of gestation (6); by midgestation, the bone marrow becomes the principal site. The

B cells leave the bone marrow and enter the periphery, where they reside in lymph nodes, spleen, and Peyer patches. At these sites, B cells are activated and undergo further differentiation.

On the basis of immunoglobulin (Ig) expression, B-lineage cells in the antigen-independent phase of development are divided into two broad classes: pre-B cells and B cells. Pre-B cells (7) express intracellular immunoglobulin μ chain, but not light chain (Fig. 1). Most of the μ chain expressed in these cells is retained in the endoplasmic reticulum and subsequently degraded. A fraction of the μ chain in pre-B cells is expressed as a complex with two invariant polypeptide chains, V_{preB} and λ5, which serve as surrogates for the variable and constant portions, respectively, of Ig light chain (8). In the mouse, pre-B cells arise by about E12; surface IgM^+ ($sIgM^+$) B cells, which synthesize both Ig heavy chain and light chain, first appear several days later (6). Engagement of the μ–V_{preB}–λ5 complex by an unknown ligand (extracellular or intracellular) has been postulated to be required for progression of pre-B cells to the B-cell stage (9).

The sequential appearance of Ig chains during B-cell ontogeny reflects an underlying developmental order in the assembly of Ig heavy and light chain genes (10). The variable regions of Ig heavy chains are encoded by three DNA segments: V_H, D, and J_H. Similarly, the variable regions of Ig light chains are encoded by two segments: V_L and J_L. During B-cell development heavy and light chain genes are assembled by a series of site-specific DNA rearrangements. The first rearrangements are D-to-J_H joinings; V_H-to-DJ_H rearrangement begins after D-to-J_H recombination has occurred on both chromosomes. Following D-to-J_H rearrangement, DJ_H transcripts are expressed (11,12). A D segment may be joined to a J_H segment in one of three reading frames: RF1, RF2, or RF3. If joining has occurred in RF2, the DJ_H transcripts encode a truncated form of μ protein, termed D_μ (11), which, like intact μ chain, can associate with V_{preB} and λ5 (13). Cells that express D_μ may be prevented from further maturation (14); mature B cells that bear DJ_H rearrangements in RF2 are rarely observed.

After assembly of a productive heavy chain gene, synthesis of μ chain, and

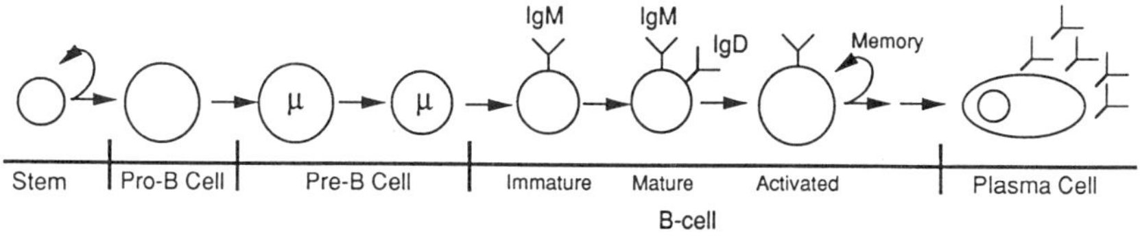

Figure 1 Outline of B-cell development. Four broad developmental compartments—pro-B, pre-B, B, and plasma cell—are indicated at bottom. The transition from large to small pre-B cell is represented schematically. For explanation, see text.

the appearance of the μ–V_{preB}–$\lambda 5$ complex, further rearrangement at the μ locus is suppressed and light chain gene rearrangement is activated (15). Rearrangement of κ genes precedes rearrangement of λ genes (16). If rearrangement at either κ allele is productive, κ-containing, surface-associated IgM (sIgM) appears, and light chain rearrangement ceases; the λ genes retain their germline configuration (16,17). If κ rearrangement has been unproductive at both alleles, λ rearrangement occurs; in this instance, successful assembly of a λ gene results in expression of λ-containing sIgM, after which rearrangement ceases. Expression of sIgM appears to be a prerequisite for progression to the antigen-dependent phase of B-cell development.

Thus, Ig heavy chain may function as a signal transductory molecule at multiple points in B-cell ontogeny (Fig. 2). The μ chain, in association with the surrogate light chain components V_{preB} and $\lambda 5$, is thought to signal cessation of heavy chain gene rearrangement and activation of light chain gene rearrangement, thereby playing a pivotal role in the establishment of allelic exclusion and in the transition from pre-B to mature B cell. The truncated D_μ chain, in association with V_{preB} and $\lambda 5$, may inhibit further development of those cells that express it. The μ–V_{preB}–$\lambda 5$ and $D_\mu V_{preB}$–$\lambda 5$ complexes, therefore, may both function in the antigen-independent phase of B-cell development, although their natural ligands, if they exist, are now unknown. Even though the $D_\mu V_{preB}$–$\lambda 5$ and μ–V_{preB}–$\lambda 5$ complexes have been found on the surfaces of some B-lymphoid cell lines, these complexes have not been observed on the surfaces of normal B-cell progenitors. It remains possible that the complexes function wholly

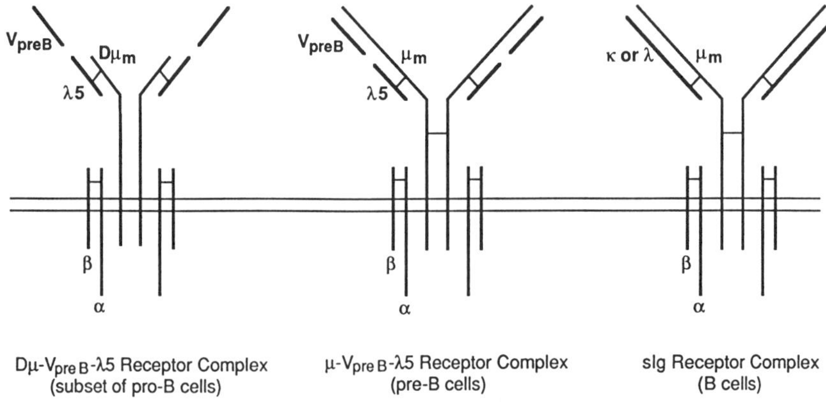

Figure 2 The μ chain-associated polypeptide complexes from B cells and B-cell progenitors. The putative receptor complexes of pro-B and pre-B cells are shown with the B-cell antigen receptor complex. Dμ, μ, V_{preB}, $\lambda 5$, Ig-α, and Ig-β chains are indicated. Interchain sulfhydryl linkages are marked. The stoichiometry of association between the Ig-α–Ig–β heterodimer and Dμ–V_{preB}–$\lambda 5$, μ–V_{preB}–$\lambda 5$ or sIg is uncertain.

intracellularly, perhaps by engaging an intracellular ligand. Immunoglobulin heavy chains, in association with κ or λ light chains, constitute the antigen receptors of B cell, and, as such, are responsible for the antigen-dependent components of B-cell activation.

III. SIGNALING THROUGH SURFACE IMMUNOGLOBULIN

A. The B-Cell Antigen–Receptor Complex

All classes of Ig—IgM, IgD, IgG, IgA, and IgE—exist in membrane-bound as well as soluble forms. The surface-associated forms of all the Ig classes have the general subunit structure H_2L_2, including IgM and IgA, which in soluble form exhibit structures of higher order. In the soluble forms, the antigen-binding sites of sIg are formed nearly exclusively by amino acid residues contributed by the complementarity-determining regions (CDRs) of the variable regions of the heavy and light chains, V_H and V_L. Thus, the membrane-bound forms of all Ig classes are divalent.

The membrane forms of the Ig heavy chains are larger than their secreted forms. As is characteristic of many so-called type I transmembrane proteins, the COOH-terminal of the membrane form of μ (μ_m), for example, contains an acidic region, followed by a 26-amino acid uncharged region and a very short cytoplasmic tail of three amino acids, two of them basic. The secreted form of μ (μ_s) differs from μ_m only in this region (Fig. 3). For each heavy chain, the membrane and secreted forms are specified by two distinct mRNAs that are generated by differential processing (18,19).

The cross-linking of sIg by polyvalent or surface-bound antigens, or by anti-Ig, results in the generation of intracellular chemical signals that initiate proliferative and differentiative responses. These signals are not produced by sIg itself, however: the cytoplasmic tails of sIg molecules are short (3 amino acids for IgM and IgD, 28 for IgG and IgE, 14 for IgA) and apparently devoid of intrinsic enzymatic activities, such as protein-tyrosine phosphorylation or GTP hydrolysis, that have been implicated in transmembrane signaling in B cells. Rather, sIg is

μ_m ...EGEVNAEEEGFENLWTTASTFIVLFLLSLFYSTTVTLFKVK-*COOH*

μ_s ...GKPTLYNVSLIMSDTGGCY-*COOH*

Figure 3 The COOH-termini of membrane and secreted forms of μ heavy chain. Top, μ_m; bottom, μ_s. The COOH-termini are indicated at right. The transmembrane region of μ_m is underlined.

noncovalently associated with transmembrane molecules that link the antigen receptor to regulatory proteins in the interior of the cell.

Under mild lysis conditions, murine sIgM is found in a noncovalent complex with a disulfide-linked heterodimer of 34- and 39-kDa transmembrane glycoproteins (20,21) and sIgD is associated with a similar disulfide-linked heterodimer of 35- and 39-kDa glycoproteins (21,22). After deglycosylation, these proteins are reduced to 21–23 kDa. The 39-kDa component of the sIgM-associated heterodimer is identical with that of the IgD-associated complex. This protein, termed Ig-β, is the product of the *B29* gene (23–25). The smaller components of the sIgM- and sIgD-associated heterodimers, collectively termed Ig-α, are both encoded by a single gene termed *mb-1*; their mobility difference arises from differences in glycosylation (24–27). These complexes are likely to include two Ig-α–Ig-β heterodimers per sIg molecule. The other classes of sIg are also associated with Ig-α–Ig-β heterodimers. The Ig-α associated with sIgE and sIgG is similar in mobility to that of sIgM; the Ig-α found with sIgA comigrates with that of sIgD (26). A similar heterodimer associated with human sIgM contains glycoproteins of 47 and 37 kDa; the larger is the homologue of murine Ig-α (28). The ensemble of sIg and Ig-α–Ig-β heterodimers, termed the B-cell antigen–receptor complex, is related in structure to complexes associated with the T-cell antigen receptor (TCR) and the Fc receptors FcεRI, FcγRIIIA, and FcγRIIIB (Fig. 4).

Surface expression of IgM requires four polypeptides: Ig-α, Ig-β, μ heavy chain, and Ig light chain (κ or λ) (20,26). Dependence on Ig-α–Ig-β for export

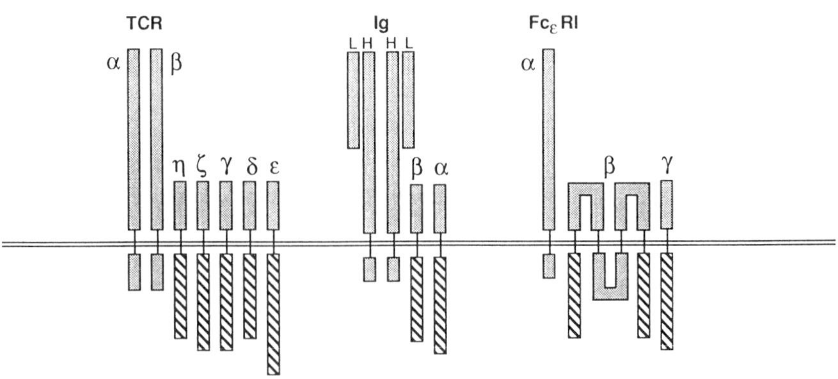

Figure 4 Schematic representation of the T-cell antigen receptor, B-cell antigen receptor, and Fc_ϵ receptor type I complexes. The plasma membrane is indicated by a double-horizontal line, with the outside of the cell at top. Identities of the constituent polypeptide chains are indicated; no attempt is made to represent the stoichiometries of association among constituent chains. Intracellular regions containing Asp/Glu-Tyr-Leu consensus repeats are indicated by crosshatching. For details, see text.

to the cell surface can be conferred on chimeric proteins by inclusion of a COOH-terminal portion of the μ chain, spanning C_H4, the transmembrane region, and the cytoplasmic region (29). The Ig-α and Ig-β are also required for efficient surface transport of IgE, IgA, and probably IgG (26,30). In some experiments, IgD has been expressed on the cell surface in the absence of Ig-α or Ig-β (26,27), but in these instances it is linked to the membrane by a glycosyl-phosphatidylinositol anchor (31). Thus, it appears that all classes of Ig require Ig-α–Ig-β for expression as membrane-spanning molecules.

B. Structures of the Immunoglobulin Accessory Chains Immunoglobulin-α and Immunoglobulin-β

Although the overall sequences of Ig-α and Ig-β are dissimilar, the proteins share several structural similarities. The processed murine Ig-α chain is a 192-amino acid polypeptide, with a 109-amino acid extracellular region, a 22-amino acid transmembrane region, and a 61-amino acid cytoplasmic domain. Mature murine Ig-β contains 199 amino acids, comprising a 129-amino acid extracellular region, a 22-amino acid transmembrane region, and a 48-amino acid cytoplasmic tail. Both are glycoproteins, and the extracellular regions of both proteins contain pairs of cysteine residues compatible with formation of an Ig-fold. Thus, Ig-α and Ig-β are members of the Ig superfamily. Expression of the genes encoding Ig-α and Ig-β, *mb-1* and *B29*, is restricted to cells of the B lineage. Expression of *mb-1* begins early in B-cell ontogeny, before μ gene assembly, and continues through the pro-B, pre-B, and B-cell stages; expression ceases upon differentiation of B cells to plasma cells (32). Expression of *B29* differs in that it continues to be expressed in terminally differentiated cells (33).

The transmembrane regions of both proteins have charged (Glu, Ig-α) or polar (Gln, Ig-β) residues in the fifth position. These residues could be shielded in the membrane if the transmembrane regions of Ig-α and Ig-β were complexed with each other or with the membrane-spanning domains of other proteins. The transmembrane regions of sIg heavy chains are also polar. The μ transmembrane region, for example, contains the intervals TTAST and YSTTVT. Conversion of the former sequence to the hydrophobic sequence VVAAV is sufficient to allow surface expression in the absence of Ig-α, suggesting that the Ig-associated molecules promote surface expression of IgM by masking polar residues (29).

The cytoplasmic regions of Ig-α and Ig-β each contain a consensus sequence motif (34) of the form

D/ExxxxxxxD/ExxYxxLxxxxxxxYxxL/I,

which, remarkably, is also found in the cytoplasmic portions of the TCR CD3 complex polypeptides γ, δ, and ϵ; the TCR accessory chains ζ and η; and the FcϵRI β and γ chains (Fig. 5). The receptors associated with these accessory chains share common features of signal transduction: engagement of each of these complexes is followed by a rapid increase in the phosphorylation of intracellular proteins on tyrosine residues and the activation of one or more *src*-family protein–tyrosine kinases, as will be discussed later. These observations have suggested that the homologous accessory chains function to couple antigen receptors to intracellular-signaling molecules, a view that has acquired some experimental support (see later discussion).

C. Biochemical Consequences of Surface Immunoglobulin Engagement

1. Immunoglobulin Cross-Linking as an Experimental Model for Receptor Engagement

The most widely employed experimental paradigm for B-cell antigen receptor signaling involves cross-linking of sIg with anti-Ig antibodies (35,36). Anti-Ig stimulation of B cells induces a sequence of responses leading to cell proliferation, but does not result in Ig secretion, except in the presence of T cells or T-cell-derived lymphokines (37); in this sense, anti-Ig stimulation resembles exposure to TD antigens. In response to low concentrations of soluble anti-Ig

```
Consensus           DXXX-XXXXDXXYXXLXXXX-----XXXYXXL
                    E        E     I              I

mouse  MB-1         DMPD-DYEDENLYEGLNLDD-----CSMYEDI
human  MB-1         DAGD-EYEDENLYEGLNLDD-----CSMYEDI
mouse  B29          DGKA-GMEEDHTYEGLNIDQ-----TATYEDI
human  B29          DSKA-GMEEDHTYEGLDIDQ-----TATYEDI

mouse  TCR-ζ/ηa     ETAANLQDPNQLYNELNLGR-----REEYDVL
mouse  TCR-ζ/ηb     KQQRRRNPQEGVYNALQKDKM----AEAYSEIKTKG
mouse  TCR-ζc       ERRRGKGH-DGLYQGLSTAT-----KDTYDALHMQTLAPR
mouse  TCR-ηc       ERRRGKGH-DGLYDSHFQAVQF---GNRRERE
mouse  CD3-γ        DKQT-LLQNEQLYQPLKDRE-----YDQYSHL
mouse  CD3-δ        EVQA-LLKNEQLYQPLRDRE-----DTQYSRL
mouse  CD3-ε        NKERPPPVPNPDYEPIRKGQ-----RDLYSGLNQRAV

mouse  FcεRI-γ      AAIASREKADAVYTGLNTRN-----QETYETL
mouse  FcεRI-β      ELESKKVPDDRLYEELNHVY-----SPIYSEL
human  FcγRIIa      ETNNDYETADGGYMTLNPRAPTDDDKNIYLTL
```

Figure 5 An amino acid sequence motif found in the intracellular regions of receptor accessory chains. Portions of the intracellular amino acid sequences of Ig-α *(mb-1)*, Ig-β *(B29)*, TCR-ζ, TCR-η, CD3-γ, CD3-δ, CD3-ε, FcεRI-γ, FcεRI-β, and FcγRIIA are aligned. In some instances, multiple copies of the consensus motif are found in a single polypeptide chain; in these cases, sequences are labeled a, b, or c according to their NH_2- to COOH-terminal order. Conserved tyrosine residues are boxed. The consensus sequence is indicated at top.

($\leq$ 5 μg/mL), resting B cells enlarge, reflecting entry into the cell cycle and progression through G_1, but fail to enter S phase. Much higher concentrations of soluble anti-Ig (25–250 μg/mL) are required for entry into S phase (38).

Anti-Ig treatment differs somewhat from stimulation of B cells by TI type 2 antigens, such as trinitrophenyl (TNP)–Ficoll: a low-density fraction of B cells (the so-called preactivated fraction) can be induced by TI type 2 antigens, but not by anti-Ig, to secrete antibody. The ability of TI type 2 antigens to induce lymphokine-independent antibody secretion by preactivated B cells can be mimicked by anti-Ig that has been insolubilized by conjugation to dextran. Conjugation to dextran also reduces the concentration of anti-Ig required to induce proliferation of resting B cells by a factor of 1000 (39).

Stimulation of B cells by anti-Ig differs from antigenic activation in that the response is polyclonal and, therefore, easier to study in heterogeneous populations (e.g., primary splenic B lymphocytes), in which the proportion of cells responsive to any one specific antigen is exceedingly small. The experimental models are simplifications, and may mask requirements for costimulatory events (for example, ligation of the complement receptor CR2; see later). In addition, the results of such assays can be confounded because the binding of anti-Ig to Fc receptors on B cells suppresses activation by sIg engagement, a problem that is obviated by the use of $F(ab')_2$ derivatives of anti-Ig as the cross-linking agent (40).

2. *The Tyrosine Phosphorylation Response*

The earliest identified response to cross-linking of sIgM or D on B cells or B-cell lines is an increase in de novo protein-tyrosine phosphorylation (41–46). This response is rapid: increased tyrosine phosphorylation of a large set of proteins is observed as early as 10 s after receptor cross-linking, and is sustained for at least 30 min–1 h before declining to basal levels. Increased tyrosine phosphorylation of at least 15 different species can be distinguished electrophoretically (43–46). A similar set of substrates is phosphorylated on tyrosine on binding of a polyvalent antigen, PC-ova, to a B-cell line expressing a PC-specific sIgM (47).

Among the proteins for which phosphotyrosine content increases after sIg engagement are the Ig-associated chains Ig-α and Ig-β (47). Analogously, engagement of the TCR and the FcεR also results in increased tyrosine phosphorylation of associated chains (48,49). By electrophoretic mobility, most proteins phosphorylated after cross-linking of IgM or IgD are similar. One important difference is that, in cells that express both IgM and IgD, the μ-associated form of Ig-α is selectively phosphorylated after ligation of IgM, and the δ-associated form of Ig-α is selectively phosphorylated after ligation of IgD (47). This suggests that the tyrosine kinase(s) activated on sIg cross-linking is associated physically with the antigen–receptor complex.

3. *Phosphatidylinositol, Calcium, and Protein Kinase C Responses*

Engagement of sIg, either by polyvalent antigen or by anti-Ig antibody, results in increased phosphatidylinositol-4,5-bisphosphate (PtdIns(4,5)P_2) hydrolysis (50), followed by a subsequent rise in the concentration of intracellular free calcium and activation of protein kinase C (PKC; 51). These events follow the tyrosine phosphorylation response and have their origin in the activation of one or more inositol phospholipid-specific phospholipase C (PLC) isozymes (52,53). The PLC isozymes comprise at least three families, distinguished by their primary sequences: β, γ, and δ; there are three known β subtypes, two γ subtypes, and three δ subtypes (54). All PLC isoforms have similar substrate specificities and optimal reaction conditions: they hydrolyze phosphatidylinositol (PtdIns), phosphatidylinositol-4-monophosphate (PtdIns(4)P) and PtdIns(4,5)P_2. The B cells express members of at least two PLC families, β and γ; both γ subtypes can be found in B cells, but the γ_2 subtype is more abundant (55).

Hydrolysis of PtdIns(4,5)P_2 by PLC enzymes yields two products with the properties of classic second-messengers (Fig. 6): inositol 1,4,5-trisphosphate (Ins(1,4,5)P_3) and diacylglycerol (DAG) (56). The Ins(1,4,5)P_3 causes release

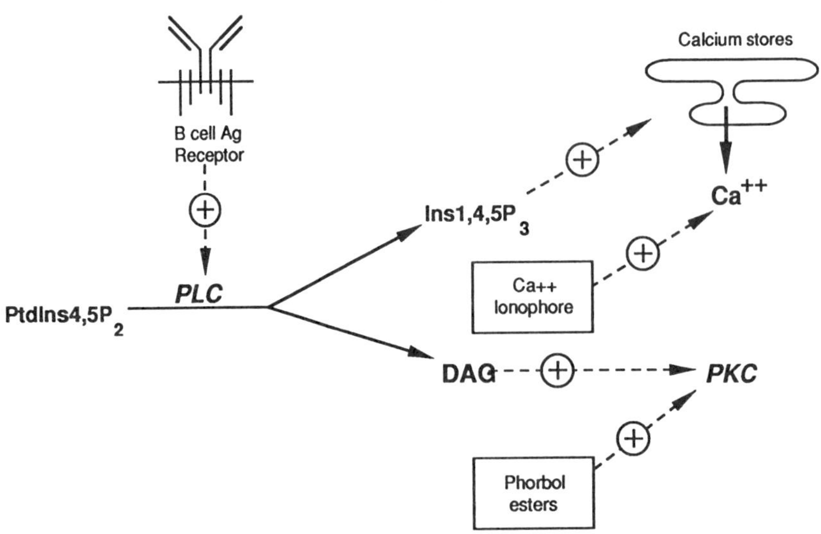

Figure 6 Inositol trisphosphate (Ins1,4,5P_3) and diacylglycerol (DG) as second-messengers. The two major intermediate physiological responses of B cells to antigen receptor stimulation, increased free intracellular calcium, and activation of protein kinase C (PKC), begin with hydrolysis of PtdIns4,5,P_2 to yield Ins1,4,5P_3, and diacylglycerol. Ins1,4,5P_3 increases the concentration of intracellular calcium by mobilizing its release from intracellular stores, and this can be mimicked by calcium ionophores. Diacylglycerol stimulates protein kinase C (PKC), which can be activated by phorbol esters.

of Ca^{2+} from an intracellular reservoir, the endoplasmic reticulum (ER), and this accounts for the initial Ca^{2+} rise observed after sIg cross-linking. The DAG binds to and activates PKC; some subtypes of PKC (α, β-I, β-II, and γ) are also Ca^{2+}-dependent, and these enzymes may respond synergistically to increases in DAG and Ca^{2+} (56,57). In several receptor-mediated signaling pathways, including those mediated by sIg and TCR, the requirement for receptor engagement can be bypassed by a combination of phorbol ester and calcium ionophore (51). This phenomenon can be understood in the context of phosphatidylinositol-mediated signaling. Phorbol esters mimic the effect of DAG by binding and activating PKC (57); calcium ionophores increase the concentration of intracellular Ca^{2+}, mimicking the effect of Ins(1,4,5)P_3.

In summary, the two major intermediate physiological responses of B cells to antigen receptor stimulation—increased free intracellular Ca^{2+} and activation of PKC—are the result of increased hydrolysis of PtdIns(4,5)P_2. Of critical importance, therefore, are the mechanisms that connect sIg engagement to increased activity of PLC isozymes.

4. *Immunoglobulin Heavy Chain Mutations That Impair Signaling*

Mutant forms of μ chain have been constructed that permit surface expression of IgM, but abolish transduction of antigenic signals, as assessed by calcium mobilization (58). Deletion of the μ cytoplasmic tail (KVK) resulted in surface expression of IgM by a glycosyl-phosphatidylinositol linkage; in this instance, it is perhaps not surprising that a defect in signaling was observed. The more interesting mutation replaces two polar amino acids in the transmembrane region with hydrophobic residues (YS to VV), resulting in loss of the intracellular calcium response to antigen or to receptor cross-linking; more subtle changes (YS to FS or YS to YA) did not impair signaling. From the example of IgD, we know that surface expression of Ig and its association with the Ig-α–Ig-β heterodimer can be separated. It seems likely, therefore, that some Ig heavy chain mutations might allow surface expression, but impair association with Ig-α and Ig-β; the YS to VV transmembrane mutation may represent a member of this class.

5. *Tyrosine Kinases as Intermediaries Between Surface Immunoglobulin and Phospholipid-Specific Phospholipase C*

In principle, activation of any of the PLC isozymes could lead to increased PtdIns(4,5)P_2 hydrolysis. Two general mechanisms for receptor-mediated activation of PLC enzymes are known. The first involves selective phosphorylation of PLC-γ isozymes on tyrosine residues (59), as occurs in signaling through growth factor receptor kinases, such as the platelet-derived growth factor (PDGF) receptor. The second involves activation of PLC-β isozymes by a class of guanine nucleotide-binding proteins called G_q (60). Although both mechanisms may function in B-cell activation, the only convincing evidence implicates the former pathway.

The rise in protein-tyrosine phosphorylation observed immediately after sIg engagement appears to be a necessary prerequisite for calcium mobilization and PKC activation. Increased tyrosine phosphorylation is not observed when B cells are stimulated with phorbol esters (43,44), and depletion of PKC by chronic treatment of B cells with phorbol dibutyrate does not impair the tyrosine kinase response to Ig cross-linking, indicating either that PKC acts downstream from protein-tyrosine phosphorylation or that it functions in a distinct pathway (46). The former is likely true, because three different, selective inhibitors of protein-tyrosine kinases were able to suppress increases in phosphatidylinositol hydrolysis and intracellular calcium in response to Ig cross-linking (52). Activation of a protein-tyrosine kinase(s), therefore, seems likely to be a primary event in Ig-mediated signal transduction.

One link between B-cell antigen receptor engagement and increased inositol phospholipid hydrolysis appears to be tyrosine phosphorylation of PLC-γ (52,53). A paradigm for this association is the role of PLC-γ1 in growth factor receptor-mediated signaling. Both platelet-derived growth factor (PDGF) and epidermal growth factor (EGF) induce the association of PLC-γ1 with their respective receptors (61), which phosphorylate PLC-γ1 in vitro and in vivo on identical tyrosine residues (771, 783, and 1254). Tyrosine kinase activity is required for the association of PLC-γ1 with these growth factor receptor kinases (62). Furthermore, phosphorylation of PLC-γ1 at tyrosine residue 783 is required for stimulation of inositol phospholipid hydrolysis by PDGF (63); thus, most or all of the increase in inositol phospholipid hydrolysis observed upon PDGF administration is due to PLC-γ1. On B-cell antigen–receptor cross-linking, increased tyrosine phosphorylation of PLC-γ is also observed (52,53). The increases in inositol phospholipid hydrolysis and intracellular calcium that follow sIgM cross-linking are suppressed by inhibitors of protein-tyrosine kinases, suggesting that these responses are the result of increased phosphorylation of PLC-γ on tyrosine (52).

6. *Role of Accessory Chains of the Surface Immunoglobulin–Receptor Complex*

Considerable evidence indicates that members of the antigen receptor accessory chain family, including the TCR-associated chain ζ and Ig-α, can themselves transduce receptor-signaling events, but that these responses are dependent on additional, lymphoid-specific components. The characteristic early and late responses to TCR engagement, including increased intracellular free calcium, inositol phospholipid hydrolysis, tyrosine phosphorylation of the cytoplasmic domain of ζ and interleukin (IL)-2 production, can be reconstituted in T cells by a chimeric protein containing the extracellular and transmembrane portions of

CD8 and the intracellular portion of ζ (64). The cytoplasmic portion of Ig-α can substitute for ζ in a similar T-cell assay (65). On the other hand, the complex of Ig, Ig-α, and Ig-β can be expressed on the surface of nonlymphoid cells, but does not reconstitute the response to receptor cross-linking (66).

Reconstitution of signaling by CD8/ζ chimeras in T cells requires the presence of several critical tyrosine residues in the ζ cytoplasmic region (64); these reside in the accessory chain consensus sequence (D/ExxxxxxxD/ExxYxxLxxxxxxx-YxxL/I; see foregoing). Tyrosine residues in a similar context are found in the sIg accessory chains and are also likely to be critical for their function. It is possible that these tyrosine residues, when phosphorylated, mediate stable interactions with tyrosine kinases or other components of the signaling pathway (see later discussion).

IV. CANDIDATE EFFECTORS OF B-CELL ACTIVATION

A. Protein-Tyrosine Kinases

1. The src Family Kinases

It is not yet certain which of the protein-tyrosine kinases are responsible for transduction of the primary sIg signal. In growth factor receptor kinases, such as the EGF receptor, ligand-binding and kinase domains are covalently joined. The members of the Ig receptor complex lack intrinsic tyrosine kinase activity, and appear to act by noncovalent association with one or more intracellular kinases, by analogy to the association between the T-cell costimulatory molecules CD4 and CD8 and the protein-tyrosine kinase *lck*, a member of the *src* family. Consistent with this idea, tyrosine kinase substrates are observed to colocalize with the B-cell antigen receptor complex after ligation of sIg (67).

Tyrosine kinases of the *src* family range in molecular weight from about 55 to 60 kDa and are similar in structure (Fig. 7). The *src* family kinases are wholly intracellular and associate with the inner leaflet of the plasma membrane. All *src* kinases contain glycine at amino acid residue 2, through which these proteins are myristoylated by an amide linkage; myristoylation is essential for membrane association. The *src* family of kinases exhibit highly divergent NH_2-terminal sequences (so-called unique region; residues 3 through 56 of *blk* and 3 through 85 of *src*). In the case of *lck*, this region mediates specific interactions with intracellular regions of CD4 and CD8 in T cells (68). It has been suggested, but not yet demonstrated, that the unique regions of the other members of the *src* family likewise mediate specific interactions with transmembrane proteins. The *src* kinases all contain two regions of homology upstream from the catalytic domain; these are termed *src* homology regions 3 and 2 (SH3 and SH2). The SH3 regions are about 50-amino acid residues long (residues 57 through 110 of *blk*); it is likely that SH3 mediates specific interactions with other proteins, including molecules that regulate tyrosine kinase activity (69). The SH2 regions are about 80-amino acid residues long (residues 118 through 199 of *blk*) and mediate specific, high-affinity interactions with phosphotyrosine-containing proteins (70). Remarkably, SH2 and SH3 regions are shared by other proteins that function in signal transductory pathways. In non-*src* proteins, SH2 and SH3 regions can be found alone or in combination and, when found together, they may appear in either order; PLC-γ, for example, contains two SH2 domains, followed by a single SH3 domain (70). These observations are further indication that SH2 and SH3 can function autonomously. The catalytic regions of *src* kinases (amino acid residues 311 to 499 in *blk*) show greater than 60% amino acid sequence identity. Within the catalytic region is a tyrosine residue (Tyr-383 of *blk*) that is a site of autophosphorylation; phosphorylation of this residue stimulates kinase activity severalfold, and may create a site for interaction with SH2 domains. The catalytic region terminates in a short regulatory region containing a critical tyrosine residue (Tyr-495 of *blk*), phosphorylation of which is correlated with decreased tyrosine kinase activity. Mutations that convert this tyrosine residue to phenylalanine confer increased kinase activity and the ability to transform fibroblasts (71).

The tyrosine kinase *blk* is the only known member of the *src* family for which expression is restricted to B-lymphoid cells (72,73), but B cells express other *src* kinases as well, including *fyn*, *lyn*, and *lck* (74–76). The *blk* kinase is expressed at all stages of B-cell ontogeny until the differentiation of B cells to

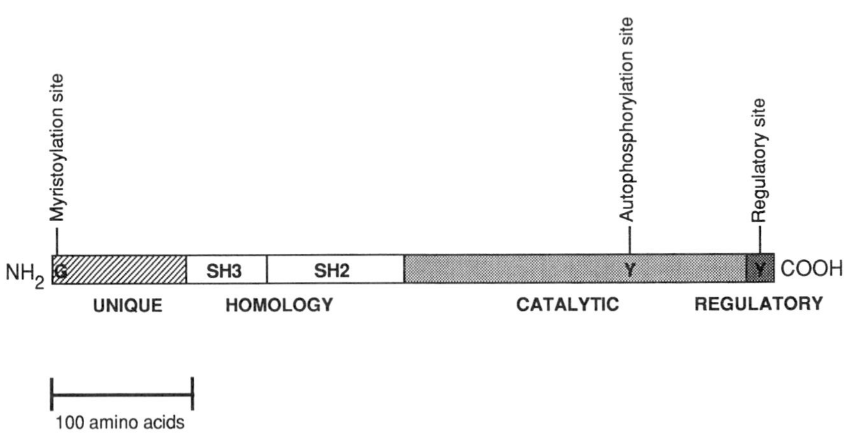

Figure 7 Structural domains of *src*-related kinases. A prototypic *src* kinase is diagrammed, with NH_2- and COOH-termini as marked. Unique, homology, catalytic, and regulatory regions are indicated by hatched, open, light shaded, and dark shaded boxes, respectively. Amino acid residues that serve as acceptors for myristoylation, autophosphorylation, and negative regulatory phosphorylation are indicated by boldface type.

plasma cells, at which time expression ceases (73). The restricted nature of *blk* expression and its coregulation with *mb-1* during B-cell ontogeny have suggested that it plays an intimate role in Ig-mediated signaling. The *fyn* and *lyn* kinases show a broader tissue distribution, although *lyn* is preferentially expressed in tissues rich in B lymphocytes and in myeloid cells. The *lck* is expressed preferentially in T cells, where it is associated with CD4 and CD8, but is also expressed at a lower level in B cells and in some B-cell lines (76,77).

All four kinases coprecipitate with the sIg–receptor complex (74–76). Unlike the *lck*–CD4/CD8 association, however, these kinase–receptor complexes are tenuous, having been observed only under conditions of mild detergent lysis, and the possibility of association after cell disruption has not been excluded. Interactions between these kinases and the sIg receptor complex are not all likely to resemble the association of *lck* with CD4 and CD8 in T cells, since their NH_2-terminal regions are dissimilar in structure. Some of the observed associations between *src* family members and the sIg complex may be mediated by their SH2 regions, particularly since the available evidence does not provide a distinction between primary kinase–receptor complexes and associations—possibly mediated by SH2—that are formed after tyrosine phosphorylation of the Ig-associated chains. In addition, a specific interaction between the interleukin-2 (IL)-2 receptor β chain and *lck* has been shown to be mediated by a portion of the catalytic domain (78), and it is possible that interactions between the sIg complex and *src* kinases are also mediated through their catalytic domains.

The *blk*, *lyn*, and *fyn* kinases all show increased tyrosine kinase activity in vitro following antigen receptor cross-linking (74,79). Within 1 min after ligation of sIg, the activity of *blk* was increased five- to eightfold, as measured by the in vitro phosphorylation of enolase by immunoprecipitated kinase; the observed increases in *lyn* and *fyn* activity were more modest (two- to fourfold). Although these results suggest that *blk*, *lyn*, and *fyn* function in Ig-mediated signaling, they do not indicate which—if any—of these kinases is the principal effector.

2. *PTK72, an Intracellular, Non-*src *Kinase*

In addition to the *src* kinases, a 72-kDa tyrosine kinase of distinctive structure (PTK72) has been purified from spleen, and a cDNA that encodes it or a related kinase has been molecularly cloned (80). The kinase encoded by this cDNA, called *syk*, is expressed preferentially in B cells, as is PTK72. The *syk* differs from *src* kinases in several ways. First, all *src* kinases are myristoylated at their NH_2-termini, a modification that contributes to their membrane localization. The *syk* kinase lacks a myristoyl acceptor sequence. Second, *syk* lacks the conserved SH3 homology region of *src* kinases and carries a duplication of the SH2 region. The *syk* kinase lacks an obvious transmembrane region and is likely to be intracellular. On cross-linking of sIgM, PTK72 (which is either identical with *syk*, or closely related to it) is phosphorylated on tyrosine and is activated (81,82), suggesting proximity to the receptor complex and a role in B-cell activation.

3. *Recruitment of Phosphatidylinositol 3-Kinase by the Surface Immunoglobulin Complex*

At least 15 different protein species are newly phosphorylated on tyrosine on B-cell antigen–receptor cross-linking. Most of these proteins are unidentified, and their roles, if any, in signal transduction are unknown. Several substrates, however, have been identified, and several of these—Ig-α, Ig-β, *blk*, *fyn*, *lyn*, PTK72, and PLC-γ—have already been discussed. Another substrate that is likely to be important in the proliferative response of B cells to antigen receptor stimulation is the 85-kDa subunit of phosphatidylinositol 3-kinase (PI3-kinase). The PI3-kinase phosphorylates PtdIns, PtdIns(4)P, and PtdIns(3,4)P_2 (83). Although it is implicated in cellular responses to several growth factors and transforming proteins, its precise function in these pathways is unknown. It binds a transforming form of *src* (p60*v-src*) and the polyoma middle T antigen, as well as to activated receptors for PDGF, colony-stimulating factor (CSF), and insulin. Mutations that abolish the association of polyoma middle T antigen with PI3-kinase impair its oncogenic potential. Similarly, association of PI3-kinase with the activated PDGF receptor is required for the mitogenic response to PDGF (83). After the sIg–receptor complex is stimulated, the 85-kDa subunit is phosphorylated on tyrosine. Coordinately, PI3-kinase is newly associated with at least one of the sIg receptor-associated *src* kinases, *lyn* (84). These observations indicate that PI3-kinase is recruited to the sIg complex after receptor engagement, and suggest a role for PI3-kinase in B-cell proliferation.

B. Guanine Nucleotide-Binding Proteins (G-Proteins)

A role for guanine nucleotide-binding proteins (G-proteins) in B-cell activation has been suggested for experiments in which G-protein activators such as GTPγS were shown to synergize with suboptimal doses of anti-Ig in stimulating phospholipid hydrolysis (85–87). In experiments with streptolysin O-permeabilized B cells, at calcium concentrations similar to those found within intact, resting cells, GTPγS stimulated inositol phosphate production with a half-maximum of 0.1–0.3 μM (85). In the same series of experiments, GTPγS and anti-IgM or anti-IgD stimulated inositol phospholipid hydrolysis synergistically; synergism was maximal at a nominal calcium concentration, similar to that achieved in intact cells on receptor activation. A non-hydrolyzable GDP analogue, GDPβS, inhibited stimulation of inositol phospholipid hydrolysis by GTPγS alone or in combination with anti-Ig. The observed effects of GTPγS on phospholipid hydrolysis were not inhibited by pertussis toxin (PTX), suggesting the involvement of one or more PTX-insensitive G-proteins (85).

A group of G-proteins, collectively called G_{ps} (for phospholipase C-stimulating G-proteins) specifically activate PLC-β isoforms. A subgroup of G_{ps}, typified by G_q, is insensitive to pertussis toxin, as is the putative G-protein activity observed to synergize with sIg. At least five distinct Gα subunits of the G_q class have been molecularly cloned. One of these, $G\alpha_{15}$, shows preferential—but not exclusive—expression in spleen and fetal liver, and in cell lines of the B and myeloid lineages (88). On the basis of these findings, one reasonable explanation for the synergy between sIg ligation and GTPγS might be that these treatments simultaneously activate PLC-γ and PLC-β isotypes through protein-tyrosine kinase and G-protein mediators, respectively.

A second group of G-proteins potentially involved in B-cell signaling is typified by p21*ras*, the product of the *ras* oncogene and one of a group of so-called small G-proteins. A role for p21*ras* in T-cell receptor signaling was suggested by experiments in which stimulation of the T-cell antigen receptor was followed by a decrease in the activity of the GTPase-activating protein (GAP), a negative regulator of p21*ras* activity, with concomitant accumulation of active p21*ras*–GTP complexes (89). The observation that p21*ras* and sIg are colocalized after antigen receptor ligation (90) suggests that p21*ras* may play a similar role in B cells.

C. Agonistic Effect of Complement Receptor Type 2 (CD21)

Activated complement component C3 serves as the ligand for complement receptors on B cells. Deficiencies of C3 or other complement deficiencies that impair the activation of C3 are associated with abnormalities in antibody production and B-cell differentiation (91). These associations suggested that B-cell responsiveness to antigens could be modified by engagement of C3 receptors, a hypothesis that has gained strong experimental support (92,93).

Two receptors for proteolytic fragments of C3 are present on B cells. C3 is activated by proteolysis to C3b, which binds the 220-kDa complement receptor type 1 (CR1; CD35). Subsequent cleavage of C3b, in a reaction that requires CR1, yields C3dg, which binds specifically to the 145-kDa complement receptor type 2 (CR2; CD21). The CR1 and CR2 are structurally homologous, type I transmembrane proteins that are found on all mature B cells. They both have relatively short intracellular regions (43 and 34 amino acids, respectively) and their extracellular domains consist of tandemly repeated consensus motifs, each about 60–70 amino acids long.

In T cells, signaling through the TCR is enhanced by coengagement of CD4, which is coupled to the intracellular protein-tyrosine kinase p56*lck*. In B cells, signaling through the sIg complex is similarly augmented by coengagement of CR2. Anti-CR2 antibody and polymeric C3dg induce B-cell proliferation in the presence of soluble T-cell factors or phorbol ester (94), and cross-linking of CR2, but not CR1, to sIgM was observed to synergistically augment the calcium mobilization response (95). In B cells, CR2 can be found in two distinct multimolecular complexes: one including CR1, the other including CD19 and the hematopoietic cell surface protein TAPA-1 (92,96).

It is the complex with CD19 that synergizes with sIg. CD19 (97) is a 95-kDa member of the Ig superfamily that is expressed throughout B-cell ontogeny. In digitonin lysates of Raji B-lymphoblastoid cells, CR2 coimmunoprecipitated with CD19 at a stoichiometric ratio of about 1:1; in addition, CR2 and CD19 were observed to associate following transfection of their cDNAs into a nonlymphoid cell line (92). The synergistic effect of CR2 ligation on the calcium response to sIg engagement may be mimicked by cross-linking CD19 to sIgM (92). Although cross-linking of CD19 alone leads to a modest increase in cytoplasmic free calcium and suppression of the proliferative response to sIg cross-linking (98), coligation of CD19 and sIgM has a profound synergistic effect on the B-cell proliferative response, resulting in a 100-fold diminution in the threshold for sIg-dependent stimulation (93).

CD19 and sIg are cocapped and comodulated, suggesting that CD19 is associated—directly or indirectly—with sIg (99). CD19 possesses a 247-amino acid cytoplasmic tail that becomes phosphorylated on tyrosine, after cross-linking of CD19 or the sIg complex. Protein-tyrosine kinase inhibitors block the rise in cytoplasmic free calcium observed after CD19 cross-linking, suggesting that tyrosine phosphorylation is required for this response. After CD19 becomes phosphorylated on tyrosine, it associates with one or more intracellular proteins; whether this association(s) is required for augmentation of the sIg response by CD19 is not yet known.

V. ROLE OF PROTEIN-TYROSINE PHOSPHATASES

Activation of B cells involves changes in the phosphorylation states of many different proteins, including Ig-α, Ig-β, *blk*, *lyn*, *fyn*, PTK72, PLCγ, and PI3-kinase. The phosphorylation state of each of these molecules is dynamic, determined by the opposing activities of kinases and phosphatases. We have examined protein-tyrosine kinases that may be responsible for de novo tyrosine phosphorylation following B-cell antigen receptor stimulation. Protein-tyrosine phosphatases (PTPases) in general oppose protein-tyrosine kinases, yet their precise physiological functions are unknown. The PTPases may act constitutively to maintain the low levels of tyrosine phosphorylation required in the basal state for proper cell function (100). A countering protein tyrosine phosphatase

(PTPase) would be required for maintenance of phosphotyrosine homeostasis following B-cell activation; CD45 is one likely candidate.

1. *Structure of CD45*

CD45 is one of a family of transmembrane glycoproteins expressed exclusively on the surface of leukocytes (101,102). CD45 isoforms, as a class, were identified as the preponderant iodinated proteins in lymphocyte membranes (103) and as the principal specificity of antilymphocyte sera (104). Approximately 10% of the lymphocyte surface is occupied by one or more CD45 isoforms (105).

The CD45 family (T200, B220, Ly-5, or L-CA), consisting of at least five isoforms ranging in relative molecular mass (M_r from 180 to 220 kDa), is encoded by a single gene that comprises 34 exons spanning approximately 110 kb of DNA on mouse and human chromosome 1 (106). The structure of CD45 family members is receptorlike, comprising a large NH_2-terminal extracellular domain, a single membrane-spanning region, and a large COOH-terminal cytoplasmic region possessing enzymatic activity. The external domain is encoded by exons 3 through 15, and distinct isoforms are generated by cell type-specific alternative splicing of exons 4, 5, and 6. Because exons 3 through 8 encode potential *O*-linked carbohydrate attachment sites, the variable use of exons 4, 5, and 6 changes not only the amino acid sequence, but the number of potential glycosylation sites as well. Exons 4–6 and their potential carbohydrate structures are likely to be functionally significant. The B-lineage cells use all three exons, resulting in the highest M_r form (220 kDa), which is designated B220 (105, 107,108).

CD45 isoforms share a common cytoplasmic region of 707 amino acids composed of two repeated domains (I and II), with 40 and 33% sequence identity to PTP1B, a cytosolic PTPase from human placenta (109,110); CD45 exhibits intrinsic PTPase activity (111). CD45, therefore, represents a family of receptor-linked PTPases. Mutational analysis of the cytoplasmic domain of CD45 suggests that PTPase domain I is active, whereas domain II is inactive (112). Domain II may have distinctive substrate specificity, or may serve a regulatory role. Because domain II is highly conserved among transmembrane phosphatases, it is likely to be of functional import.

2. *CD45 as a Potential Receptor Protein-Tyrosine Phosphatase*

The structural similarity between CD45 and receptor tyrosine kinases suggests that the activity of the cytoplasmic domain might be similarly regulated by an extracellular ligand. Cloning by homology has led to the identification of ten additional PTPases that, with CD45 and PTPase 1B, fall into three groups (Fig. 8). Class III phosphatases, including CD45, contain a large extracellular region, a transmembrane peptide, and duplicated cytoplasmic PTPase domains. The extracellular domains are quite diverse: LAR and DLAR phosphatases are both

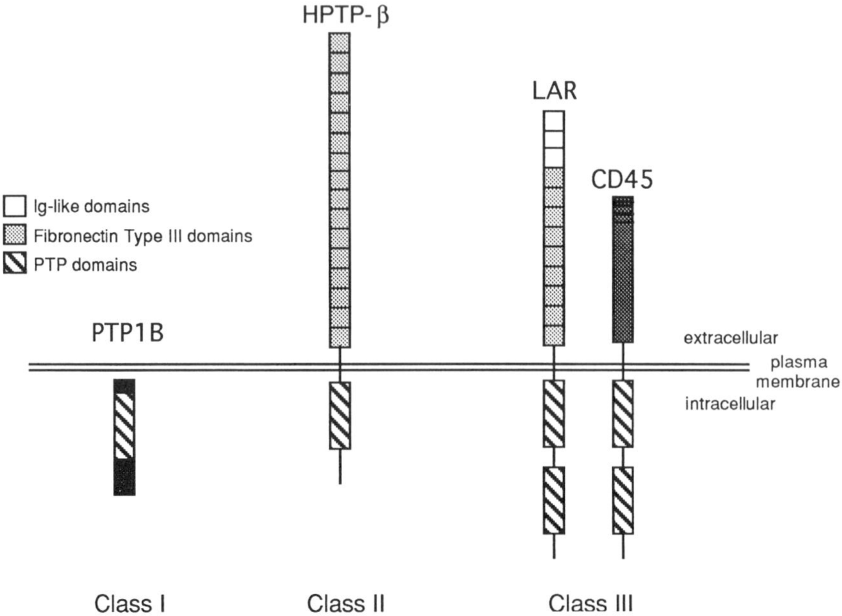

Figure 8 Schematic representation of members of the protein-tyrosine phosphatase (PTP) family. Representative members of PTP classes I, II, and III are diagrammed with their NH_2-termini oriented toward the tope of the figure. Ig-like domains, fibronectin type II domains, and PTP domains are indicated by open boxes, light shading, and crosshatching, respectively. The plasma membrane is indicated by a double line.

composed of three immunoglobulinlike (Ig) domains and eight (LAR) or nine (DLAR) fibronectin type-III (FN-III) domains; DPTP contains two Ig domains and two FN-III domains. Multiple Ig domains and FN-III domains are common to cell adhesion molecules, such as the neural-cell adhesion molecule, N-CAM. The FN-III repeats are also found in the extracellular regions of the receptors for IL-2, IL-4, IL-6, GM–CSF, prolactin, erythropoietin, and growth hormone. These structural similarities suggest that LAR, DLAR, and DPTP may also be cell adhesion-type receptors. Although the extracellular region of CD45 is not related to N-CAM or any other known receptor, these observations, together with the tissue-specific expression of CD45 isoforms, suggests that CD45 is a receptor with tissue-specific ligands that modulate its phosphatase activity.

Little is known about the regulation of CD45 activity. A tyrosine-containing region found near the COOH-terminus of CD45 resembles the COOH-terminal negative regulatory region of *src* family kinases. If CD45 were negatively regulated by tyrosine phosphorylation, it could autoactivate by dephosphorylating itself—a process analogous to autophosphorylation by receptor protein-tyrosine kinases (100). Most receptor PTKs, such as the PDGF receptor or EGF receptor, are activated by ligand-dependent dimerization. Dimerization may also regulate CD45 activity.

3. The B-lineage-Specific Isoform of CD45, B220

The B-lineage isoform of CD45, B220, is detectable on the cell surface very early in B-cell differentiation (113). Its expression, in fact, is one of the earliest indications of commitment to the B lineage, and precedes expression of cytoplasmic μ heavy chain. B220 is expressed on the surface of pro-B, pre-B, and mature B cells; differentiation into antibody-secreting plasma cells is accompanied by decreased B220 expression. Few plasma cells or plasmacytoma cell lines express B220.

B-cell activation does not alter B220 expression or induce expression of additional CD45 isoforms, but does lead to changes in the glycosylation state of B220 (114). This is manifested by differential ability to bind various plant lectins. The B lymphocytes undergoing activation in lymph node germinal centers gain the ability to bind peanut agglutinin (114), which does not bind to resting and memory B cells. Interleukin-2, a factor important in B-cell activation and proliferation, can induce a 5- to 15-fold increase in the number of peanut agglutinin and soybean lectin sites per B220 molecule (115).

Some insight into the functional significance of these changes in carbohydrate structure has come from B220 cross-linking studies using anti-CD45 monoclonal antibodies (MAb). Different anti-CD45 MAbs, having distinct epitope specificities, exhibit different effects on anti-Ig-induced B-cell proliferation. Although some anti-CD45 antibodies inhibit proliferation, one antibody (136-4B5), which recognizes a neuraminidase-sensitive epitope of CD45, enhances proliferation; this enhancement could be achieved through either increased or diminished B220 phosphatase activity. Even though these observations have suggested involvement of a specific carbohydrate epitope of B220 in the generation of a proliferative signal (116), it seems equally likely that responsiveness to CD45 oligomerization may be a function of the relative orientation of CD45 monomers in the complex, which would differ with differing MAbs. Thus, the role of carbohydrate in CD45-mediated signaling remains an open question.

4. Potential Function of B220 in B-Cell Activation

Evidence for a direct role of B220 in sIg-mediated signaling is now limited. A B220 B-cell line (J558μm3), which failed to mobilize calcium in response to sIg cross-linking, acquired this ability following transfection with B220 cDNA (117). In these experiments, the molar ratio of sIg and B220 on maximally responsive transfectants was approximately 1:1, and treatment with anti-B220 MAb resulted in the loss of B220 and IgM from the surface of a proportion of cells. Comodulation of B220 and sIg has suggested a specific association between these molecules. The function of B220 associated with the sIg complex might be to dephosphorylate Ig-α and Ig-β following receptor activation. In this scheme, a hyperphosphorylated receptor complex would be refractory to a second round of signaling, requiring dephosphorylation to return to a signaling-competent state.

Alternatively, or in addition, B220 could activate sIg-associated tyrosine kinases of the *src* type by dephosphorylation of tyrosine within the COOH-terminal regulatory region. Precedent for such a mechanism exists for p56*lck* in T-cell activation. Cross-linking of CD4 leads to increased tyrosine phosphorylation of p56*lck* and an increase in the activity of CD4-associated p56*lck* in vitro. Relative to p56*lck* from CD45$^+$ T-cell lines, p56*lck* from CD45$^-$ lines was hyperphosphorylated, and a greater proportion of phosphorylation was at Tyr-505, in the COOH-terminal regulatory region. Signaling through CD4 was augmented by cocross-linking with CD45 and, under these conditions, p56*lck* became dephosphorylated at Tyr-505 (118).

5. *Kinase–Phosphatase Interactions: A Summary*

Changes in tyrosine phosphorylation associated with B-cell activation are likely to involve regulated interactions among protein-tyrosine kinases, B220, and the sIg–receptor complex. Within seconds after ligation of the B-cell antigen–receptor complex, activation of protein-tyrosine kinases is likely accompanied by assembly of a transient, submembrane-signaling complex including, in addition to the Ig accessory chains and tyrosine kinases themselves, PLC-γ, PI3-kinase, and probably the p21*ras* GTPase-accelerating protein (GAP) or a related protein (119). Such a complex would be specified in part by specific interactions between phosphotyrosine-containing protein sequences and SH2 domains. Following signal transduction, the receptor complex might be restored to its resting state by tyrosine dephosphorylation, perhaps mediated by B220.

VI. LYMPHOKINES AND B-CELL ACTIVATION

Ligation of the B-cell antigen receptor induces a sequence of responses leading to cell proliferation, but does not result in Ig secretion except in the presence of various soluble mediators called cytokines. Cytokines regulate both antigen-independent and antigen-dependent B-cell growth and differentiation. Interleukin-7, critical in the antigen-independent phase of B-cell development, is produced by bone marrow stromal cells. A large subset of cytokines, including IL-2, IL-4, IL-5, and IL-6, are produced principally by T lymphocytes. In the mouse, CD4$^+$ T cells can be divided into two subsets according to the lymphokines they produce: T helper 1 (Th1) cells produce IL-2, interferon (IFN)-γ, IL-3, and GM-CSF; Th2 cells produce IL-3, IL-4, IL-5, IL-6, IL-10, and GM–CSF (120). A clear distinction between T-helper-cell subsets has not been observed in humans. The effects of most cytokines are pleiotrophic, and we will briefly focus on their primary roles in B-cell development and activation.

A. Interleukin-5

Interleukin-5 induces the growth of a subset of $CD5^+$ B cells and controls differentiation of B cells into antibody-secreting plasma cells.

The development of B-lineage cells can be divided into two pathways (Fig. 9). The first is dependent on IL-7 and produces $CD5^-$ B cells. The second is independent of IL-7 and dependent on IL-5; it produces $CD5^+$ B cells. The $CD5^+$ lineage cells differ from conventional, $CD5^-$ B-lineage cells by their restricted repertoire, distinctive tissue localization, and secretion of polyreactive IgM antibodies. Interleukin-5 selectively supports the early development of this $CD5^+$ subpopulation before rearrangement of IgH genes (121).

In combination with antigen, IL-5 induces polyclonal IgM production (122). In addition, it induces antigen-specific and polyclonal IgA production in antigen-primed B cells and lipopolysaccharide-stimulated B cells, respectively (123). In these experiments, IL-5 acts on $sIgA^+$ B cells, but not on $sIgA^-$ B cells, to induce IgA production; that is, IL-5 appears to act on B cells committed to become IgA-secreting cells.

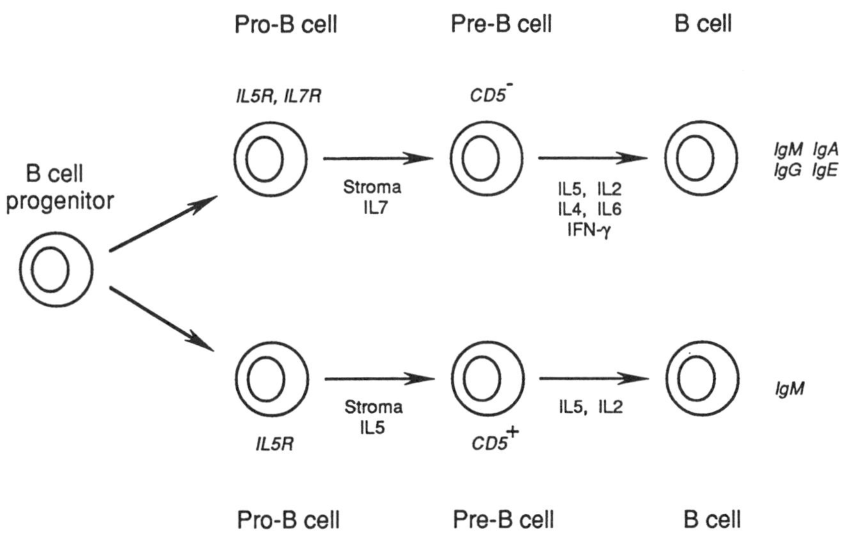

Figure 9 Schematic representation of interleukin-5 (IL-5)- and interleukin-7 (IL-7)-dependent B-cell development. The IL-5- and stromal cell-dependent pro-B cell and pre-B cells express both CD5 and IL-5 receptor (IL5R), and appear to represent a different B-cell lineage from that which preferentially responds to IL-7. Cell surface markers and other protein products of B-lymphoid cells are indicated in italics. For details, see text.

The phenotype of IL-5 transgenic mice is consistent with these observations: levels of serum IgM and IgA are elevated, an unusual $CD5^+$, $B220^+$ population appears in spleen, and polyreactive IgM autoantibodies are synthesized against ssDNA, dsDNA, cardiolipin, poly-(ADP-ribose) and trinitrophenyl (124).

B. Interleukin-7

Interleukin-7 is a growth factor secreted by bone marrow stromal cells that stimulates proliferation of $CD5^-$ B-lymphoid progenitors (125) (see Fig. 9). Interleukin-7 is required in cell culture systems that allow long-term, continuous development of bone marrow-derived B-lineage cells from the precursor to sIg^+ stage (125,126).

C. Interleukin-2

Interleukin-2 acts, at least in vitro, as the principal auto- and paracine T-cell growth factor (127). It does not, however, act exclusively on T cells, but also has been observed to induce antigen-activated B cells to progress through the cell cycle and to differentiate into antibody-secreting plasma cells (128). Interestingly, mice incapable of expressing IL-2 because of targeted disruption of the IL-2 gene showed normal thymopoiesis and normal numbers of peripheral T cells and B cells, but exhibited increased serum levels of IgG1, IgG2a, and IgG2b. In addition, T cells from these mice responded poorly to polyclonal mitogens and were unable to help B cells secrete IgM in response to polyclonal activation by anti-IgM (129).

D. Interleukin-4

Interleukin-4 possesses a broad spectrum of biological activities throughout B-cell development. In resting mature B cells, IL-4 induces an increase in the expression of class II MHC molecules (Ia) (130,131) and low-affinity Fc receptors for IgE (CD23) (132) and primes them to proliferate in response to anti-Ig (133,134). B cells stimulated with mitogens, anti-Ig (135), or T cells and antigen (136–138) go on to secrete IgG1 (139) and IgE (160) in the presence of IL-4. Interleukin-4 appears to be required to generate and sustain in vivo IgE responses by inducing heavy-chain class switching to IgE. Similar results were obtained in IL-4 transgenic mice: B cells from mice expressing IL-4 under the control of the Ig heavy chain promoter and enhancer showed a significant increase in expression of Ia molecules, and increased serum levels of IgG1 and IgE were observed (141,142).

E. Interleukin-6

Interleukin-6 induces terminal differentiation of B cells into plasma cells (143). In addition, dysregulated expression of IL-6 may be instrumental in the pathogenesis of various polyclonal and monoclonal plasma cell abnormalities associated with such diseases as rheumatoid arthritis (144). Similarly, IL-6 can serve as an autocrine growth factor for human myeloma cells (145). In the mouse, expression of the human IL-6 gene in the B lineage resulted in massive plasmacytosis. Profound mesangial cell proliferation was observed in the kidney as well, resembling mesangioproliferative glomerulonephritis. The serum level of IgG1 was increased 120- to 400-fold, and the levels of IgG2b, IgG3, and IgA were increased approximately threefold, compared with age-matched control normal mice (146).

VII. CONCLUSIONS AND FUTURE DIRECTIONS

B-cell activation is regulated by receptors for five main classes of ligand: antigens, lymphokines, antibody (Fc), complement, and cell surface adhesion molecules. The antigen receptor complex shares several structural and functional features with some Fc receptor complexes, including the presence of homologous polypeptide chains and activation of intracellular protein-tyrosine kinases. The means by which lymphokine and cell surface adhesion receptors transduce extracellular signals is less well understood, but protein-tyrosine phosphorylation has also been implicated here. The tyrosine kinases involved, the molecular mechanisms by which receptor engagement is coupled to kinase activation, and the immediate consequences of kinase activation are important problems that are as yet unanswered. These questions will need to be answered before more complex phenomena—such as modulation of antigenic signaling by other B-cell surface molecules—can be fully understood.

REFERENCES

1. I. Scher, The CBA/N mouse strain: An experimental model illustrating the influence of the X-chromosome on immunity, *Adv. Immunol.*, *33*: 1 (1982).
2. D. F. Amsbaugh, C. T. Hansen, B. Prescott, P. W. Stashak, D. R. Barthold, and P. J. Baker, Genetic control of the antibody response to type III pneumococcal polysaccharide in mice. Evidence that an X-linked gene plays a decisive role in determining responsiveness, *J. Exp. Med.*, *136*: 931 (1972).
3. D. E. Mosier, I. Scher, and W. E. Paul, In vitro response of CBA/N mice: Spleen cells of mice with an X-linked defect that precludes immune response to several thymus-independent antigens can respond to TNP-lipopolysaccharide, *J. Immunol.*, *117*: 1336 (1976).
4. J. J. Mond, P. K. A. Mongini, D. Sieckmann, and W. E. Paul, Role of T lymphocytes in the response to TNO-AECM-Ficoll, *J. Immunol.*, *125*: 1066 (1980).
5. R. W. Dutton, R. Falkoff, J. A. Hirst, et al., Is there evidence for a non-antigen specific diffusable chemical mediator from the thymus derived cell in the initiation of the immune response? *Prog. Immunol.*, *1*: 355 (1971).

6. M. D. Cooper, and P. D. Burrows, B-cell differentiation, *Immunoglobulin Genes* (F. W. Alt and T. Honjo, eds.), Academic Press, New York (1989).
7. D. Levitt, and M. D. Cooper, Mouse pre-B cells synthesize and secrete mu heavy chains but not light chains, *Cell, 19*: 617 (1980).
8. H. Karasuyama, A. Kudo, and F. Melchers, The proteins encoded by the V_{preB} and 5 pre-B cell-specific genes can associate with each other and with μ heavy chain, *J. Exp. Med., 172*: 969 (1990).
9. F. Melchers, The many roles of immunoglobulin molecules in the development and growth control of the B-lymphocyte lineage, *Immunoglobulin Genes* (F. W. Alt, and T. Honjo, eds.), Academic Press, New York, p. 23 (1989).
10. S. V. Desiderio, Organization and assembly of immunoglobulin genes, *Developmental Immunology* (E. Cooper, and E. Nesbitt-Brown, eds.), Oxford University Press, New York, (in press).
11. M. G. Reth, and F. W. Alt, Novel immunoglobulin heavy chains are produced from DJ_H gene segment rearrangements in lymphoid cells, *Nature, 312*: 418 (1984).
12. A. Alessandrini, and S. V. Desiderio, Coodination of immunoglobulin DJ_H transcription and D-to-J_H rearrangement by promoter-enhancer approximation, *Mol. Cell. Biol., 11*: 2096 (1991).
13. T. Tsubata, R. Tsubata, and M. G. Reth, Cell surface expression of the short immunoglobulin μ chain (Dμ protein) in murine pre-B cells is differently regulated from that of the intact μ chain, *Eur. J. Immunol., 21*: 1359 (1991).
14. H. Gu, D. Kitamura, and K. Rajewsky, B cell development regulated by gene rearrangement: Arrest of maturation by membrane-bound Dμ protein and selection of D_H element reading frames, *Cell, 65*: 47 (1991).
15. F. W. Alt, N. Rosenberg, E. Enea, E. Siden, and D. Baltimore, Organization and reorganization of immunoglobulin genes in A-mu-LV-transformed cells: Rearrangement of heavy but not light chain genes, *Cell, 27*: 381 (1981).
16. P. A. Hieter, S. J. Korsmeyer, T. A. Waldmann, and P. Leder, Human immunoglobulin κ light chain genes are deleted or rearranged in λ-producing B-cells, *Nature, 290*: 368 (1981).
17. S. J. Korsmeyer, P. A. Hieter, J. V. Ravetch, D. G. Poplack, T. A. Waldmann, and P. Leder, Developmental hierarchy of immunoglobulin gene rearrangements in human leukemic pre-B-cells, *Proc. Natl. Acad. Sci. USA, 78*: 7096 (1981).
18. P. Early, J. Rogers, M. Davis, K. Calame, M. Bond, R. Wall, and L. Hood, Two mRNAs can be produced from a single immunoglobulin μ gene by alternative RNA processing pathways, *Cell, 20*: 313 (1980).
19. J. Rogers, P. Early, C. Carter, K. Calame, M. Bond, L. Hood, and R. Wall, Two mRNAs with different 3′ ends encode membrane-bound and secreted forms of immunoglobulin μ chain, *Cell, 20*: 303 (1980).
20. J. Hombach, T. Takeshi, L. Leclercq, H. Stappert, and M. Reth, Molecular components of the B-cell antigen receptor complex of the IgM class, *Nature, 343*: 760 (1990).
21. K. S. Campbell, and J. C. Cambier, B lymphocyte antigen receptors (mlg) are non-covalently associated with a disulfide linked, inducibly phosphorylated glycoprotein complex, *EMBO J., 9*: 441 (1990).

22. J. Wienands, J. Hombach, A. Radbruch, C. Riesterer, and M. Reth, Molecular components of the B cell antigen receptor complex of class IgD differ partly from those of IgM, *EMBO J.*, *9*: 449 (1990).
23. G. G. Hermanson, D. Eisenberg, P. W. Kincade, and R. Wall, B29: A member of the immunoglobulin gene superfamily exclusively expressed on B-lineage cells, *Proc. Natl. Acad. Sci. USA*, *85*: 6890 (1988).
24. J. Hombach, F. Lottspeich, and M. Reth, Identification of the genes encoding the IgM-α and Ig-β components of the IgM antigen receptor complex by amino-terminal sequencing, *Eur. J. Immunol.*, *20*: 2795 (1990).
25. K. S. Campbell, E. J. Hager, R. J. Friedrich, and J. C. Cambier, IgM antigen receptor complex contains phosphoprotein products of *B29* and *mb1* genes, *Proc. Natl. Acad. Sci. USA*, *88*: 3982 (1991).
26. A. R. Venkitaraman, G. T. Williams, and P. Dariavac, The B-cell antigen receptor of the five immunoglobulin classes, *Nature*, *352*: 777 (1991).
27. J. Wienands, and M. Reth, The B cell antigen receptor of class 1gD can be expressed on the cell surface in two different forms, *Eur. J. Immunol.*, *21*: 2373 (1991).
28. C. J. M. Van Noesel, R. A. W. Van Lier, J. L. Cordell, A. G. D. Tse, G. M. W. Van Schijndel, E. F. R. De Vries, D. Y. Mason, and J. Borst, The membrane IgM-associated heterodimer on human B cells is a newly defined B cell antigen that contains the protein product of the *mb-1* gene, *J. Immunol.*, *146*: 1 (1991).
29. G. T. Williams, A. R. Wenkitaraman, D. J. Gilmore, and M. S. Neuberger, The sequence of the μ transmembrane segment determines the tissue specificity of the transport of immunoglobulin M to the cell surface, *J. Exp. Med.*, *171*: 947 (1990).
30. A. K. Bachhawat, and S. Pillai, Distinct intracellular fates of membrane and secretory immunoglobulin heavy chains in a pre-B cell line, *J. Cell Biol.*, *115*: 619 (1991).
31. J. Wienands, and M. Reth, Glycosyl-phosphatidylinositol linkage as a mechanism for cell surface expression of immunoglobulin D, *Nature*, *356*: 236 (1992).
32. S. Kashiwamura, T. Koyama, T. Matsuo, M. Steinmetz, M. Kimoto, and N. Sakaguchi, Structure of the murine *mb-1* gene encoding a putative sIgM-associated molecule, *J. Immunol.*, *145*: 337 (1990).
33. G. G. Hermanson, D. Eisenberg, P. W. Kincade, and R. Wall, A member of the immunoglobulin gene superfamily exclusively expressed on B-lineage cells, *Proc. Natl. Acad. Sci. USA*, *86*: 7341 (1989).
34. M. Reth, Antigen receptor tail clue, *Nature*, *338*: 383 (1989).
35. S. Sell, and P. G. H. Gell, Studies on rabbit lymphocytes in vitro. I. Stimulation of blast transformation with an antiallotypic serum, *J. Exp. Med.*, *122*: 423 (1965).
36. D. Parker, Induction and suppression of polyclonal antibody responses by anti-Ig reagents and antigen-nonspecific helper factors, *Immunol. Rev.*, *52*: 115 (1980).
37. D. C. Parker, J. J. Fothergill, and D. C. Wadsworth, B lymphocyte activation by insoluble anti-immunoglobulin: Induction of immunoglobulin secretion by a T cell-dependent soluble factor, *J. Immunol.*, *123*: 931 (1979).
38. A. L. DeFranco, E. S. Raveche, and W. E. Paul, Separate control of B lymphocyte early activation and proliferation in response to anti-IgM antibodies, *J. Immunol.*, *135*: 87 (1985).

39. L. M. T. Pecanha, C. M. Snapper, F. D. Finkelman, and J. J. Mond, Dextran-conjugated anti-Ig antibodies as a model for T cell-independent type 2 antigen-mediated stimulation of Ig secretion in vitro. I. Lymphokine dependence, *J. Immunol.*, *146*: 833 (1991).
40. M. K. Bijsterbosch, and G. D. B. Klaus, Crosslinking of surface immunoglobulin and Fc receptors on B lymphocytes inhibits stimulation of inositol phospholipid breakdown via the antigen receptor, *J. Exp. Med.*, *162*: 1825 (1985).
41. R. D. Klausner, and L. E. Samuelson, Cell antigen receptor activation pathways: The tyrosine kinase connection, *Cell*, *64*: 875 (1991).
42. A. E. Nel, G. E. Landreth, P. J. Goldschmidt-Clermont, H. E. Tung, and R. M. Galbraith, Enhanced tyrosine phosphorylation in B lymphocytes upon complexing of membrane immunoglobulin, *Biochem. Biophys. Res. Comm.*, *125*: 859 (1984).
43. M. R. Gold, D. A. Law, and A. L. Defranco, Stimulation of protein tyrosine phosphorylation by the B-lymphocyte antigen receptor, *Nature*, *345*: 810 (1990).
44. M.-A. Campbell, and B. M. Sefton, Protein tyrosine phosphorylation is induced in murine B lymphocytes in response to stimulation with anti-immunoglobulin, *EMBO J.*, *9*: 2125 (1990).
45. P. J. L. Lane, F. M. McConnell, G. L. Schieven, E. A. Clark, and J. A. Ledbetter, The role of class II molecules in human B cell activation: Association with phosphatidylinositol turnover, protein tyrosine phosphorylation, and proliferation, *J. Immunol.*, *144*: 3684 (1990).
46. M. Brunswick, L. E. Samuelson, and J. J. Mond, Surface immunoglobulin crosslinking activates a tyrosine kinase pathway in B cells that is independent of protein kinase C, *Proc. Natl. Acad. Sci. USA*, *88*: 1311 (1991).
47. M. R. Gold, L. Matsuuchi, R. B. Kelly, and A. L. Defranco, Tyrosine phosphorylation of components of the B-cell antigen receptors following receptor crosslinking, *Proc. Natl. Acad. Sci. USA*, *88*: 3436 (1991).
48. M. Baniyash, P. Garcia-Morales, E. Luong, L. E. Samuelson, and R. D. Klausner, *J. Biol. Chem.*, *263*: 18225 (1988).
49. J. J. O'Shea, A. M. Weissman, I. C. Kennedy, and J. R. Ortaldo, Engagement of the natural killer cell IgG Fc receptor results in tyrosine phosphorylation of the zeta chain, *Proc. Natl. Acad. Sci. USA*, *88*: 350 (1991).
50. M. K. Bijsterbosch, J. C. Meade, G. A. Turner, and G. D. B. Klaus, B lymphocyte receptors and polyphosphoinositide degradation, *Cell*, *41*: 999 (1985).
51. J. C. Cambier, and J. T. Ransom, Molecular mechanisms of transmembrane signaling in B lymphocytes, *Annu. Rev. Immunol.*, *5*: 175 (1987).
52. R. H. Carter, D. J. Park, S. G. Rhee, and D. T. Fearon, Tyrosine phosphorylation of phospholipase C induced by membrane immunoglobulin in B lymphocytes, *Proc. Natl. Acad. Sci. USA*, *88*: 2745 (1991).
53. S. Padeh, A. Levitzki, A. Gazit, G. B. Mills, and C. M. Roifman, Activation of phospholipase C in human B cells is dependent on tyrosine phosphorylation, *J. Clin. Invest.*, *87*: 1114 (1991).
54. S. G. Rhee, P.-G. Suh, S.-H. Ryu, and S. Y. Lee, Studies of inositol phospholipid-specific phospholipase C, *Science*, *244*: 546 (1989).
55. W. M. Hempel, and A. L. Defranco, Expression of phospholipase C isozymes by murine B lymphocytes, *J. Immunol.*, *146*: 3713 (1991).

56. M. J. Berridge, Inositol triphosphate and diacylglycerol as second messengers, *Biochem. J.*, *220*: 345 (1984).
57. R. M. Bell, and D. J. Burns, Lipid activation of protein kinase C, *J. Biol. Chem.*, *266*: 4661 (1991).
58. A. C. Shaw, R. N. Mitchell, Y. K. Weaver, J. Campos-Torres, A. K. Abbas, and P. Leder, Mutations of immunoglobulin transmembrane and cytoplasmic domains: Effects on intracellular signaling and antigen presentation, *Cell*, *63*: 381 (1990).
59. S. Nishibe, M. I. Wahl, S. M. T. Hernandez-Sotomayor, N. K. Tonks, S. G. Rhee, and G. Carpenter, Increase of the catalytic activity of phospholipase C-γ1 by tyrosine phosphorylation, *Science*, *250*: 1253 (1990).
60. S. J. Taylor, H. Z. Chae, S. G. Rhee, and J. H. Exton, Activation of the b1 isozyme of phospholipase C by a subunits of the G_q class of G proteins, *Nature*, *350*: 516 (1991).
61. J. Meisenhelder, P.-G. Suh, S. G. Rhee, and T. Hunter, Phospholipase C-γ is a substrate for the PDGF and EGF receptor protein-tyrosine kinases in vivo and in vitro, *Cell*, *57*: 1109 (1989).
62. B. Margolis, F. Bellot, A. M. Honegger, A. Ullrich, J. Schlessinger, and A. Zilberstein, Tyrosine kinase activity is essential for the association of phospholipase C-γ with the epidermal growth factor receptor, *Mol. Cell. Biol.*, *10*: 435 (1990).
63. H. K. Kim, J. W. Kim, A. Zilberstein, B. Margolis, J. G. Kim, J. Schlessinger, and S. G. Rhee, PDGF stimulation of inositol phospholipid hydrolysis requires PLC-γ1 phosphorylation on tyrosine residues 783 and 1254, *Cell*, *65*: 435 (1991).
64. B. A. Irving, and A. Weiss, The cytoplasmic domain of the T cell receptor ζ chain is sufficient to couple to receptor-associated signal transduction pathways, *Cell*, *64*: 891 (1991).
65. T. E. Costa, R. R. Franke, M. Sanchez, Z. Misulovin, and M. C. Nussenzweig, Functional reconstitution of an immunoglobulin antigen receptor in T cells, *J. Exp. Med.*, *175*: 1669 (1992).
66. L. Matsuuchi, M. R. Gold, A. Travis, R. Grosschedl, A. L. DeFranco, and R. B. Kelly, The membrane IgM-associated proteins MB-1 and Ig-β are sufficient to promote surface expression of a partially functional B-cell antigen receptor in a nonlymphoid cell line, *Proc. Natl. Acad. Sci. USA*, *89*: 3404 (1992).
67. S. Takagi, M. Daibata, T. J. Last, R. E. Humphreys, D. C. Parker, and T. Sairenji, Intracellular localization of tyrosine kinase substrates beneath crosslinked surface immunoglobulins in B cells, *J. Exp. Med.*, *174*: 381 (1991).
68. A. S. Shaw, K. E. Amrein, C. Hammond, D. F. Stern, B. M. Sefton, and J. K. Rose, The *lck* tyrosine protein kinase interacts with the cytoplasmic tail of the CD4 glycoprotein through its unique amino-terminal domain, *Cell*, *59*: 627 (1989).
69. P. Cicchetti, B. J. Mayer, G. Thiel, and D. Baltimore, Identification of a protein that binds to the SH3 region of *Abl* and is similar to *Bcr* and GAP-rho, *Science*, *257*: 803 (1992).
70. C. A. Koch, D. Anderson, M. F. Moran, C. Ellis, and T. Pawson, SH2 and SH3 domains: Elements that control interactions of cytoplasmic signaling proteins, *Science*, *252*: 668 (1991).

71. T. E. Kmiecik, and D. Shalloway, Activation and suppression of pp60c-*src* transforming ability by mutation of its primary sites of tyrosine phosphorylation, *Cell*, *49*: 65 (1987).
72. S. M. Dymecki, J. E. Niederhuber, and S. V. Desiderio, Specific expression of a tyrosine kinase gene, *blk*, in B lymphoid cells, *Science*, *247*: 332 (1990).
73. S. M. Dymecki, P. Zwollo, K. Zeller, F. P. Kuhajda, and S. V. Desiderio, Structure and developmental regulation of the B-lymphoid tyrosine kinase gene *blk*, *J. Biol. Chem.*, *267*: 4815 (1992).
74. A. L. Burkhardt, M. Brunswick, J. B. Bolen, and J. J. Mond, Anti-immunoglobulin stimulation of B lymphocytes activates *src*-related protein-tyrosine kinases, *Proc. Natl. Acad. Sci. USA*, *88*: 7410 (1991).
75. Y. Yamanashi, T. Kakiuchi, J. Mizuguchi, T. Yamamoto, and K. Toyoshima, Association of B cell antigen receptor with protein tyrosine kinase *lyn*, *Science*, *251*: 192 (1991).
76. M.-A. Campbell, and B. M. Sefton, Association between B-lymphocyte membrane immunoglobulin and multiple members of the *src* family of protein tyrosine kinases, *Mol. Cell. Biol.*, *12*: 2315 (1992).
77. J. D. Marth, D. B. Lewis, C. B. Wilson, M. E. Gearn, E. G. Krebs, and R. M. Perlmutter, Regulation of pp56*lck* during T-cell activation: Functional implications for the *src*-like protein tyrosine kinases, *EMBO J.*, *6*: 2727 (1987).
78. M. Hatakeyama, T. Kono, N. Kobayashi, A. Kawahara, S. D. Levin, R. M. Perlmutter, and T. Taniguchi, Interaction of the IL-2 receptor with the *src*-family kinase p56*lck*: Identification of novel intermolecular association, *Science*, *252*: 1523 (1991).
79. Y. Yamanashi, Y. Fukui, B. Wongsasant, Y. Kinoshita, Y. Ichimori, K. Toyoshima, and T. Yamamoto, Activation of *src*-like protein-tyrosine kinase *lyn* and its association with phosphatidylinositol 3-kinase upon B-cell antigen receptor-mediated signaling, *Proc. Natl. Acad. Sci. USA*, *89*: 1118 (1992).
80. T. T. Taniguchi, T. Kobayashit, J. Kondo, K. Takahaski, H. Nakamura, J. Suzuki, K. Nagai, T. Yamada, S. Nakamura, and H. Yamamura, Molecular cloning of a porcine gene *syk* that encodes a 72-kDa protein-tyrosine kinase showing high susceptibility to proteolysis, *J. Biol. Chem.*, *266*: 15790 (1991).
81. J. E. Hutchcroft, M. L. Harrison, and R. L. Geahlen, B lymphocyte activation is accompanied by phosphorylation of a 72-kDa protein-tyrosine kinase, *J. Biol. Chem.*, *266*: 14846 (1991).
82. J. E. Hutchcroft, M. L. Harrison, and R. L. Geahlen, Association of the 72-kDa protein-tyrosine kinase PTK72 with the B cell antigen receptor, *J. Biol. Chem.*, *267*: 8613 (1992).
83. P. W. Majerus, T. S. Rosa, T. W. Cunningham, K. K. Caldwell, A. B. Jefferson, and V. S. Bansal, Recent insights in phosphatidylinositol signaling, *Cell*, *63*: 459 (1990).
84. Y. Yamanashi, Y. Fukui, B. Wongsasant, Y. Kinoshita, Y. Ichimori, K. Toyoshima, and T. Yamamoto, Activation of *src*-like protein-tyrosine kinase *lyn* and its association with phosphatidylinositol 3-kinase upon B-cell antigen receptor-mediated signaling, *Proc. Natl. Acad. Sci. USA*, *89*: 1118 (1992).

85. M. M. Harnett, and G. G. B. Klaus, G protein coupling of antigen receptor-stimulated polyphosphoinositide hydrolysis in B cells, *J. Immunol.*, *140*: 3135 (1988).
86. M. R. Gold, J. P. Jakway, and A. L. DeFranco, Involvement of a guanine nucleotide-binding component in membrane IgM-stimulated phosphoinositide breakdown, *J. Immunol.*, *139*: 3604 (1987).
87. J. G. Monroe, and S. Haldar, Involvement of a specific guanine nucleotide binding protein in receptor immunoglobulin stimulated inositol phospholipid hydrolysis, *Biochim. Biophys. Acta*, *1013*: 273 (1989).
88. T. M. Wilkie, P. A. Scherle, M. P. Strathmann, V. Z. Slepak, and M. I. Simon, Characterization of G-protein α subunits in the G_q class: Expression in murine tissues and in stromal and hematopoietic cell lines, *Proc. Natl. Acad. Sci. USA*, *88*: 10049 (1991).
89. J. Downward, J. D. Graves, P. H. Warne, S. Rayter, and D. A. Cantrell, Stimulation of p21*ras* upon T-cell activation, *Nature*, *346*: 719 (1990).
90. L. Graziadei, K. Riabowol, and D. Bar-Sagi, Co-capping of *ras* proteins with surface immunoglobulins in B lymphocytes, *Nature*, *347*: 396 (1990).
91. R. D. Campbell, M. C. Carroll, and R. R. Porter, The molecular genetics of components of complement, *Adv. Immunol.*, *38*: 203 (1986).
92. A. K. Matsumoto, J. Kopicky-Burd, R. H. Carter, D. A. Tuveson, T. F. Tedder, and D. T. Fearon, Intersection of the complement and immune systems: A signal transduction complex of the B lymphocyte-containing complement receptor type 2 and CD19, *J. Exp. Med.*, *173*: 55 (1991).
93. R. H. Carter, and D. T. Fearon, CD 19: Lowering the threshold for antigen receptor stimulation of B lymphocytes, *Science*, *256*: 105 (1992).
94. F. Melchers, A. Erdei, T. Schulz, and M. P. Dierich, Growth control of activated, synchronized murine B cells by the C3d fragment of human complement, *Nature*, *317*: 264 (1985).
95. R. H. Carter, M. O. Spycher, Y. C. Ng, R. Hoffmann, and D. T. Fearon, Synergistic interaction between complement receptor type 2 and membrane IgM on B lymphocytes, *J. Immunol.*, *141*: 457 (1988).
96. D. A. Tuveson, J. M. Ahearn, A. K. Matsumoto, and D. T. Fearon, Molecular interactions of complement receptors on B lymphocytes: A CR1/CR2 complex distinct from the CR2/CD19 complex, *J. Exp. Med.*, *173*: 1083 (1991).
97. I. Stamenkovic, and B. Seed, CD19, the earliest differentiation antigen of the B cell lineage, bears three extracellular immunoglobulin-like domains and an Epstein–Barr virus-related cytoplasmic tail, *J. Exp. Med.*, *168*: 1205 (1988).
98. A. Pezzutto, B. Dorken, P. S. Rabinovitch, J. A. Ledbetter, G. Moldenhauer, and E. A. Clark, CD19 monoclonal antibody HD37 inhibits anti-immunoglobulin-induced B cell activation and proliferation, *J. Immunol.*, *138*: 2793 (1987).
99. J. M. Pesando, L. S. Bouchard, and B. E. McMaster, CD19 is functionally and physically associated with surface immunoglobulin, *J. Exp. Med.*, *170*: 2159 (1989).
100. T. Hunter, Protein-tyrosine phosphatases: The other side of the coin, *Cell*, *58*: 1013 (1989).

101. A. N. Barclay, D. I. Jackson, A. C. Willis, and A. F. Williams, Lymphocyte specific heterogeneity in the rat leucocyte common antigen (T200) is due to differences in polypeptide sequences near the NH_2-terminus, *EMBO J.*, *6*: 1259 (1987).
102. P. Johnson, L. Greenbaum, K. Bottomly, and I. S. Trowbridge, Identification of the alternatively spliced exons of murine CD45 (T200) required for reactivity with B220 and other T200-restricted antibodies, *J. Exp. Med.*, *169*: 1179 (1989).
103. I. S. Trowbridge, P. Ralph, and M. J. Bevan, Differences in the surface proteins of mouse B and T cells, *Proc. Natl. Acad. Sci. USA*, *72*: 157 (1975).
104. J. W. Fabre, and A. F. Williams, Quantitative serological analysis of a rabbit anti-rat lymphocyte serum and preliminary biochemical characterization of the major antigen recognized, *Transplantation*, *23*: 349 (1977).
105. M. L. Thomas, P. J. Reynolds, A. Chain, Y. Ben-Neriah, and I. S. Trowbridge, Evidence for alternative mRNA splicing, *Proc. Natl. Acad. USA*, *84*: 5360 (1987).
106. M. L. Thomas, The leukocyte common antigen family, *Annu. Rev. Immunol.*, *7*: 339 (1989).
107. Y. Saga, J.-S. Tung, F.-W. Shen, and E. A. Boyse, Alternative use of 5′ exons in the specification of LY05 isoforms distinguishing hematopoietic cell lineages, *Proc. Natl. Acad. Sci. USA*, *84*: 5364 (1987).
108. M. Streuli, L. R. Hall, Y. Saga, S. F. Schlossman, and H. Saito, Differential usage of three exons generates at least five different mRNAs encoding human leukocyte common antigens, *J. Exp. Med.*, *166*: 1548 (1987).
109. H. Charbonneau, N. K. Tonks, K. A. Walsh, and E. H. Fischer, *Proc. Natl. Acad. Sci. USA*, *85*: 7182 (1988).
110. H. Charbonneau, N. K. Tonks, S. Kumar, C. D. Diltz, M. Harrylock, D. E. Cool, E. G. Krebs, E. H. Fischer, and K. A. Walsh, *Proc. Natl. Acad. Sci. USA*, *86*: 5252 (1989).
111. J. K. Tonks, H. Charbonneau, C. D. Diltz, E. H. Fischer, and K. A. Walsh, Demonstration that the leukocyte common antigen is required for antigen-induced T lymphocyte proliferation, *Biochemistry*, *27*: 8695 (1988).
112. M. Streuli, N. X. Krueger, A. Y. M. Tsai, and H. Saito, A family of receptor-linked protein tyrosine phosphatase in humans and *Drosophila*, *Proc. Natl. Acad. Sci. USA*, *86*: 8698 (1989).
113. P. W. Kincase, Experimental models for understanding B lymphocyte formation, *Adv. Immunol.*, *41*: 181 (1987).
114. E. C. Butcher, R. V. Rouse, R. L. Coffman, C. N. Nottenburg, R. R. Hardy, and I. L. Weissman, Surface phenotype of Peyer's patch germinal center cells: Implications for the role of germinal centers in B cell differentiation, *J. Immunol.*, *129*: 2698 (1982).
115. R. G. Cook, N. F. Landolfi, V. Mehta, J. Leone, and D. Hoyland, Interleukin 2 mediates an alteration in the T200 antigen expressed on activated B lymphocytes, *J. Immunol.*, *139*: 991 (1987).
116. E. Alsinet, J. Inglesesteve, R. Bilella, F. Lozano, J. Mila, I. Rojo, J. Martorell, J. Vives, and A. Gaya, Differential effects of anti-CD45 monoclonal antibody on human B-cell proliferation—a monoclonal antibody recognizing a neuraminidase-sensitive epitope of the T200 molecule enhances anti-immunoglobulin-induced proliferation, *Eur J. Immunol.*, *20*: 2801 (1990).

117. L. B. Justement, K. S. Campbell, N. C. Dhien, and J. C. Cambier, Regulation of B cell antigen receptor signal transduction and phosphorylation by CD45, *Science*, *252*: 1839 (1991).
118. H. L. Ostergaard, and I. S. Trowbridge, Coclustering CD45 with CD4 or CD8 alters the phosphorylation and kinase activity of p56*lck*, *J. Exp. Med.*, *172*: 347 (19).
119. D. Bar-Sagi, personal communication.
120. T. R. Mosmann, H. Cherwinski, M. W. Bond, M. A. Giedlin, and R. L. Coffman, Two types of murine helper T cell clone. I. Definition according to profiles of lymphokine activities and secreted proteins, *J. Immunol.*, *136*: 2348 (1986).
121. K. Takatsu, Interleukin-5, *Curr. Opin. Immunol.*, *4*: 299 (1992).
122. S. F. Webb, C. Das, S. Eeton, K. Calame, and P. W. Tucker, Novel protein–DNA interactions associated with increased immunoglobulin transcription in response to antigen plus interleukin-5, *Mol. Cell. Biol.*, *11*: 5197 (1991).
123. E. Sonoda, Y. Hitoshi, N. Yamaguchi, T. Ishii, A. Tominaga, S. Araki, and K. Takatsu, Differential regulation of IgA production by TGF-β and IL-5: TGF-β induces surface IgA-positive cells bearing IL-5 receptor, whereas IL-5 promotes their survival and maturation into IgA-secreting cells, *Cell. Immunol.*, *140*: 158 (1992).
124. Y. Hosoya, S. Yamauchi, Y. Kanai, et al., Transgenic mice expressing a B cell growth and differentiation factor gene (interleukin 5) develop eosinophilia and autoantibody production, *J. Exp. Med.*, *173*: 429 (1991).
125. A. E. Namen, A. E. Schmierer, C. J. March, R. W. Overell, L. S. Park, D. L. Urdal, and D. Y. Mochizuki, B cell precursor growth-promoting activity, purification and characterization of a growth factor active on lymphocyte precursors, *J. Exp. Med.*, *167*: 988 (1988).
126. C. A. Whitlock, and O. N. Witte, Long-term culture of B lymphocytes and their precursors from murine bone marrow, *Proc. Natl. Acad. Sci. USA*, *79*: 3608 (1982).
127. K. Smith, Interleukin-2: Inception, impact, and implications, *Science*, *240*: 1169 (1988).
128. M. A. Tigges, L. S. Casey, and M. E. Koshland, Mechanism of interleukin-2 signaling: Mediation of different outcomes by a single receptor and transduction pathway, *Science*, *243*: 781 (1989).
129. H. Schorle, T. Holtschke, T. Hunig, A. Schimpl, and I. Horak, Development and function of T cells in mice rendered interleukin-2 deficient by gene targeting, *Nature*, *352*: 621 (1991).
130. R. Noelle, P. H. Krammer, J. Ohara, J. W. Uhr, and E. S. Vitetta, Increased expression of Ia antigens on resting B cells: An additional role for B-cell growth factor, *Proc. Natl. Acad. Sci. USA*, *81*: 6149 (1984).
131. N. W. Roehm, H. J. Liebson, A. Slotnik, J. Kappler, P. Marrack, and J. C. Cambier, Interleukin-induced increase in Ia expression by normal mouse B cells, *J. Exp. Med.*, *160*: 679 (1984).
132. S. A. Hudak, S. O. Gollnick, D. H. Conrad, and M. R. Kehry, Murine B-cell stimulatory factor 1 (interleukin 4) increases expression of the Fc receptor for IgE on mouse B cells, *Proc. Natl. Acad. Sci. USA*, *84*: 4606 (1987).

133. K. Oliver, R. J. Noelle, J. W. Uhr, P. H. Krammer, and E. S. Vitetta, B-cell growth factor (B-cell growth factor I or B-cell-stimulating factor, provisional 1) is a differentiation factor for resting B cells and may not induce cell growth, *Proc. Natl. Acad. Sci. USA*, *82*: 2465 (1985).
134. E. M. Rabin, J. Ohara, and W. E. Paul, B-cell stimulatory factor 1 activates resting B cells, *Proc. Natl. Acad. Sci. USA*, *82*: 2935 (1985).
135. J. M. Purkerson, and I. C. Isakson, Isotype switching in anti-immunoglobulin-activated B lymphoblasts: Differential requirements for interleukin 4 and other lymphokines to elicit membrane vs. secreted IgG1, *Eur. J. Immunol.*, *21*: 707 (1991).
136. T. L. Stevens, A. Bossie, V. M. Sanders, R. Fernandez-Botran, R. L. Coffman, T. R. Mosmann, and E. S. Vitetta, Regulation of antibody isotype secretion by subsets of antigen-specific helper T cells, *Nature*, *334*: 255 (1988).
137. W. H. Boom, D. Liano, and A. K. Abbas, Heterogeneity of helper/inducer T lymphocytes, *J. Exp. Med.*, *167*: 1350 (1988).
138. D. A. Lebman, and R. L. Coffman, Interleukin 4 causes isotype switching to IgE in T cell-stimulated clonal B cell cultures, *J. Exp. Med.*, *168*: 853 (1988).
139. P. C. Isakson, E. Pure, E. S. Vitetta, and P. H. Kramer, T cell-derived B cell differentiation factor(s). Effect on the isotype switch of murine B cells, *J. Exp. Med.*, *155*: 734 (1982).
140. R. L. Coffman, and J. Carty, A T cell activity that enhances polyclonal IgE production and its inhibition by interferon-γ, *J. Immunol.*, *136*: 949 (1986).
141. R. I. Tepper, D. A. Levinson, B. Z. Stanger, J. Compos-Torres, A. K. Abbas, and P. Leder, *Cell*, *62*: 347 (1990).
142. W. Muller, R. Kuhn, and K. Rajewsky, Major histocompatibility complex class II hyperexpression on β cells in interleukin 4-transgenic mice does not lead to β cell proliferation and hypergammaglobulinemia, *Eur. J. Immunol.*, *21*: 921 (1991).
143. T. Kishimoto, and T. Hirano, Molecular regulation of B lymphocyte response, *Annu. Rev. Immunol.*, *6*: 485 (1988).
144. T. Hirano, T. Matsuda, M. Turner, N. Miyasaka, G. Buchan, B. Tang, K. Sato, M. Shimizu, R. Maini, M. Feldman, and T. Kishimoto, *Eur. J. Immunol.*, *18*: 1797 (1988).
145. M. Kawano, T. Hirano, T. Matsuda, T. Taga, Y. Horii, K. Iwato, H. Asaoku, B. Tang, O. Tanabe, H. Tanaka, M. A. Kuramoto, and T. Kishimoto, Autocrine generation and requirement of BSF-2/IL-6 for human multiple myelomas, *Nature*, *332*: 83 (1988).
146. S. Suematsu, et al., IgG1 plasmacytosis in interleukin 6 transgenic mice, *Proc. Natl. Acad. Sci. USA*, *86*: 7547 (1989).

3

Major Histocompatibility Antigens and Genes

David M. Robinson
Virginia Mason Medical Center, Seattle, Washington

Gerald T. Nepom
Virginia Mason Research Center, Seattle, Washington

I. INTRODUCTION

One of the fundamental controls of the immune response is provided by the gene products of the major histocompatibility complex (MHC). The MHC is composed of a limited set of genes, which are variable among individuals. This variability of MHC genes and their protein products allows the broad range and diversity of the immune response to foreign antigens. This has important implications for determining the response to allergens as well as the generation of autoimmune disorders. This chapter will serve as an introduction to the human major histocompatibility complex—the human leukocyte antigen (HLA) complex—and the antigens that it encodes. The terms MHC and HLA are used somewhat interchangeably throughout.

The MHC proteins on antigen-presenting cells serve to bind antigenic peptide fragments and present them to T cells. The specific interactions between MHC protein, peptide fragment, and T-cell receptor determine the character of the subsequent immune response. Since the MHC molecules differ among individuals, these interactions are the structural basis for HLA-linked genetic control of individual variation in immune response. Several HLA associations have been made with allergen responsiveness (1–3). Virtually all of these associations have been found with HLA class II genes (or class II gene products). Nonresponsiveness to specific allergen has also been shown to be associated with a particular

HLA type (4). The functional participation of *HLA-DR* gene products has also been demonstrated in the recognition of allergen by helper T cells (5).

Studies of large families indicate that the development of allergic disease not only involves the MHC, but also other factors outside the MHC (6). The tentative assignment of an "atopy gene" has been made to chromosome 11q by the use of molecular genetic techniques (7). The same situation exists for HLA associations with autoimmune diseases, for which a specific HLA type associated with disease is a necessary, but not a sufficient, factor determining disease susceptibility (8). The respective contributions of other genetic and environmental susceptibility factors remain areas of intense interest in both the allergic and autoimmune diseases. Allergic diseases have one distinct advantage, in that the environmental factors (allergens) are often well defined, some at the molecular level (9,10).

The terminology for specific HLA loci and allelic variants has undergone significant revision in recent years. Synonyms and frequently used terminology will be supplied where appropriate.

II. MAJOR HISTOCOMPATIBILITY COMPLEX GENE ORGANIZATION

The MHC complex is a region of 40–50 genes that, in humans, is arrayed, along with a variety of immunologically related and seemingly unrelated genes, within a contiguous stretch of approximately 4000 kilobases (kb) of DNA found 15 centimorgans (cM) (recombination map distance) from the centromere on the short arm of chromosome 6 (Fig. 1). The MHC is referred to as the HLA complex in humans. In mice it is called the H-2 complex and occupies about 2500 kb on chromosome 17. A comprehensive map of gene loci within the HLA has recently been published (11), and this chapter will concentrate on the human MHC antigen and genes. However, since so much basic immunological research is performed with the mouse model, a brief overview of the corresponding organization and nomenclature in the murine system is also included.

A. *HLA* Class I Genes

The *HLA* class I genes are located most telomeric and include at least six known *HLA* class I loci. Three of these loci encode the highly polymorphic class I histocompatibility antigens or "classic transplantation antigens." The remaining *HLA* class I genes are felt to possibly play some functional role during development (12). Several other genes also lie within this region. Some appear to be intact genes, but are not known to be expressed. Others are pseudogenes, which are similar to typical class I genes, but lack some important structural or regulatory elements.

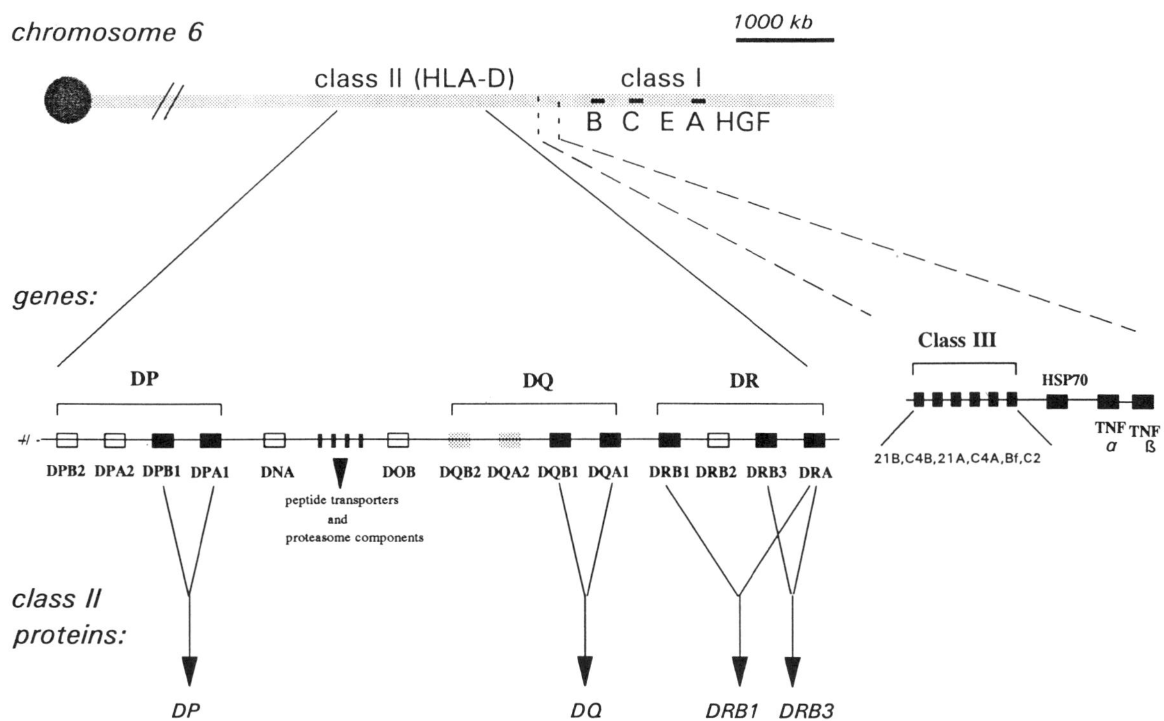

Figure 1 Gene organization of the major histocompatibility complex (MHC) located on chromosome 6. The *HLA* genes segregate into the class I and class II clusters, separated by the non-*HLA* class III cluster. Several additional non-*HLA* genes that map into the MHC are also shown; susceptibility associated with a particular *HLA* gene may be due to an *HLA* structural gene or to a non-*HLA* gene in close linkage to other genes in the complex. Vertical arrows denote *HLA* class II expressed genes that encode the major DR, DQ, and DP molecules. Other class II genes are pseudogenes of unknown function. The centromere lies to the left of the HLA complex as shown here. *21A* and *21B*, 21-hydroxylase genes; *C4B*, *C4A*, *C2*, complement component genes; *Bf*, properdin factor B; *TNF*, tumor necrosis factor; *HSP70*, heat-shock protein 70. (Adapted from Ref. 11.)

The *HLA* class I gene loci are arranged as shown in Figure 1. Three loci, termed *HLA-A, HLA-B,* and *HLA-C* encode the classic transplantation antigens and are expressed on all nucleated cells. These loci encode the gene products that form the major histocompatibility barrier to allogenic transplantation. The remaining class I genes are known as the nonclassic *HLA* class I genes.

The typical class I gene consists of seven exons, which code for specific domains of the final protein product, such as the signal sequence, three external domains (two of which bind peptide), the transmembrane and intracytoplasmic tail of the HLA class I protein. Each gene also has a variety of regulatory elements, which will be covered briefly later. There is only a single type of *HLA* class I gene, that encoding the single heavy chain. The *HLA* class I loci are highly polymorphic with over 80 total alleles and over 30 known alleles defined at some loci (Table 1).

Centromeric to the *HLA* class I genes are several important but unrelated, genes encoding such things as tumor necrosis factor (TNF)-α and TNF-β; complement components C2, C4, and Bf; as well as the structural gene for 21-hydroxylase (13–19). This diverse region is often referred to as the HLA class III region.

B. *HLA* Class II Genes

The *HLA* class II genes are located centromeric to the class III and class I regions (Fig. 1). The number of *HLA* class II genes varies, but averages approximately 14, some of which are pseudogenes. There are seven or eight II gene loci, which encode class II polypeptides that are functionally important and are expressed on the cell surface. As seen with the class I region, a variety of non-*HLA* genes are found in the class II region. Some of these non-*HLA* gene products may play a role in antigen processing and transport and, thus, are potentially important in terms of regulating the immune response and, possibly, in disease susceptibility (18,20). Other non-*HLA* candidate genes are also found within the HLA class II region and are of unknown import.

There are two types of genes in the *HLA* class II region, those coding for α-polypeptides and those coding for β-polypeptides. The mature class II molecule consists of an α–β heterodimer. Although there are multiple class II genes, the gene organization is such that there are certain clusters of genes that correspond to discrete patterns of α–β pairs. Three major subregions of the *HLA* class II complex are described as *HLA-DR, HLA-DQ,* and *HLA-DP* (see Fig. 1). Previous designations of the *HLA* class II region included "HLA-D" and HLA-DR, DC, and SB (for DR, DQ, and DP, respectively).

In the *HLA-DR* region, a single nonpolymorphic α gene, *HLA-DRA,* is linked to one or more β-genes known as *HLA-DRB1, DRB3, DRB4,* or *DRB5*. On most haplotypes, most of these *DRB* genes are nonfunctional or pseudogenes (as is *HLA-DRB2*). Generally, only two *DRB* genes are expressed per haplotype (four per individual), and the resulting HLA-DR proteins are heterodimeric products of two such gene combinations (i.e., *HLA-DRA/DRB1* and *HLA-DRA/DRB3*). The *DRB1* locus encodes the β-chains, comprising the alleles *DR1* (currently *DRB1 *0101*), *DR2, DR3,* whereas the second expressed gene varies on different haplotypes, such that most *DR3* haplotypes, for example, also express *DRw52* (currently, *DRB3*0101*) (Table 2).

The *HLA-DQ* region consists of two pairs of linked α- and β-genes. The *HLA-DQA1* and *DQB1* genes encode the expressed HLA-DQ heterodimer. The other pair of *HLA-DQ* genes, *DQA2* and *DQB2* encode an as of yet unidentified product, often designated as "DX."

The *HLA-DP* region lies most centromeric within the class II region. One set of functional α- and β-genes, *HLA-DPA1* and *DPB1* encode the expressed DP molecule. The *HLA-DPA2* and *DPB2* loci are nonexpressed pseudogenes.

Additional class II genes lie between the *HLA-DQ* and *DP* regions, including an α-gene, *HLA-DNA* (formerly *HLA-DZA*) and a β-gene, *HLA-DOB*. Transcripts for each have been found, but the function of the molecules encoded by these loci is not yet unknown.

A typical *HLA* class II gene consists of five or six exons, each encoding a different functional domain of the HLA class II α- or β-polypeptide. The different *HLA* class II loci differ in terms of their tissue sites of expression. This is presumed to indicate functional differences among their products. The *HLA-DR* and *DP* genes are constitutively expressed on antigen-presenting cells such as B lymphocytes and monocyte–macrophage lineage cells. The *HLA-DR* and *DP* expression is inducible in other cells, including activated T lymphocytes and somatic tissues, such as vascular endothelium at sites of cytokine-mediated inflammation. The HLA-DQ molecules have a more restricted constitutive range, but likewise may be induced on a variety of lymphoid and nonlymphoid cell types (21).

As seen with *HLA* class I genes, allelic polymorphism is common to most *HLA* class II genes. The exception is the *HLA-DRA* locus, which is present as a single allele. All other expressed class II genes are highly polymorphic with from 4 to over 40 alleles identified thus far at each locus (see Table 2).

C. Nonclassic Genes

Within the *HLA* class I region, there are several other class I genes, designated *HLA-E, HLA-F,* and *HLA-G,* plus other class I-like loci of unknown function, often referred to as nonclassic class I genes. The *HLA-E* gene has a much more restricted range of expression than *HLA-A, B,* and *C* (22). Expression of the *HLA-F* and *HLA-G* genes has not yet been thoroughly studied. Transcription of the *HLA-F* gene has been detected in cell lines, although a protein product has not yet been identified (23). Expression of the *HLA-G* gene has been found only in the placental trophoblast, an observation that is intriguing with potential functional consequences (12).

Table 1 Current Nomenclature of the *HLA-A*, *HLA-B*, and *HLA-C* Alleles Recognized by the WHO Committee for Factors of the HLA System, as of 1990

HLA-A Alleles		*HLA-B* Alleles		*HLA-C* Alleles	
1990 Nomenclature	Previous Designations	1990 Nomenclature	Previous Designations	1990 Nomenclature	Previous Designations
*A*0101*	*A1*	*B*0701*	*B7, B7.1*	*Cw*0101*	*Cw1*
*A*0201*	*A2, A2.1*	*B*0702*	*B7, B7.2*	*Cw*0201*	*Cw2, Cw2.1*
*A*0202*	*A2, A2.2F*	*B*0801*	*B8*	*Cw*02021*	*Cw2, Cw2.2*
*A*0203*	*A2, A2.3*	*B*1301*	*B13, B13.1*	*Cw*02022*	*Cw2, Cw2.2*
*A*0204*	*A2*	*B*1302*	*B13, B13.2*	*Cw*0301*	*Cw3*
*A*0205*	*A2, A2.2Y*	*B*1401*	*B14*	*Cw*0501*	*Cw5*
*A*0206*	*A2, A2.4a*	*B*1402*	*Bw65 (14)*	*Cw*0601*	*Cw6*
*A*0207*	*A2, A2.4b*	*B*1501*	*Bw62 (15)*	*C2*0701*	*Cw7*
*A*0208*	*A2, A2.4c*	*B*1801*	*B18*	*Cw*0702*	*Cw7, JY328*
*A*0209*	*A2, A2-OZB*	*B*2701*	*B27, 27f*	*Cw*1101*	*Cw11*
*A*0210*	*A2, A2-LEE*	*B*2702*	*B27, 27e, 27K, B27.2*	*Cw*1201*	*Cx52*
*A*0301*	*A3, A3.1*	*B*2703*	*B27, 27d, 27J*	*Cw*1202*	*Cb-2*
*A*0302*	*A3, A3.2*	*B*2704*	*B27, 27b, 27C, B27.3*	*Cw*1301*	*CwBL18*
*A*1101*	*A11, A11E*	*B*2705*	*B27, 27a, 27W, B27.1*	*Cw*1401*	*Cb-1*

*A*1102*	*A11, A11K*	*B*2706*	*B27, 27D, B27.4*
*A*2401*	*A24 (9)*	*B*3501*	*B35*
*A*2501*	*A25 (10)*	*B*3502*	*B35*
*A*2601*	*A26 (10*	*B*3701*	*B37*
*A*2901*	*A29 (w19)*	*B*3801*	*B38 (16), B16.1*
*A*3001*	*A30 (w19), A30.3*	*B*3901*	*B39 (16), B16.2*
*A*3101*	*A31 (w19)*	*B*4001*	*Bw60 (40)*
*A*3201*	*A32 (w19)*	*B*4002*	*B40, B40**
*A*3301*	*Aw33 (w19), Aw33.1*	*B*4101*	*Bw41*
*A*6801*	*Aw68 (28), Aw68.1*	*B*4201*	*Bw42*
*A*6802*	*Aw68 (28), Aw68.2*	*B*4401*	*B44 (12), B44.1*
*A*6901*	*Aw69 (28)*	*B*4402*	*B44 (12), B44.2*
		*B*4601*	*Bw46*
		*B*4701*	*Bw47*
		*B*4901*	*B49 (21)*
		*B*5101*	*B51 (5)*
		*B*5201*	*Bw52 (5)*
		*B*5301*	*Bw53*
		*B*5701*	*Bw57 (17)*
		*B*5801*	*Bw58 (17)*
		*B*7801*	*B'SNA'*

Source: Ref. 54.

Table 2 Current Nomenclature of the *HLA-DR*, *HLA-DQ*, and *HLA-DP* Alleles Recognized by the WHO Committee for Factors of the HLA System, as of 1990

DRBI Alleles		*DQA1* Alleles	
1990 Nomenclature	Previous Designations	1990 Nomenclture	Previous Designations
*DRB1*0101*	*DR1, Dw1*	*DQA1*0101*	*DQA 1.1, 1.9*
*DRB1*0102*	*DR1, Dw20*	*DQA1*0102*	*DQA 1.2, 1.19, 1.AZH*
*DRB1*0103*	*DR'BR', Dw'BON'*	*DQA1*0103*	*DQA 1.3, 1.18, DRw8-DQw1*
*DRB1*1501*	*DRw15 (2), Dw2*	*DQA1*0201*	*DQA 2, 3.7*
*DRB1*1502*	*DRw15 (2), Dw12*	*DQA1*03011*	*DQA 3, 3.1, 3.2*
*DRB1*1601*	*DRw16 (2), Dw21*	*DQA1*03012*	*DQA 3, 3.1, 3.2, DR9-DQw3*
*DRB1*1602*	*DRw16 (2), Dw22*	*DQA1*0302*	*DQA 3, 3.1, 3.2, DR9-DQw3*
*DRB1*0301*	*DRw17 (3), Dw3*	*DQA1*0401*	*DQA 4.2, 3.8*
*DRB1*0302*	*DRw18 (3), Dw'RSH'*	*DQA1*0501*	*DQA 4.1, 2*
*DRB1*0401*	*DR4, Dw4*	*DQA1*0501*	*DQA 4.1, 2*
*DRB1*0402*	*DR4, Dw10*	*DQA1*05012*	*DQA 4.1, 2*
*DRB1*0403*	*DR4, Dw13, 13.1*	*DQA1*05013*	*DQA 4.1, 2*
*DRB1*0404*	*DR4, Dw14, 14.1*	*DQA1*0601*	*DQA 4.3*
*DRB1*0405*	*DR4, Dw15*		
*DRB1*0406*	*DR4, Dw'KT2'*		
*DRB1*0407*	*DR4, Dw13, 13.2*	*DQB1* Alleles	
*DRB1*0408*	*DR4, Dw14, Dw14.2*		
*DRB1*0409*	*DR4*	*DQB1*0501*	*DQw5 (w1), DQB 1.1, DRw10-DQw1.1*
*DRB1*0410*	*DR4*	*DQB1*0502*	*DQw5 (w1), DQB 1.2, 1.21*
*DRB1*0411*	*DR4*	*DQB1*05031*	*DQw5 (w1), DQB 1.3, 1.9, 13.1*
*DRB1*1101*	*DRw11 (5), Dw5, DRw11.1*	*DQB1*05032*	*DQw5 (w1), DQB 1.3, 1.9, 13.2*
*DRB1*1102*	*DRw11 (5), Dw'JVM', DRw11.2*	*DQB1*0504*	*DQB 1.9*
*DRB1*1103*	*DRw11 (5), DRw11.3*	*DQB1*0601*	*DQw6 (w1), DQB 1.4, 1.12*
*DRB1*1104*	*DRw11 (5), Dw'FS'*	*DQB1*0602*	*DQw6 (w1), DQB 1.5, 1.2*
*DRB1*1201*	*DRw12 (5), Dw'DB6'*	*DQB1*0603*	*DQw6 (w1), DQB 1.6, 1.18*

*DRB1*1202*	*DRw12 (5), DRw12b*	*DQB1*0604*	*DQw6 (w1), DQB 1.7, 1.19*
*DRB1*1301*	*DRw13 (w6), Dw18, DRw6a*	*DQB1*0605*	*DQw6 (w1), DQB 1.8, 1.19b*
*DRB1*1302*	*DRw13 (w6), Dw19, DRw6c*	*DQB1*0201*	*DQw2, DQB 2*
*DRB1*1303*	*DRw13 (w6), Dw'HAG'*	*DQB1*0301*	*DWw7 (w3), DQB 3.1*
*DRB1*1304*	*DRw13 (w6)*	*DQB1*0302*	*DQw8 (w3), DQB 3.2*
*DRB1*1305*	*DRw13 (w6)*	*DQB1*03031*	*DQw9 (w3), DQB 3.3*
*DRB1*1401*	*DRw14 (w6), Dw9, DR26b, DRB1*1401*	*DQB1*03032*	*DQw9 (w3), DQB 3.3*
*DRB1*1402*	*DRw14 (w6), Dw16, DRB1*1401*	*DQB1*0401*	*DQw4, DQB 4.1, Wa*
*DRB1*1403*	*DRw14 (w6)*	*DQB1*0402*	*DQw4, DQB 4.2, Wa*
*DRB1*1404*	*DRw6b.2*		
*DRB1*1405*	*DRw14 (w6)*		
*DRB1*0701*	*DR7, Dw17*	*DPA1* Alleles	
*DRB1*0702*	*DR7, Dw'DB1'*		
*DRB1*0801*	*DRw8, Dw8.1*	*DPA1*0101*	*LB14/LB24, DPA1*
*DRB1*08021*	*DRw8, Dw8.2*	*DPA1*0102*	*pSBα-318*
*DRB1*08022*	*DRw8, Dw8.2*	*DPA1*0103*	*DPw4α1*
*DRB1*08031*	*DRw8, Dw8.3*	*DPA1*0201*	*DPA2, pDAα13B*
*DRB1*08032*	*DRw8, Dw8.3*		
*DRB1*0804*	*DRw8*		
*DRB1*09011*	*DR9, Dw23*		
*DRB1*09012*	*DR9, Dw23*		
*DRB1*1001*	*Drw10*		

Source: Ref. 54.

D. Murine Major Histocompatibility Complex

The *H-2* locus in the mouse shares many common features with the *HLA* locus in humans, although the nomenclature and overall organization of the genes is somewhat different. Thus, in both mammals, the *MHC* genes are organized into three classes of genes with the same names, class I, II, and III, which encode molecules with essentially similar structures and functions. In the mouse, the three class I molecules H-2K, H-2D, and H-2L are encoded by two regions of DNA K (*H-2K*) and D (which includes the *H-2D* and *H-2L* genes) that, unlike in the human genome (*HLA-B, HLA-C,* and *HLA-A*), are not contiguous. In addition to these classic class I molecules, which serve the same function in antigen presentation as HLA-A, B, and C, two nonclassic class I molecules, Qa and Tla, the precise function of which is not yet known, are also encoded in the class I region of the mouse. Unlike the *HLA* class II genes, which have three loci that encode DP, DQ, and DR molecules the mouse has only two class II molecules, designated I-A and I-E.

Polymorphic variants of the *H-2* gene products have been named from inbred strains of mice that express the same allele at each locus. The most common of these are $H\text{-}2^{k}$, $H\text{-}2^{b}$, $H\text{-}2^{s}$, and $H\text{-}2^{d}$. Thus, a mouse that is homozygous for one of these alleles, for example, $H\text{-}2^{k}$ will express I-A^{k} and I-E^{k} class II molecules and $H\text{-}2K^{k}$, H-2d^{k} and H-2L^{k} class I molecules. Several inbred strains of differing genetic background can have the same *H-2* haplotype, for example B10.Br, C3H, and A/J mice are syngeneic at the *H-2* locus ($H\text{-}2^{k}$), but differ in genes outside the MHC complex. The F_1 progeny of two inbred strains of mice with different *H-2* haplotypes will inherit both haplotypes and express both parental alleles at each *H-2* locus. For example, F_1 (Balb/c $\times$ A/J) mice will express I-A^{d}, I-A^{k}, I-E^{d}, and I-E^{k} class II molecules and K^{k}, K^{d}, D^{k}, D^{d}, L^{k}, and L^{d} molecules and will be histocompatible with both parental strains. As with the *HLA* class II genes, the murine class II molecules code for discrete patterns of α–β chain pairs within each subregion. Thus, mixed class II molecules will be expressed by the F_1 generation described such as I-$A\alpha^{d}A\beta^{k}$, I-$A\alpha^{k}A\beta^{d}$ and similar heterologous I-E molecules, but in general I-A α- or β-chains are not believed to associate with I-E α- or β-chains.

III. MAJOR HISTOCOMPATIBILITY COMPLEX POLYMORPHISM

A. Identification of Alleles

Class I alleles were originally defined using heterotypic serum from multiparous women and multiply transfused patients. Later, monoclonal antibodies (MAb) were available and allowed increased precision in typing, but little increased discrimination among alleles. Many of the original specificities defined, such as

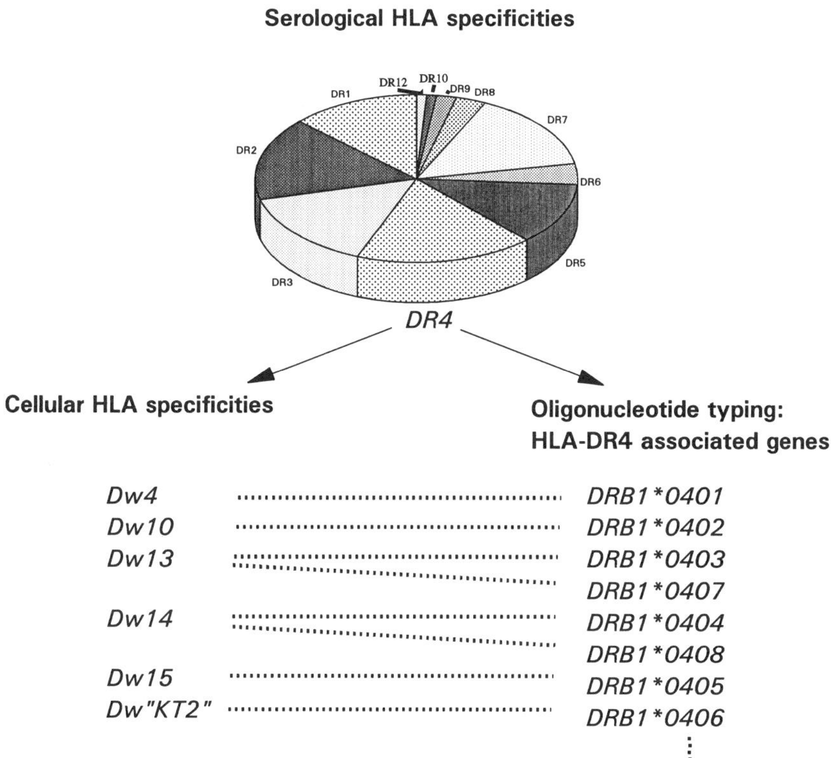

Figure 2 The spectrum of HLA class II typing. Serologic reactivity patterns identify traditional antibody-defined HLA "types," which represent public or shared specificities present on the HLA class II molecule. The major allogeneic determinants on HLA molecules are defined by mixed lymphocyte culture or HLA-D typing and represent an additional set of cellularly defined specificities on the same class II molecules. The relation between the serologically and cellularly defined determinants can be quite complicated. This is illustrated for the *HLA-DR4* family of alleles: *HLA-DR4* is defined by antibody reactivity; this type can be split into six identifiable specificities by T-cell recognition in mixed lymphocyte cultures. Synthetic oligonucleotide typing takes advantage of individual nucleotide differences to identify each gene on the basis of its DNA sequence, distinguishing further alleles. (Adapted from Ref. 53).

HLA-B27, have been shown by biochemical and DNA sequence studies to be families of several related alleles; *HLA-B27* consists of six alleles (24,25).

As with class I molecules, many class II molecules were originally serologically recognized specificities (Fig. 2). Mixed lymphocyte culture (MLC) allowed finer distinctions to be made by using cellular recognition of foreign HLA antigens by cultured lymphocytes of known specificities. By using MLC, the serologically defined specificity, DR4, could be shown to consist of many more subtypes. Nucleotide-sequencing is available for most class II alleles now and this makes it possible to design oligonucleotide probes that can be used for extremely sensitive typing techniques (such as the polymerase chain reaction; PCR) to identify alleles (26). Amino acid sequences of several class II *DRB1* alleles are shown in Figure 3.

B. Haplotype Organization

The high degree of allelic variability among the various class I and class II loci make for an astounding number of possible unique combinations in a given individual. The assortment of alleles within *HLA-DP, DQ, DR, HLA-B, C,* and *A* loci is not actually random, however. The alleles at each *MHC* locus on any given individual's single chromosome is termed their *haplotype*. This lack of randomness between the *MHC* loci is termed *linkage disequilibrium* and still allows a large, but finite, number of distinct haplotypes. The observed rearrangement frequency between fairly distant genetic loci is often less than the actual genetic distance would predict. For instance, one of the alleles of the *HLA-B* locus, *HLA-B8,* is most often found on haplotypes with *HLA-DR3, DQ2,* and *DP1*. This and many other examples of linkage disequilibrium are important to consider when interpreting the *HLA* genetic contributions to disease. Any assignment must be done cautiously in persons in whom there may be linkage on the same haplotype to other *HLA* alleles. Another problem raised by linkage disequilibrium is that "extended haplotypes" may also include other polymorphic non-HLA genes that could play a primary role in disease and the *HLA* gene merely functions as a marker gene.

C. Regulation of *HLA* Genes

Transcription of *HLA* genes is tightly regulated. An ordered set of promoter and enhancer elements controls basal as well as inducible expression (27,28). The HLA class I (*HLA-A, B,* and *C*) genes appear to be coordinately regulated, and their expression is constitutive on all nucleated cells. The *HLA* class II genes, however, appear to have multiple levels of control. The coordinate control of class II genes has been demonstrated by the rare disorder known as "bare lymphocyte syndrome" in which expression of class I genes is normal, but class II genes are not transcribed, and severe immunodeficiency results (29–31). Fusion of bare lymphocyte cells with normal lymphocytes restores HLA class II gene expression indicating that the deficiency was for a key coordinated transcription factor (32).

Differential regulation of class II genes is implied by the variable levels of constitutive and inducible expression of *HLA-DR, HLA-DQ,* and *HLA-DP* in various tissues. The nuclear proteins that bind to promoter elements of each of these loci appear to provide the tissue specificity and differential expression (28,33). DNA sequence allelic polymorphism has been observed in *HLA-DQB*

```
                   10         20         30         40         50         60         70         80         90
DRB1*0101   GDTRPRFLWQ LKFECHFFNG TERVRLLERC IYNQEESVRF DSDVGEYRAV TELGRPDAEY WNSQKDLLEQ RRAAVDTYCR HNYGVGESFT VQRR
DRB1*0102   ---------- ---------- ---------- ---------- ---------- ---------- ---------- ---------- ----AV---- ----
DRB1*1501   --------Q- D-Y------- -----F-H-D ------DL-- ---------- ---------- ------F--D ---------- ---------- ----
DRB1*1502   --------Q- D-Y------- -----F-H-G ------N--- ---------- ---------- ------F--D ---------- ---------- ----
DRB1*0301   --------EY STS------- -----Y-D-Y FH----N--- ------F--- ---------- ---------- K-GR--N--- -----V---- ----
DRB1*0401   --------E- V-H------- -----F-D-Y F-H---Y--- ---------- ---------- ---------- K--------- ---------- ----
DRB1*0402   --------E- V-H------- -----F-D-Y F-H---Y--- ---------- ---------- ------I--D E--------- -----V---- ----
DRB1*0403   --------E- V-H------- -----F-D-Y F-H---Y--- ---------- ---------- ---------- ---E------ -----V---- ----
DRB1*0404   --------E- V-H------- -----F-D-Y F-H---Y--- ---------- ---------- ---------- ---------- -----V---- ----
DRB1*0405   --------E- V-H------- -----F-D-Y F-H---Y--- ---------- ------S--- ---------- ---------- ---------- ----
DRB1*0406   --------E- V-H------- -----F-D-Y F-H------- ---------- ---------- ---------- ---E------ -----V---- ----
DRB1*1101   --------EY STS------- -----F-D-Y F-----Y--- ------F--- -------E-- ------F--D ---------- ---------- ----
DRB1*1102   --------EY STS------- -----F-D-Y F-----Y--- ------F--- -------E-- ------I--D E--------- -----V---- ----
DRB1*1201   --------EY STG--Y---- ---------H FH----LL-- ------F--- ------V--S ------I--D ---------- ----AV---- ----
DRB1*1301   --------EY STS------- -----F-D-Y FH----N--- ------F--- ---------- ------I--D E--------- -----V---- ----
DRB1*1401         --EY STS------- -----F-D-Y FH----F--- ---------- ------A--H ---------R ---E------ -----V---- ----
DRB1*1402   --------EY STS------- -----F---Y FH----N--- ---------- ---------- ---------- ---------- ---------- ----
DRB1*0701   ---Q------ G-YK------ ----QF---L F-----F--- ---------- ------V--S ------I--D --GQ---V-- ---------- ----
DRB1*0801   --------EY STG--Y---- -----F-D-Y F-----Y--- ---------- ------S--- ------F--D ---L------ ------
DRB1*08031  --------EY STG--Y---- -----F-D-Y F-----Y--- ---------- ------S--- ------I--D ---L------
DRB1*0901   ---Q----K- D--------- -----Y-H-G ------N--- ---------- ------V--S ------F--R ---E---V-- ---------- ----
DRB1*1001   --------EE V--------- ---------R VH----YA-Y ---------- ---------- ---------R ---------- ---------- ----
```

Peptide-interaction residues: "floor" of antigen-binding groove (see Fig. 4)

T cell and peptide recognition residues: α-helical logs (see Fig. 4)

Figure 3 Amino acid sequences, using the single-letter amino acid code, encoded by several of the most common alleles within the *HLA-DRB1* locus. The sequence shown spans the first domain of the β-chain of the HLA-DR class II molecule, demonstrating regions of "hypervariability" interspersed among conserved regions. The variable sequences are responsible for functional differences among different alleles, primarily by restricting the binding of antigenic peptides and the T-cell receptor.

promoter elements, with resulting differences in both the nuclear protein binding to different alleles at the same *DQBI* locus and corresponding effects on the transcriptional activity of the promoter (34).

IV. HUMAN LEUKOCYTE ANTIGEN MOLECULES

The genetic organization and polymorphism of the *HLA* genes have direct structural repercussions that are critical for function. The HLA class II molecules and classic class I molecules are expressed as membrane-bound surface proteins, with their polymorphic features oriented outward.

A. Class I Molecules

The HLA class I molecule is present on the cell surface as a dimeric protein consisting of a relative molecular mass (M_r) 44,000 heavy chain glycoprotein encoded by the class I gene. The heavy chain is complexed with a smaller 12,000 M_r subunit, β_2-microglobulin, which is encoded outside the MHC. The structure of a class I molecule has been derived from an x-ray crystallographic analysis of the HLA-A2 molecule (Fig. 4). There are several structural features of the class I molecule that relate to its function. The class I protein is folded into three structural domains, referred to as α_1, α_2, and α_3. The presentation of peptide antigens to T cells is accomplished by the α_1- and α_2-domains, which form the peptide-binding groove. The external loops of the α_1- and α_2-domains, as well as the bound peptide, provide the recognition surface (contact residues) for specific T-cell receptors.

The polymorphism that characterizes the class I genes is reflected in the prc'ein structure by several amino acid substitutions that tend to cluster in the regions of the α_1- and α_2-domains, both in the peptide-binding groove as well as the contact residues for the T-cell receptor (36). The amino acid substitutions determine whether a peptide will bind to a particular HLA molecule and the conformation the peptide will assume when bound. Since T-cell recognition requires peptide-binding by specific HLA molecules, the interaction is said to be *restricted* by the nature of the polymorphic variation of the HLA molecule.

Several conserved sites are found on the HLA class I molecule that serve to maintain the architecture of the molecule. One region, the α_3-domain serves to interact with the CD8 coreceptor molecule of the T cell (37). It is this interaction that is responsible for the functional dichotomy of T cells whereby CD8-positive T cells are restricted by HLA class 1 molecules, whereas CD4-positive T cells are restricted by class II molecules.

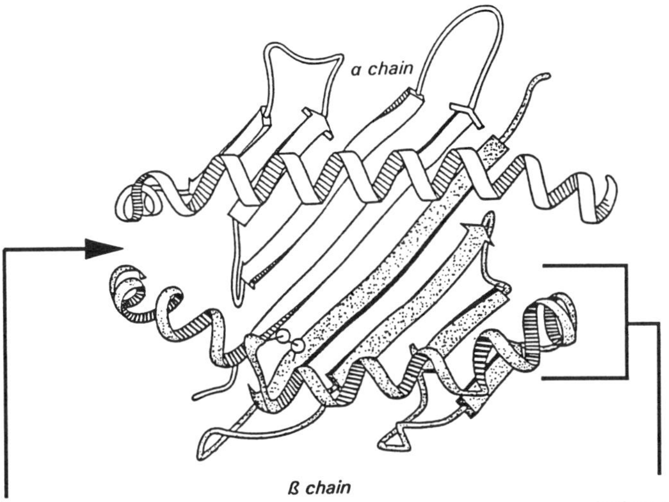

Figure 4 Three-dimensional view of the antigen-recognition domains of an HLA molecule, looking down into the antigen-binding groove, with α-helices forming the walls and β-pleated sheets forming the floor of the groove. The antigen-recognition domains are covalently linked to additional membrane-proximal domains in both class I and class II molecules, which provide for signal transduction and corecognition properties of the molecule. HLA polymorphisms, the amino acids that uniquely identify each allele, consist of structural variation primarily focused along the inner faces of the antigen-binding groove in positions suitable for direct contact with the antigenic peptides. (Adapted from the x-ray crystallographic analysis of *HLA-A2*, Ref. 35.)

B. Class II Molecules

The HLA class II molecules are glycoprotein dimers that consist of an α-chain and a β-chain. Both are encoded by the *A* and *B* genes within the class II *MHC* region. The 31,000 M_r α-chain and the 27,000 M_r β-chain assemble as an intracellular dimer, with the assistance of a third polypeptide known as the invariant chain, which is encoded outside of the *MHC* (38). The invariant chain also helps transport the class II heterodimer through various compartments to the cell surface (39). The invariant chain does not bind class I molecules, which may be one of the more important differences between class I- and class II-mediated immune recognition.

The precise structure of the class II molecule is not yet known; however, a high degree of homology with the class I structure is presumed, given the similarities in a number of conserved residues important for the tertiary structure of the protein (40). The peptide-binding cleft in the HLA class II molecule consists of separate domains contributed by both the α- and β-polypeptides. The α_1-domain of the class II α-molecule and the β_1-domain of the class II β-molecule (both NH_2-terminals) form the peptide recognition and binding cleft.

Several structural studies involving site-directed mutagenesis have been performed around the binding site for peptide and suspected contact residues between the HLA class II molecules and T-cell receptor, using the predicted location of critical residues from the known HLA class I structure (41–44). All are consistent with the concept that T-cell recognition of peptide bound to class II molecule occurs by direct contact between residues of the α_1-domain, the β_1-domain, bound peptide, and the T-cell receptor (Fig. 5). There is an important coreceptor relation between the HLA class II molecule and the CD4 molecule

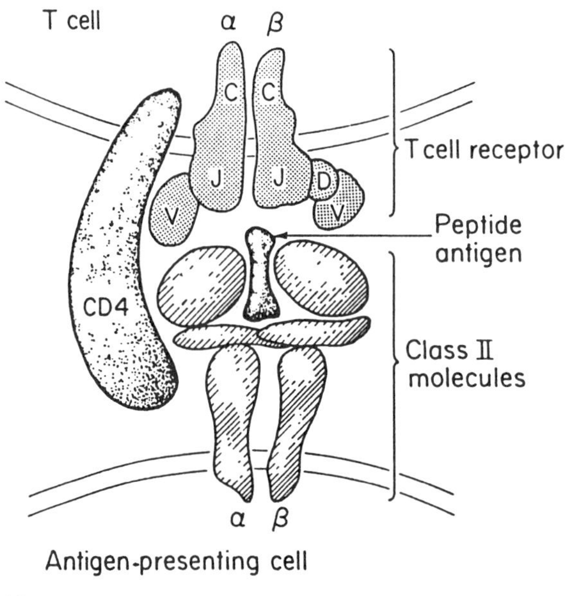

Figure 5 Molecular interactions in the triggering of antigen-specific immune responses. Illustrated are the components of the trimolecular complex essential for T-cell activation—the HLA class II molecule, a peptide antigen, and the T-cell antigen-specific receptor—as well as the CD4 molecule. The HLA molecule is anchored in the membrane of an antigen-presenting cell with the antigen-recognition domain, and its bound peptide accessible to the T-cell receptor. The T-cell receptor recognizes the processed peptide antigen in the context of its HLA class II molecule, and a CD4 molecule enhances the binding interaction and assists with signal transduction into the T-cell compartment. The orientation of the variable segments of the T-cell receptor and the structural interactions of the CD4 molecule are modeled after indirect experimental evidence and are not yet precisely known.

found on a subpopulation of T cells. This increases the affinity of the T-cell receptor–HLA class II interaction and assists the intercellular communication of activation signals (45). This special affinity of the class II molecule for the CD4 molecule is responsible for the preference of the CD4-positive T cell to recognize (and be "restricted by") HLA class II-positive cells.

V. HISTOCOMPATIBILITY ANTIGEN ASSOCIATIONS WITH ALLERGEN REACTIVITY AND AUTOIMMUNITY

The invariant chain associates with the HLA class II α- and β-chain and facilitates transport of the molecule to the cell surface. The invariant chain is thought to occupy or block the peptide-binding pocket during transport, preventing interaction of self-peptides from binding with the class II molecule at this stage (46). Endosomes containing class II molecules appear to intersect with endosomes that have carried external antigens into the intracellular compartment (47). Thus, it appears that the class II molecules first contact external antigens within endosomal compartments. The HLA class I molecules demonstrate a somewhat different pattern of activity in monitoring potential antigen (covered in much greater detail in Chap. 7 on antigen processing and presentation; also see 48).

A. Histocompatibility Antigen Association with Allergen Reactivity

Virtually all HLA associations with allergen reactivity would be predicted to be with class II alleles since, as cited earlier, sensitization to exogenous antigens is primarily restricted to the class II compartment. Allergens often show association with more than one HLA class II specificity (or no association with any specificity), presumably reflecting the complex nature of these proteins that may contain multiple antigenic sites or epitopes. Older studies using serologically or mixed lymphocyte culture-defined HLA types should be considered potentially unreliable in view of the high degrees of precision available using molecular-typing techniques. Extended haplotypes and the linkage disequilibrium that are common features of the *HLA* region will continue to complicate the analysis of HLA association with reactivity toward a specific allergen. Another potential complication not previously mentioned concerns the potential for trans-combinatorial assortment between class II α- and β-chains of differing allotype (49), although the degree to which this mechanism operates is unknown.

Despite these concerns, several studies have been published that do indeed demonstrate highly significant associations between specific HLA types and specific allergen reactivity. Some of the best-studied examples include short ragweed allergens Amb a V (HLA-DR2, Dw2; currently DRB1*1501) (1), Amb a VI (HLA-DR5, Dw11; currently designated by a group of "splits" DRB1*1101-1104) (3), and perennial ryegrass allergens Lol p I, II, and III (all are HLA-DR3; currently DRB1*0301) (2). Direct functional participation by *DRB* gene products in the recognition of dust mite antigen mix *(Dermatophagoides farinae)* by helper T cells has also been demonstrated (5,50). The presentation of antigen was optimal with constructed antigen-presenting cells displaying the Dw52b specificity (currently "split" as DRB3*0201 and 0202 by nucleotide sequencing). Unfortunately, these otherwise excellent studies were not performed with highly purified allergen. Preliminary results from a large collaborative study have recently been published examining HLA and specific immune response to allergens as part of the 11th International Histocompatibility Workshop (51), and should continue to provide a source of patient material for future studies. A negative association has been described between *Cryptomeria japonicum* (Japanese cedar) pollen allergy and HLA class II type *(DQw3)* (4).

B. Histocompatibility Antigen Association with Autoimmunity

The best-studied associations of HLA and autoimmune diseases are found for type-1 diabetes mellitus (DR3, DR4, DQw3.2), rheumatoid arthritis (DR4, Dw4, Dw14), seropositive juvenile rheumatoid arthritis (DR4, Dw4, Dw14, often heterozygous), gluten-sensitive enteropathy (celiac disease; DR3, DR7), and pemphigus vulgaris (*DR4, Dw10,* and *DR6, Dw9* haplotypes). A comprehensive review is available dealing with this area of continued interest and debate (52).

C. Mechanisms of Genetic Susceptibility and Resistance

Current mechanistic models to account for the genetic associations between immunological disease and HLA fall into three broad categories (53). Tolerance models presume that physiological mechanisms that normally prevent autoreactivity are defective in disease. Antigen presentation models propose that exogenous antigen triggers disease in an MHC-restricted fashion. Molecular mimicry models require an exogenous event, such as a viral infection, to activate lymphocytes with receptors that cross-react with self-antigens. Although all three of the foregoing mechanisms may operate in autoimmune or rheumatic diseases, reactivity to specific allergen is most compatible with the model in which MHC-type acts as a restrictive or permissive element controlling immune responsiveness. It is this that makes the study of allergic diseases an attractive way to study fine details of the human immune response.

ACKNOWLEDGMENTS

We thank members of the Immunology Program at Virginia Mason Research Center who contributed to the original studies cited in this chapter, and thank Dr. Barbara Nepom for permission to adapt and reprint Figures and Tables that appear in another work (53).

REFERENCES

1. S.-K. Huang, P. Zwollo, and D. G. Marsh, Class II major histocompatibility complex restriction of human T cell responses to short ragweed allergen, Amb a V, *Eur. J. Immunol., 21:*1469 (1991).
2. A. A. Ansari, L. R. Freidhoff, and D. G. Marsh, Molecular genetics of human immune responsiveness to *Lolium perenne* (rye) allergen, Lol p III, *Int. Arch. Allergy Appl. Immunol., 88:* 164 (1989).
3. D. G. Marsh, L. R. Freidhoff, E. Ehrlich-Kautzky, W. B. Bias, and M. Roebber, Immune responsiveness to *Ambrosia artemesiifolia* (shore ragweed) pollen allergen Amb a VI (Ra6) is associated with HLA-DR5 in allergic humans, *Immunogenetics, 26:* 230 (1987).
4. S. Matsushita, M. Muto, M. Sucmura, Y. Salto, and T. Sasazuki, HLA-linked non-responsiveness to *Cryptomeria japonica* pollen antigen, I. Nonresponsiveness is mediated by antigen-specific suppressor T cell, *J. Immunol., 138:* 109 (1987).
5. R. E. O'Hehir, B. Mach, C. Berte, R. Greenlaw, J.-M. Tiery, V. Bal, R. I. Lechler, J. Trowsdale, and J. R. Lamb, Direct evidence for a functional role of HLA-DRβ1 and -DRβ3 gene products is the recognition of *Dermatophagoides* spp. (house dust mite) by helper T lymphocytes, *Int. Immunol., 2:* 885 (1990).
6. D. A. Myers, W. B. Bias, and D. G. Marsh, A genetic study of total IgE levels in the Amish, *Hum. Hered., 32:* 15 (1982).
7. W. O. Cookson, P. A. Sharp, J. A. Faux, and J. M. Hopkins, Linkage between immunoglobulin E responses underlying asthma and rhinitis and chromosome 11q, *Lancet, 1:* 1292 (1989).
8. G. T. Nepom, and D. M. Robinson, HLA-DQ and diabetes mellitus: A genetic and structural paradigm for models of disease susceptibility, *The Molecular Biology of Autoimmune Disease* (A. G. Demaine, et al., eds.), Springer-Verlag, Berlin, p. 251 (1990).
9. A. A. Ansari, P. Shenbagamurthi, and D. G. Marsh, Complete primary structure of a *Lolium perenne* (perennial rye grass) pollen allergen, Lol p III: Comparison with known Lol p I and II sequences, *Biochemistry, 28:* 8665 (1980).
10. L. Goodfriend, A. M. Choudhury, D. G. Klapper, K. M. Coulter, G. Durval, J. Delcarpio, and C. K. Osterland, Ra5G, a homologue of Ra5 in giant ragweed pollen: Isolation, HLA-DR associated activity and amino acid sequence, *Mol. Immunol., 22:* 899 (1985).
11. J. Trowsdale, J. Ragoussis, and R. D. Campbell, Map of the human MHC, *Immunol. Today, 12:* 443 (1991).
12. D. E. Geraghty, B. H. Koller, and H. T. Orr, A human major histocompatibility complex class I gene that encodes a protein with a shortened cytoplasmic segment, *Proc. Natl. Acad. Sci. USA, 84:* 9145 (1987).

13. M. C. Carrol, P. Katzman, Z. M. Alicot et al., Linkage map of the human major histocompatibility complex including the tumor necrosis factor genes, *Proc. Natl. Acad. Sci. USA, 84:* 8535 (1987).
14. C. A. Sargent, I. Dunham, J. Trowsdale, et al., Human major histocompatibility complex contains genes for the major heat shock protein HSP70, *Proc. Natl. Acad. Sci. USA, 86:* 1968 (1989).
15. T. Spies, G. Blank, M. Breshnahan, et al., A new cluster of genes within the human major histocompatibility complex, *Science, 243:* 214 (1989).
16. E. V. Deverson, I. R. Gow, W. J. Coadwell, et al., MHC class II region encoding proteins related to multidrug resistance family of transmembrane transporters, *Nature, 348:* 738 (1990).
17. J. Trowsdale, I. Hanson, I. Mockridge, et al., Sequences encoded in the class II region of the MHC related to the "ABC" superfamily of transporters, *Nature, 348:* 741 (1990).
18. T. Spies, M. Bresnahan, S. Bahram, et al., A gene in the human major histocompatibility complex class II region controlling the class I antigen presentation pathway, *Nature, 348:* 744 (1990).
19. P. C. White, Molecular genetics of the class III region of the HLA complex, *Immunobiology of HLA. Immunogenetics and Histocompatibility* (B. Dupont, ed.), Springer-Verlag, New York, p. 62 (1989).
20. M. Robertson, Antigen processing. Proteasomes in the pathway, *Nature, 353:* 300 (1991).
21. J. S. Lee, Regulation of HLA class II gene expression, *Immunobiology of HLA. Immunogenetics and Histocompatibility* (B. Dupont, ed.), Springer-Verlag, New York, p. 49 (1989).
22. Y. Shimuzu, D. E. Geraghty, B. H. Koller et al., Transfer and expression of three cloned human non-HLA-A, B, C, class I major histocompatibility complex genes in mutant lymphoblastoid cells, *Proc. Natl. Acad. Sci. USA, 85:* 227 (1988).
23. B. H. Koller, D. E. Geraghty, R. DeMars, et al., Chromosomal organization of the human major histocompatibility complex class I gene family, *J. Exp. Med., 169:* 469 (1989).
24. S. Y. Choo, P. Antonelli, B. Nisperos, et al., Six variants of HLA-B27 identified by isoelectric focusing, *Immunogenetics, 23:* 23 (1986).
25. S. Rojo, and J. A. Lopez de Castro, Structure of HLA-B27 subtypes: Evolutionary implications, *Immunobiology of HLA. Immunogenetics and Histocompatibility* (B. Dupont, ed.), Springer-Verlag, New York, p. 111 (1989).
26. S. J. Scharf, R. L. Griffith, and H. A. Erlich, Rapid typing of DNA sequence polymorphism at the *HLA-DR*β*1* locus using the polymerase chain reaction and nonradioactive oligonucleotide probes. *Hum. Immunol., 30:* 190 (1991).
27. K. E. Sullivan, A. F. Calman, M. Nakanishi, et al., A model for the transcriptional regulation of MHC class I genes, *Immunol. Today, 8:* 289 (1987).
28. C. Benoist, and D. Mathis, Regulation of major histocompatibility complex class-II genes: X, Y and other letters of the alphabet, *Annu. Rev. Immunol., 8:* 681 (1990).
29. B. Lisowka-Grospierre, D. J. Charron, C. dePreval, et al., Defect of expression of MHC genes responsible for an abnormal HLA class I phenotype and the class II

negative phenotype of lymphocytes from patients with combined immunodeficiency. *Histocompatability Testing 1984* (E. D. Albert, et al., eds.), Springer-Verlag, Berlin, p. 650 (1984).
30. J.-L. Tourane, G.-L. Marseglia, and H. Betuel, Thirty international cases of bare lymphocyte syndrome: Biological significance of HLA antigens, *Exp. Hematol., 13:* 86 (1985).
31. C. R. Hume, L. A. Shookster, W. Collins, et al., Bare lymphocyte syndrome: Altered HLA class II expression in B cell lines derived from two patients, *Hum. Immunol., 21*: 1 (1989).
32. Z. Yang, R. S. Accolla, D. Pious, et al., Two distinct loci regulatory class II gene expression are defective in human mutant and patient cell lines, *EMBO J.*, 7: 1965 (1988).
33. M. Boothby, H. C. Liou, and L. H. Glimcher, Differences in DNA sequence specificity among MHC class II X box binding proteins, *J. Immunol., 142:* 1005 (1989).
34. L. C. Andersen, J. S. Beaty, J. W. Nettles, et al., Allelic polymorphism in transcriptional regulatory regions of *HLA-DQB*, *J. Exp. Med., 173:* 181 (1991).
35. P. J. Bjorkman, M. A. Saper, B., Samaouri, et al., Structure of the human class I histocompatibility antigen, HLA-A2, *Nature, 329:* 506 (1987).
36. P. J. Bjorkman, M. A. Saper, B. Samroui, et al., The foreign antigen binding site and T cell recognition region of class I histocompatibility antigens, *Nature, 329:* 512 (1987).
37. Y. Rosenstein, S. Ratnotsky, S. J. Barakoff, et al., Direct evidence for binding of CD8 to HLA class I antigens, *J. Exp. Med., 169:* 149 (1989).
38. L. Claesson-Welsh, P. E. Barker, O. Larhammar, et al., The gene encoding the human class I antigen-associated γ chain is located on chromosome 5, *Immunogenetics, 20:* 89 (1984).
39. V. Lotteau, L. Teyton, A. Peleraux, et al., Intercellular transport of class II MHC molecules directed by invariant chain, *Nature, 348:* 600 (1990).
40. J. H. Brown, T. Jardetzky, M. A. Saper, et al., A hypothetical model of the foreign antigen binding site of class II histocompatibility molecules, *Nature, 322:* 845 (1988).
41. W. W. Kwok, C. Lotshaw, E. C. Milner, et al., Mutational analysis of the *HLA-DQ3.2* IDDM susceptibility gene, *Proc. Natl. Acad. Sci. USA, 87:* 8051 (1990).
42. A. Hiraiwa, K. Yamanaka, W. W. Kwok, et al., Structural requirements for recognition of the HLA-Dw14 class II epitope—a key determinant associated with rheumatoid arthritis, *Proc. Natl. Acad. Sci. USA, 87:* 8051 (1990).
43. J. Peccoud, P. Dellabonna, P. Allen, et al., Delineation of antigen contact residues on an MHC class II molecule, *EMBO J., 9:* 4215 (1990).
44. J. I. Krieger, R. W. Karr, H. M. Grey, et al., Single amino acid changes in DR and antigen define residues critical for peptide—MHC binding and T cell recognition, *J. Immunol., 146:* 2331 (1991).
45. R. S. Mittler, S. J. Goldman, G. L. Spitalny, et al., T-cell receptor–CD4 physical association in a murine T-cell hybridoma: Induction by antigen receptor ligation, *Proc. Natl. Acad. Sci. USA, 86:* 8531 (1989).

46. P. A. Roche, and P. Cresswell, Invariant chain association with HLA-DR molecules inhibits immunogenic peptide binding, *Nature, 345:* 615 (1990).
47. P. Cresswell, Intracellular class II HLA antigens are accessible to transferrin–neuraminidase conjugates internalized by receptor-mediated endocytosis, *Proc. Natl. Acad. Sci. USA, 82:* 8188 (1985).
48. T. J. Brachiale, and V. L. Brachiale, Antigen presentation: Structural themes and functional variations, *Immunol. Today, 12:* 124 (1991).
49. W. W. Kwok, D. Schwarz, B. Nepom, et al., HLA-DQ molecules form α–β heterodimers of mixed allotype, *J. Immunol., 141:* 3123 (1988).
50. R. E. O'Hehir, R. D. Garman, J. L. Greenstein, and J. R. Lamb, The specificity and regulation of T-cell responsiveness to allergens, *Annu. Rev. Immunol., 9:* 67 (1991).
51. D. G. Marsh, M. N. Blumenthal, T., Ishikawa, et al., HLA and specific immune responsiveness to allergens, *HLA 1991,* Oxford University Press, vol. 1, pp. 765–71 (1992).
52. G. T. Nepom, and H. Erlich, MHC class-I molecules and autoimmunity, *Annu. Rev. Immunol., 9:* 493 (1991).
53. Immunogenetics and rheumatic diseases, *Textbook of Rheumatology,* 4th ed. (W. N. Kelley, et al., eds.), pp. 89–104 (1993).
54. J. G. Bodmer, S. G. E. Marsh, E. D. Albert, et al., Nomenclature for factors of the HLA system, 1990, *Tissue Antigens, 37:* 97 (1991).

4

T-Cell Receptor Recognition of Antigen: Genes and Proteins

Elwyn Loh
University of Pennsylvania, Philadelphia, Pennsylvania

I. THE ROLE OF THE T-CELL ANTIGEN RECEPTOR IN THE ACTIVATION OF T CELLS

A. The Decision-Making Process

T cells play a central role in the decision-making process of the immune system. In the vernacular of information science, the T cell serves multiple functions in the storage and processing of analog information (Fig. 1). The input information that the T cell receives comes through many routes: receptors specific for protein or peptide molecules; receptors for other messengers, such as glucocorticoids and prostaglandins; and other changes in the environment not recognized by specific receptors, such as electrolytes. The past information that the T cell has received is stored as the number, localization, and differentiation state of the T cell. Each T cell operates as a parallel processor for information such that the output of the entire immune system is the sum of the ouputs of each individual cell. Thus eventually, it will be necessary to understand not only how individual cells work, but also how the entire system operates as an integrated entity.

B. Information Transfer

The T-cell receptor for antigen (TCR) is one of many surface glycoproteins of the T cell that transfers information to the T cell. The uniqueness of its role is not that

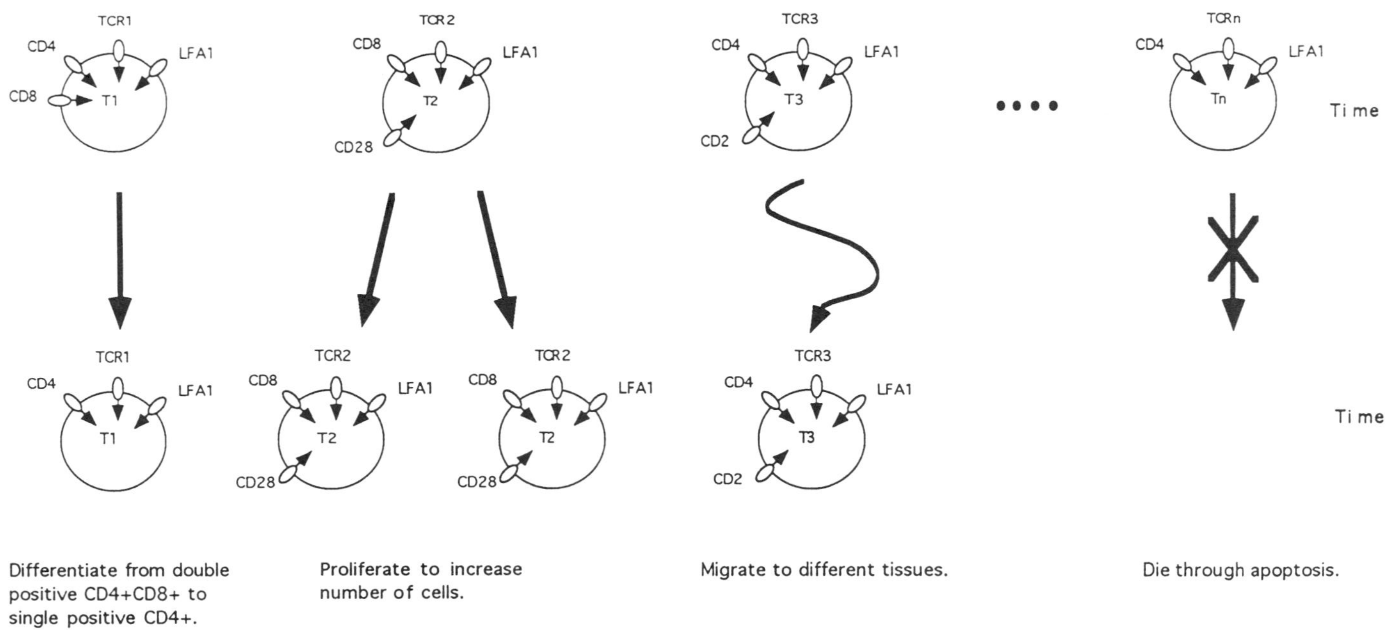

Figure 1 T cells serve as parallel processors of information. Each T cell has a unique set of input devices including many cell surface receptors of which the TCR, LFA1, CD2, and CD4 are examples. Each T cell also has a set of output mechanisms, which are not shown, including secretion of interleukins, adhesion receptors, and alterations of morphologic appearance and of mobility functions. In a continual process of change, the cell integrates the input information and can differentiate, replicate, die, or migrate.

a signal through the TCR is always necessary for activation or that the strength of that signal is greater or different from that of other inputs, but rather, that only the TCR can differentiate one antigen from another. Thus, the discrimination between self and nonself must be performed by the T cell through the TCR. The ability of the TCR to recognize a vast universe of foreign antigens requires that it, itself, be equally complex, and the description and understanding of that complexity will be the challenge of this chapter. One fundamental generalization that can be made is that, just as each B cell expresses only one species of immunoglobulin receptor, each T cell expresses only one species of TCR [with roughly 2×10^4–2×10^5 identical copies per cell (1)] thereby allowing the decision to respond or not to respond to a specific antigen to be made at the cellular level of a T cell. This is illustrated in Figure 1 by showing each T cell expressing different TCR, but similar T-cell differentiation antigens. It is also a receptor that, for most purposes, defines a T cell, in that its expression is necessary and sufficient for a cell to be called a T cell. This definition, rather than the older definition of a T cell, that it is a lymphocyte derived from the thymus, thus includes some cells that are not thymically derived, but that do express TCR. Because of this unique role, the nature of the TCR and how and what it recognizes play a central role in immunology. The other general point to emphasize about the role of the TCR in T-cell activation is the dynamic state of the T cell. What happens when antigen is presented to the T cell depends not only on which other of the surface receptors of the T cell are engaged, but also on the history of the T cell in terms of what is its differentiation state. This chapter will not discuss the events that occur after the antigen–receptor interaction has taken place; a description of these will be found in Chapter 6 in this volume. Instead, we will summarize the current state of knowledge about the functional protein structure and genetics of the T-cell receptor. In the context of medical applications of this knowledge, there are several possible implications: (1) mutations or polymorphisms in the genes that code for the structural genes or regulatory genes of the TCR may alter susceptibility of persons to autoimmune or infectious diseases; and (2) steps in the pathophysiology of a disease that requires T cells may be manipulatable in a therapeutic way through the TCR. Although neither of these situations has been proved in humans, animal models exist that hold promise for the application of this knowledge.

II. THE BIOCHEMICAL CHARACTERIZATION OF THE ANTIGEN RECOGNITION COMPLEX OF THE T CELL

A. The T-Cell Receptor for Antigen

The TCR for antigen is a complex of at least six different proteins. This is shown schematically in Figure 2. The antigen specificity of the receptor is determined by an $\alpha\beta$ or $\gamma\delta$ heterodimer, referred to in this chapter as the T-cell receptor

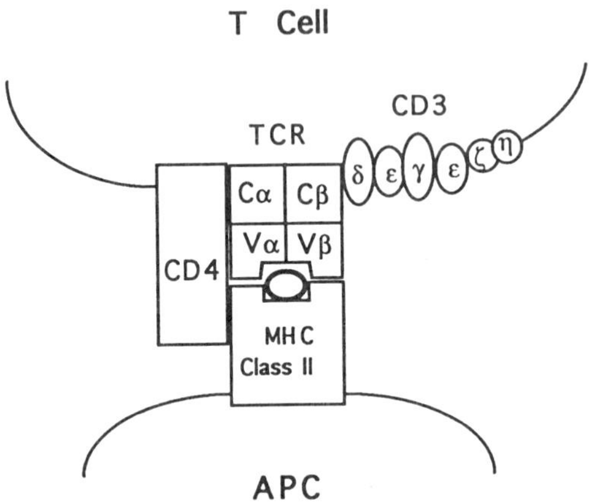

Figure 2 Schematic diagram of the T-cell receptor complex. The complex includes molecules that are physically associated, namely TCRα (shown as a C region and V region domain), TCRβ, CD3γ, CD3δ, CD3ε, CD3ζ, CD3η, as well as other molecules functionally and possibly physically associated including CD4 or CD8. The two cell types involved are the T cell and the antigen-presenting cell (APC).

(TCR), as contrasted with the T-cell receptor complex, which includes the other polypeptide chains. (An alternative nomenclature refers to the heterodimer as Ti, and the entire complex as the TCR). The determination of the antigen specificity by the αβ heteroduplex has been demonstrated by several lines of evidence, including the transfer of specificity for an peptide–major histocompatibility complex (MHC) by the molecular transfer of the TCR (2). In those experiments, a gene representing the TCR from a T cell that recognizes a specific peptide–MHC molecule was cloned and transfected into a recipient T cell. It was then shown that the recipient cell could now respond to a cell expressing the appropriate peptide–MHC complex.

In general, an individual T cell has only αβ or γδ receptors. Under unusual circumstances, there may be other combinations of the heterodimer such as a δβ or ββ, but whether these combinations are biologically significant is unknown (3,4). The distinct and overlapping features of αβ and γδ TCR will be discussed later in this chapter. Physically and functionally associated with the TCR are the CD3 molecules that are involved in the signaling process of transferring the information of antigen recognition into the cell. The tight physical association between TCR and CD3 is shown by the observation that antibodies that precipitate the TCR in nonionic detergents also precipitate the CD3 complex. Conversely, antibodies that precipitate the CD3 molecules generally precipitate the TCR. Not only are the TCR and the CD3 molecules physically associated but, in most circumstances, one cannot appear on the cell surface without the other. When mutations are introduced in the TCR polypeptide chains that abrogate their expression, the CD3 molecule is also not expressed on the cell surface. Similarly, when CD3 mutants are made that affect CD3 expression, the TCR no longer appears at the cell surface. The stoichiometry of the associated complex is not precisely known, but is likely to occur in several forms, possibly even within a single cell. These possible combinations include TCR-αβ,CD3δεγεξ2 and TCR-αβ,CD3δεγεξη in the mouse (5), whereas the situation in humans is less clear (6). The proteins that are included in the entire complex may differ, depending on the type of T cell. For example, some large granular lymphocytes have a FcεRIγ as part of the complex, instead of CD3ξη (5).

B. Contribution of CD4 and CD8 Molecules

The CD4 or CD8 molecules also contribute to the recognition specificity of the complex. The expression of the αβ TCR in some transfection experiments is not sufficient to confer antigen recognition on a cell, without the concomitant expression of a CD4 or CD8 molecule on the T cell (7,8). Consequently, these molecules have been termed coreceptors (9). In general, the CD4 molecule is important to MHC class II-restricted recognition, whereas CD8 recognizes MHC class I molecules. The association between the TCR and the CD4 and CD8 molecules is neither covalent nor physically tight, since even under gentle dissociation conditions, antibodies directed against the TCR do not coprecipitate CD4 or CD8 molecules. Thus, the association of the TCR with the CD4 or CD8 molecule may be a functional and transient one that may require the presence of the antigenic molecule. The role of the CD3, CD4, and CD8 molecules and their role in activation are discussed in detail elsewhere in this book. We will focus on the structure of the antigen receptor that comprises either an αβ or γδ heterodimer.

C. The Function of the TCR has been Studied in Detail Only in the Presence of the Entire Complex on the Surface of a Cell

Many attempts have been made to produce soluble TCR that could be used to specifically recognize antigen, just as the immunoglobulin molecule in its secreted form recognizes its antigen (10,11). Although substantial amounts of soluble receptor have been produced through molecular manipulations, the ability of these molecules to recognize peptide–MHC has been difficult to convincingly demonstrate, possibly because the affinity of the receptor for antigen is too low. A recent paper has described a soluble TCR–immunoglobulin chimeric molecule that could be used to block peptide–T-cell interactions specifically (12). This may be an unusual TCR with high affinity. It is possible that only with the increased avidity of many cell surface receptors can specific recognition take place. It is of potential interest that there have been reports of soluble TCR that occur naturally, associated with biologically important functions, such as suppression and delayed-type hypersensitivity (13). What the molecular form of the soluble receptors may be and what their role is in the immune system remains to be demonstrated.

D. Functional Regions of the T-Cell Receptor

The TCR is a protein that can be divided into regions of functional importance. Given its primary protein structure, the TCR is a member of the immunoglobulin supergene family, with close homology to immunoglobulins. All of the TCR chains can be divided into presumptive functional domains consisting of a leader peptide (L) that is cleaved off during synthesis, an immunoglobulinlike V region and a C region domain, each of approximately 90 amino acids, followed by a connecting peptide (CP), a transmembrane region (TM), and a short cytoplasmic tail (CY) (Fig. 3a). The V region domain interacts with antigen and can be sudivided into functional regions, based on patterns of diversity similar to the hypervariable regions described for immunoglobulins. The four regions of increased variability are designated CDR1, 2, and 3, representing the complementarity-determining regions 1, 2, and 3, as well as HV4 for hypervariable region 4 (14–16). We will describe potential functions for these hypervariable regions and how they are derived genetically later in this chapter. Figure 3b shows one set of V regions, the Vγl family of γ-chains to illustrate the diversity patterns (17). As one can see, the diversity within this family is found are preponderantly within the CDRs and HV4; the rest of the V region, which can be termed framework regions, is remarkably conserved. When more sequences are compared, the clustering of the differences into CDR becomes obscured because the framework regions vary. Only the first two CDRs and the HV4 is shown because CDR3 is coded for by a separate genetic region, with far greater diversity.

E. Structural Relation of T-Cell Receptor to Antigen

A hypothetical model exists that describes the structural relation of the TCR to antigen. The discovery of MHC restriction of antigens raised the key question as to whether the TCR was, in fact, two molecular structures, one interacting with the antigen and one interacting with the presenting molecule. After the characterization of the TCR, experiments proved that the TCR heterodimer was indeed responsible both for peptide and MHC recognition. The first crystal structure of an MHC suggested a model that reconciled what was known about the structure of MHC and peptide with what was known about the TCR (18). The assumptions of the model were that the three-dimensional structure of the TCR was essentially similar to that of an Fab of an immunoglobulin molecule and, therefore, the surface of the TCR that interacts with the peptide–MHC

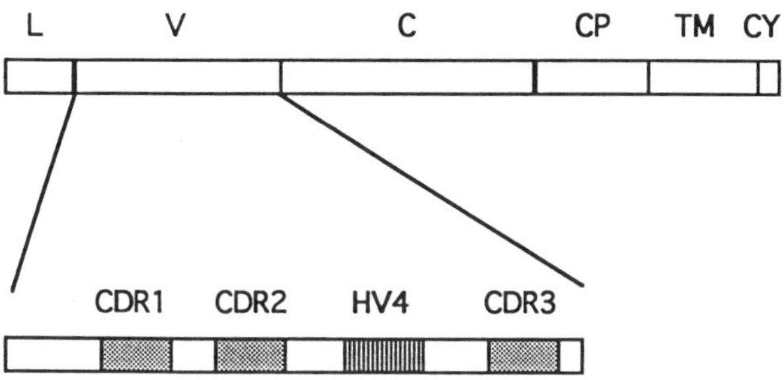

(a)

```
                                                  CDR1
      1                    20                                40
Vγ1   SQKSSNLEGRTKSVIRQTGSSAEITCDL  AEGSNGY      IHWYLHQEGKAPQRLQYY
Vγ3   ----------------------------  TVTNTF-      ---------------L--
Vγ4   ----------------------------  ----T--      ---------------L--
Vγ5   ---------G----T-P-R---------  TVINAF-      ---------------L--
Vγ8   ----------------P------V----  PVENAV-      ---------------L--

            CDR2                    HV4
             60                      80
Vγ1   DSYNSKVVLESGV  SPGKY  YTYASTRNNLRLI  LRNLIENDSGVYYCATWDG
Vγ3   -VSTARD-----L  -----  --HTPR-WSWI-R  -Q-----------------R
Vγ4   ---T-S------I  -----  D--G---K---M-  -------------------
Vγ5   VS---D------L  -----  --HTPR-WSWI--  -------------------
Vγ8   -----R------I  -----  H-----GKS-KF-  -E----R-----------R
```

(b)

Figure 3 Structure of the TCR protein. (a) Schematic illustration of the division of the primary structure of the TCR into functional domains. The leader sequence (L) is cleaved off during protein synthesis, The V region domain recognizes antigen and can be subdivided into complementarity-determining regions (CDR) and HV4 (hypervariable region 4). The C region contains the first domain, which interacts with the other chain of the heterodimer, followed by the connecting peptide (CP), the transmembrane domain (TM), and the cytoplasmic tail (CY). (b) Hypervariable regions in the TCR γ-chain family, *Vγl*. Five V regions are included. The single-letter code is used for the amino acids and dashes indicate identity with the sequences above. CDR3 sequences follow the sequences shown and are not included because they are not coded by germline genes.

complex is composed of the CDR1, CDR2, and CDR3 of the two polypeptide chains (Fig. 4, upper panels). It was further proposed that the linear pocket of the MHC molecule in which the peptide resides could orient beneath the linear structure of the CDR3s of the TCR, as in models A and C in the figure. Since the model has been published, it has become widely used in the conceptualization of the ternary structure of peptide–MHC–TCR structure. A more general application of the model would allow the rotation of the structure in other angles, such as 90° in models B and D. Even more general, would be the possibility of an infinite number of other rotations and translations.

One set of TCR that recognizes peptides from the model protein antigen cytochrome C has been experimentally examined to try to test the model.

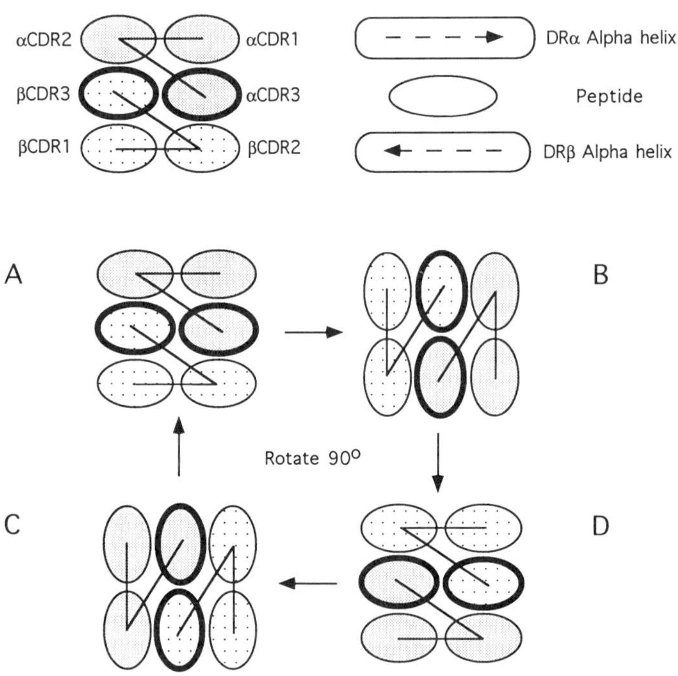

Figure 4 Model of the interaction between the TCR and the MHC–peptide. The upper panel shows a schematic diagram of the domains of the surface of the TCR and MHC. The lines connecting the domains in the TCR show the three domains that belong to the same polypeptide chain. The CDR3s are in bold and are in the middle of the TCR surface. The orientation of the TCR domains is actually from behind the molecule, so that to see the interaction of the domains with MHC one needs only to place the TCR over the NHC. The four simplest possible orientations of the TCR to the MHC are illustrated in the bottom panel. In reality, the TCR could be rotated to any angle and translated laterally or vertically. The orientation of the TCR to the MHC is a testable hypothesis currently under study.

Although the details of the approach are complex, the idea was to mutate the peptide and see where adaptive changes in the TCR would occur. By aligning the changes, it was interpreted that the α-chain of the TCR was oriented above the α-helix of the polymorphic β-chain of the MHC molecule (see Fig. 4, model C) (19). Importantly, even if one TCR is oriented in a particular fashion over one MHC–protide complex, other TCR–MHC combinations may be oriented in very different ways. No known general constraints exist that dictate that all TCR have the same alignment relative to the MHC molecule. The predicted model of the three-dimensional structure of the TCR–MHC–peptide complex remains to be validated by the actual structure, as determined by x-ray crystallography.

F. Binding Affinity of T-Cell Receptor for Antigen

The binding affinity of the interaction between the TCR and peptide antigen has been measured by two approaches. One measured the amount of soluble MHC–peptide required to block the binding of an antibody to TCR; and the other measured the amount of soluble TCR required to block T-cell responses (1,12). Both gave values of approximately 10^{-4}–10^{-5}M. That level of affinity can be viewed in the context of antibody–antigen affinities that are usually in the range of 10^{-5}–10^{-10}M.

G. T-Cell Receptor Recognition of Superantigens

TCR can recognize a second class of molecules that has been loosely termed superantigens (20). These differ from conventional peptide antigens in several remarkable ways: (1) The antigens do not require processing, but instead stimulate T cells when they are intact. (2) Presentation of superantigens by class II molecules is necessary, but there is not a strict MHC restriction. This is interpreted to mean that superantigens are bound to class II molecules when they stimulate T cells, but they can associate with a wide range of class II molecules, unlike peptides that can fit into the grooves of a narrow choice of MHC molecules. (3) Recognition of superantigens is determined predominantly by their interaction with the β-chain of the TCR. Thus, when a complex mixture of T cells are assayed for stimulation by a superantigen, T cells with certain Vβs respond and others do not, suggesting that recognition is a function of the Vβ. More recent data, however, suggest that the α-chain also contributed to the recognition of superantigens (21). (4) The region of the TCR that interacts with superantigen has been examined in two mouse models (22,23) and one human example (24). The observation was made that polymorphic forms of mouse Vβ8 varied in their ability to recognize a known superantigen. Replacement of some of those amino acids that were different in the polymorphic form, unable to recognize the superantigen, restored the recognition.

This work and the other examples have suggested that the region around HV4 of the Vβ is important for the interaction with superantigen.

III. THE GERMLINE GENES CODING FOR THE T-CELL RECEPTOR

The general description of the germline genes that code for the TCR has been a major success of molecular immunology. Eventually, through research related to the human genome project, the complete sequence of the loci, involving several megabases, will be completed. The following description, therefore, is incomplete, but includes the major features. Beyond the completion of one genomic individual sequence, which actually will be a compilation of several persons' genes, the description of the polymorphic nature of the loci is unfolding. The clinical implications of such polymorphisms may be important, as we discuss later. It is important to emphasize that, in this chapter, the human genes are described and those of other species will differ.

A. The Structure of the β-Locus

Essential features of the β-locus are illustrated in the central structure of Figure 5. The locus is at band q35-36 of chromosome 7. The figure also illustrates the rearrangements that will be discussed in the next section.

A cluster of *V* regions lies centromeric to the *C* regions. Various attempts have been made to estimate the number of *V* regions that are present in the genome. The most straightforward approach, which is to isolate and sequence the entire region, has been partially accomplished. Because this region covers on the order of a megabase, the final results of this effort have not been published, although it has been partially mapped with genetic and physical techniques (27). A second approach has been to sequence randomly selected cDNAs from the peripheral blood of given individuals. By using that approach, four recent papers have analyzed more than 700 β-chain cDNA sequences (28–31). Interpretation of this data has led to the conclusion that there exist 24 families of *V* regions and a total of approximately 60 functional V region genes.* The number of members within

---

*This interpretation involves some problems of nomenclature and ambiguity. Repetitions of identical V region sequences are interpreted to represent repeat sequences of one V region. Sequences that differ by only a few bases can be due to separate V regions, sequencing errors, or polymorphisms. Sequences that differ by up to 25% are felt to be different genes and have been grouped within a family (some authors have called these subfamilies). The V gene family concept developed for historical and technical reasons. Workers used individual V region probes for hybridization studies, and sequences that differed by less than 25% generally could cross-hybridize. Thus, by probing Southern digests of unrearranged DNA, the number of members of a given family could be estimated. With this criteria of defining families, 24 β-families have been described.

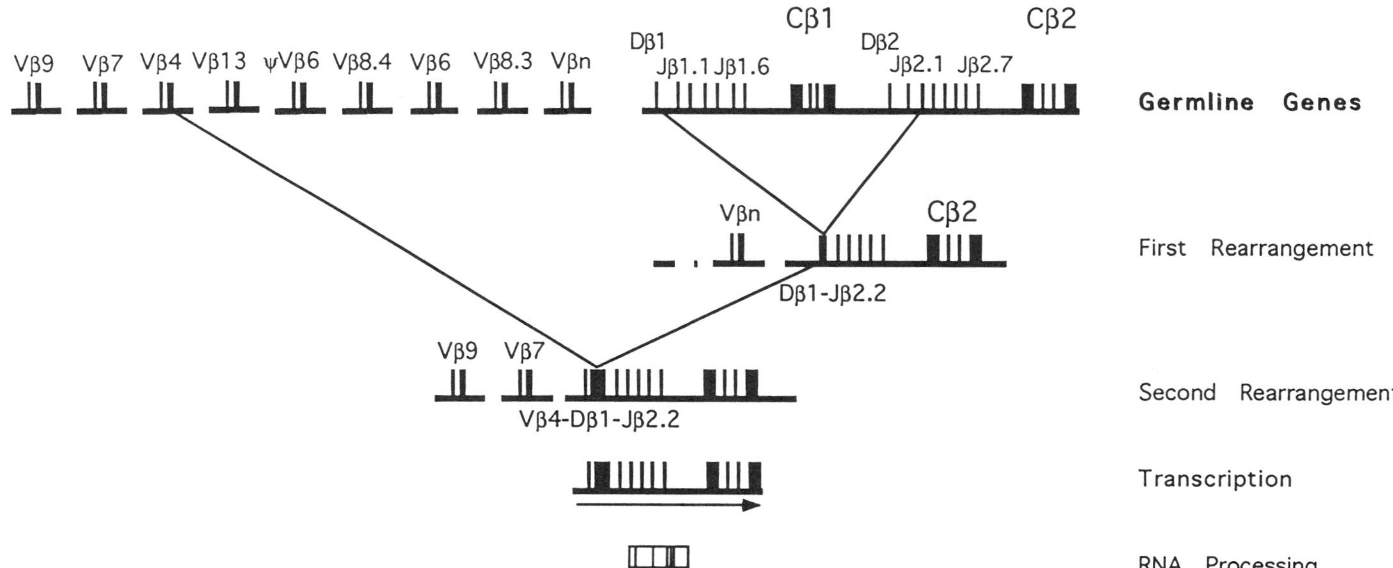

Figure 5 TCR β-locus: At the top are shown the germline configuration of the loci. Note the set of *V* regions on one side and two *J-C* clusters on the other. Each *V* region consists of two exons, one corresponding to the leader peptide and one to the *V* region. Each rearrangement deletes the region of the chromosome between the two pieces. After two rearrangements, transcription creates an mRNA containing the intronic sequences that are spliced out during RNA processing. (Adapted from multiple sources, including Refs. 25 and 26.)

cases, multiple pseudogenes for the TCR were found. (*Pseudogenes* are homologous sequences to the functional TCR genes, but appear to have mutations that render them nonfunctional, such as internal stop codons.) This suggests that, similar to other clusters of homologous genes, the *V*β*s* are undergoing homologous, but unequal, crossovers that can expand and contract the set during evolution, with gradual divergence of related genes. The related members of a family have not been found grouped together, suggesting that these homologous, but unequal, cross-overs can occur throughout the region. The orientation of the *V* regions that have been examined are all in the same translational direction, a finding that may have unknown functional importance. One exception is a mouse *V* region, *V*β*17*, which has been shown to be located 3′ to the *C* region cluster and oriented in the opposite direction (34). A human counterpart may also exist. Although the rearrangement of that V region necessarily needs to occur by an inversional, rather than a deletional mechanism, the significance of the position is unknown. For immunoglobulin heavy chain genes there is the suggestion that *V* regions proximal to the *C* region rearrange first. No such data have been found for the human β-locus.

Centromeric to the V region cluster are the genes coding for the D-J-C region of the molecule. The sequence of the entire region has been published (25). Two homologous clusters exist, each has one *D* region followed by either six *J* regions and one *C* region for the first cluster, or seven *J* regions and one *C* region for the second cluster. The formal possibility remains that other *D* regions may exist, but they have not yet been described. The two *C* regions are closely related and have no known functional differences.

Numerous polymorphisms of the β-locus are certain to exist, and a few have been described (33,35–37).

B. The Structure of the γ-Locus

The overall structure of the γ-locus, on chromosome 7p15, resembles that of the β-locus, with multiple V regions and two clusters of *J-C* genes (Fig. 6) (38,39). Although the βγ*V* region locus has not been described in detail, the γ-chain locus has been characterized, and the following description is likely to be complete, although again, the entire region has not been sequenced.

A cluster of *V* regions precedes the rest of the locus with the *V* regions oriented in one direction. Unlike the β-locus, there is a structural organization of the *V* regions in that the nine most closely related sequences come first, and the more divergent sequences follow. This suggests that homologous, but unequal, crossover events have not occurred recently between the first set and the others. The first set of nine *V* regions contains several pseudogenes that may not code for functional genes. Five can clearly code for functional proteins and are designated *V*γ*1*, *V*γ*3*, *V*γ*4*, *V*γ*5*, and *V*γ*8*; these are closely related and have been given a family name of Vγ*I* (See Fig. 3b). Five other γ-chain *V* regions follow, including two more

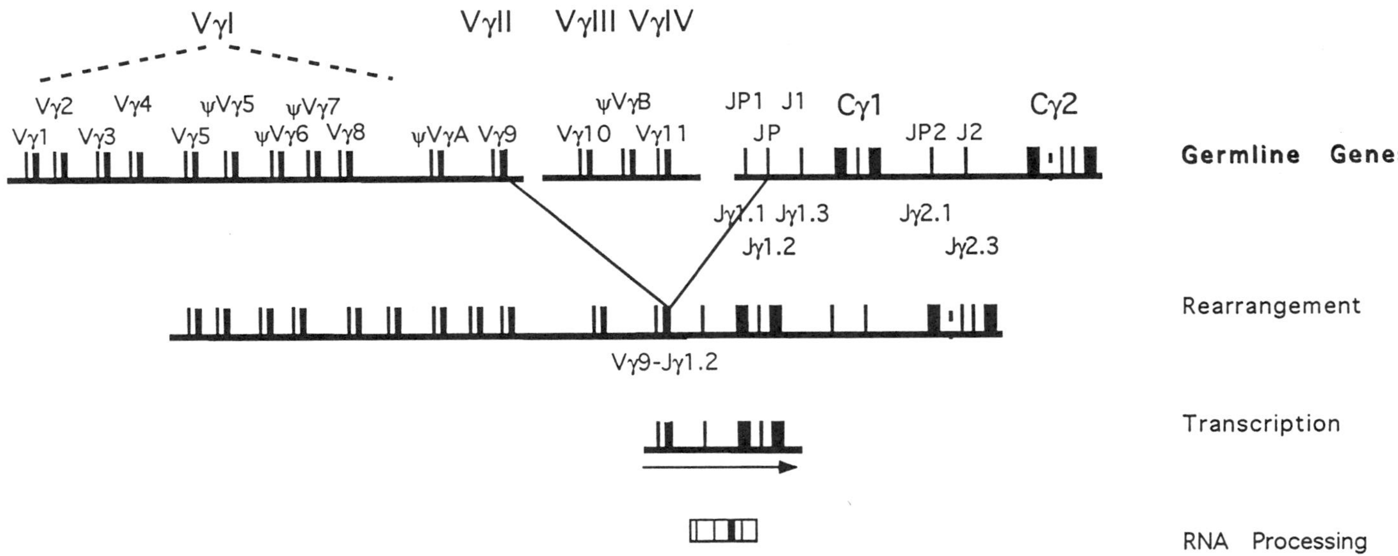

Figure 6 TCR γ-locus: At the top are the germline genes for the γ locus. The two notations for the *V* regions are shown. Pseudogenes are indicated with a ψ. An example of a rearrangement is shown followed by transcription and splicing.

pseudogenes. Three functional *V* regions not in the *VγI* family are widely divergent from each other and have been designated *Vγ9 (VγII)*, *Vγ10 (VγIII)*, and *Vγ11 (VγIV)*.

Centromeric to the *V* regions are two clusters of *J* regions, each with a *C* region following the *J* regions. The first cluster of *J* regions has three members, whereas the second has two members that closely resemble the first and third of the first set, explaining the notation of not having a Jγ2.2. The two *C* regions have a marked difference, in that *Cγ1* has one second exon, which codes for the connecting peptide of the constant region, while *Cγ2* has a duplication of the second exon. Even more remarkable is that the second exons of *Cγ2* have lost the cysteine that forms the disulfide bond between the C regions of the γ- and δ-chains. There is probably a functional difference between *Cγ1* and *Cγ2* because of the observation that most of the γδ cells in the peripheral blood are Cγ1, whereas those of the thymus are Cγ2. The possibility exists however that J region or V region differences may account for the presence of Cγ1 or Cγ2 in certain populations of cells, since specific V or J regions are associated with the C regions.

The most striking polymorphism of the γ-chain is the presence of either two or three second exons. The form with three tandem second exons occurs in up to 68% of the black African population, but only in 16% of the French and 13% of the Chinese population (each group was sampled in their native countries) (40,41). This polymorphism leads to a signifcant measurable difference in the molecular mass of the γ-chain, in part because of the asparagine-linked carbohydrate on the connecting peptide. No functional difference or disease association is known, but no systematic attempts have been made to find a functional difference. For example, it has not been determined if the distribution of the genotypes in a population is in Hardy–Weinberg equilibrium. Two polymorphic forms of the *VγI* set of genes have been described. One haplotype has seven instead of nine genes with the loss of *Vγ4* and *Vγ5*; this occurs in approximately 21% of the French population. Another haplotype has ten *VγI* genes, with the addition of a pseudo *Vγ3* gene. A functional difference in the protein products of these polymorphic genes has not yet been determined.

C. The Structure of the α–δ Locus

The most complex TCR locus is the α–δ locus at band 14q11 (Fig. 7) (42,43). The overall structure of the locus resembles the γ- and β-locus in having a cluster of *V* regions, followed by two *J-C* clusters. A major difference is that the two *J-C* regions are very divergent genetically and functionally. Related to the wide divergence of α and δ constant regions, some *V* regions are rearranged and expressed with the δ *J* and *C* region and others are consistently found as α-chains. The mechanism for that separation is unknown. It may simply be due to the physical of *V* regions, followed by two *J-C* clusters. A major difference is that the two *J-C* regions are very divergent genetically and functionally. Related to the wide divergence of α and δ constant regions, some *V* regions are rearranged and expressed with the δ *J* and *C* region and others are consistently found as α-chains. The mechanism for that separation is unknown. It may simply be due to the physical clustering of *V*δ, since they are mostly found near the *J-C* region. It may also relate to unknown regulation of rearrangements and/or transcription.

The *V* regions of the cluster have again been divided into families, based on criteria similar to that used for the β-chains. The most extensive characterization of a population of T cells has recently been published in which sequences of 300 α-chains were determined. A total of 29 families were defined (28). Most of the families had few members, such that the current estimate of the number of *V* region families is 30, with a total number of *V* regions being 50. This does not include the three *V* region subfamilies that have been associated primarily with δ chains. The order of the *V* regions has been only partially determined, and the orientation is exclusively in one direction, with the exception of one *V*δ which lies 3′ of the δ constant region in an inverted orientation (44).

The δ-chain cluster consists of three *D* regions, three *J* regions, and the constant region (44). No polymorphisms have been described for the δ-chains, although some have been described for the α-chain (37,45–47).

The α-chain cluster consists of a large domain of *J* regions that number approximately 100, covering 75 kb between the inverted *V*δ*3* amd the α *C* region (48). A single α-chain *C* region exists. Many polymorphisms are likely to exist in the α *J* regions, but have yet to be described.

IV. THE GENERATION OF T-CELL RECEPTOR DIVERSITY

In the germline and in early T-committed cells, the TCR genes are unrearranged with separate *V*, *D*, *J*, and *C* region segments as described in the last section. Early in T-cell development the rearrangement mechanism is initiated, leading to formation of the mature repertoire of the TCR loci. The rearrangement process in T cells is similar to that in B cells, as described in Chapter 1 in this book. Evidence for the similarity includes the observations that (1) the types of joints produced by rearrangement appear to be the same in TCR and immunoglobulins, (2) the SCID mutation of the mouse affects immunoglobulin and TCR gene rearrangement equally (49), and (3) transgenic mice with nonfunctional *RAG1* and *RAG2* genes, which are required to activate *VDJ* recombination, are incapable of rearranging either TCR or immunoglobulin genes (50,51). Representative products of rearrangement are shown in Figure 8, using the γ chain as the example. The diversity of the junctional region corresponds functionally to the *CDR3* mentioned earlier (see Figs. 3 and 4). Thus the tremendous diversity in

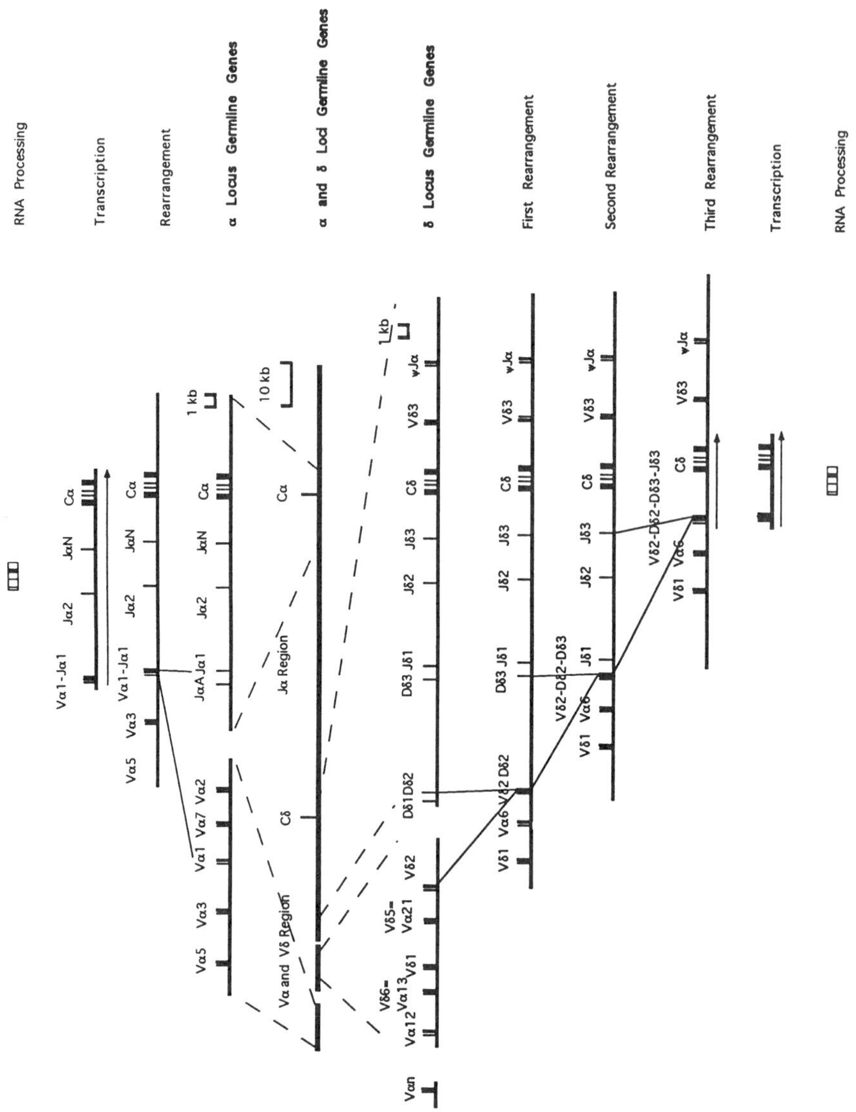
RNA Processing
Transcription
Rearrangement
α Locus Germline Genes
α and δ Loci Germline Genes
δ Locus Germline Genes
First Rearrangement
Second Rearrangement
Third Rearrangement
Transcription
RNA Processing
1 kb
10 kb
Cα
Cδ
Jα Region
Vα and Vδ Region

this region allows the recognition of a corresponding diversity of peptides, assuming that the model of recognition is true. The simplest rearrangement is the direct ligation of *V* to *J*, without modification of the connected ends (line 1). The choice of *V* and *J*, thus, allows the formation of as many new chains as the product of the number of *V* regions times the number of *J* regions. A second mechanism for generation of diversity is the addition of *P* sequences between the *V* and *J*. The notable feature of these sequences is that they are added onto the *V* or *J* region ends and are inverted repeats (palindromes) of the ends of the *V* or *J* region (line 2). After endonuclease cleavage of the signal sequence, the ends can also be modified by an exonuclease, resulting in shorter ends (line 3), or by the enzyme terminal deoxynucleotidyl transferase, which adds random nucleotides (line 4). The rearrangement process involves a series of enzymatic steps that have only recently begun to be dissected experimentally (52). The most notable difference between the generation of diversity of TCR chains as compared with immunoglobulins is the lack of somatic mutation of TCR genes.

The *V*, *D*, and *J* genes in the germline define the *germline diversity* of a locus. If all the germline genes that are functional can rearrange randomly using the mechanisms of rearrangement, those rearrangements constitute the *potential diversity* or potential repertoire of a receptor chain. The heteroduplex nature of the TCR expands diversity further by allowing each chain to associate with different partners, termed combinatorial association. The potential diversity of the TCR loci, therefore, is an enormous number that is well beyond the number of cells in the body (Table 1). The repertoire that is actually expressed as a functional receptor is a tiny fraction of the potential repertoire. How that repertoire matures is the topic of the next section.

V. ORDERED AND REGULATED REARRANGEMENTS IN T-CELL DEVELOPMENT

A. Developmental Process

Every fully functional T cell must necessarily have undergone a developmental process that includes multiple steps and decision points. A major goal of modern immunology has been the delineation of that process, which is described in Chapter 5. As a part of the early developmental pathways of the T cell, every

---

Figure 7 TCR α–δ locus: The middle three lines show germline genes. Because the two sets of genes are imbedded, two separate lines show the expanded α- and δ-loci. One rearrangement can create a functional α-gene, but up to four rearrangements are necessary to make a δ-gene.

Germline genes	Vγ9	TGT GCC TTG TGG GAG GTG ↓ **CACAGCA**GCAGACAGTTTGAGCCATCCCAT**TCAATAAAT**
	Jγ1.2	**GATTCTTAT**AAAGGCTTCTCA**GGTGGTG** ↓ GG CAA GAG TTG GGC AAA AAA
1. No modified ends		TGT GCC TTG TGG GAG GTG GG CAA GAG TTG GGC AAA AAA
2. P additions		TGT GCC TTG TGG GAG GTG GTG G GG CAA GAG TTG GGC AAA AAA
3. Exonuclease		TGT GCC TTG TGG GA A GAG TTG GGC AAA AAA
4. N additions		TGT GCC TTG TGG GA C GAA TTC CG A GAG TTG GGC AAA AAA

Figure 8 Junctional diversity of TCR: The germline genes of a *Vg* and *Jg* are shown in the first two lines. In boldface are the heptamer and nonamer signals. The germline genes are cut between the heptamer and the coding sequence by an endonuclease, as shown by arrows. The coding ends are rejoined either after no modification, one, or several varieties of modification, giving many possible junctional sequences, 2–4. The sequences that are added include P (palindromic) sequences, 2, or N sequences, 4, and are underlined.

Table 1 Sequence Diversity in T-Cell Receptor and Immunoglobulin Genes

	IG		α:β TCR		γ:δ TCR	
Gene locus	*H*	κ	α	β	γ	δ
Variable segments	250–1000	250	50	60	9	6
Diversity segments	10(?)	0	0	2	0	3
Ds read all frames	Rarely	None	None	Often	None	Often
N-region addition	V-D, D-J	None	V-J	V-D,D-J	V-J	V-D1,D1-D2,D2-D3,D3-J
Joining segments	6	5	90	13	5	3
Variable region combinations	62,500–250,000		3,000		54	
Junctional combinations	10^{11}		$>10^{15}$		$>10^{20}$	

Source: Adapted from Ref. 18.

fully functional TCR on the surface of a T cell must have undergone a series of developmental steps, described in the following and outlined in Figure 9.

B. Cell Lineages

One of the fundamental concepts in immunology is the description of lineages of cells that have specialized functions. A lineage of cells can be traced to a developmental decision that led to a single lineage dividing into multiple lineages. Cell lineages are distinguished by cell surface markers or by intracellular markers, such as rearrangements; for example, the CD4 and CD8 lineage division. Unfortunately, many of the markers are not perfect indicators, in that the same molecules are expressed on other lineages, such as some of the CD3 components. The first locus in a potential T cell to rearrange is probably either the β or δ locus. Following the rearrangement of one locus, a second locus must rearrange, the α for the αβ lineage and the γ for the γδ lineage. The commitment of a progenitor cell to a specific lineage probably dictates which rearrangements take place. Thus, the cell with the rearrangement machinery turned on must selectively rearrange the TCR loci and not the immunoglobulin loci and more precisely, it must correctly select a specific TCR locus and not others. The details of how that selection is made remains an important unanswered question, but it may be related to transcriptional regulation (i.e., the same mechanisms that open up a locus to transcription may open it up for rearrangement). All of the possible rearrangements that occur within a lineage generate the potential repertoire of that lineage. In other words, if in αβ T cells, the γ and δ loci do not rearrange, then the potential repertoire of the αβ cell does not include any receptors of the

State of the TCR in T cells:	Developmental process:	Regulatory decision affecting TCR repertoire:
Germline receptor genes with germline repertoire.	Commitment of pleuripotential stem cell to T cell lineage.	
One receptor gene locus rearranged.	Commitment to either the αβ or γδ lineage.	Selection of V, D, J to rearrange.
A second rearrangement occurs; the potential repertoire is generated.		Selection of V, D, J for second locus.
A cell surface receptor is expressed; the expressed repertoire is generated.	Maturation to CD3 low, CD4, and CD8 positive cell.	TCR must be able to be expressed as a functional complex.
TCR on cells that recognize self-MHC are selected; other cells die.	Positive Selection.	Cells recognizing self-MHC differentiate and proliferate.
TCR on cells that recognize self-peptide MHC are eliminated; the naive mature repertoire is generated.	Negative Selection; maturation. to CD3 high, single positive cell.	Cells recognizing self-peptide-MHC die.
Specialized lineages of T cells with limited repertoires may populate specific tissues.	Cells leave the thymus and migrate to peripheral organs or stay in the blood.	Different lineages of cells express different cell surface molecules that control migration.
T cells that recognize antigens replicate and differentiate; the effector and memory repertoire is generated.	Antigen stimulation through TCR.	Antigenic exposure dictates which cells change.

Figure 9 Developmental steps involved in TCR repertoire differentiation. On the left are the progressive states of the TCR during development of a mature immune system. The middle column describes the developmental process that matures the repertoire from one state to the next. The right column describes the decision the cell makes during that developmental process.

γδ type. These events are not mutually exclusive. In many, if not most, αβ T cells there are rearrangements of the γ locus and sometimes of the δ locus. The α or β rearrangements have not been commonly found in γδ T cells.

C. Nonproductive Rearrangements

Many rearrangements cannot produce a functional molecule. The following situations result in nonproductive rearrangements: (1) a pseudogene is rearranged; (2) the junction between the segments does not preserve the translational frame of the sequence; (3) the addition of *N* region sequences introduces a stop codon; (4) the exonuclease activity deletes an essential part of the V region or J region, such as the cysteine residue. By far the most frequent event is the generation of out-of-frame products, which have been found in approximately two thirds of the junctions that are taken from cells that have not been selected for function. If the first allele of the locus produces an aberrant rearrangement, the second allele may rearrange. If the first rearranges correctly, a second rearrangement of the same locus on the other allele may be prohibited. Similar mechanisms may account for the phenomena of allelic exclusion, originally described for B cells, in which a given B lymphocyte expresses only one allele of a polymorphic locus (such as for the immunoglobulin κ-chain) while having two ostensible alleles coding for each of the chains involved. Multiple mechanisms probably exist for the phenomenon which has been studied to a limited extent for T cells. No general studies have been done because no allele-specific C region antibodies have been available. On a cellular level, no T-cell clones have been described that express two different β-chains, two δ-chains, or two γ-chains. Multiple instances have been described for which two α-chain cDNAs are expressed in the same cell (53). The general rule is that allelic exclusion does not exist for α-chains, whereas it does exist for β-chains. The role of allelic exclusion in the expression of γ- and δ-chains remains unclear.

A further barrier to the successful development of a functional TCR is the association between the two chains. For example, certain combinations of γ and δ TCR have not been observed. One explanation of that phenomenon is that structural constraints exist between the two chains of a heterodimer that prevents certain combinations existing as a stable heterodimer. Such a phenomenon has been observed for the two chains of an immunoglobulin molecule. Thus, the next level of repertoire of T cells is the expressed repertoire, which would include all receptors that can actually appear on the surface of T cells.

When a functional TCR appears on the surface of the cell, functional selection of the T cell can now occur on at least four levels: thymic-positive selection for self-MHC, thymic-negative selection against self-peptide–MHC; positive- or negative-stimulation with superantigens in the periphery; and positive- or negative-stimulation by peptide antigens in the periphery. Thus, the selected repertoire of T cells will shrink further. These mechanisms of T cell maturation are discussed in Chapter 5.

Although each T cell has only one receptor, the biological effect of a T-cell response is dictated by a population of T cells with a heterogeneous array of antigen receptors. Attempts to describe and measure the diversity of the receptors that make up the T-cell receptor repertoire are incomplete. Thus, the repertoire is a complex issue that combines theory and data. Aspects of the repertoire can be discussed on the level of a species, an individual, a population of cells, or by the response to a single peptide–MHC complex.

D. Repertoire Expression

Lineages of T cells may express limited repertoires of TCR. Examples in the mouse and, to a more limited extent, in humans, have provided a model whereby certain lineages of T cells with specialized function express TCR with limited diversity. This model implies that some diseases may be associated with the over- or underexpansion of these subsets of cells and, more importantly, that some diseases may be associated with antigens that may have an etiological role in the growth of these lineages. An analogy is with the CD5-positive B cells, which have unique immunoglobulin receptors and are expressed on B-cell lymphomas and in chronic lymphocytic leukemia (CLL). The CD4- and CD8-positive lineages are examples for which it is known that the MHC molecules that stimulate the two sets of cells are different. Recent evidence suggests that the TCR repertoire may differ between those two populations of cells (54). An example of a lineage of cells with a unique receptor comes from studies of dendritic epithelial cells (DEC) of the mouse, which were found to be γδ T cells, with a nearly homogeneous TCR with very limited junctional diversity (55). Remarkably, these γδ cells were one of the first to develop in mouse embryogenesis in the thymus, leaving the thymus to populate the skin before the appearance of other T cells. Following those cells, other waves of γδ T cells, each with its own set of receptors, appear in the thymus (56).

E. Regulation of Rearrangements

Regulation of rearrangements may determine the TCR repertoire of separate lineages of T cells. Data have been presented that T cells make an early commitment to either an αβ or a γδ lineage (57). One suggestion has been that the decision to rearrange the δ or α loci is also made very early. The alternative theory would be that the cells can rearrange both loci, but the cells that successfully make an αβ receptor become αβ T cells, and those that make a successful γδ receptor become γδ cells. Because the rearrangements are not reversible, one can obtain evidence of a cell's developmental history by examining the rearranged products in mature T cells. One study found no evidence, in mature cells of the αβ lineage, that the δ locus had also rearranged. This would be expected if the decision to commit to a lineage is made after the rearrangement step. [Some conflicting data have also been described, such that the relation of the lineage determination to the rearrangement process does not remain entirely clear (58).] In humans, when the early fetal rearrangements were examined, it was found that *Vδ2–Dδ3* rearrangements appeared first, with other rearrangements *Vδ1–Dδ2* appearing later (59,60). These early rearrangements have not yet been associated with any adult lineage of T cells. They may have a special role in fetal immunity and disappear, or they may be in a compartment of the adult immune system that has not been carefully examined.

VI. T-CELL RECEPTOR REPERTOIRE AND DISEASE ASSOCIATIONS

A. Association with Human Disease

There have been many reports in which differences in the *TCR* genes or repertoire have been associated with human diseases. The reports have fallen in two categories. First are reports of diseases that appear to be associated with TCR locus polymorphisms. Prominent among these diseases are autoimmune disorders that are thought to be T-cell–mediated, such as multiple sclerosis (61–63). These studies find increased frequency of certain TCR alleles in groups of unrelated patients when compared with control groups. These suggest that structural or regulatory polymorphisms in the TCR genes may play a role in the susceptibility to immunological diseases, analogous to MHC associations. Precedent for these studies was established by the use of the approach in finding disease associations between *HLA* polymorphisms and diseases of the immune system. To date, none of these associations with the TCR have been documented in multiple studies from different laboratories. It is unclear if those associations are real or if technical problems and disease heterogeneity have obscured real biologically important effects (64). Family or twin studies of disease associations would be more convincing, but are not yet available. Importantly, this type of study makes no direct link between a gene and a disease. It could be any genetic locus that is linked to the polymorphism studied that is really associated mechanistically to the disease.

The second category are reports of diseases that have an altered TCR repertoire. A strong impetus to those studies was the finding, in rodents, that the TCRs that recognized myelin basic protein (MBP), an immunogen that could cause experimental allergic encephalomyelitis (EAE), a demyelinating disease, were of a very limited heterogeneity (65–67). For example, TCR sequences were obtained from T-cell clones that react to a MBP peptide in PL/J mice. All of the clones used the same Vβ, many used the same Vα. Not only were the V regions of limited heterogeneity, but also the junctional sequences were closely related in terms of the amino acid structure, suggesting that the disease process that produces anti-MBP T cells that mediate demyelination expanded out T cells of a very narrow spectrum. Furthermore, it has been possible to treat the clinical disease with antibodies against the disease-associated T-cell receptors (68). If human syndromes could be found with similar pathophysiology, the prospects for treatment are exciting.

Since the early reports of altered repertoire in animal models of autoimmune diseases, many attempts have been made to ask if human diseases were similar. Reports of that type have been common in many diseases, most prominently multiple scleroses (69), but also sarcoidosis (70), rheumatoid arthritis (71,72), inflammatory bowel disease (73), human immunodeficiency virus (HIV) infection, autoimmune thyroiditis (74), and many others. These studies have not been repeated in large well-controlled patient populations, and it is too early to know if any of them will show that the TCR repertoire selection has any fundamental role in the evolution of a disease. Key to the interpretation of these studies is the realization that basic immunological knowledge is not yet available to do the studies correctly or to interpret them clearly. The T cell repertoire can be very different depending on where and how it is investigated. Superantigens can expand in vitro populations of T cells with certain Vβ. The TCR repertoire peptide–antigen stimulation in the mouse can be very homogeneous, as with the MBP example, or highly diverse, as with a viral peptide (75). Different tissues can have a limited TCR repertoire, as with the increase in Vα12 in the human intestine (76). Thus, when one finds that the repertoire is altered in a disease, one may be looking at tissue or developmental differences, rather than disease-associated changes. Furthermore, technical difficulties in measuring TCR repertoire may have a great influence on the observations. In the mouse, most studies of V region use of β chains were made possible through the use of monoclonal antibodies against specific V regions. In humans, those antibodies have not been widely available. The PCR techniques have largely replaced Southern blot analyses as a molecular method to examine the TCR gene repertoire. The quantitative PCR with specific primers has become the preponderant technology, even though the reproducibility, accuracy, and specificity of the techniques remain to be clearly demonstrated (77).

In conclusion, the role of the T cell in a range of human disease is unquestioned. It remains a fascinating possibility that alterations in the T-cell receptor repertoire caused by genetic polymorphisms by altered developmental steps, or by interactions with the environment could be causal in human disease. If that proves to be true, it could have very important implications both diagnostically and therapeutically.

REFERENCES

1. K. Matsui, J. J. Boniface, P. A. Reay, H. Schild, B. Fazekas de St. Groth, and M. M. Davis, Low affinity interaction of peptide–MHC complexes with T cell receptors, *Science*, *254*: 1788 (1991).
2. T. Saito, and R. N. Germain, The molecular basis of MHC-restricted antigen recognition by T cells, *Int. Rev. Immunol.*, *3*: 147 (1988).
3. F. Hochstenbach, and M. B. Brenner, Newly identified gamma delta and beta delta T-cell receptors, *J. Clin. Immunol.*, *10*: 1 (1990).
4. M. Groettrup, A. Baron, G. Griffiths, R. Palacios, and H. von Boehmer, T cell receptor (TCR) β chain homodimers on the surface of immature but not mature α, gamma, δ chain deficient T cell lines, *EMBO J.*, *11*: 2735 (1992).
5. S. Koyasu, L. DAdamio, A. R. Arulanandam, S. Abraham, L. K. Clayton, and

E. L. Reinherz, T cell receptor complexes containing Fc epsilon RI gamma homodimers in lieu of CD3 zeta and CD3 eta components: A novel isoform expressed on large granular lymphocytes. *J. Exp. Med.*, *175*: 203 (1992).

6. J. P. Jensen, D. Hou, M. Ramsburg, A. Taylor, M. Dean, and A. M. Weissman, Organization of the human T cell receptor zeta/eta gene and its genetic linkage to the *FcgammaRII–FcgammaRIII* gene cluster, *J. Immunol.*, *148*: 2563 (1992).
7. J. Gabert, C. Langlet, R. Zamoyska, J. R. Parnes, A. M. Schmitt-Verhulst, and B. Malissen, Recognition of MHC class I specificity by transfer of the T cell receptor and *Lyt-2* genes, *Cell*, *50*: 545 (1987).
8. W. G. Ballhausen, A. B. Reske-Kunz, B. Tourvieille, P. S. Ohashi, J. R. Parnes, and T. W. Mak, Acquisition of an additional antigen specificity after mouse CD4 gene transfer into a T helper hybridoma, *J. Exp. Med*, *167*: 1493 (1988).
9. C. A. Janeway, The T cell receptor as a multicomponent signalling machine: CD4/CD8 coreceptors and CD45 in T cell activation, *Annu. Rev. Immunol.*, *10*: 645 (1992).
10. C. Grégoire, N. Rebaï, F. Schweisguth, A. Necker, G. Mazza, N. Auphan, A. Millward, A.-M. Schmitt-Verhulst, and B. Malissen, Engineered secreted T-cell receptor αβ heterodimers, *Proc. Natl. Acad. Sci. USA*, *88*: 8077 (1991).
11. B. Devaux, P. J. Bjorkman, C. Stevenson, W. Greif, J. F. Elliott, C. Sagerstrom, C. Clayberger, A. M. Krensky, and M. M. Davis, Generation of monoclonal antibodies against soluble human T cell receptor polypeptides, *Eur. J. Immunol.*, *21*: 2111 (1991).
12. S. Weber, A. Traunecker, F. Oliveri, W. Gerhard, and K. Karjalainen, Specific low-affinity recognition of major histocompatibility complex plus peptide by soluble T-cell receptor. *Nature*, *356*: 793 (1992).
13. B. R. Bloom, R. L. Modlin, and P. Salgame, Stigma variations: Observations on suppressor T cells and leprosy, *Annu. Rev. Immunol.* *10*: 453 (1992).
14. J. Novotny, S. Tonegawa, H. Saito, D. M. Kranz, and H. N. Eisen, Secondary, tertiary, and quaternary structure of T-cell–specific immunoglobulin-like polypeptide chains, *Proc. Natl. Acad. Sci. USA*, *83*: 742 (1986).
15. C. Chothia, D. R. Boswell, and A. M. Lesk, The outline structure of the T-cell αβ receptor, *EMBO J.*, *7*: 3745 (1988).
16. R. Jores, P. M. Alzari, and T. Meo, Resolution of hypervariable regions in T-cell receptor beta chains by a modified Wu–Kabat index of amino acid diversity, *Proc. Natl. Acad. Sci. USA*, *87*: 9138 (1990).
17. S. Huck, P. Dariavach, and M. P. Lefranc, Variable region genes in the human T-cell rearranging gamma (TRG) locus: V-J junction and homology with the mouse genes, *EMBO J.*, *7*: 719 (1988).
18. M. M. Davis, and P. J. Bjorkman, T-cell antigen receptor genes and T-cell recognition [published erratum appears in *Nature* 1988 Oct 20;*335*(6192):744], *Nature*, *334*: 395 (1988).
19. J. L. Jorgensen, P. A. Reay, E. W. Ehrich, and M. M. Davis, Molecular components of T-cell recognition, *Annu Rev. Immunol.*, *10*: 835 (1992).
20. P. Marrack, and J. Kappler, The staphylococcal enterotoxins and their relatives [published erratum appears in *Science* 1990 Jun 1;*248*(4959):1066; see comments], *Science*, *248*: 705 (1990).

21. H. P. Smith, P. Le, D. L. Woodland, and M. A. Blackman, T cell receptor α-chain influences reactivity to Mis-1 in Vβ8.1 transgenic mice, *J. Immunol.*, *149*: 887 (1992).
22. A. M. Pullen, J. Bill, R. T. Kubo, P. Marrack, and J. W. Kappler, Analysis of the interaction site for the self superantigen Mls-1a on T cell receptor V beta, *J. Exp. Med.*, *173*: 1183 (1991).
23. P. Dellabona, J. Peccoud, J. Kappler, P. Marrack, C. Benoist, and D. Mathis, Superantigens interact with MHC class II molecules outside of the antigen groove, *Cell*, *62*: 1115 (1990).
24. Y. W. Choi, B. Kotzin, L. Herron, J. Callahan, P. Marrack, and J. Kappler, Interaction of *Staphylococcus aureus* toxin "superantigens" with human T cells, *Proc. Natl. Acad. Sci. USA*, *86*: 8941 (1989).
25. B. Toyonaga, and T. W. Mak, Genes of the T-cell antigen receptor in normal and malignant T cells, *Annu. Rev. Immunol.*, *5*: 585 (1987).
26. M. H. Klein, P. Concannon, M. Everett, L. D. Kim, T. Hunkapiller, and L. Hood, Diversity and structrure of human T-cell receptor alpha-chain variable region genes, *Proc. Natl. Acad. Sci. USA*, *84*: 6884 (1987).
27. E. Lai, R. K. Wilson, and L. E. Hood, Physical maps of the mouse and human immunoglobulin-like loci, *Adv. Immunol.*, *46*: 1 (1989).
28. S. Roman-Roman, L. Ferradini, J. Azocar, C. Genevee, T. Hercend, and F. Triebel, Studies on the human T cell receptor alpha/beta variable region genes. I. Identification of 7 additional V alpha subfamilies and 14 J alpha gene segments, *Eur. J. Immunol.*, *21*: 927 (1991).
29. W. M. C. Rosenberg, P. A. H. Moss, and J. I. Bell, Variation in human T cell receptor V_β and J_β repertoire: Analysis using anchor polymerase chain reaction, *Eur. J. Immunol.*, *22*: 541 (1992).
30. M. A. Robinson, The human T cell receptor beta-chain gene complex contains at least 57 variable gene segments. Identification of six V beta genes in four new gene families, *J. Immunol.*, *146*: 4392 (1991).
31. A Plaza, D. H. Kono, and A. N. Theofilopoulos, New Human V beta genes and polymorphic variants, *J. Immunol.*, *147*: 4360 (1991).
32. G. Siu, E. C. Strauss, E. Lai, and L. E. Hood, Analysis of a human *V beta* gene subfamily, *J. Exp. Med.*, *164*: 1600 (1986).
33. Y. Li, P. Szabo, and D. N. Posnett, The genomic structure of human Vβ6 T cell antigen receptor genes, *J. Exp. Med.*, *174*: 1537 (1991).
34. M. Malissen, C. McCoy, D. Blanc, J. Trucy, C. Devaux, A. M. Schmitt Verhulst, F. Fitch, L. Hood, and B. Malissen, Direct evidence for chromosomal inversion during T-cell receptor beta–gene rearrangements, *Nature*, *319*: 28 (1986).
35. P. Charmley, A. Chao, P. Concannon, L. Hood, and R. A. Gatti, Haplotyping the human T-cell receptor beta-chain gene complex by use of restriction fragment length polymorphisms, *Proc. Natl. Acad. Sci. USA*, *87*: 4823 (1990).
36. S.-J. Gahm, B. J. Fowlkes, S. C. Jameson, N. R. J. Gascoigne, M. M. Cotterman, O. Kanagawa, R. H. Schwartz, and L. A. Matis, Profound alteration in an αβ T-cell antigen receptor repertoire due to polymorphism in the first complementarity-determining region of the β chain, *Proc. Natl. Acad. Sci. USA*, *88*: 10267 (1991).
37. S. C. Jameson, P. B. Nakajima, J. L. Brooks, W. Heath, O. Kanagawa, and N.

R. J. Gascoigne, the T cell receptor *Vα11* gene family: Analysis of allelic sequence polymorphism and demonstration of Jα region-dependent recognition by allele-specific antibodies, *J. Immunol.*, *147*: 3185 (1991).
38. M. P. Lefranc, and T. H. Rabbitts, The human T-cell receptor gamma (TRG) genes, *Trends Biochem. Sci*, *14*: 214 (1989).
39. M. P. Lefranc, P. Chuchana, P. Dariavach, C. Nguyen, S. Huck, F. Brockly, B. Jordan, and G. Lefranc, Molecular mapping of the human T cell receptor gamma (TRG) genes and linkage of the variable and constant regions, *Eur. J. Immunol.*, *19*: 989 (1989).
40. N. Ghanem, C. Buresi, J. P. Moisan, M. Bensmana, P. Chuchana, S. Huck, G. Lefranc, and M. P. Lefranc, Deletion, insertion, and restriction site polymorphism of the T-cell receptor gamma variable locus in French, Lebanese, Tunisian, and black African populations, *Immunogenetics*, *30*: 350 (1989).
41. C. Buresi, N. Ghanem, S. Huck, G. Lefranc, and M. P. Lefranc, Exon duplication and triplication in the human T-cell receptor gamma constant region genes and RFLP in Franch, Lebanese, Tunisian, and black African populations [published erratum appears in *Immunogenetics* 1989; *30*: 148], *Immunogenetics*, *29*: 161 (1989).
42. H. Griesser, E. Champagne, D. Tkachuk, Y. Takihara, M. Lalande, E. Baillie, M. Minden, and T. W. Mak, the Human T cell receptor alpha–delta locus: A physical map of the variable, joining and constant region genes, *Eur. J. Immunol.*, *18*: 641 (1988).
43. K. Satyanarayana, S. Hata, P. Devlin, M. G. Roncarolo, J. E. De Vries, H. Spits, J. L. Strominger, and M. S. Krangel, Genomic organization of the human T-cell antigen-receptor alpha/delta locus, *Proc. Natl. Acad. Sci. USA*, *85*: 8166 (1988).
44. E. Y. Loh, S. Cwirla, A. T. Serafini, J. H. Phillips, and L. L. Lanier, Human T-cell-receptor delta chain: Genomic organization, diversity, and expression in populations of cells, *Proc. Natl. Acad. Sci. USA*, *85*: 9714 (1988).
45. A. Chan, R. P. Du, M. Reis, E. Baillie, L. M. Meske, M. Sheehy, and T. W. Mak, Polymorphism of the human T cell receptor alpha chain variable genes: Identification of a highly polymorphic V gene probe, *Int. Immunol.*, *1*: 267 (1989).
46. A. Chan, R. P. Du, M. Reis, L. M. Meske, M. Sheehy, E. Baillie, and T. W. Mak, Human T cell receptor V alpha gene polymorphism, *Exp. Clin. Immunogenet.*, *7*: 26 (1990).
47. J. A. Wright, L. Hood, and P. Concannon, Human T-cell receptor *Vα* gene polymorphism, *Hum. Immunol.*, *32*: 277 (1991).
48. E. Champagne, U. Sagman, A. Biondi, W. H. Lewis, T. W. Mak, and M. D. Minden, Structure and rearrangement of the T cell receptor J alpha locus in T cells and leukemic T cell lines, *Eur. J. Immunol.*, *18*: 1033 (1988).
49. A. M. Carroll, and M. J. Bosma, T-lymphocyte development in *scid* mice is arrested shortly after the initiation of T-cell receptor δ gene recombination, *Genes Dev.*, *5*: 1357 (1991).
50. Y. G. Shinkai, G. Rathbun, K.-P. Lam, E. M. Olyz, V. Stewart, M. Mendelsohn, J. Charron, M. Datta, F. Young, A. M. Stall, and F. W. Alt, *RAG-2* deficient mice lack mature lymphocytes owing to inability to initiate V(D)J rearrangement, *Cell*, *68*: 855 (1992).

51. P. Mombaerts, J. Iacomini, R. S. Johnson, K. Herrup, S. Tonegawa, and V. E. Papaioannou, *RAG-1*-deficient mice have no mature B and T lymphocytes, *Cell*, *68*: 869 (1992).
52. M. R. Lieber, The mechanism of V(D)J recombination: A balance of diversity, specificity, and stability, *Cell*, *70*: 873 (1992).
53. M. Malissen, J. Trucy, F. Letourneur, N. Rebai, D. E. Dunn, F. W. Fitch, L. Hood, and B. Malissen, A T cell clone expresses two T cell receptor alpha genes but uses one alpha beta heterodimer for allorecognition and self MHC-restricted antigen recognition, *Cell*, *55*: 49 (1988).
54. M. P. Davey, M. M. Meyer, D. D. Munkirs, D. Babcock, M. P. Braun, J. B. Hayden, and A. C. Bakke, T-cell receptor variable beta genes show differential expression in CD4 and CD8 T cells, *Hum. Immunol.*, *32*: 194 (1991).
55. J. P. Allison, D. M. Asarnow, M. Bonyhadi, A. Carbone, W. L. Havran, D. Nandi, and J. Noble, lambdaδ T cells in murine epithelia: Origin, repertoire, and function, *Adv. Exp. Med. Biol.*, *292*: 63 (1991).
56. W. Haas, and S. Tonegawa, Development and selection of gammaδ T cells, *Curr. Opin. Immunol.*, *4*: 147 (1992).
57. A. Winoto, and D. Baltimore, Separate lineages of T cells expressing the alpha beta and gamma delta receptors, *Nature*, *338*: 430 (1989).
58. K. Okazaki, and H. Sakano, Thymocyte circular DNA excised from T cell receptor alpha–delta gene complex, *EMBO J.*, *7*: 1669 (1988).
59. H. Spits, H. Yssel, C. Brockelhurst, and M. Krangel, Evidence for controlled gene rearrangements and cytokine production during development of human TCR-gammaδ$^+$ lymphocytes, *Curr. Top. Microbiol. Immunol.*, *173*: 47 (1991).
60. M. S. Krangel, H. Yssel, C. Brocklehurst, and H. Spits, A distinct wave of human T cell receptor gamma/delta lymphocytes in the early fetal thymus: Evidence for controlled gene rearrangement and cytokine production, *J. Exp. Med.*, *172*: 847 (1990).
61. P. Charmley, S. S. Beall, P. Concannon, L. Hood, and R. A. Gatti, Further localization of a multiple sclerosis susceptibility gene on chromosome 7q using a new T cell receptor beta-chain DNA polymorphism, *J. Neuroimmunol.*, *32*: 231 (1991).
62. V. Kumar, D. H. Kono, J. L. Urban, and L. Hood, The T-cell receptor repertoire and autoimmune diseases, *Annu. Rev. Immunol.*, *7*: 657 (1989).
63. L. L. Hashimoto, T. W. Mak, and G. C. Ebers, T. cell receptor α chain polymorphisms in multiple sclerosis, *J. Neuroimmunol.*, *40*: 41 (1992).
64. J. Hillert, and O. Olerup, Debate: Do TCR genes influence susceptibility to autoimmune disease in humans? Germ-line polymorphism of TCR genes and disease susceptibility—factor hypothesis? *Immunol. Today*, *13*: 47 (1992).
65. H. Acha-Orbea, Limited heterogeneity of autoantigens and T cells in autoimmune diseases? *Res. Immunol.*, *142*: 487 (1991).
66. T. Esch, L. Clark, X.-M. Zhang, S. Goldman, and E. Heber-Katz, Observations, legends, and conjectures concerning restricted T-cell receptor usage and autoimmune disease, *Crit. Rev. Immunol.*, *11*: 249 (1992).
67. H. Acha-Orbea, D. J. Mitchell, L. Timmermann, D. C. Wraith, G. S. Tausch, M. K. Waldor, S. S. Zamvil, H. O. McDevitt, and L. Steinman, Limited heterogeneity

of T cell receptors from lymphocytes mediating autoimmune encephalomyelitis allows specific immune intervention, *Cell, 54*: 263 (1988).
68. D. M. Zaller, G. Osman, O. Kanagawa, and L. Hood, Prevention and treatment of murine experimental allergic encephalomyelitis with T cell receptor V beta-specific antibodies, *J. Exp. Med., 171*: 1943 (1990).
69. L. Hood, V. Kumar, G. Osman, S. S. Beall, C. Gomez, W. Funkhouser, D. H. Kono, D. Nickerson, D. M. Zaller, and J. L. Urban, Autoimmune disease and T-cell immunologic recognition, *Cold Spring Harbor Symp. Quant. Biol., 54* (part 2): 859 (1989).
70. N. Tamura, K. J. Holroyd, M. K. Banks, H. Okayama, and R. G. Crystal, Diversity in junctional sequences associated with the common human *Vg9* and *Vd2* gene segments in normal blood and lung compared with the limited diversity in a granulomatous disease, *J. Exp. Med., 172*: 169 (1990).
71. G. Pluschke, G. Ricken, H. Taube, S. Kroninger, I. Melchers, H. H. Peters, K. Eichmann, and U. Krawinkel, Biased T cell receptor V_α region repertoire in the synovial fluid of rheumatoid arthritis patients, *Eur. J. Immunol., 21*: 2749 (1991).
72. T. L. Whiteside, D. S. Heo, S. Takagi, and R. B. Herberman, Tumor-infiltrating lymphocytes from human solid tumors: Antigen-specific killer T lymphocytes of activated natural killer lymphocytes, *Immunol. Ser., 48*: 139 (1989).
73. D. N. Posnett, I. Schmelkin, D. A. Burton, A. August, H. McGrath, and L. F. Mayer, T cell antigen V region usage: Increases in $Vb8^+$ T cells in Crohn's disease, *J. Clin. Invest., 85*: 1770 (1990).
74. T. F. Davies, A. Martin, E. S. Concepcion, P. Graves, L. Cohen, and A. Ben Nun, Evidence of limited variability of antigen receptors on intrathyroidal T cells in autoimmune thyroid disease, *N. Engl. J. Med., 325*: 238 (1991).
75./ A. H. Taylor, A. M. Haberman, W. Gerhard, and A. J. Caton, Structure–function relationships among highly diverse T cells that recognize a determinant from influenza virus hemagglutinin, *J. Exp. Med., 172*: 1643 (1990).
76. H. DerSimonian, H. Band, and M. B. Brenner, Increased frequency of T cell receptor V alpha 12.1 expression on $CD8^+$ T cells: Evidence that V alpha participates in shaping the peripheral T cell repertoire, *J. Exp. Med., 174*: 639 (1991).
77. B. L. Hall, and O. J. Finn, PCR-based analysis of the T-cell receptor Vβ multigene family: Experimental parameters affecting its validity, *Biotechniques, 13*: 248 (1992).

5

Differentiation of T Cells in the Thymus

I. Nicholas Crispe
Yale University School of Medicine, New Haven, Connecticut

I. INTRODUCTION

The human thymus is a multilobular organ located in the superior mediastinum, which is prominent in neonates and small children, reaches a maximum size at about 5 years of age, and slowly involutes thereafter. In healthy old age, much of the organ is replaced by fat, but islands of functional thymic tissue remain. The thymus is the source of virtually all T cells, which is most clearly demonstrated by "experiments of nature": athymic mice and athymic humans lack T cells and are deficient in T-cell–mediated immune responses. The thymus originates from pharyngeal pouch material, which gives rise to the thymic stroma; and from hematopoietic progenitor cells, which give rise to thymic macrophages, dendritic cells, and developing T cells. Small numbers of progenitor cells enter the thymus from the surrounding tissue during embryogenesis, and bone marrow-derived progenitor cells continue to colonize the thymus during adult life. Inside the thymus, these cells proliferate, undergo gene rearrangements that generate antigen-specific T-cell receptors (TCR) of two types ($\alpha\beta$ and $\gamma\delta$ receptors), and are then subject to selection processes based on their specificity. The outcome of these processes is the production of $\alpha\beta$ T cells with a specificity repertoire that is tolerant of most self-proteins, and competent to recognize foreign antigenic peptides presented by molecules encoded by the major histocompatibility complex (MHC). In this chapter, the current understanding of

T-cell development in the thymus is summarized. Much of this understanding is derived from experiments in mice, for which in vivo experiments are possible, but the information available suggests that thymic differentiation is similar in most mammals and that reliable extrapolation from mouse to humans is possible.

II. MORPHOLOGY AND CELLULAR COMPOSITION

A. Anatomy

The human thymus is a multilobular organ consisting of a left and a right lobe. Each lobe is subdivided by fibrous septa into several lobules. Each lobule has two main anatomical zones: an outer cortex and an inner medulla. In the mouse, the entire thymus consists of only two lobes, each organized like a human thymus lobule. Outside the cortex is a connective tissue capsule, beneath which is a fluid-filled subcapsular sinus. The septa that separate the lobules of the human thymus are invaginations of the capsule.

The thymus consists of thymocytes (developing T cells) and stromal cells (everything else). The thymocytes are mainly small, lymphoid cells, but approximately 20% of them are blast cells, and a similar number are in cycle, based on measurements of DNA content (1). The stroma cells are mainly epithelial cells, which form a reticulum in the cortex and are more densely packed in the medulla. Also found in the thymus are dendritic cells and macrophages. The dendritic cells are concentrated in the medulla and at the corticomedullary junction.

The blood supply of the thymus gains access to the tissue by the main arterioles, which pass through the interlobular septa; these break up into a network of capillaries. The venules in the inner cortex have specialized endothelium, similar to the "high endothelial venules" through which lymphocytes leave the circulation to enter the parenchyma of lymph nodes. These structures are particularly seen near the corticomedullary junction and are probably the principal route by which cells enter the thymus. The site of egress of mature thymocytes is unknown.

The thymus also has a sympathetic and parasympathetic nerve supply, which forms a plexus around the septal blood vessels; and a peptidergic supply, which includes nerve terminals containing oxytocin and vasopressin. However, hemi-denervation experiments (in which the nerve supply to one thymic lobe is surgically interrupted) do not lead to identifiable changes in thymus function, at least in the short term; thus, the physiological importance of the thymic nerve supply is unclear.

B. Thymocytes

Almost all thymocytes are immature T cells, which differ from peripheral T cells in that they are functionally incompetent. In the mouse, they express a distinctive pattern of cell surface markers: low MHC class I, high Thy-1, and binding to the lectin peanut agglutinin (1). In the human, they express a high level of CD1, a molecule structurally homologous to major histocompatibility complex (MHC) class I (HLA-A, B, and C) molecules, but encoded on chromosome 1, unlinked to the human leukocyte antigen *(HLA)* complex on chromosome 6 (2).

Thymocytes exhibit a pattern of expression of the principal T-cell antigen recognition molecules, the T-cell antigen receptor (TCR), CD4, and CD8, which is strikingly different from that of peripheral T cells. Although peripheral T cells express approximately 30,000–40,000 TCR molecules, and either CD4 or CD8, but not both, most thymocytes express both CD4 and CD8, and tenfold lower levels of TCR (3). However, there are minor subsets of thymocytes that share the phenotypic and functional characteristics of peripheral T cells (4,5). The mature thymocytes are found mainly in the medulla and are usually interpreted as being cells at the end of their intrathymic differentiation process and ready to exit to the periphery.

C. Intrathymic Precursors

About 2–5% of thymocytes express neither CD4 nor CD8. These "double-negative" cells mostly lack TCR and contain most of the T-cell precursor activity (6,7). Cells in this population will home back to the thymus after intravenous injection. The rearrangement of the TCR β-locus seems to occur in this cell population, and the rearrangement of α starts, but is completed later (8). It has recently been found that the very earliest cells in the mouse thymus express a low level of CD4, which is lost early in development and then regained as the cells mature (9). Early human thymocytes express the marker CD7, which is also observed on cells in the connective tissue surrounding the developing fetal thymus (10). These have been interpreted as progenitors in the process of seeding the thymus. During the double-negative stage, developing mouse thymocyte cells transiently express the low-affinity (p55) chain of the interleukin (IL)-2 receptor (11), which has led to the speculation that IL-2 plays a role early in T-cell maturation. However, treatment of thymocytes with IL-2 in vitro favors the expansion of mature cells, rather than the maturation of progenitors. In addition, mice in which the IL-2 gene has been disrupted by homologous recombination in embryonic stem cells, resulting in no IL-2 production, have normal αβ-T-cell differentiation in the thymus (12). Thus, it seems likely that the role of the IL-2R p55 chain on early thymocytes has little to do with αβ-T-cell development (although an effect on γδ cells is not excluded); but it is possible that there is a set of cytokines with overlapping functions, such that, in IL-2-deficient animals, another growth factor is able to substitute. Alternatively, the molecule may have a function that does not involve IL-2. One possibility is that it forms part of a different receptor. Other examples of this kind of multiple function of a single receptor chain are the occurrence of the CD3ζ chain in the FcεR (13), and the sharing of the cytokine receptor p130 chain between the IL-3, IL-5, and granulocyte–macrophage colony-stimulating factor (GM-CSF) receptors (14).

The maturation stage of progenitor cells when they reach the thymus is a matter of debate. In the mouse, CD4-low early progenitor cells have been reported to express progenitor activity for both T- and B-cell lineages, but not for myeloid cells (9). Although purified bone marrow stem cells deposited in the thymus will give rise to a burst of T-cell differentiation, this is preceded by a transient wave of myeloid cells; the same is not true for early thymocytes (15). Such evidence suggests that stem cells become partly committed to the T-cell lineage before migrating to the thymus; but the prethymic progenitor T cells, the existence of which is implied by this line of argument, have not yet been identified.

III. THE THYMOCYTE MATURATION PATHWAY

In the mouse, the earliest thymocytes are CD4-low, $CD8^-$, and TCR^-, whereas, in humans, the expression of CD4 on early intrathymic progenitor cells has not been documented. Within the $CD4^-$,$CD8^-$ thymocytes, several distinct progenitor subsets have been identified and, in the mouse, the sequence of subsets is known. The earliest $CD4^-$,$CD8^-$ cells express the marker CD44 (Ly-24, Pgp-1), which is a receptor for hyaluronic acid (16,17). As the cells mature, this cell surface molecule is down-regulated, and the IL-2R p55 chain is expressed (11,18). As these phenotypic changes occur, TCR gene rearrangement is initiated. The β, γ, and δ gene segments appear to rearrange synchronously, whereas rearrangements of TCRα are delayed (see following section). A little later, the cells begin to express CD8 and CD4, usually, but not always, in that order. This results (in sheep, rat, and mouse) in a short-lived population of immature $CD8^+$,$CD4^-$ cells (19–21); in the mouse a similar, but smaller, population of $CD8^-$,$CD4^+$ immature cells has been described (22). These cell should not be confused with the very early progenitor cells, from which they differ in other surface markers, developmental kinetics, and TCR gene rearrangement status. As the cells transit from $CD4^-$,$CD8^-$ to $CD4^+$,$CD8^+$, αβTCR appears at very low density on the cell surface. The proportion of $CD4^+$,$CD8^+$ cells estimated to express αβTCR is increasing steadily as the sensitivity of detection methods improves. Some of the uncertainty results from the use of monoclonal antibodies against different components of the TCR; it is no longer assumed that the β chain of the TCR, the ε chain of CD3, and a single α chain are always coexpressed on all categories of thymocytes. Nevertheless, it is clear that most of the cells in the thymus have the $CD4^+$,$CD8^+$, αβTCR-low phenotype. These cells are subject to selective events that shape the T-cell repertoire; after 2–3 days they either undergo cell death by apoptosis, or they mature into either $CD4^+$ or $CD8^+$ single-positive T cells. As this final maturation step occurs, the surface density of αβTCR increases approximately tenfold. Intermediate cells between typical $CD4^+$, $CD8^+$ cells and mature thymocytes have been described; these cells have either a $CD4^+$,CD8-low or a CD4-low, $CD8^+$ phenotype, and express a higher level of TCR (23,24). It appears that the high level of TCR expression characteristic of medullary thymocytes is achieved before the single-positive coreceptor phenotype. The main features of thymocyte development are summarized in Figure 1, and the early maturation processes are shown in more detail in Figure 2.

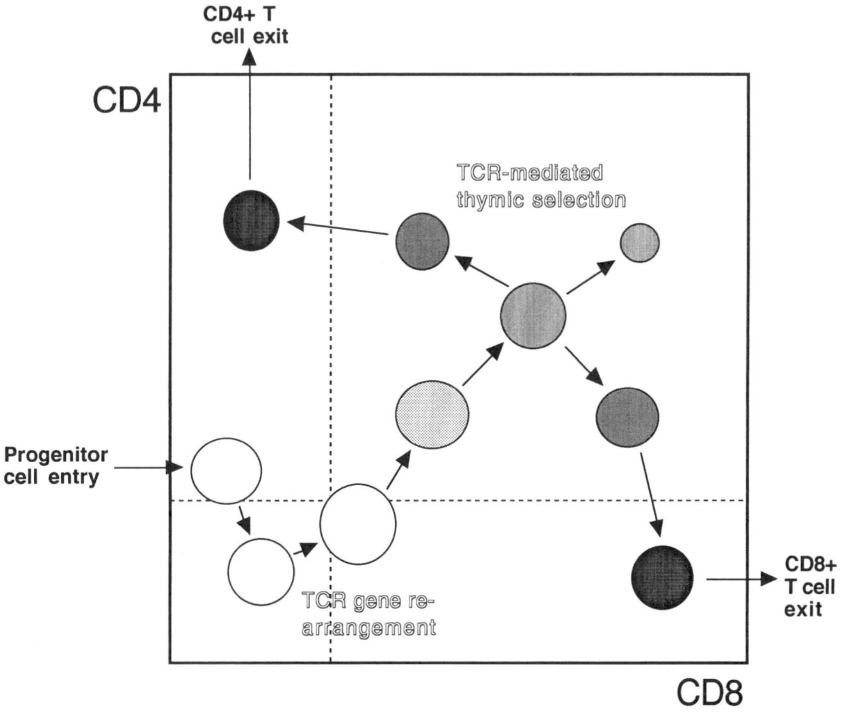

Figure 1 A schematic that illustrates the progression of thymocytes from a CD4-low early progenitor stage, through $CD4^-$,$CD8^-$ cells, and then to the "selectable" cortical thymocytes which are $CD4^+$,$CD8^+$ and express low levels of the T-cell antigen receptor. This phenotypic maturation is accompanied by rearrangement of TCR genes, which starts at the $CD4^-$,$CD8^-$ stage. Selection acts on $CD4^+$,$CD8^+$ cells. After selection, many of the cells become committed to die in the thymus by apoptosis, whereas a few become either $CD4^+$ or $CD8^+$ mature T cells. The density of shading indicates the surface level of αβTCR expression.

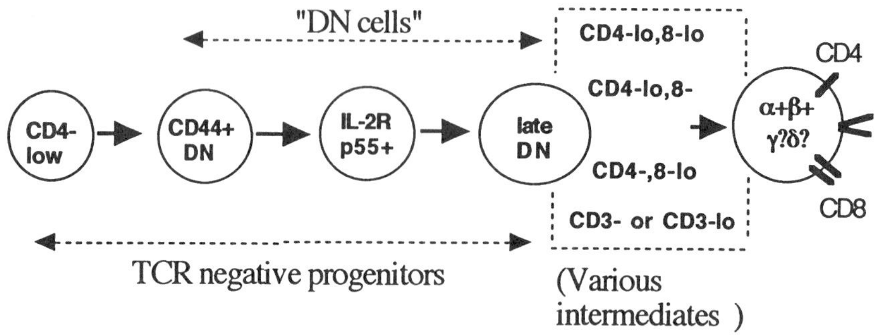

Figure 2 The MHC-independent early maturation processes in thymocyte differentiation. The phenotypic transition from CD4-low progenitor cells to $CD4^+$,$CD8^+$ cortical thymocytes is accompanied by transient expression of the p55 chain of the IL-2 receptor (CD25), and includes a number of transitional cell types, the existence of which argues that the order of expression of CD4,CD8 and the αβTCR is not fixed.

A. T-Cell Receptor Diversity

Most T cells in the thymus express TCR of the αβ type. These are the TCR expressed on peripheral $CD4^+$ and $CD8^+$ T cells found in the circulation, in lymph nodes, and in the spleen. However, a minor population of thymocytes and of circulating T cells expresses a distinct set of TCR, termed γδ receptors, which are assembled by the recombination of different genes. T cells with γδ receptors are mostly $CD4^-$,$CD8^-$, although $CD8^+$ peripheral γδ cells and thymic $CD4^+8^+$ γδ cells occur. They are rare in blood, lymph nodes, and spleen; but they are more abundant in the fetal thymus, at early developmental stages, and in a variety of epithelia: the skin (at least in mice), the uterus, and the gastrointestinal tract (25). The relative proportion of αβ and γδT cells differs dramatically among species; sheep, cattle, and chickens have many more circulating γδT cells than mice and humans. The function of γδT cells is unknown. Evidence is accumulating that some of them may recognize stress-induced proteins, such as heat-shock proteins, and that they interact with non-classical MHC class 1b molecules (CD1 in human, TL in mouse) (26–29). One suggestion is that they recognize not foreign antigens, but cellular proteins, which are up-regulated in response to various types of injury. These cells are prominent in the intestinal epithelium in celiac disease; and in lesions associated with several intracellular bacterial parasites. The pathogenic or protective role of γδ cells in any of these situations remains to be demonstrated.

B. Intrathymic γδ-Cells

A small subset of adult $CD4^-$,$CD8^-$ thymocytes express γδTCR (30,31). The γδ cells found in the thymus differ throughout life. In the mouse, a population of

$\gamma\delta$ cells are the earliest TCR-bearing cells to be seen in the fetal thymus; these cells express a monomorphic antigen receptor that is not seen in the adult thymus, but they migrate to the skin where they take up residence as dendritic epidermal (dEC) T cells (32,33). A second wave of $\gamma\delta$-cells is seen a few days later, followed by the expansion of $\alpha\beta$ T cells at about fetal day 16. In the adult thymus, $\gamma\delta$ cells express a subset of TCR γ-genes distinct from those expressed in the fetus, and also distinct from those $\gamma\delta$-cells found in epithelia or peripheral lymphoid organs (25). These adult intrathymic $\gamma\delta$-cells are competent to synthesize IL-2 and can be induced to kill target cells when linked to them through their $\gamma\delta$-antigen receptor; but their physiological function in the thymus is completely unknown.

C. The $\alpha\beta/\gamma\delta$ Lineage Split

Since both $\alpha\beta$ and $\gamma\delta$T cells express CD3-associated receptors, whereas B cells do not, it is a reasonable proposal that the two types of T cells share a common progenitor that is not also a B-cell progenitor. The earliest cell population in the mouse thymus gives rise to $\alpha\beta$, $\gamma\delta$, and B cells in adoptive transfer (9). Later populations of thymocytes give rise to $\gamma\delta$ cells in vitro and $\alpha\beta$ cells in vivo (34); but these experiments relate to cell populations, and the problem of whether individual cells within these populations are multipotential is still unresolved.

One possible mechanism for the control of the differentiation of progenitors into either $\alpha\beta$ or $\gamma\delta$ cells is the regulation of rearrangement of the α/δ locus (31). This complex genetic locus encodes the TCR δ chain, the gene segments for which are nested inside the genes for the TCR α locus (see the preceding chapter). In transfection experiments, the α locus has been shown to contain "silencer" elements, the control of which may determine the course of α/δ locus rearrangement. An early hypothesis proposed that the developmental fate of progenitor cells was controlled by the success or failure of attempts to rearrange the γ and δ loci; in this model, successful rearrangements would commit the cells to a $\gamma\delta$-developmental fate, whereas failure would lead on to β- and α-rearrangement (31). After the structural organization of the α/δ locus was understood, attempts were made to test one version of this model by examining the δ gene segments excised as circular DNA fragments during the process of α rearrangement. Two studies of this kind gave discordant results, so the truth remains in doubt (35,36).

D. Rearrangement of β, γ, and δ Genes

The CD4-low very early thymocytes identified in the adult mouse thymus have both γ and β genes in germline configuration. Several subsets of $CD4^-$,$CD8^-$ thymocytes have been examined for full-length (1.3-kb) mRNA for TCRβ; the IL-2R $p55^+$ cells express both 1.0 and 1.3-kb mRNA, suggesting that some of the cells are transcribing from *DJ* rearrangements, while others have complete *VDJ* rearrangements. How many of these rearrangements are in-frame and translatable is not known; nor is the fraction of these cells with at least one β gene segment still in germline configuration.

E. Rearrangement and Expression of α Genes

In the mouse fetus, expression of mRNA for the TCR γ, δ, and β chains is found from the earliest stages of thymus development, but the appearance of TCRα mRNA is delayed by several days (37). Similarly, in the adult, some early thymocyte subsets have β, but not α, rearrangement (8,38). This temporal arrangement is reminiscent of the situation in pre-B cells, in which rearrangement of the immunoglobulin heavy (H) chain precedes rearrangement of the light (L) chain, and a complex of H chain and "surrogate L chains" is expressed transiently and at low levels on the surface of pre-B cells (39). It has been proposed that this complex is involved in a signaling event that controls the termination of H chain rearrangement, and initiates L chain rearrangement (40).

In T cells, a population of early thymocytes has been described in TCR transgenic mice, which comes somewhere between the $CD4^-$,$CD8^-$ precursors and the TCRαβ-low, $CD4^+$,$CD8^+$ cortical thymocytes in the developmental sequence and expresses CD3-associated TCR β chain without α (41). Cells of this kind have recently been described even in mice in which the *C*α gene has been inactivated by genetic manipulation. Although it is too early to be sure of the function of these atypical T-cell receptors, they could serve to transduce a signal that a successful β rearrangement has occurred, preventing further β rearrangement or promoting TCR α chain rearrangement.

Fetal thymocytes express cytoplasmic CD3ε, and it seems that all of the other components of the αβTCR are synthesized before cells complete the rearrangement of the α locus. Once α is rearranged and expressed, αβTCR can be assembled and expressed on the thymocyte surface.

IV. SELECTION OF THE T-CELL REPERTOIRE

A. Immunological Tolerance

The random recombination events that generate both B- and T-cell repertoires create a problem of self-tolerance: since the repertoire is *a priori* competent to recognize anything, what happens to the cells with the potential to recognize self-components?

Antigens may be sequestered from the immune system, so that the possibility of antigen recognition does not occur. This might apply to antigens in tissues without a vascular supply, such as the lens of the eye. If T cells are able to encounter self-molecules, the result may be not T-cell activation, but the induction of a form of nonresponsiveness (anergy) that renders the T cell unable to respond to the same antigen, even if it is subsequently presented under the most favorable conditions.

Antigen sequestration and the induction of anergy both allow self-reactive precursor T cells to coexist with self-antigens. An alternative model to account for self-tolerance is the physical removal of self-reactive cells; this is termed *clonal deletion* (42). In the thymus, self-tolerance results partly from a process of clonal deletion, in which self-reactive cells are physically removed from the repertoire and die in the thymus by apoptosis. This process occurs when nonmature cortical-type thymocytes interact with antigen presented by bone marrow-derived dendritic cells, macrophages, or T cells (Fig. 3). Intrathymic clonal deletion is described in more detail later.

It has also been proposed that some self-tolerance is mediated by suppressor T cells, a T-cell subset that has been supposed to actively suppress antigen-specific responses in an antigen- or idiotype-specific way (43). Although a voluminous literature on suppressor T cells exists, many of the experiments have been impossible to accommodate within the modern synthesis, and the field is very confused. Many of the experiments, particularly the in vitro assays that have been used to justify claims for the involvement of suppressor T cells in human disease, are explained by mechanisms of doubtful in vivo significance, such as the consumption of cytokines by preactivated cells. A detailed critical review of this field is out of place here, but the debate has been lively in the columns of *Immunology Today* and can be followed there (44,45). One suppressive mechanism, with better credentials than most, is the mutually inhibitory activity of the two functionally distinct subsets of $CD4^+$ T cells. Inflammatory $CD4^+$ T cells secrete IL-2 and interferon gamma (INF-γ), and tend to inactivate helper $CD4^+$ cells (46). Conversely, $CD4^+$ helper cells, which secrete IL-4 and IL-5, tend to inactivate inflammatory cells. Much of the in vivo literature on T-cell suppression may be explained by activation of the inappropriate functional subset of $CD4^+$ T cells. Evidence for this kind of suppression seems very clear in experimental infectious diseases, such as leishmaniasis (47), and it may also operate in human infectious diseases.

B. The Thymus Mediates Experimental Tolerance

The contribution of the thymus to the induction of tolerance has been studied by thymus grafting in mice. Such grafting experiments show that transfer of an $(A\times B)F_1$ (MHC heterozygous) neonatal thymus graft into an irradiated, bone marrow-reconstituted A host gives an animal that is fully tolerant to B antigens as well as A antigens. However, a bone marrow transplant of $(A\times B)F_1$ marrow into an A host also leads to tolerance to antigens of B type (48). Grafting of a

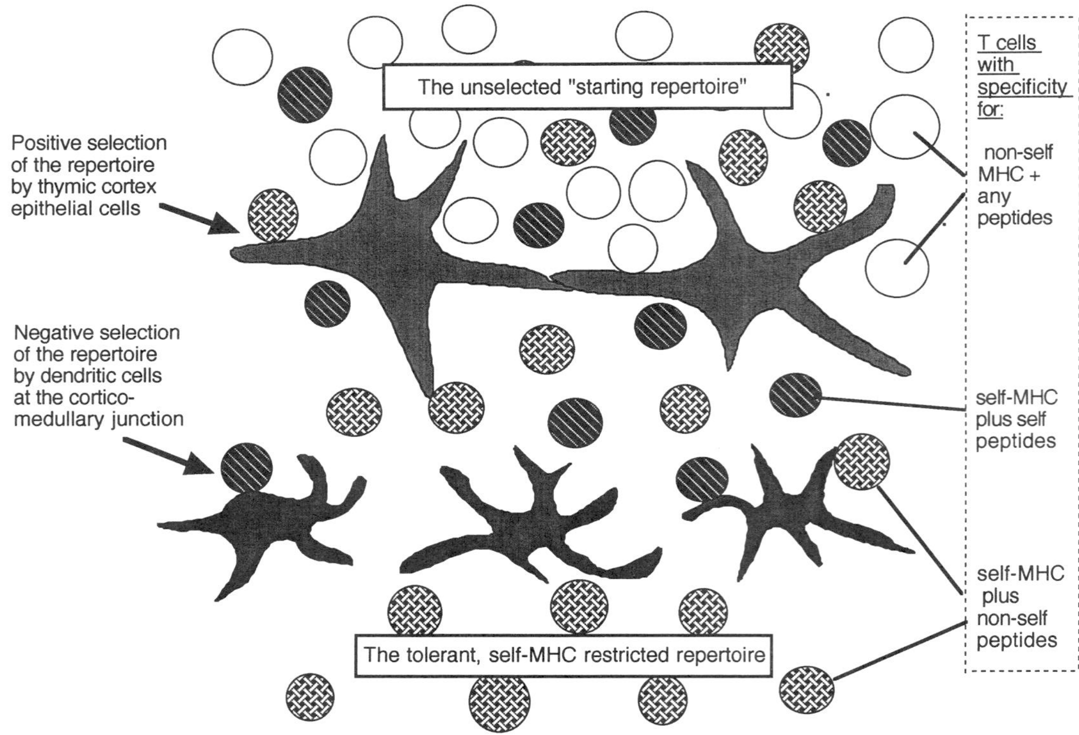

Figure 3 Selection of the T-cell repertoire in the thymus depends on confrontations between $CD4^+$,$CD8^+$ thymocytes and selecting elements of the thymic stroma, principally thymus epithelial cells and dendritic cells. In the diagram, the mixture of thymocytes at the top represents the unselected repertoire of receptors generated by gene recombination. The implication that positive selection precedes negative selection is likely, but is unproven.

thymus depleted of bone marrow-derived cells does not induce tolerance to graft antigens (49). This has led to the current consensus that bone marrow-derived cells, acting in the thymus, are prime mediators of tolerance.

Tissue grafts of thymic epithelium alone are not rejected, but do not induce systemic tolerance measured by mixed lymphocyte reaction (MLR) and cytotoxic T-lymphocytes (CTL) (50). Nevertheless, thymic epithelium grafts express MHC class I and class II. Furthermore, isolated thymic epithelial cells can present antigens to T cells in vitro (51). The antigen-presenting properties of thymic epithelial cells are under intense scrutiny because of the importance of these cells in positive selection (see later discussion).

C. Direct Evidence for Clonal Deletion

Several T-cell receptor Vβ gene products can be identified using monoclonal antibodies. One of these, Vβ17a, is expressed on peripheral T cells only in mice that do not express the *I-E* locus of the MHC. However, I-E positive mice express Vβ17a at low level on immature thymocytes (52,53). The implication is that Vβ17a T cells have a high affinity for I-E molecules, and are deleted as they mature in I-E$^+$ animals. Several other TCR Vβ regions are associated with reactivity to I-E, or occasionally I-A, in association with an "endogenous superantigen," a molecule encoded by an open-reading frame in the long-terminal repeat (LTR) of an endogenous retrovirus (54–57). These viruses are widespread in both inbred laboratory mice and wild mice, and dramatically shape the T-cell repertoire. Similar effects on the rat or human T-cell repertoires, for which there are some but fewer anti-TCR antibodies, have not been documented. If endogenous superantigens are, in fact, found to modify the human T-cell pool by clonal deletion, this could be an inherited determinant of disease susceptibility.

The details of clonal deletion differ between experimental systems, but in general, self-reactive TCR-expressing cells disappear at the transition between TCR-low immature thymocytes and TCR-high mature cells.

D. Clonal Deletion Depends on T-Cell Receptor, CD4, and CD8

Saturation of the thymus with anti-TCR antibodies mimics clonal deletion by self-antigens and causes global deletion of all developing T cells (58). Deletion is characterized by nuclear pyknosis and early DNA fragmentation (apoptosis). When T cells that express a particular Vβ are deleted from the mature thymocyte pool, they disappear from both CD4$^+$ and CD8$^+$ mature cells, arguing that deletion happens before the cells are committed to become either CD4$^+$ or CD8$^+$ single-positive cells. In support of this model, saturation with anti-CD4 blocks clonal deletion at the CD4$^+$,CD8$^+$ stage; CD4$^+$ single-positive cells do not appear, but TCRs that are normally deleted become visible in the $CD8^+$ single-positive cells (59).

E. Clonal Deletion in Transgenic Models

Until the late 1980s, functional analysis of the control of the T-cell repertoire had been limited by the problem that antigen-reactive and self-restricted cells could be identified only by functional assays. These experiments were always open to several interpretations. For example, the lack of T cells reactive to self-antigens could be due to either the absence of such self-reactive cells, or to their suppression by another T-cell population.

This deadlock was broken by the analysis of mice that express a single TCR as a result of the introduction of rearranged TCR α and β chain genes into the germline. The presence of rearranged genes inhibits the rearrangement of endogenous TCR genes; and the small number of cells that, nevertheless, do rearrange endogenous TCR genes can be removed by back-crossing the TCR transgenes onto a homozygous *scid* (severe combined immunodeficiency) background. In these mice, all T cells behave in the same way, and their developmental fate is determined by the thymus in which they differentiate.

In TCR transgenic mice the fate of developing thymocytes may be studied in mice that either lack or express the antigen for which the transgenic TCR is specific. Antigen-negative mice often have normal-sized thymi, but the thymocyte phenotypic pattern is skewed toward mature TCR-high cells of the $CD4^+$ or $CD8^+$ single-positive phenotype. Antigen-positive mice typically have very small thymi, with thymocytes strongly skewed toward the $CD4^-$,$CD8^-$ or TCR^- phenotype (60,61). Antigen-negative mice have a very high T-cell precursor frequency toward the antigen in question; antigen-positive mice are antigen-tolerant in functional experiments. An experiment that illustrates clonal deletion of thymocytes is shown in Figure 4.

F. Split Tolerance

Antigens expressed on bone marrow-derived cells typically cause complete tolerance of MHC class I-restricted cells (i.e., CTLp) and class II-restricted cells (the major contributors to the MLR). However, expression of antigens on nonhemopoietic tissues or on T cells may cause a different pattern, in which the MLR proliferative response is intact, but the CTL response is abrogated. This is known as split tolerance (62). This form of tolerance is commonly accompanied by inability to reject skin grafts and may be induced by the expression of foreign MHC on various tissue cells in transgenic mice (63), or by the grafting of leukocyte-depleted tissue. The mechanisms are unknown, but one hypothesis is that nonlymphoid tissues induce incomplete tolerance of $CD4^+$ cells, because the deletion mechanism (like T-cell activation) depends on a “second signal”

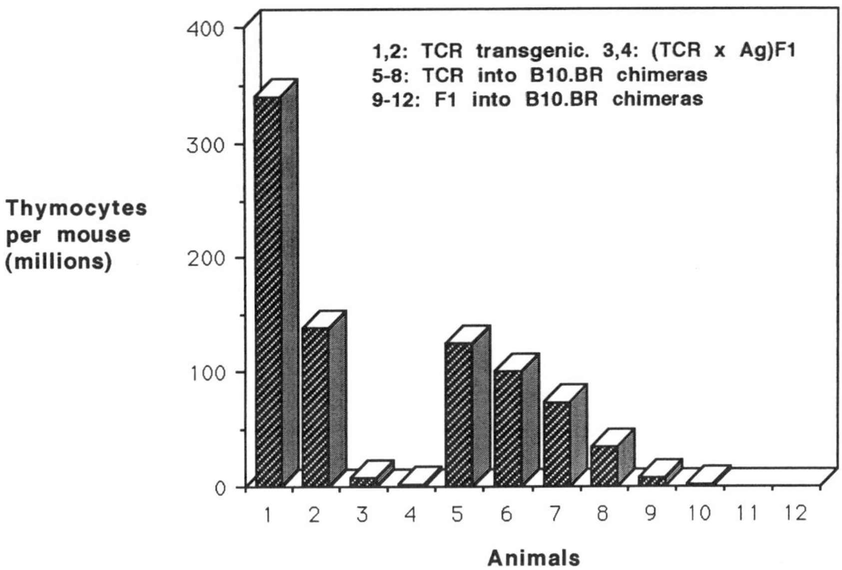

Figure 4 An experiment that illustrates clonal deletion in the thymus, caused by the expression of antigen on bone marrow cells. Radiation chimeras were prepared in nontransgenic hosts, using different sources of donor bone marrow. Thymocyte numbers are shown for TCR transgenics that recognize SV40 large-T antigen (animals 1,2); F_1 hybrids between TCR-transgenics and SV40 large-T transgenics (animals 3,4); chimeras in a positively selecting host, and TCR-transgenic bone marrow (animals 5–8); and chimeras made with bone marrow from double transgenics that expressed both the anti-large-T TCR and the large-T antigen (animals 9–12). The data are from a collaborative project with T. Geiger and R. A. Flavell.

alongside TCR ligation. A number of candidate molecules have been identified that may transmit the second-signal; one of the most promising candidates is CD28, a T-cell molecule that interacts with a known ligand, B7, on antigen-presenting cells (64). Anti-CD28 antibodies have costimulatory activity and can abrogate the induction of anergy.

G. The Major Histocompatibility Complex Restriction of T-Cell Antigen Recognition

T cells interact with antigens only in the form of antigenic peptides anchored in the peptide-binding groove of MHC molecules. T cells that express the CD8 coreceptor typically recognize nine amino acid peptides bound to class I molecules of MHC, whereas T cells that express CD4 recognize longer peptides of 11–13 (occasionally more) amino acids bound to MHC class II molecules (65). The TCR makes contact with a surface consisting of amino acid side chains of the peptide, flanked by α-helical regions of the MHC molecule (66). The specificity of TCRs is determined both by the peptide and the MHC molecule. Thus, a single amino acid substitution at a critical residue on the peptide may destroy the antigenic epitope recognized by a T cell; but so may the allelic difference between two MHC alleles. T-cell specificity is described as MHC-restricted because a single T cell recognizes an antigenic peptide only in the context of one, or several closely related MHC molecules. There is evidence that the peripheral T-cell precursor pool is skewed toward the recognition of antigen in the context of those allelic forms of MHC molecules expressed in the organism (self-MHC), rather than in other members of the species; and that the imposition of this skewing is one of the functions of the thymus.

H. The Thymus Controls Major Histocompatibility Complex Restriction

Both $CD4^+$ and $CD8^+$ T cells emerge from the thymus with the capacity to recognize antigen in association with self-MHC. In the late 1970s, thymus-grafting experiments in chimeric mice suggested that the thymus imposes restriction-specificity on T cells as they develop. In such experiments, F_1 mice were lethally irradiated, thymectomized, grafted with a parental thymus, and reconstituted with F_1 bone marrow. T-cell restriction-specificity was skewed toward the MHC haplotype of the thymus graft (67,68). This skewing is termed *positive selection* of the repertoire.

The importance of thymus epithelium in positive selection was first demonstrated in experiments grafting isolated thymus epithelium. Mouse fetal thymus lobes, isolated from day 14–15 embryos, may be depleted of all hemopoietic cells by a 5-day culture in 2-deoxyguanosine (69). The surviving lobe consists entirely of thymus epithelium. Such lobes may be grafted into nude or thymectomized mice, where they are colonized by hemopoietic cells and support thymocyte differentiation. The T cells that develop in the grafted lobe are functionally tolerant to the graft (i.e., it is not rejected), but fully responsive to graft antigens in MLR, and will differentiate to form cytotoxic T cells against target cells of graft type. The MHC restriction of T cells, both in the graft and in the periphery of grafted animals, is controlled by the MHC of the epithelium (70). This argues that in the normal intact thymus, epithelial cells, and not bone marrow-derived cells, are dominant in imposing positive selection. This principle is illustrated in Figure 3.

I. Positive Selection in T-Cell Receptor Transgenic Mice

As in the study of clonal deletion, TCR transgenic mice have greatly clarified the process of positive selection by allowing the unambiguous identification of T cells of known MHC restriction-specificity in different environments. The phenomenology of positive selection differs in detail between TCR transgenic systems, but the following general rules emerge. T cells that are neither deleted nor positively selected develop to a stage at which they express both CD4 and CD8, and a level of TCR intermediate between that of cortical and medullary cells in normal mice. The thymus is typically of near-normal size, suggesting that most thymocyte proliferation is independent of positive selection. However, the cells do not make the transition to the TCR-high, $CD4^+$,$CD8^+$ stage, and die by apoptosis (61,71,72).

If the TCR transgenic thymocytes develop in a thymus that expresses the MHC molecule that is the restriction element for the transgenic antigen receptor, the thymocytes develop beyond the $CD4^+$,$CD8^+$ stage; essentially all of the mature cells are either $CD4^-$,$CD8^+$, TCR-high if the receptor is restricted by MHC class I, or $CD4^+$,$CD8^-$, TCR-high if the receptor is restricted by class II. These cells do not die, but are exported to the periphery where the mouse exhibits a very high precursor frequency for the antigen recognized by the transgenic TCR (60). In line with older functional experiments, transfer of TCR transgenic bone marrow into lethally irradiated recombinant inbred hosts has shown that positive selection is controlled by radioresistant components of the thymus (71).

J. Molecular Models of Positive Selection

The main conceptual problem in positive selection is to explain how a cell can be selected for the capacity to recognize previously unknown antigens in the context of self-MHC, and at the same time selected against reactivity with self-antigens plus self-MHC. Several models have been proposed.

The *affinity model* proposes that developing T cells are positively selected by self-molecules across a very wide range of affinity and that only the higher-affinity cells are subsequently removed by clonal deletion. In this model, the intermediate- to low-affinity cells are not functionally self-reactive in the periphery. They are able to respond to exogenous antigens because of a heteroclitic interaction with exogenous antigenic peptides plus self-MHC. This model implies that positive and negative selection use the recognition of the same set of self-peptides in two different ways, with different affinity requirements. To test these ideas, much effort has been put into developing soluble forms of the T-cell receptor so that affinities can be measured. These molecules do not as yet exhibit detectable binding to MHC plus antigen, and the affinity of the T-cell receptor remains unknown. One problem with the affinity model is that the cells that are supposed to be selected at very low affinity express a very low surface density of TCR. How can very few receptors signal, if the affinity is so small?

An alternative model proposes that thymic epithelium displays a unique set of *selecting peptides* that are not part of peripheral self. These peptides might be inert in the selection process, with positive selection occurring on the flanking α-helical regions only; or they might be involved in the interaction. If so, it has been proposed that they act as templates that select at a high affinity cells that will interact with exogenous antigens. Staining of sections of mouse thymus with an antibody against a peptide derived from I-E and presented by I-A, which recognizes an epitope differentially expressed between the thymic cortex and the medulla (69), supports this view.

V. ROLE OF THE THYMUS IN DISEASE STATES

A. Defects of Self-Tolerance

Autoimmune diseases are, by definition, caused by an immune response against self-components. Since T cells control B cells in most immune responses, one might expect both antibody-mediated autoimmune diseases (such as Graves' thyroiditis) and diseases in which the primary abnormality is T-cell-mediated (such as juvenile-onset diabetes) to result from a failure of effective self-tolerance in the thymus. However, concrete evidence for diseases caused by this mechanism has not been forthcoming. In the murine systemic autoimmune disease that occurs in MRL-Mp-+/+ mice, self-reactive T cells have been described in peripheral lymphoid tissues (73), but the origin of these cells is not yet clarified. In humans, the problem of identifying self-reactive T cells is much more acute; there are now no monoclonal antibodies that identify such cells. As more monoclonal anti-human TCR reagents become available, this situation may improve. With current reagents and techniques, self-reactive T cells must be detected by functional assays. If such cells are found, it is unclear whether they are due to a failure of thymic tolerance, or to defective regulation in the periphery.

B. Pathogens Masquerading as Self

Developing T cells that encounter antigen in the thymus become tolerant. It follows that pathogens that can gain access to the thymus can tolerize the T-cell pool at source, a dangerous situation, since no new pathogen-specific T cells will be produced. No examples of this mechanism are known in humans, but in mice, the lymphocytic choriomeningitis virus (LCMV) evidently gains access to the thymus. This has been shown in TCR transgenic mice, in which all of the T cells express a TCR specific for an LCMV antigen (74). In these animals, neonatal LCMV infection results in clonal deletion in the thymus. Likewise, in other TCR transgenic mice, systemic injection of the antigenic peptide for which the transgenic TCR is specific results in deletion of thymocytes (75).

Since thymus biopsy is not indicated for diagnostic reasons, there is no information on the effects of acute infections on the thymus in humans. The situation is complicated by the thymic atrophy that results from the stress of severe illness; but in principle, it should be possible to obtain samples of thymocytes from victims of fatal acute infections and to test them for specific unresponsiveness to the pathogen. Animal experiments with large tracer molecules have shown that these molecules are excluded from the thymic cortex, whereas antibodies are not (76). This has given rise to the notion of a selectively permeable blood–thymus barrier, analogous to the blood–brain barrier. To understand fully their interactions with the T-cell system, it will be important to determine whether human pathogens respect or ignore this barrier.

C. Myesthenia, Thymomas, and Thymectomy

The autoimmune disease myesthenia gravis, which is caused by autoantibodies directed against motor end plates on muscle cells, particularly against the acetylcholine receptor (AChR), has a poorly understood association with thymic pathology. Many myesthenic patients have a nonmalignant hyperplasia of the thymus ("thymitis") (77); a smaller percentage develop thymic lymphomas. Strangely, complete surgical removal of the thymus frequently results in remission of the myesthenia. The effectiveness of thymectomy varies with patient age and sex; the operation is most often beneficial in young female patients. In contrast, patients with thymoma benefit from thymectomy only because of the removal of the tumor; their myesthenia does not regress.

This association has led to the proposal that the thymus either produces anti-AChR antibodies, or it is the site of the expression of the antigen that incites a systemic autoimmune response. In favor of the former proposal, the thymus of myesthnic patients contains germinal centers full of proliferating B cells. It remains to be shown that these B cells are making anti-AChR antibodies. The human thymus contains a small population of cells of unknown function, which resemble smooth muscle. These are termed myoid cells. Current research is aimed at the identification of myoid cell antigens which cross-react with the AChR, and might be inducing the autoimmunity (78).

Whatever the mechanism of the therapeutic effect of thymectomy, the surgical treatment of myesthenia has created a large population of patients in whom the peripheral T-cell pool is no longer replenished by emigrants from the thymus. In the short-term, these patients retain immunological competence, which argues that precursor, as well as memory T cells can be long-lived in this situation. Detailed long-term immunological analysis of thymectomized myesthenic patients would provide fascinating information on the importance of the thymus in adult life, the long-term life history of peripheral T cells, and the role of extrathymic T-cell maturation in humans.

FURTHER READING

M. A. Ritter, and I. N. Crispe, *The Thymus*. IRL Books, Oxford University Press (1992).

REFERENCES

1. R. Scollay, P. Bartlett, and K. Shortman, T cell development in the adult murine thymus: Changes in the expression of the antigens Ly2, L3T4 and B2A2 during development from early precursors to emigrants, *Immunol. Rev.*, *82*: 79 (1984).
2. F. Cabali, and C. Milstein, A novel family of major histocompatibility complex related genes not mapping to chromosome 6, *Nature*, *323*: 540 (1986).
3. J. A. Bluestone, D. Pardoll, S. O. Sharrow, and B. J. Fowlkes, Characterization of murine thymocytes with CD3 associated T cell receptor structures, *Nature*, *326*: 82 (1987).
4. R. Ceredig, A. L. Glasebrook, and H. R. MacDonald, Phenotypic and functional properties of murine thymocytes. 1. Precursors of cytolytic T lymphocytes and interleukin-2 producing cells are all contained within a population of mature thymocytes as analyzed by monoclonal antibodies and flow microfluorimetry, *J. Exp. Med.*, *155*: 358 (1982).
5. R. Ceredig, D. P. Dialynas, F. W. Fitch, and H. R. MacDonald, Precursors of T cell growth factor producing cells in the thymus: Ontogeny, frequency and quantitative recovery in a subpopulation of phenotypically mature thymocytes defined by monoclonal antibody GK1.5, *J. Exp. Med.*, *158*: 1654 (1983).
6. B. J. Fowlkes, L. Edison, B. J. Mathieson, and T. M. Chused, Early T lymphocytes: Differentiation in vivo of adult intrathymic precursor cells, *J. Exp. Med.*, *162*: 802 (1985).
7. I. N. Crispe, M. W. Moore, L. A. Husmann, L. Smith, M. J. Bevan, and R. P. Shimonkevitz, Differentiation potential of subsets of $CD4^-$,$CD8^-$ thymocytes, *Nature*, *329*: 336 (1987).
8. M. Pearse, P. Gallagher, A. Wilson, L. Wu, N. Fisicaro, J. F. A. P. Miller, R. Scollay, and K. Shortman, Molecular characterization of T-cell antigen receptor expression by subsets of $CD4^-$,$CD8^-$ murine thymocytes, *Proc. Natl. Acad. Sci. USA*, *85*: 6082 (1988).
9. L. Wu, R. Scollay, M. Egerton, M. Pearse, G. J. Spangrude, and K. Shortman, CD4 expressed on earliest T-lineage cells in the adult murine thymus, *Nature*, *349*: 71 (1991).
10. B. F. Haynes, M. E. Martin, H. H. Kay, and J. Kurtzberg, Early events in human T cell ontogeny. Phenotypic characterization and immunohistologic localization of T cell precursors in early human fetal tissues, *J. Exp. Med.*, *168*: 1061, (1988).
11. R. P. Shimonkevitz, L. A. Husmann, M. J. Bevan, and I. N. Crispe, Transient

expression of IL-2 receptor precedes the differentiation of immature thymocytes, *Nature*, *329*: 157 (1987).
12. H. Schorle, T. Holtschke, T. Hunig, A. Schimpl, and I. Horak, Development and function of T cells in mice rendered interleukin-2 deficient by gene targeting, *Nature*, *352*: 621 (1991).
13. U. Blank, C. Ra, L. Miller, H. Metzger, and J. P. Kinet, Complete structure and expression in transfected cells of high affinity IgE receptor, *Nature*, *337*: 187 (1989).
14. N. A. Nicola, and D. Metcalf, Subunit promiscuity among hemopoietic growth factor receptors, *Cell*, *67*: 1 (1991).
15. G. J. Spangrude, and R. Scollay, Differentiation of hematopoietic stem cells in irradiated mouse thymic lobes: Kinetics and phenotype of progeny, *J. Immunol.*, *145*: 3661 (1990).
16. R. Hyman, J. Lesley, R. Schulte, and J. Trotter, Progenitor cells in the thymus: Most thymus-homing progenitor cells in the adult mouse thymus bear Pgp-1 glycoprotein but not interleukin-2 receptor on their cell surface, *Cell. Immunol.*, *101*: 320 (1986).
17. A. Aruffo, I. Stamenkovic, M. Melnick, C. B. Underhill, and B. Seed, CD44 is the principal cell surface receptor for hyaluronidate, *Cell*, *61*: 1303 (1990).
18. M. Pearse, W. Li, M. Egerton, A. Wilson, K. Shortman, and R. Scollay, A murine early thymocyte developmental sequence is marked by transient expression of the interleukin-2 receptor, *Proc. Natl. Acad. Sci. USA.*, *86*: 1614 (1989).
19. L. A. Husmann, R. P. Shimonkevitz, I. N. Crispe, and M. J. Bevan, Thymocyte subpopulations during early fetal development in the BALB/c mouse, *J Immunol.*, *141*: 736 (1988).
20. D. J. Paterson, and A. F. Williams, An intermediate cell in thymocyte differentiation that expresses CD8 but not CD4 antigen, *J. Exp. Med.*, *166*: 1603 (1987).
21. C. R. MacKay, J. F. Maddox, and M. R. Brandon, Thymocyte subpopulations during early fetal development in sheep, *J. Immunol.*, *136*: 1592 (1986).
22. P. Hugo, G. A. Waanders, R. Scollay, H. T. Petrie, and R. L. Boyd, Characterization of immature $CD4^+CD8^-CD3^-$ thymocytes, *Eur. J. Immunol.*, *21*: 835 (1991).
23. C. J. Guidos, J. S. Danska, C. G. Fathman, and I. L. Weissman, T cell receptor-mediated negative selection of autoreactive T lymphocyte precursors occurs after commitment to the CD4 or CD8 lineages, *J. Exp. Med.*, *172*: 835 (1990).
24. P. Hugo, R. L. Boyd, G. A. Waanders, H. T. Petrie, and R. Scollay, Timing of deletion of autoreactive V beta 6^+ cells and down-modulation of either CD4 or CD8 on phenotypically distinct $CD4^+8^+$ subsets of thymocytes expressing intermediate or high levels of T cell receptor, *Int. Immunol.*, *3*: 265 (1991).
25. S. Tonegawa, A. Berns, M. Bonneville, A. G. Farr, I. Ishida, K. Ito, S. Itohara, C. A. Janeway, Jr., O. Kanagawa, M. Katsuiki, R. T. Kubo, J. J. Lafaille, P. Mombaerts, D. B. Murphy, N. Nakanishi, Y. Takagaki, L. Van Kaer, and S. Veebeek, Diversity, development, ligands and probable functions of gamma delta T cells, *Prog. Immunol.*, *7*: 243 (1989).
26. J. Holoshitz, F. Koning, J. Coligan, J. De Bruyn, and S. Strober, Isolation of $CD4^-CD8^-$ T lymphocyte clones from rheumatoid arthritis synovial fluid, *Nature*, *339*: 226 (1989).

27. R. O'Brien, M. H. Happ, A. Dallas, E. Palmer, R. Kubo, and W. K. Born, Stimulation of a major subset of lymphocytes expressing a T cell receptor gamma-delta by an antigen derived from *Mycobacterium tuberculosis*, *Cell*, *57*: 667 (1989).
28. M. Bonneville, K. Ito, E. G. Krecko, S. Itohara, D. Kappes, O. Kanagawa, C. A. Janeway, D. B. Murphy, and S. Tonegawa, Recognition of a major histocompatibility complex TL region product by gamma-delta T cell receptors, *Proc. Natl. Acad. Sci. USA*, *86*: 5928 (1989).
29. S. Porcelli, M. B. Brenner, J. L. Greenstein, S. P. Balk, C. Terhorst, and P. A. Bleicher, Recognition of cluster of differentiation 1 antigens by human $CD4^-CD8^-$ cytolytic T lymphocytes, *Nature*, *341*: 115 (1989).
30. A. M. Lew, D. M. Pardoll, W. L. Maloy, B. J. Fowlkes, A. Kruisbeek, S.-F. Cheng, R. N. Germain, J. A. Bluestone, R. H. Schwartz, and J. E. Coligan, Characterization of T cell receptor gamma expression in a subset of murine thymocytes, *Science*, *234*: 1401 (1986).
31. D. M. Pardoll, B. J. Fowlkes, J. A. Bluestone, A. Kruisbeek, L. M. Maloy, J. E. Coligan, and R. H. Schwartz, Differential expression of two distinct T-cell receptors during thymocyte development, *Nature*, *326*: 78 (1987).
32. W. L. Havran, and J. P. Allison, Developmentally ordered appearance of thymocytes expressing different T-cell antigen receptors, *Nature*, *335*: 443 (1988).
33. W. L. Havran, and J. P. Allison, Origin of Thy-1^+ dendritic epidermal cells of adult mice from fetal thymic precursors, *Nature*, *344*: 68 (1990).
34. H. T. Petrie, R. Scollay, and K. Shortman, Commitment to the TCR alpha-beta or gamma-delta lineages can occur just prior to the onset of CD4 and CD8 expression amongst immature thymocytes, *Eur. J. Immunol.*, (in press; 1992).
35. A. Winoto, and D. Baltimore, Separate lineages of T cells expressing the alpha beta and gamma delta receptors, *Nature*, *338*: 430 (1989).
36. K. Okazaki, and H. Sakano, Thymocyte circular DNA excised from T cell receptor alpha-delta gene complexes, *EMBO J.*, *7*: 1669 (1988).
37. D. H. Raulet, R. D. Garam, H. Saito, and S. Tonegawa, Developmental regulation of T cell receptor gene expression, *Nature*, *314*: 103 (1985).
38. M. L. K. Collins, G. Tanigawa, A. M. Kissonerghis, M. Ritter, K. M. Price, S. Tonegawa, and M. J. Owen, Regulation of T cell receptor gene expression in human T cell development, *Proc. Natl. Acad. Sci. USA*, *82*: 4503 (1985).
39. T. Tsubata, and M. Reth, The products of pre-B cell-specific genes ($\lambda 5$ and *VpreB*) and the immunoglobulin μ chain form a complex that is transported onto the cell surface, *J. Exp. Med.*, *172*: 973 (1990).
40. D. Kitamura, J. Roes, R. Kuhn, and K. Rajewsky, A B cell-deficient mouse by targeted disruption of the membrane exon of the immunoglobulin mu chain gene, *Nature*, *350*: 423 (1991).
41. H. Kishi, P. Borgulya, B. Scott, K. Karjalainen, A. Traunecker, J. Kaufman, and H. von Boehmer, Surface expression of the beta T cell receptor (TCR) chain in the absence of other TCR or CD3 proteins on immature T cells, *EMBO J.*, *10*: 93 (1991).
42. M. F. Good, K. W. Pyke, and G. J. V. Nossal, Functional clonal deletion of cytotoxic T lymphocyte precursors in chimeric thymus produced in vitro from embryonic anlagen, *Proc. Natl. Acad. Sci. USA*, *80*: 3045 (1983).

43. A. J. Cunningham, Self tolerance maintained by active suppressor mechanisms, *Transplantation*, *31*: 23 (1976).
44. H.-G. Rammensee, M. J. Bevan, and P. J. Fink, Antigen-specific suppression of T cell responses—the veto concept, *Immunol. Today*, *6*: 41 (1985).
45. I. N. Crispe, Mechanisms of tolerance, *Immunol. Today*, *9*: 329 (1989).
46. T. Mosmann, and R. L. Coffman, Th1 and Th2 cells: Different patterns of lymphokine secretion lead to different functional properties, *Annu. Rev. Immunol.*, *7*: 145 (1989).
47. A. Sher, and R. L. Coffmann, Regulation of immunity to parasites by T cells and T cell derived cytokines, *Annu. Rev. Immunol.*, *10*: 385 (1992).
48. J. Sprent, H. von Boehmer, and M. Nabholz, Association of immunity and tolerance to host H-2 determinants in irradiated F_1 hybrid mice reconstituted with bone marrow cells from one parental strain, *J. Exp. Med.*, *142*: 321 (1975).
49. H. von Boehmer, and K. Schubiger, Thymocytes appear to ignore class I major histocompatibility antigens expressed on thymus epithelial cells, *Eur. J. Immunol.*, *14*: 1048 (1984).
50. H. von Boehmer, and K. Hafen, Minor but not major histocompatibility antigens of thymus epithelium tolerize precursors of cytotoxic cells, *Nature*, *320*: 626 (1986).
51. R. G. Lorenz, and P. M. Allen, Thymic cortical epithelial cells lack full capacity for antigen presentation, *Nature*, *340*: 557 (1989).
52. J. Kappler, T. Wade, J. White, E. Kushnir, M. Blackman, J. Bill, N. Roehm, and P. Marrack, A T cell receptor Vβ segment that imparts reactivity to a class II major histocompatibility complex product, *Cell*, *49*: 263 (1987).
53. J. Kappler, N. Roehm, and P. Marrack, T cell tolerance by clonal elimination in the thymus, *Cell*, *49*: 273 (1987).
54. J. W. Kappler, U. Staerz, J. White, and P. C. Marrack, Self-tolerance eliminates T cells specific for Mls-modified products of the major histocompatibility complex, *Nature*, *332*: 35 (1988).
55. H. R. MacDonald, R. Schneider, R. K. Lees, R. C. Howe, H. Acha-Orbea, H. Festenstein, R. M. Zinkernagel, and H. Hengartner, T-cell receptor Vβ use predicts reactivity and tolerance to *Mls-a* encoded antigens, *Nature*, *332*: 40 (1988).
56. D. Woodland, M. Happ, J. Bill, and E. Palmer, Requirement for cotolerogenic gene products in the clonal deletion of I-E reactive cells, *Science*, *247*: 964 (1990).
57. J. Bill, O. Kanagawa, D. Woodland, and E. Palmer, The MHC molecule I-E is necessary but not sufficient for the clonal deletion of Vβ11 bearing T cells, *J. Exp. Med.*, *169*: 1405 (1989).
58. M. MacDuffie, W. Born, J. Kappler, and P. Marrack, The role of the T cell receptor in thymocyte maturation: Effects in vivo of antireceptor antibody, *Proc. Natl. Acad. Sci. USA*, *83*: 8728 (1986).
59. B. J. Fowlkes, R. H. Schwartz, and D. M. Pardoll, Deletion of self-reactive thymocytes occurs at a $CD4^+$,$CD8^+$ precursor stage, *Nature*, *334*: 620 (1988).
60. P. Kisielow, H. Bluthmann, U. D. Staerz, M. Steinmetz, and H. von Boehmer, Tolerance in T cell receptor transgenic mice involves deletion of nonmature $CD4^+8^+$ thymocytes, *Nature*, *333*: 742 (1988).
61. W. C. Sha, C. A. Nelson, R. D. Newberry, D. M. Kranz, J. H. Russell, and D. Y. Loh, Positive and negative selection of an antigen receptor on T cells in transgenic mice, *Nature*, *336*: 73 (1988).

62. R. P. Shimonkevitz, and M. J. Bevan, Split tolerance induced by the intrathymic adoptive transfer of thymocyte stem cells, *J. Exp. Med.*, *168*: 143 (1988).
63. L. Burkly, D. Lo, O. Kanagawa, R. L. Brinster, and R. A. Flavell, T cell tolerance by clonal anergy in transgenic mice with nonlymphoid expression of MHC class II I-E, *Nature*, *342*: 564 (1989).
64. F. A. Harding, J. G. McArthur, G. A. Gross, D. H. Raulet, and J. P. Allison, CD-28 mediated signalling co-stimulates murine T cells and prevents induction of anergy in T cell clones, *Nature*, *356*: 607 (1992).
65. R. N. Germain, The second class story, *Nature*, *353*: 605 (1991).
66. M. M. Davis, and P. J. Bjorkman, T-cell antigen receptor genes and T-cell recognition, *Nature*, *334*: 395 (1988).
67. P. J. Fink, and M. J. Bevan, H-2 antigens of the thymus determine lymphocyte specificity, *J. Exp. Med.*, *148*: 766 (1978).
68. R. M. Zinkernagel, G. N. Callahan, A. Althage, S. Cooper, P. A. Klein, and J. Klein, On the thymus in the differentiation of H-2 self-recognition by T cells: Evidence for dual recognition, *J. Exp. Med.*, *147*: 882 (1978).
69. E. J. Jenkinson, L. L. Franchi, R. Kingston, and J. J. T. Owen, Effect of deoxyguanosine on lymphopoiesis in the developing thymus rudiment in vitro: Application in the production of chimeric thymus rudiments, *Eur. J. Immunol.*, *12*: 583 (1982).
70. D. Lo, and J. Sprent, Identity of cells that imprint H-2 restricted T cell specificity in the thymus, *Nature*, *319*: 672 (1986).
71. P. Kisielow, H-S. Teh, H. Bluthmann, and H. von Boehmer, Positive selection of antigen-specific T cells in the thymus by restricting MHC molecules, *Nature*, *335*: 730 (1988).
72. L. J. Berg, A. M. Pullen, B. Fazekas de St. Groth, D. Mathis, C. Benoist, and M. M. Davis, Antigen/MHC specific T cells are preferentially exported from the thymus in the presence of their ligand, *Cell*, *58*: 1035 (1989).
73. L. Smyth, M. Howell, and I. N. Crispe, Abnormal $CD4^+$ T cell populations and defective clonal deletion vary independently in MRL–$Mp^{+/+}$ and MRL–Mp–lpr/lpr mice. *Dev. Immunol.*, *3*: (in press; 1992).
74. H. Pircher, K. Burki, R. Lang, H. Hengartner, and R. M. Zinkernagel, Tolerance induction in double specific T cell receptor transgenic mice varies with antigen, *Nature*, *342*: 559 (1989).
75. K. M. Murphy, A. B. Heimberger, and D. Y. Loh, Induction by antigen of intrathymic apoptosis of CD4/CD8/TCR-lo thymocytes in vivo, *Science*, *250*: 1720 (1990).
76. P. Nieuwenhuis, Self-tolerance induction and the blood-thymus barrier, *Thymus Update*, *3*: 31 (1990).
77. T. Kirchner, B. Schalke, A. Melms, T. Kugelagen, and H. K. Muller-Hermelink, Immunohistochemical patterns of non-neoplastic change in the thymus in myasthenia gravis, *Virchows Archiv. Cell. Pathol.*, *52*: 237 (1986).
78. K. Geuder, A. Marx, V. Witzeman, B. Schalke, K. Toyka, T. Kirchner, and H.-K. Muller-Hermelink, Pathogenic significance of fetal-type acetylcholine receptors on thymic myoid cells in myasthenia gravis, *Dev. Immunol.*, *2*: 69 (1992).

6

T-Cell Activation and Signaling

Joanne C. Pratt and Glen N. Gaulton
University of Pennsylvania, Philadelphia, Pennsylvania

I. INTRODUCTION

Activation of resting T lymphocytes is initiated by the binding of antigen, in the context of products of the major histocompatibility complex (MHC), to specific, clonally distributed and highly polymorphic receptor molecules on the surface of T cells. The T-cell antigen receptor (TCR) acts as the dominant signaling complex that defines immunity and self. Signaling through the TCR complex is a complicated, and only partially understood, biochemical process that links ligand binding on the outer membrane surface into intramembrane, and then, cytoplasmic and nuclear changes that drive lymphocyte proliferation and differentiation. Signal transduction in the TCR system is also characterized by the utilization of multiple associated molecules. These molecules include integral membrane proteins, nonintegral membrane proteins, and cytosolic proteins that transiently translocate to different intracellular compartments during TCR engagement. The immediate consequences of this association are the stimulation of multiple enzymatic activities that serve to amplify and diversify intracellular signals. The most notable, and perhaps central, of these is the activation of a TCR-associated tyrosine kinase function.

Immediate activation events are followed by the induction of new gene expression. Some of these products directly control T-cell replication and others, such as the interleukins (IL), the communication of T cells with different lymphoid cell components. The most well characterized of these events is the synthesis and expression of interleukin-2 (IL-2) and its cellular receptor (IL-2R). In the absence of IL-2 or IL-2R induction, most T cells accumulate in the G_1 phase of the cell cycle, and fail to initiate DNA synthesis. Interleukin-2 is essential for autocrine cell cycle progression and for paracrine and endocrine integration of immune responses. The structure of the nonpolymorphic IL-2R is quite distinct from that of the TCR; however, molecular organization of IL-2R-signaling components bears striking similarities to the TCR. Both receptor systems use multisubunit intermolecular associations that include, among others, essential engagement of tyrosine kinase function. This chapter presents a review of the signaling elements and function of these two receptor systems and their respective roles in the regulation of T-cell growth.

II. STRUCTURE OF THE T-CELL RECEPTOR COMPLEX

A. T-Cell Receptor Molecules

The TCR is a multisubunit complex that contains molecular components of antigen binding, MHC restriction, and intracellular signaling. The antigen–MHC-binding component of the TCR, termed Ti, is a highly polymorphic, immunoglobulinlike 90-kDa heterodimer. The Ti component is formed by the covalent association of either α–β or γ–δ protein chains. Although the organization of the genes that encode Ti chains is diverse, the overall structures of these molecules are quite similar. The Ti proteins have a large extracellular ligand-binding domain and a short cytoplasmic domain that limits potential for intrinsic enzymatic signaling activity. A thorough review of Ti components is presented in the accompanying chapter by Loh (see Chap. 4). These proteins are held in noncovalent association with three nonpolymorphic chains, termed CD3 γ, δ, ϵ (2). In contrast with the Ti proteins, CD3 molecules possess relatively short extracellular regions and long cytoplasmic tails. The CD3 components share a high level of homology and may have evolved from a single gene. The two additional transmembrane proteins, ζ and η, are also found in association with CD3 molecules. Associated ζ and η molecules exist as either homo (ζ–ζ) or heterodimers (ζ–η). These proteins are derived from alternative RNA splice products of a single gene (64). The Ti, and CD3 γ and δ components of the TCR are extensively glycosylated. Lectins that bind to these sugars act as polyclonal T-cell mitogens, and thus have been useful for biochemical and molecular genetic analyses of TCR aggregation and signaling. Several recent studies have illustrated the spatial and stochiometric relations of the TCR complex. It appears that a functional TCR complex is composed of two α–β dimers, in association with CD3γ and CD3δ proteins, a CD3ϵ dimer, and either ζ–ζ or ζ–η dimers. An illustration of this complex is shown in Figure 1.

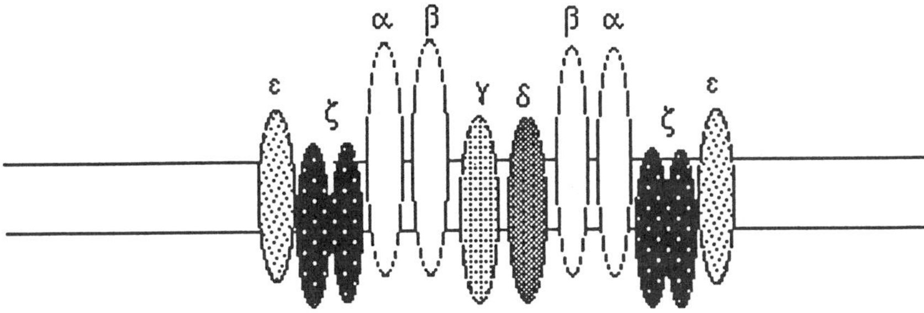

Figure 1 A schematic of the transmembrane proteins that compose functional TCR complexes.

The roles of each component of the TCR have been under study for several years. The large cytoplasmic domains of the CD3 chains, and particularly of the ζ and η proteins, have implicated these molecules in the transduction of antigen-induced activation signals. It has been proposed that ζ–ζ and ζ–η TCR exhibit differential coupling to cellular signal transduction systems (90,91). The role of the ζ component of the TCR in signal transduction has been carefully analyzed by using ζ chain mutants. These studies demonstrated that truncation of up to 40% of the cytosolic portion of ζ resulted in normal TCR assembly and cell surface expression, but defective antigen-induced lymphokine secretion and growth inhibition (34). Interestingly, hybrid molecules composed of the extracellular and transmembrane domains of the nonpolymorphic accessory molecule CD8, fused to the cytoplasmic domain of the ζ chain were capable of transmitting a full activation signal when dissociated from the other components of the TCR (62). More detailed analysis of TCR signaling is presented in Section III.

Exit of the TCR from the endoplasmic reticulum and optimal expression on the cell surface requires complete assembly of all seven protein chains. The Ti heterodimer components that are not in association with the CD3 complex undergo degradation. Transfection of the products encoding these proteins into Chinese hamster ovary (CHO) cells has demonstrated that these proteins are both necessary and sufficient for cell surface expression of the TCR (17).

B. Accessory Molecules

There are an increasingly large number of T-cell surface molecules that modulate signaling through the TCR. For convenience, these accessory molecules may be divided into two categories: those that are believed to interact directly with, and augment, primary signaling of the TCR complex; and those that provide necessary, but TCR-independent, costimulatory signals for cell competence and cycle progression. In this chapter we briefly present only the most thoroughly characterized examples, including, in the former category, CD4 and CD45 molecules, and in the latter, CD28 molecules.

The CD4 molecules are nonpolymorphic glycoproteins expressed on subpopulations of immature and mature T lymphocytes that function as accessory molecules in T-cell recognition and activation. The CD4 molecules bind MHC class II molecules, and become physically and functionally associated with the TCR complex during activation (63). The participation of CD4 enhances and is sometimes requisite for activation (32). Although it is reasonable to speculate that the participation of CD4 in MHC binding increases overall TCR–antigen affinity or cell adhesion, it is also apparent that CD4 serves other roles in activation. For example, chimeric CD4 molecules that contain only the extracellular CD4 domain linked to glycosylated phosphatidylinositol molecules fail to enhance TCR signaling (138). Thus, CD4 molecules may directly participate in TCR signaling. This effect may be coupled through the association of the cytosolic domain of CD4 with the tyrosine protein kinase $p56^{lck}$ (134). Recent studies by Haughn et al., who used mutant cell clones that lack CD4 expression, have demonstrated that CD4 molecules directly affect the capacity of TCR components to deliver positive growth signals (51). Activation of tyrosine protein kinase is requisite for optimal TCR signaling (see Sec. III.A).

Interestingly, dephosphorylation events elicited by tyrosine phosphatases may also serve as mediators of TCR signal transduction. T cells contain a family of transmembrane proteins collectively termed CD45 that are intimately associated with the TCR and play critical roles in activation (39,163). When the CD45 molecule is moved into close physical proximity to other key T-cell surface molecules, such as the TCR, by the use of antibody heteroconjugates, normal activation events are inhibited (76). In contrast, mutants of T-cell tumor lines that lack CD45 fail to proliferate in response to TCR stimulation (115). Cocapping and cross-linking studies have indicated that CD45 can form a complex with the TCR and the CD4 molecule in T cells. It has been proposed that possible substrates of CD45 are the tyrosine kinases *lck*, *fyn*, and *yes*, and that dephosphorylation of these proteins by CD45 may activate their kinase function (14,73,114).

The CD28 molecule plays an important role in T-cell activation by synergizing with early TCR-signaling events. Anti-CD28 antibodies costimulate T-cell proliferation when added in combination with suboptimal doses of phytohemagglutinin (PHA) or antibodies that bind to the TCR (66). However, anti-CD28 antibodies alone do not elicit a proliferative response. The ligand for CD28 has been identified as an activation antigen, termed B7/BB1 that is expressed on several antigen-presenting cells, including B cells (82). Incubation with anti-CD28 or B7/BB1 in the presence of phorbol ester induces tyrosine phosphorylation, but does not augment either inositol phosphate production or calcium accumulation (84,158). One of the main consequences of CD28 engagement is to increase the stability of the mRNA of several T-cell–derived cytokines, including those for IL-2 (81). This results in enhanced and prolonged cytokine production and, accordingly, the augmentation of normal activation events (152). A distinguishing feature of CD28 costimulation, in contrast with TCR-mediated activation, is that it is not inhibited by the immunosuppressive agent cyclosporine.

C. Alternative Activation Pathways

In addition to the TCR, several other cell surface molecules have been identified that are capable of inducing T-cell activation. The CD2 molecule is one of the earliest T-cell differentiation antigens expressed during thymic ontogeny. Its cell surface expression precedes that of the TCR. The ligand for this protein is the lymphocyte function-associated antigen LFA-3, which is distributed on a wide variety of tissues, including thymic epithelial cells. The interaction of these two proteins contributes to adhesion between the cell types on which they are expressed. In addition, a combination of monoclonal antibodies directed against CD2 are mitogenic for T cells (97). These antibodies induce events that are similar to, or identical with, those elicited by TCR engagement. Studies that have compared intracellular signaling responses in T cells induced by either TCR or CD2 stimulation suggest that these molecules represent two independent receptors that work through a common signaling pathway that requires the CD3ζ molecule (35,55).

The cell surface proteins Thy-1 and T-cell-activating protein (Ly-6/TAP) have also been implicated in T-cell activation. These molecules have in common an unusual means of linkage to the cell surface, that utilizes a glycosyl phosphatidylinositol (GPI) lipid anchor in contrast with a protein transmembrane domain (120). Antibodies directed against either of these proteins stimulate lymphokine production and initiate cell proliferation (47,124). Activation by the Thy-1 molecule requires coordinate expression of the TCR (46). The role of the GPI moiety in this process is unknown; however, recent studies suggest that Thy-1 and Ly-6 molecules may associate with *src* kinase family members following activation (141). It has also been proposed that the lipid moieties of these types of molecules may, themselves, serve as secondary messengers of signal transduction. The physiological ligands of these proteins have not been identified.

III. BIOCHEMISTRY OF T-CELL RECEPTOR SIGNALING

The requirements for maximal activation of T cells can be measured in several ways. These include biological assays, such as lymphokine production; biochemical assays, such as activation of tyrosine kinases; and molecular genetic assays, such as induced gene expression. An illustration of the signal transduction cascade activated by TCR engagement is presented in Figure 2. A two-signal hypothesis has been proposed for the activation of resting, naive T lymphocytes (13,87,170,173). The first signal is delivered by antigen binding to the Ti components of the TCR. This is mimicked by binding of immobilized antibodies that recognize the Ti or the CD3 portion of the TCR, or by mitogenic lectins (95,96). The second signal, normally delivered by antigen-presenting cells (APC), is believed to involve binding of one or more accessory molecules such as CD28. In cells that are not at a complete resting stage, such as T-cell clones that have been in culture for extended periods, immobilized anti-CD3 alone may be capable of inducing maximal activation. Kinetic studies have indicated that sustained occupancy of the TCR for 2–4 h is required for the production of IL-2 in the T-cell leukemic line Jurkat (172).

A. Activation of Cellular Tyrosine Kinases

The phosphorylation of several proteins on tyrosine residues is an important initial signal of T-cell activation (65). Addition of the selective inhibitor of

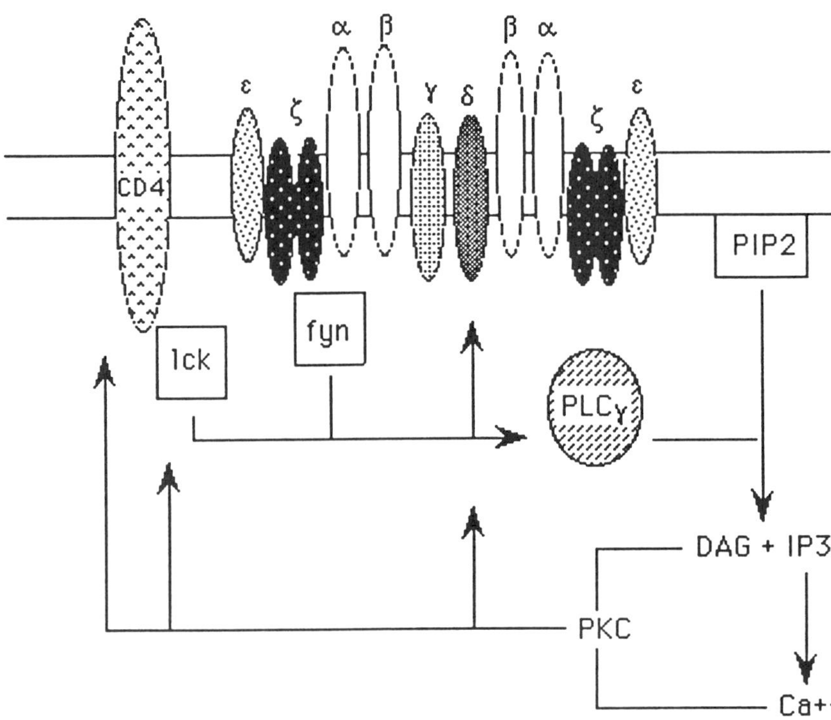

Figure 2 Signal transduction through the TCR complex.

tyrosine phosphorylation, genistein, to T cells inhibits TCR-induced inositol phosphate formation and proliferation, suggesting that phosphorylation events are critical to the progression of activation (107). Tyrosine phosphorylation has been detected as early as 15 s following T-cell stimulation. Tyrosine kinase substrates include the ζ chain of the TCR complex (4), phosphatidylinositol-specific phospholipase Cγ (PI-PLCγ) (117,131,171), and a limited number of additional unidentified proteins. Although the function of ζ chain phosphorylation is unclear, phosphorylation of PI-PLC-γ is a critical factor in the regulation of its enzymatic activity in A421 and 3T3 cells (see later discussion).

In addition to the identification of tyrosine kinase substrates, the kinases themselves have received a great deal of attention. Analysis of TCR primary structure revealed that none of the TCR proteins contain an inherent tyrosine kinase catalytic domain. This led investigators to search for enzymes that are physically or functionally associated with the TCR. Several tyrosine kinases including *lck*, *fyn*, *yes*, *csk*, and Zap70 have been identified in T cells. The product of the *lck* gene, $p56^{lck}$, is a 56-kDA tyrosine kinase that is distributed in both T-cell cytosol and in association with CD4 and CD8 molecules (160). The importance of $p56^{lck}$ in TCR-mediated signal transduction was demonstrated using antigen-specific T cells into which mutant CD4 molecules were introduced. Cells expressing mutant CD4 molecules that were unable to associate with $p56^{lck}$ failed to produce IL-2 in response to antigen (38). In addition, cells defective in *lck* expression were also unable to respond to stimulation through the TCR (144). Although $p56^{lck}$ was a primary candidate for the ζ-chain-associated kinase, subsequent studies revealed that *lck* is not activated following the direct triggering of the TCR, making it unlikely that *lck* is the sole kinase coupled to the TCR (161). Additional support for this belief is that a different pattern of protein tyrosine phosphorylation occurs in response to activation of *lck* by anti-CD4 cross-linking and TCR triggering, and that cells that lack CD4 and CD8 molecules remain coupled to a tyrosine kinase pathway through the TCR (85,159). Thus, the coupling of *lck* regulation to TCR function remains unclear. The protein tyrosine kinase *csk* was recently shown to phosphorylate *lck* on a single tyrosine residue, resulting in suppression of *lck* activity (6). The linkage of *csk* activation to TCR engagement is currently under investigation.

The protein tyrosine kinase *fyn* has also been identified in T cells (129). Although expression of *fyn* is ubiquitous, T cells contain a specific form of *fyn*. Activation of *fyn* by TCR triggering has not been directly demonstrated; however, immunoprecipitation studies have shown that up to 25% of *fyn* in T cells is physically associated with the ζ chain of the TCR. In addition the abrogation of *fyn* expression in mutant mice caused defective TCR signaling and proliferation in thymocytes, whereas peripheral T-cell signaling remained normal (1,142). Another member of the *src* family of tyrosine kinases expressed in T cells is *yes*. This protein is also expressed in many other cell types, and its involvement in signal transduction in T cells has not been well characterized. Finally, a protein tyrosine kinase termed Zap70 was described in the leukemic T-cell line Jurkat (19). This kinase was shown to specifically associate with TCRζ in the absence of other TCR molecules (19).

These data indicate that one component of activation through the TCR proceeds through a tyrosine kinase pathway. However, not all receptors, nor components of the TCR-signaling cascade need be coupled to this kinase system. For example, it is possible to elicit IL-2 secretion in Jurkat lymphoblastoma cells without induction of tyrosine kinase activity. These studies were conducted using Jurkat cells that had been transfected with the cDNA for the human muscarinic receptor HM1, a non-tyrosine kinase-coupled receptor system. Stimulation of these cells with muscarinic agonists resulted in the production of IL-2 and IL-2Rα subunits, without any detectable increases in tyrosine phosphorylation (28).

B. T-Cell Receptor-Coupled Phosphoinositide Hydrolysis

The T-cell membrane is composed of a mixture of phospholipids, for which the roles in activation have only just begun to be elucidated. Both intact phospholipids and their metabolites appear to be involved in the response of T cells to TCR stimulation. The major membrane lipids include phosphatidylcholine (PC), phosphatidylethanolamine (PE), phosphatidylserine (PS) and the phosphoinositides (PI). The phosphoinositides represent about 1% of the total membrane phospholipids and are present in several forms that differ in phosphate content. Phosphoinositide cleavage is catalyzed by a phosphatidylinositol-specific phospholipase C (PI-PLC) that is activated within 15 s after TCR triggering (58). The cleavage products of PI provide critical second-messengers of activation. These include diacylglycerol (DAG) and the inositol phosphates (IP_1, IP_2, IP_3, IP_4). The importance of these second-messengers in the progression of T-cell activation is strengthened by the observation that pharmacological reagents that mimic these molecules, such as phorbol esters and ionophores, are capable of inducing T-cell activation and proliferation (155). Inositol 1,4,5-trisphosphate (IP_3) and inositol 1,3,4,5-tetrakisphosphate (IP_4) regulate the concentration of intracellular calcium by eliciting its release from intracellular (7) and extracellular (61) sources, respectively. The role of IP_1 and IP_2 in activation is unclear; however, they may sustain steady-state levels of the polyphosphoinositides during activation by providing a reservoir of PI that may serve as kinase substrates. Diacylglycerol functions as a cofactor of the serine/threonine kinase, protein kinase C (PKC) (8). The role of PKC in T-cell activation is discussed in detail in the following section. Despite these observations, there is some evidence in T-cell hybridoma systems that these early biochemical events can be dissociated from downstream effects, such as IL-2 production (146). These data have not, however, been repeated in normal peripheral blood T cells. Activation of several other cell surface receptors, including CD2 and Thy-1, is also coupled to phosphoinositide hydrolysis in T cells.

The PI-PLCs that catalyze the hydrolysis of the phosphoinositides have been best characterized in nonlymphoid tissues. Antibodies prepared against these lipases reveal that there are at least six immunologically distinct forms and that a single tissue can express several different PI-PLC isozymes. Two distinct PI-PLC isozymes, PI-PLC-γ_1 and -γ_2, have been reported in T cells. The PI-PLC-γ_1 was originally identified in brain and fibroblastoid cell lines, whereas the related PI-PLC-γ_2 form was originally cloned from transformed human T lymphocytes (112). Antibody blots have failed to detect the presence of other PI-PLC isozymes (α, β, and δ) in T cells; however, it is conceivable that the presence of very small amounts of these isozymes in T cells would not be detected by these analyses.

The expression of multiple isoforms of T cell PI-PLC is consistent with reported heterogeneity in the regulation of T-cell PI-PLC by phorbol esters (168). In some T-cell lines, PKC activation inhibited both inositol phosphate production and calcium-induced PI metabolism, whereas in others, PKC activation had no effect on TCR-stimulated inositol phosphate production. These observations suggest that either regulation of PI-PLC-γ is mediated by PKC in T cells, or that some T cells contain PI-PLC isoforms, such as PI-PLC-α or PI-PLC-β, in levels that are below detection by Western blot analysis and that are affected by PKC.

It has been proposed that either GTP-binding (G) proteins or tyrosine kinase activation couple the TCR to PI-PLC activation. The G-protein hypothesis is based on observations that inositol phosphate production can be stimulated with G-protein activators such as GTPγS and $AlFl_4$ that bypass the TCR (110,130). The GTP-binding proteins are involved in phosphorylation of TCRζ and a 68-kDa GTP-binding protein was coprecipitated with the TCR (18,111). Data supporting the involvement of tyrosine kinase activation in PI-PLC regulation are derived from studies demonstrating that the tyrosine kinase inhibitor, genistein, inhibits TCR-mediated phosphoinositide hydrolysis and proliferation (107), and that PI-PLC-γ is tyrosine-phosphorylated following TCR engagement (117,131,171). Although it has not been directly demonstrated, this phosphorylation event is believed to regulate enzymatic activity, as the kinetics of phosphorylation are consistent with those of phosphoinositide hydrolysis (131,171).

The PI-PLC-γ_1 becomes phosphorylated under several conditions. Wahl et al. first showed that phospholipase C activity from epidermal growth factor (EGF)-stimulated A431 cell lysates, but not unstimulated cell lysates, was retained on a phosphotyrosine immunoaffinity matrix (164). Meisenhelder et al. and Margolis et al. subsequently demonstrated that the platelet-derived growth factor receptor (PDGFR) and epidermal growth factor receptor (EGFR) coprecipitate with PI-PLC-γ_1 in 3T3 and A431 cells, respectively, indicating that these receptors can phosphorylate PI-PLC-γ_1 directly (88,89). Both groups detected tyrosine phosphorylation of PI-PLC-γ within 30 s, and this phosphorylation lasted up to 1 h following stimulation. Kinase-negative mutants of the EGFR were not able to couple with PI-PLC-γ, and the addition of the tyrosine kinase inhibitor, tyrphostin, RG50864, blocked EGF-induced tyrosine phosphorylation, PI-PLC-γ_1 association with the EGFR, and EGF-induced CA^{2+} mobilization and proliferation (88). In addition, purified EGF or PDGF receptor phosphorylated PI-PLC-γ in vitro. Four tyrosine residues on PI-PLC-γ_1 have been identified as kinase substrates in vitro, at positions 472, 771, 783, and 1254. Purification of PI-PLC-γ_1 and tryptic digestion following treatment of HSC-1 cells with EGF have shown that two of these residues, 771 and 1254 are also phosphorylated in vivo (165). Two thirds of the total PI-PLC-γ_1 pool is phosphorylated on tyrosine residues within 2 min of EGF addition (165). This level of phosphorylation of PI-PLC-γ_1 correlates with the amount of PI-PLC that translocates from the cytosol to the membrane during cellular stimulation.

Reconstitution studies have also increased our understanding of the role of tyrosine phosphorylation in the regulation of PI-PLC activity. Goldschmidt-Clermont et al. have proposed that tyrosine phosphorylation of PI-PLC up-regulates enzymatic activity that is detectable when using a PIP_2-profilin substrate (41,42). Profilin is a 15-kDa protein that binds to actin and inhibits its polymerization. One protein that has been shown to directly interact with PI-PLC-γ_1 is cAMP-dependent protein kinase A (72,113). Reagents that increase the intracellular concentration of cAMP in rat glioma cells and BALB/c-3T3 cells induce phosphorylation on serine residues, with a concomitant decrease in the formation of inositol phosphates. A slight increase in the phosphoserine content of PI-PLC-γ in T cells as been detected following PHA stimulation (117), and reagents that increase intracellular cAMP inhibit both CD2 and TCR-mediated phosphoinositide hydrolysis and proliferation (10,80).

Several PI-PLC isozymes (e.g., PI-PLC-β) are not substrates of tyrosine kinases, and their regulation is believed to be linked to the activation of a unique guanine nucleotide-binding protein termed Gp (150). In T cells, GTPγS and $AlFl_4$ induce PI turnover, elevation of cytoplasmic calcium, and phosphorylation of the CD3γ chain (102,110). Additionally, other early T-cell activation events are inhibited by the selective depletion of guanine nucleotides and by treatment of cells with mycophenolic acid (106).

The cDNA sequences for five PI-PLC (α, β, γ_1, γ_2, and δ) have been determined. The three isozymes isolated from bovine brain, PI-PLC-β, γ, and δ, as well as PI-PLC-γ_2, share two regions of sequency homology. Removal of either of these regions results in loss of enzymatic activity, suggesting that they contain the catalytic site of the enzyme. Isoenzymes PI-PLC-γ_1 and PI-PLC-γ_2 also contain a region that shares sequence similarity with the NH_2-terminal regulatory domain of nonreceptor tyrosine kinases of the *src* family. This region contains three subdomains: two adjacent copies of a domain termed SH2 (*src* homology 2), and one copy of SH3. The SH2 domains are also present in the GTPase-activating protein (GAP) of *ras* p21 (162), α-spectrin (169), and PI 3-kinase (PI3K) (33,68). One possible function of this region is to mediate interactions between the lipase and other phosphorylated proteins, such as tyrosine kinases. This interaction has been reported for both EGF and PDGF receptor systems, but has not been described in T cells.

C. The Role of Calcium and Protein Kinase C in Activation

Stimulation of T cells by antibodies that are reactive with either Ti or CD3 components of the TCR induce an increase in intracellular calcium concentration to more than 300 nM from resting levels of approximately 100 nM (60, 170). This increase in intracellular calcium is due to both the release of calcium from intracellular stores and to its influx from the external medium (58). In normal Jurkat cells, calcium increases are sustained for up to 2 h before IL-2 gene induction. It is generally believed that IP_3 elicits the release of Ca^{2+} from intracellular, nonmitochondrial stores, and IP_4 mediates the influx of external Ca^{+2} (61). It has been suggested that the ability of IP_4 to induce a calcium influx may be dependent on its prior depletion from intracellular stores by an IP_3-mediated release, as IP_4 alone is not sufficient to induce an increase in intracellular Ca^{+2} (59). Voltage-independent IP_3-mediated calcium channels that respond to both TCR and CD2 triggering have been described in T cells (35) and localized to the plasma membrane (71). Other patch-clamping studies have suggested that IP_3 itself is the second-messenger that initiates the influx of calcium.

The sustained elevation (>2 h) in calcium levels seen in normal activation is necessary for the induction of later cell cycle progression events (43). This was demonstrated by the abolition of downstream signaling events when sustained intracellular Ca^{2+} levels were depleted by the addition of extracellular calcium chelators, such as EGTA. In addition, transient increases in Ca^{2+} (< 20 min) have been detected in mutant Jurkat cell lines, without concomitant IL-2 gene expression. However, TCR-dependent production of inositol phosphates was also significantly diminished in these mutant cells.

One characteristic of PI-PLC isolated from nonlymphoid cells is a strong Ca^{2+} requirement for activation in vitro; yet in T cells, activation of PI-PLC precedes increases in intracellular calcium. In calcium titration studies and kinetic analyses, purified T-cell PI-PLC had a requirement for Ca^{2+} for optimal hydrolysis of both PI and PIP_2, however, 50% of maximal phosphatidylinositol 4,5-bisphosphate (PIP_2) hydrolysis was seen in the absence of calcium (117). It is reasonable to assume that calcium would not play a role in the initial activation of PI-PLC in vivo, as the concentrations found to be necessary for in vitro activation (>10 μM) are considerably higher than resting intracellular levels (~150 nM). In activated cells, calcium levels increase to ~300 nM, a concentration still well below that necessary for in vitro PIP_2 cleavage.

The other major product of phosphoinositide hydrolysis is DAG. The interaction of DAG with PKC enables this enzyme to bind phosphatidylserine and increases its affinity for Ca^{2+}. The production of this trimolecular complex (PKC, PS, and Ca^{2+}) induces the kinase activity of PKC. Activation of PKC results in its association with the cell membrane. In T cells, a transient activation of PKC results in incomplete activation signals, whereas a sustained association of PKC with the membrane is correlated with cellular proliferation (86). The importance of PKC activation is underscored by the observation that the addition of the PKC inhibitor, staurosporine, to T cells abrogates IL-2 secretion and cellular proliferation (103). Transient or prolonged activation of PKC can be generated by the stimulation of cells with soluble or immobilized anti-CD3 antibodies. Substrates of PKC include the γ and δ chains of CD3, CD4, MHC class I molecules, CD2, IL-2Rα subunit, the transferrin receptor, and a number of unidentified proteins. The PKC-mediated phosphorylations of CD3, CD4, IL-2Rα subunit, and the transferrin receptor result in their modulation from the cell surface, and thus, it appears that PKC events may function to dampen the immune response by negative-feedback (15,27,56,75).

D. T-Cell Receptor Regulation of Gene Expression

Lymphocyte activation is characterized by a pattern of molecular events. Immediate activation events involve alteration of the activities of preformed transcription factors, such as NFκB and AP-1. Perhaps the best-described nuclear regulatory effects of TCR activation are coupled to the activation of PKC. This enzyme controls the phosphorylation of two proteins that play prominent roles as regulators of gene transcription, including the lymphokine IL-2 and its cellular receptor (IL-2R) (3,37). The protein IκB serves as an inhibitor of the positive regulator of transcription NFκB. Phosphorylation of IκB following PKC activation results in the dissociation of IκB from NFκB, thereby allowing the translocation of NFκB to the nucleus and enhanced gene transcription (3). Similar effects have been reported for the regulation of the activity of the transcription factor AP-1, a product of the c-*jun* protooncogene.

The activation of preformed elements controls the expression of several additional genes, such as c-*fos*, that also act as transcription regulatory factors (25). Production of these proteins further regulates the production of so-called early genes such as c-*myc* and the lymphocyte selective proteins IL-2 and IL-2Rα subunit (25,69,118). This temporally coordinated sequence of gene expression seems to be essential for the progression of antigen- and mitogen-activated T cells through the cell cycle. The *fos* protein regulates expression of its own and other growth-associated genes (20), the c-*myc* gene product p65 appears essential for the initiation of cellular DNA synthesis (52,145), and the expression of IL-2 and IL-2Rα gene products are essential for the metabolic commitment of T cells to cycle progression (93,139).

IV. STRUCTURE OF INTERLEUKIN-2 AND THE INTERLEUKIN-2 RECEPTOR COMPLEX

A. Interleukin-2 Structure

Interleukin-2 is the major regulatory cytokine of the immune system (140). It induces the proliferation and differentiation of a variety of lymphocytic subsets, including T and B cells (139,148,153). The biological effects of IL-2 are mediated through binding to specific membrane receptors that are expressed in an autocrine, paracrine, and perhaps endocrine fashion (139). The production of IL-2 is very tightly linked to lymphocyte activation, predominantly within the $CD4^+$ T-cell subset. Interleukin-2 is a single polypeptide of 15 kDa (153 amino acids), folded into six independent helical domains, with a single internal disulfide bond (12). The receptor contact domains have been mapped to the first two NH_2-terminal helices, which are nonlinear (21). The IL-2 protein is only slightly modified at the posttranslational phase (74,121,123,139,148). Interleukin-2 is variably *O*-glycosylated, phosphorylated, and sialyated, but contains no *N*-glycosylation sites. The biological activity of nonglycosylated recombinant IL-2 is indistinguishable from that of native mammalian IL-2; however, glycosylation may affect serum half-life (148).

The human IL-2 gene is present in single copy and organized in four exons (53,149). Gene transcription is controlled primarily by a ~500-base pair region that lies upstream (5′) of the transcription start signal. Several established enhancer sequences have been identified within this promoter. These include: a κB-like site (57); a phorbol-responsive element—AP-1-binding site (104); at least two T-cell-specific enhancer–antigen-responsive regulatory elements (25,29,30); and two possible negative regulatory sites (67). The induction of IL-2 synthesis is a hallmark of TCR activation in $CD4^+$ T cells, and is governed by the cascade of signaling events outlined in the previous section.

B. Interleukin-2 Receptor Structure

The biological effects of IL-2 are mediated through binding to specific cellular membrane receptors (IL-2R). The ligand-binding component of the IL-2R is composed of two subunits, termed α (p55) and β (p75) (79,121,133,167). These proteins are encoded by distinct and structurally unrelated genes (50,167). Both α and β IL-2R subunits are capable of independently binding IL-2, albeit with differing affinities. The IL-2Rα subunit binds IL-2 with low affinity (equilibrium dissociation constant = 0.7×10^{-9} M) whether expressed in lymphoid or nonlymphoid cells (122). This reaction is characterized by slow association (3.8×10^{5}/M·s), and dissociation rates ($2.5 \times 10^{-4}s^{-1}$), and a relatively long occupancy time ($t_{1/2}$ = 46 min). The IL-2Rβ subunit displays a higher independent affinity (equilibrium dissociation constant = 2.9×10^{-8} M) that is characterized by both faster association (1.4×10^{7}/M·s) and dissociation rates ($4.0 \times 10^{-1}s^{-1}$), and a short occupancy time ($t_{1/2}$ = 1.7 s). The dimeric IL-2Rαβ complex utilizes the independent properties of the α and β chains to full advantage: The on rate is reflective of rapid β subunit trapping of IL-2 molecules (3.1×10^{7}/M·s), while the dissociation rate is representative of the slower dissociation of IL-2Rα chains ($2.3 \times 10^{-4}s^{-1}$). The net effect is to increase total receptor occupancy time ($t_{1/2}$ = 50 min) and secure a vastly more sensitive aggregate affinity (equilibrium dissociation constant = 7.0×10^{-12} M) (167). Expression of both IL-2Rα and β subunits fails to reconstitute a high-affinity complex in some nonlymphoid cell lines (50). This has led to speculation that one, or several, additional molecules that are constitutively expressed in lymphoid cells, participate in binding. One likely candidate, termed the p64γ subunit, has been recently cloned (147).

Up-regulation of IL-2R expression is another hallmark of lymphocyte activation. Antibodies prepared to the IL-2Rα subunit, termed anti-TAC, showed that low affinity IL-2R numbers rose from undetectable levels to >100,000 per cell within 24–48 h after mitogen stimulation (157,166). The rise in high-affinity receptors showed similar kinetics, but only reached 5,000–10,000 receptors per cell. Thus, in activated cells, the number of high-affinity receptors is limited by the availability of the IL-2Rβ subunit. These are produced in small amounts, and most likely exist in preformed complexes with the more abundant IL-2Rα molecules. Resting cells are devoid of both IL-2Rα protein and mRNA transcripts (70,118). The induction of IL-2Rα gene expression is positively and negatively regulated by several elements including NFκB (57,83,116). The IL-2Rα gene is also induced by the human immunodeficiency virus (HIV) *tat* and human T-lymphotropic virus (HTLV)-1 *tax* gene products, by addition of phorbol esters or calcium ionophores (70,132), and is largely independent of extracellular calcium or protein kinase C activation (99,103). Thus, expression of the IL-2Rα chain is not as tightly controlled as that of IL-2. Regulation of IL-2Rβ chain expression is less clear. There is still debate about whether resting T cells express low levels of IL-2Rβ mRNA and protein (9,137). This issue is confounded because natural killer cells, which constitutively express IL-2Rβ molecules, copurify with T cells in most protocols. It is also possible that different T-cell subsets or memory populations may express different IL-2Rβ levels. There is no doubt, however, that activation up-regulates IL-2Rβ mRNA and protein levels, and that this increase in receptor number is a key element in the increased expression of high-affinity IL-2R (40).

V. INTERLEUKIN-2 RECEPTOR SIGNAL TRANSDUCTION

A. Interleukin-2 Receptor Biochemical Signaling

The signaling mechanisms of the IL-2R remained an enigma until recent years. Early reports linked IL-2 binding with activation of phosphoinositide hydrolysis, the accumulation of intracellular calcium, and activation of protein kinase C (11,22,125). These results are reminiscent of those described in the foregoing following TCR engagement and, as such, made little sense in the context of the need for separate and patterned IL-2 binding. Indeed, we now know that these results were erroneous, and that IL-2Rs are coupled to their own unique-signaling pathways that complement, rather than duplicate TCR signaling (77,98,99,101). An illustration of the signal transduction processes that are believed to control IL-2-dependent events are presented in Figure 3.

Elegant studies by Taniguichi and colleagues (49,50) have demonstrated that the IL-2Rβ chain is essential for signal coupling. Cell mutants or transfectants

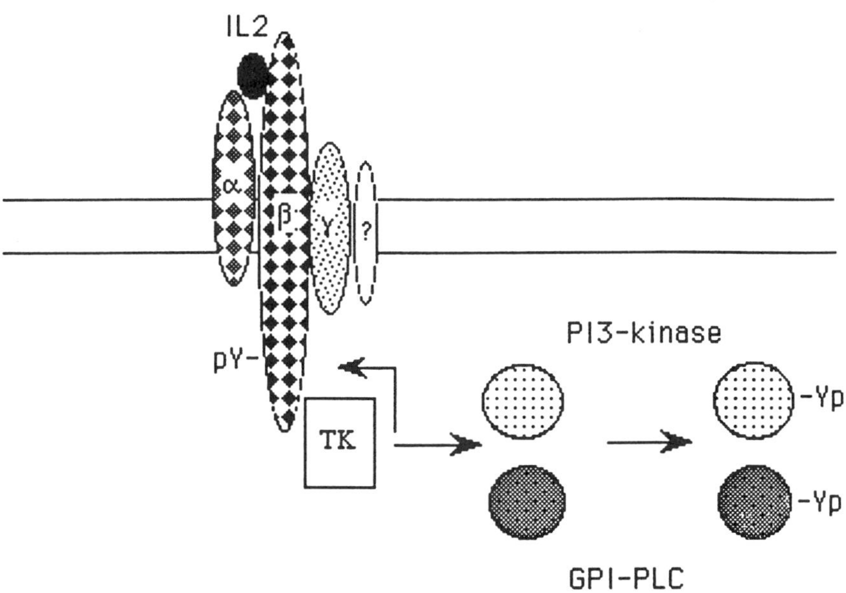

Figure 3 Potential pathways of IL-2R signal transduction.

that lack IL-2Rβ do not transmit growth signals, even in the presence of the IL-2Rα subunit. This is not surprising, as the IL-2Rα molecule contains only 13 cytoplasmic amino acid residues (78). The IL-2Rβ molecule, in contrast, contains 286 amino acids in its cytosolic domain (50). Sequence comparisons indicate that the extracellular domain of the IL-2Rβ molecule shares extensive homology with a number of other hematopoietic growth factor receptors, including erythropoietin. The reasons for this are unclear, but may indicate that these receptors use common or related molecules in ligand binding. The cytoplasmic domain of the IL-2Rβ chain bears no obvious homology to any other reported receptors, but does contain distinctive serine (S)- and acidic (A)-rich regions, and is generally rich in proline. Mutations of either the A domain or the 145 COOH-terminal residues do not affect growth signal delivery; however, deletion of the entire cytoplasmic domain or the S region inhibits cell growth. These results suggest that some signaling functions are encoded by distinct domains within this molecule. Interestingly, all of the cytoplasmic mutants, including full truncations, may associate with IL-2Rα molecules to reconstitute high-affinity-binding molecules (49). The IL-2R p64γ subunit contains only 86 cytosolic residues and is also without homology to tyrosine kinases or other proteins with known enzymatic functions. A portion of the cytosolic domain does, however, bear homology to two of the five well-defined subdomains of SH2 regions. This suggests that p64γ may play a role in IL2R signal coupling, perhaps by linking IL2Rβ subunits to tyrosine kinases.

The overall scheme of IL-2R signaling is similar to the TCR in that ligand-binding activates a complicated and branching series of enzymatic steps that use protein tyrosine kinase as an essential component. The first observation of IL-2R-associated protein kinase activation was reported in 1986 by Gaulton and Eardley (36). This was followed by the description of specific tyrosine kinase activation (100,128) and, most recently, by the coprecipitation of IL-2-induced protein tyrosine kinase in the IL-2–IL2R-binding complex (2). The IL-2Rβ subunit is rapidly tyrosine-phosphorylated following IL-2 binding and is believed to be a direct substrate of this kinase (100). The nature of the IL-2R-associated tyrosine kinase remains in question, but recent reports have indicated that the lymphocyte-specific tyrosine kinase *lck* may be an important component. As described previously, *lck* is a member of the *src* tyrosine kinase family, and has been detected in noncovalent association with CD4 and CD8 molecules (134). The data in favor of a role for *lck* in IL-2 signaling are multifold. Interleukin-2 binding rapidly activates *lck* kinase activity. This binding induces phosphorylation of the *lck* kinase protein and *lck* coprecipitates with the IL-2Rβ subunit when coexpressed at high levels in COS cells (48,54). The problems with assignment of *lck* as the dominant IL-2R-associated tyrosine kinase are twofold. First, mutants of the IL-2Rβ chain that fail to associate with *lck*, such as mutants of the acidic A region of the cytoplasmic domain, are capable of growth signal transduction (49). Second, several cell types, including some T cells, that fail to express *lck* remain IL-2 responsive (154).

More recent reports have indicated that the IL-2R associates with additional signaling molecules. The first to be described was IL-2-dependent hydrolysis of glycosylated phosphatidylinositol (GPI) molecules (31,93). The GPI molecules have been reported to function as important regulators of cellular metabolism in several hormonal systems, including insulin and nerve growth factor. Interleukin-2 has also been shown to upregulate lymphocyte metabolism, probably as a prerequisite of both the energetic and biosynthetic requirements of cell replication. The GPI metabolites are suggested as regulators of this effect. A second molecule that has recently been identified in association with the IL-2R is a phosphatidylinositol 3-kinase (PI3K) (94). The IL-2-dependent PI3K activation results in the in vivo membrane accumulation of a subclass of polyphosphoinositides that are phosphorylated on the 3-position of the inositol ring. Although the function of these lipids is unknown, they have been postulated to participate in reorganization of the cytoskeletal network. The IL-2 activation of PI3K is linked to coordinate activation of tyrosine kinase, although it is not known if PI3K is a direct substrate. Interleukin-2 binding activates the *raf*-1 kinase, the $p21^{ras}$ protein, and the Na^+/H^+ antiport and inhibits adenylate cyclase activity (5,44,45,156). The mechanisms for regulation of these factors and their importance in IL-2-induced growth and differentiation awaits further study. Incorporation of these signal transduction pathways into a model of IL-2R action is presented in Figure 3.

B. Interleukin-2-Induced Gene Expression

The binding of IL-2 to activated T lymphocytes induces the expression of several genes that serve to regulate subsequent cell proliferation and differentiation. These include well-characterized nuclear proteins encoded by the protooncogenes c-*fos*, c-*myc*, and c-*myb*, receptors for the essential cycle progression factors transferrin and prolactin, and numerous genes that encode metabolic enzymes, transport systems, and replicative machinery. The protooncogenes c-*fos*, c-*myc*, and c-*myb* show a unique pattern of induction during T-cell activation. Protooncogenes c-*fos* and c-*myc* are up-regulated within minutes to hours following stimulation with competence growth factors such as PDGF, EGF, interleukin-3, and TCR mitogens (118,143). In contrast, the up-regulation of c-*myb* expression is generally associated with signals for cell cycle progression, such as delivered by IL-2 (119,143). This pattern is not, however, universal, as some T-cell lines maintained in culture may coordinately up-regulate c-*fos*, c-*myc*, and c-*myb* following IL-2 binding (23,26). Interpretation of these data should be cautious, since most of these cell lines constitutively express IL-2R and never truly return to the G_0 state at cycle completion. The regulation of c-*myb* expression is different from that previously described for c-*fos* and c-*myc*. Control of c-*myb* mRNA levels is primarily posttranscriptional (26,151). Cyclohexamide treatment of activated cells results in superinduction of c-*myb*, suggesting that mRNA levels are regulated by a labile protein that enhances mRNA degradation (151).

Interleukin-2 also induces the expression of several genes that encode essential membrane receptors. The most notable of these are receptors for prolactin and transferrin. Interleukin-2 induces a sevenfold increase in prolactin receptor expression (24). The mechanism of this effect is unknown. Prolactin, in part, functions to further up-regulate several growth-associated genes and proteins, including the transferrin receptor. Transferrin binding and iron transport function are obligatory for the growth of most eukaryotic cells including T lymphocytes (108). The expression of transferrin receptors is dependent on IL-2R induction and follows IL-2R expression (109). Although the precise regulation of transferrin receptor gene expression is not fully understood, this event is affected by the concentration of intracellular iron. Decreased intracellular iron induces the binding of a cytosolic protein to a cisacting element within the 3′-untranslated region of the transferrin mRNA, thereby repressing mRNA degradation (105).

Interleukin-2 binding also induces the expression of genes that encode proteins with essential functions in cellular metabolism, cytoskeletal structure, and integrity. The most common metabolic genes are those for glyceraldehyde-3-phosphate dehydrogenase (GAPDH), pyruvate dehydrogenase, triosephosphate isomerase, enolase, lactate dehydrogenase, and phosphoglycerate mutase (126). Up-regulation of these genes and proteins appears essential for the generation of increased demands on ATP pools during replication. The transcription rate of the GAPDH gene increases 1.5-fold, GAPDH mRNA half-life increases 3.8-fold, and overall, the steady-state levels of GAPDH mRNA increase two to fivefold following IL-2 binding (126). This results in a fivefold increase in GAPDH enzymatic activity. Dramatic changes in cell volume and shape accompany these metabolic events, and it is not surprising that IL-2 coordinately up-regulates the expression of mRNA encoding the cytoskeletal components α-tubulin, β-actin, γ-actin, and the intermediate filament protein vimentin (127). Finally, IL-2 also up-regulates the expression of enzymes that direct DNA synthesis. The most thoroughly characterized is an auxiliary protein for DNA polymerase δ, termed proliferating cell nuclear antigen (PCNA). The PCNA mRNA levels increase 25- to 30-fold, and protein levels increase 20-fold following IL-2 binding (135). This results from a combination of increased PCNA gene transcription and mRNA stability (136).

VI. SUMMARY

The selection and clonal expansion of T lymphocytes is an essential component in both the development of protective immunity and the coordinate preservation of nonresponsiveness to self. Recognition and responsiveness are controlled by the binding and subsequent activation of a complicated molecular cascade that begins with TCR complex engagement. Unlike most eukaryotic growth factor systems, the regulation of T-lymphocyte replication is split into two overlapping, yet separately controlled phases, that may be thought of as the competence and progression steps. Competence for replication is mediated by binding of the TCR complex in association with secondary signals delivered by accessory molecules such as CD28. These events result in transit from the G_0 to G_1 phase of the cell cycle and are accompanied by lymphokine secretion and the acquisition of several progression factor receptors. Progression from G_1 into S, and ultimately M phase of the cell cycle is linked predominantly to the binding of IL-2. The extent of T-cell clonal expansion is dependent on the concentration of IL-2, the density of functional IL-2R, and the duration of the interaction of IL-2 with IL-2R (16). The unique aspects of this split system are most probably linked to the regulatory roles of T-cell cytokines that extend beyond direct lymphoid activation events and, possibly, to mechanistic requirements for the induction of nonresponsiveness to self.

REFERENCES

1. M. W. Appleby, J. A. Gross, M. P. Cooke, S. D. Levin, X. Qian, and R. M. Perlmutter, Defective T cell receptor signaling in mice lacking the thymic isoform of $p59^{fyn}$, *Cell*, *70*: 751 (1992).
2. J. D. Ashwell, and R. D. Klausner, Genetic and mutational analysis of the T cell antigen receptor, *Annu. Rev. Immunol.*, *8*: 139 (1990).
3. P. Baeuerle, and D. Baltimore, IκB: A specific inhibitor of the NF-κB transcription factor, *Science*, *242*: 540 (1988).
4. M. Baniyash, P. Garcia-Morales, E. Luong, L. E. Samuelson, and R. D. Klausner, The T cell antigen receptor ζ is tyrosine phosphorylated upon activation, *J. Biol. Chem.*, *263*: 18225 (1988).
5. S. K. Beckner, and W. L. Farrar, Interleukin 2 modulation of adenylate cyclase, *J. Biol. Chem.*, *261*: 3043 (1986).
6. M. Bergman, T. Mustelin, C. Oetken, J. Partanen, N. A. Flint, K. E. Amrein, M. Autero, P. Burn, and K. Alitalo, The human $p50^{csk}$ tyrosine kinase phosphorylates $p56^{lck}$ at Tyr-505 and down regulates its catalytic activity, *EMBO J.*, *11*: 2919 (1992).
7. M. J. Berridge, and R. F. Irvine, Inositol trisphosphate, a novel second messenger in cellular signal transduction, *Nature*, *312*: 315 (1984).
8. M. J. Berridge, and R. F. Irvine, Inositol phosphate and cell signalling, *Nature*, *341*: 197 (1989).
9. L. T. Bich-Thuy, M. Dukovich, N. J. Peffer, A. S. Fauci, J. H. Kehrl, and W. C. Greene, Direct activation of human resting T cells by IL2: The role of an IL2 receptor distinct from the *Tac* protein, *J. Immunol.*, *139*: 1550 (1987).

10. G. Bismuth, J. Theodorou, H. Gouy, S. L. Gouvello, A. Bernard, and P. Debre, Cyclic AMP-mediated alteration of the CD2 activation process in human T lymphocytes. Preferential inhibition of the phosphoinositide cycle-related transduction pathway, *Eur. J. Immunol.*, *18*: 1351 (1988).
11. E. Bonvini, F. W. Ruscetti, M. Ponzoni, T. Hoffman, and W. L. Farrar, Interleukin 2 rapidly stimulates synthesis and breakdown of polyphosphoinositides in interleukin 2-dependent, murine T-cell lines, *J. Biol. Chem.*, *262*: 4160 (1987).
12. B. J. Brandhuber, T. Boone, W. C. Kenney, and D. B. McKay, Three-dimensional structure of interleukin-2, *Science*, *238*: 1707 (1987).
13. P. Bretscher, The two-signal model of lymphocyte activation twenty-one years later, *Immunol. Today*, *13*: 74 (1992).
14. L. C. Cantley, K. R. Auger, C. Carpenter, B. Duckworth, A. Graziani, R. Kapeller, and S. Soltoff, Oncogenes and signal transduction, *Cell*, *64*: 281 (1991).
15. D. A. Cantrell, A. A. Davies, and M. J. Crumpton, Activators of protein kinase C down-regulate and phosphorylate the T3/T-cell antigen receptor complex of human T lymphocytes, *Proc. Natl. Acad. Sci. USA*, *82*: 8158 (1985).
16. D. A. Cantrell, and K. A. Smith, The interleukin-2 T-cell system: A new cell growth model, *Science*, *224*: 1312 (1984).
17. G. R. Carson, R. E. Kuestner, A. Ahmed, C. L. Pettey, and M. F. Concino, Six chains of the human T cell antigen receptor · CD3 complex are necessary and sufficient for processing the receptor heterodimer to the cell surface, *J. Biol. Chem.*, *266*: 7883 (1991).
18. C. Ceniarelli, R. J. Hohman, T. P. Atkinson, F. Gusovsky, and A. M. Weissman, Evidence for GTP-binding protein involvement in the tyrosine phosphorylation of the T cell receptor ζ chain, *J. Biol. Chem.*, *267*: 14527 (1992).
19. A. C. Chan, B. A. Irving, J. D. Fraser, and A. Weiss, The ζ chain is associated with a tyrosine kinase and upon T-cell antigen receptor stimulation associates with ZAP-70, a 70-kDa tyrosine phosphoprotein, *Proc. Natl. Acad. Sci.*, *88*: 9166 (1991).
20. R. Chiu, W. J. Boyle, J. Meek, T. Smeal, T. Hunter, and M. Karin, The c-*fos* protein interacts with c-*jun*/AP-1 to stimulate transcription of AP-1 responsive genes, *Cell*, *54*: 541 (1988).
21. T. L. Ciardelli, H. Holley, K. A. Smith, F. E. Cohen, L. Butler, and R. Gadski, Interleukin 2: Approaches to structure determination, Interleukin 2 (K. A. Smith, ed.), Academic Press, Inc., New York (1988).
22. R. B. Clark, J. T. Love, D. Sgroi, E. G. Lingenheld, and R. I. Sha'afi, The protein kinase inhibitor, H-7, inhibits antigen and IL-2-induced proliferation of murine T cell lines, *Biochem. Biophys. Res. Commun.*, *145*: 666 (1987).
23. J. L. Cleveland, U. R. Rapp, and W. L. Farrar, Role of c-*myc* and other genes in interleukin-2-stimulated proliferation of T lymphocytes, *J. Immunol.*, *138*: 3495 (1987).
24. C. V. Clevenger, D. H. Russell, P. M. Appasamy, and M. B. Prystowsky, Regulation of interleukin 2-driven T-lymphocyte proliferation by prolactin, *Proc. Natl. Acad. Sci. USA*, *87*: 6460 (1990).
25. G. R. Crabtree, Contingent genetic regulatory events in T lymphocyte activation, *Science*, *243*: 355 (1989).
26. F. Dautry, D. Weil, J. Yu, and A. Dautry-Varsat, Regulation of *pim* and *myb* mRNA

26. F. Dautry, D. Weil, J. Yu, and A. Dautry-Varsat, Regulation of *pim* and *myb* mRNA accumulation by interleukin 2 and interleukin 3 in murine hematopoietic cell lines, *J. Biol. Chem.*, *263*: 176 (1988).
27. L. Davis, R. Vida, and P. E. Lipsky, Regulation of human T lymphocyte mitogenesis by antibodies to CD3, *J. Immunol.*, *137*: 3758 (1986).
28. D. M. Desai, M. E. Newton, T. Kadlecek, and A. Weiss, Stimulation of the phosphatidylinositol pathway can induce T-cell activation, *Nature*, *348*: 66 (1990).
29. D. B. Durand, M. R. Bush, J. G. Morgan, A. Weiss, and G. R. Crabtree, A 275 basepair fragment at the 5′ end of the interleukin 2 gene enhances expression from a heterologous promotor in response to signals from the T cell antigen receptor, *J. Exp. Med.*, *165*: 395 (1987).
30. D. B. Durand, J.-P. Shaw, M. R. Bush, R. E. Relogle, R. Belagaje, and G. R. Crabtree, Characterization of antigen receptor response elements within the interleukin-2 enhancer, *Mol. Cell. Biol.*, *8*: 1715 (1988).
31. D. D. Eardley, and M. E. Koshland, Glycosylphosphatidylinositol: A candidate system for interleukin-2 signal transduction, *Science*, *251*: 78 (1991).
32. K. Eichman, J.-I. Jonsson, I. Falk, and F. Emmrich, Effective activation of resting mouse T lymphocytes by cross-linking submitogenic concentrations of the T cell antigen receptor with either Lyt-2 or L3T4, *Eur. J. Immunol.*, *17*: 643 (1987).
33. J. A. Escobedo, and L. T. Williams, A PDGF receptor domain essential for mitogenesis but not for other responses to PDGF, *Nature*, *335*: 85 (1988).
34. S. J. Frank, B. B. Niklinska, D. G. Orloff, M. Mercep, J. D. Ashwell, and R. D. Klausner, Structural mutations of the T cell receptor ζ chain and its role in T cell activation, *Science*, *249*: 174 (1990).
35. P. Gardner, A. Alcover, M. Kuno, P. Moingeon, C. M. Weyand, J. Goronzy, and E. L. Reinherz, Triggering of T-lymphocytes via either T3-Ti or T11 surface structures opens a voltage-insensitive plasma membrane calcium-permeable channel: Requirement for interleukin-2 gene function, *J. Biol. Chem.*, *264*: 1068 (1989).
36. G. N. Gaulton, and D. D. Eardley, Interleukin 2-dependent phosphorylation of interleukin 2 receptors and other T cell membrane proteins, *J. Immunol.*, *136*: 2470 (1986).
37. S. Ghosh, and D. Baltimore, Activation in vitro of NF-κB by phosphorylation of its inhibitor Iκ-B, *Nature*, *344*: 678 (1990).
38. N. Glaichenhaus, N. Shastri, D. R. Litman, and J. M. Turner, Requirement for association of $p56^{lck}$ with CD4 in antigen-specific signal transduction in T cells. *Cell*, *64*: 511 (1991).
39. H. Gmunder, and W. Lesslauer, A 45-kDa human T-cell membrane glycoprotein functions in the regulation of cell proliferative responses, *Eur. J. Biochem.*, *142*: 153 (1984).
40. J. R. Gnarra, H. Otani, M. G. Wang, W. McBride, M. Sharon, and W. J. Leonard, Human interleukin 2 receptor β-chain gene: Chromosomal localization and identification of 5′ regulatory sequences, *Proc. Natl. Acad. Sci. USA*, *87*: 3440 (1990).
41. P. J. Goldschmidt-Clermont, J. W. Kim, L. M. Machesky, S. G. Rhee, and T. D. Pollard, Regulation of phospholipase C-γ1 by profilin and tyrosine phosphorylation, *Science*, *251*: 1231 (1991).

42. P. J. Goldschmidt-Clermont, L. M. Machesky, J. J. Baldassare, and T. D. Pollard, The actin-binding protein profilin binds to PIP_2 and inhibits its hydrolysis by phospholipase C, *Science*, *247*: 1575 (1990).
43. M. A. Goldsmith, and A. Weiss, Early signal transduction by the antigen receptor without commitment to T cell activation, *Science*, *240*: 1029 (1988).
44. J. D. Graves, J. Downward, M. Izquierdo-Pastor, S. Rayter, P. H. Warne, and D. A. Cantrell, The growth factor IL-2 activates p21*ras* proteins in normal human T lymphocytes, *J. Immunol.*, *148*: 2417 (1992).
45. S. Grinstein, E. Mack, and G. B. Mills, Osmotic activation of the Na^+/H^+ antiport in protein kinase C-depleted lymphocytes, *Biochem. Biophys. Res. Commun.*, *134*: 8 (1986).
46. K. C. Gunter, R. N. Germain, R. A. Korczek, T. Saito, W. M. Yokoyama, C. Chan, A. Weiss, and E. M. Shevach, Thy-1-mediated T-cell activation requires co-expression of CD3/Ti complex, *Nature*, *326*: 505 (1987).
47. K. C. Gunter, T. R. Malek, and E. M. Shevach, T cell activating properties of an anti-Thy-1 monoclonal antibody, *J. Exp. Med.*, *159*: 716 (1984).
48. M. Hatakeyama, T. Kono, N. Kobayashi, A. Kawahara, S. D. Levin, R. M. Perlmutter, and T. Taniguchi, Interaction of the IL-2 receptor with the *src*-family kinase p56*lck*: Identification of novel intermolecular association, *Science*, *252*: 1523 (1991).
49. M. Hatakeyama, H. Mori, T. Doi, and T. Taniguchi, A restricted cytoplasmic region of IL-2 receptor β chain is essential for growth signal transduction but not for ligand binding and internalization, *Cell*, *59*: 837 (1989).
50. M. Hatakeyama, M. Tsudo, S. Minamoto, T. Kono, T. Doi, T. Miyata, M. Miyasaka, and T. Taniguchi, Interleukin-2 receptor β chain gene: Generation of three receptor forms by cloned human α and β chain cDNAs, *Science*, *244*: 551 (1989).
51. L. Haughn, S. Gratton, L. Caron, R.-P. Sekaly, A. Veillette, and M. Julius, Association of tyrosine kinase p56*lck* with CD4 inhibits the induction of growth through the αβ T-cell receptor, *Nature*, *358*: 328 (1992).
52. R. Heikkila, G. Schwab, E. Wickstrom, S. L. Loke, D. H. Pluznik, R. Watt, and L. M. Neckers, A c-*myc* antisense oligodeoxynucleotide inhibits entry into S phase but not progress from G_0 to G_1, *Nature*, *328*: 445 (1987).
53. N. J. Holbrook, K. A. Smith, A. J. F. C. M. Comeau, Jr, R. L. Wiskocil, and G. R. Crabtree, T-cell growth factor: Complete nucleotide sequence and organization of the gene in normal and malignant cells, *Proc. Natl. Acad. Sci. USA*, *81*: 1634 (1984).
54. I. D. Horak, R. E. Gress, P. J. Lucas, E. M. Horak, T. A. Waldmann, and J. B. Bolen, T-lymphocyte interleukin 2-dependent tyrosine protein kinase signal transduction involves the activation of p56*lck*, *Proc. Natl. Acad. Sci. USA*, *88*: 1996 (1991).
55. F. D. Howard, P. Moingeon, U. Moebius, D. J. McConkey, B. Yandava, T. E. Gennert, and E. L. Reinherz, The CD3 ζ cytoplasmic domain mediates CD2-induced T cell activation, *J. Exp. Med.*, *176*: 139 (1992).
56. J. A. Hoxie, J. L. Rackowski, B. S. Haggarty, and G. N. Gaulton, T4 endocytosis and phosphorylation induced by phorbol esters but not by mitogen or HIV infection, *J. Immunol.*, *140*: 786 (1988).

57. B. Hoyos, D. W. Ballard, E. Bohnlein, M. Siekevitz, and W. Greene, kappa B-Specific DNA binding proteins: Role in the regulation of human interleukin-2 gene expression, *Science*, *244*: 457 (1989).
58. J. B. Imboden, and J. D. Stobo, Transmembrane signalling by the T cell antigen receptor, *J. Exp. Med.*, *161*: 446 (1985).
59. J. B. Imboden, and A. Weiss, The T-cell antigen receptor regulates sustained increases in cytoplasmic free Ca^{2+} through extracellular Ca^{2+} influx and ongoing intracellular Ca^{2+} mobilization, *Biochem. J.*, *247*: 695 (1987).
60. J. B. Imboden, A. Weiss, and J. D. Stobo, The antigen receptor on a human T cell line initiates activation by increasing cytoplasmic free calium, *J. Immunol.*, *134*: 663 (1985).
61. R. F. Irvine, and R. M. Moor, Micro-injection of inositol 1,3,4,5-tetrakisphosphate activates sea urchin eggs by a mechanism dependent on external Ca^{2+}, *Biochem. J.*, *240*: 917 (1986).
62. B. A. Irving, and A. Weiss, The cytoplasmic domain of the T cell receptor ζ chain is sufficient to couple to receptor-associated signal transduction pathways, *Cell*, *64*: 891 (1991).
63. C. A. Janeway, Accessories or coreceptors, *Nature*, *335*: 208 (1988).
64. Y.-J. Jin, L. K. Clayton, F. D. Howard, S. Koyasu, M. Sieh, R. Steinbrich, G. E. Tarr, and E. L. Reinherz, Molecular cloning of the CD3η subunit identifies a CD3ζ-related product in thymus-derived cells, *Proc. Natl. Acad. Sci.*, *87*: 3319 (1990).
65. C. H. June, M. C. Fletcher, J. A. Ledbetter, and L. E. Samelson, Increases in tyrosine phosphorylation are detectable before phospholipase C activation and after T cell receptor stimulation, *J. Immunol.*, *144*: 1591 (1990).
66. C. H. June, J. A. Ledbetter, P. S. Linsley, and C. B. Thompson, Role of the CD28 receptor in T-cell activation, *Immunol. Today*, *11*: 211 (1990b).
67. J. A. Kant, M. Kamoun, J. E. Burlein, and T. M. Williams, Positive and negative regulatory elements in the human IL2 gene promotor, *Coll. INSERM*, *179*: 53 (1989).
68. A. Kazlauskas, and J. A. Cooper, Autophosphorylation of the PDGF Receptor in the kinase insert region regulates interactions with cell proteins, *Cell*, *58*: 1121 (1989).
69. K. Kelly, B. H. Cochran, C. D. Stiles, and P. Leder, Cell-specific regulation of the c-*myc* gene by lymphocyte mitogens and platelet-derived growth factor, *Cell*, *35*: 603 (1983).
70. J. A. Kern, J. C. Reed, R. P. Daniele, and P. C. Nowell, The role of the accessory cell in mitogen-stimulated human T cell gene expression, *J. Immunol.*, *137*: 764 (1986).
71. A. A. Khan, J. P. Steiner, M. G. Klein, M. F. Schneider, and S. H. Snyder, IP_3 receptor: Localization to plasma membrane of T cells and cocapping with the T cell receptor, *Science*, *257*: 815 (1992).
72. U.-H. Kim, J. W. Kim, and S. G. Rhee, Phosphorylation of phospholipase C-γ by cAMP-dependent protein kinase, *J. Biol. Chem.*, *264*: 20167 (1989).
73. R. D. Klausner, and L. E. Samelson, T cell antigen receptor activation pathways: The tyrosine kinase connection, *Cell*, *64*: 875 (1991).

74. H. F. Kung, I. Calvert, E. Bekesi, F. R. Kahn, K. P. Huang, S. Oroszlan, L. E. Henderson, T. D. Copeland, R. C. Sowder, and S. J. Wei, Phosphorylation of human interleukin-2 (IL-2), *Mol. Cell. Biochem.*, *89*: 29 (1989).
75. C. S. Larsen, N. O. Christiansen, and V. Esmann, Modulation of high-affinity interleukin 2 receptors on activated human T lymphocytes by activators of protein kinase C, *Scand. J. Immunol.*, *28*: 285 (1988).
76. J. A. Ledbetter, N. K. Tonks, E. H. Fischer, and E. A. Clark, CD45 regulates signal transduction and lymphocyte activation by specific association with receptor molecules on T or B cells, *Proc. Natl. Acad. Sci.*, *85*: 8628 (1988).
77. S. J. LeGrue, Does interleukin 2 stimulus-response coupling result in generation of intracellular second messengers? *Lymphokine Res.*, *7*: 187 (1988).
78. W. J. Leonard, J. M. Depper, G. R. Crabtree, S. Rudikoff, J. Pumphrey, R. J. Robb, M. Kronke, P. B. Svetlik, N. J. Peffer, T. A. Waldmann, and W. C. Greene, Molecular cloning and expression of cDNAs for the human interleukin-2 receptor, *Nature*, *311*: 626 (1984).
79. W. J. Leonard, J. M. Depper, R. J. Robb, T. A. Waldmann, and W. C. Greene, Characterization of the human receptor for T-cell growth factor, *Proc. Natl. Acad. Sci. USA*, *80*: 6957 (1983).
80. A. Lerner, B. Jacobson, and R. A. Miller, Cyclic AMP concentrations modulate both calcium flux and hydrolysis of phosphatidylinositol phosphates in mouse T lymphocytes, *J. Immunol.*, *140*: 936 (1988).
81. T. Lindsten, C. H. June, J. A. Ledbetter, G. Stella, and C. B. Thompson, Regulation of lymphokine messenger RNA stability by a surface-mediated T cell activation pathway, *Science*, *244*: 339 (1989).
82. P. S. Linsley, E. A. Clark, and J. A. Ledbetter, The cell antigen, CD28, mediates adhesion with B cells by interacting with the activation antigen, B7/BB-1, *Proc. Natl. Acad. Sci. USA*, *87*: 5031 (1990).
83. J. W. Lowenthal, E. Bohnlein, D. W. Ballard, and W. C. Greene, Regulation of interleukin 2 receptor α subunit (*Tac* or CD25 antigen) gene expression: Binding of inducible nuclear proteins to discrete promoter sequences correlates with transcriptional activation, *Proc. Natl. Acad. Sci. USA*, *85*: 4468 (1988).
84. Y. Lu, A. Granelli-Piperno, J. M. Bjorndahl, C. A. Phillips, and J. M. Trevillyan, CD28-induced T cell activation. Evidence for a protein-tyrosine kinase signal transduction pathway, *J. Immunol.*, *149*: 24 (1992).
85. K. Luo, and B. M. Sefton, Cross-linking of T-cell surface molecules CD4 and CD8 stimulates phosphorylation of *lck* tyrosine protein kinase at the autophosphorylation site, *Mol. Cell. Biol.*, *10*: 5305 (1990).
86. B. Manger, A. Weiss, J. Imboden, T. Laing, and J. D. Stobo, The role of protein kinase C in transmembrane signaling by the T cell antigen receptor complex, *J. Immunol.*, *139*: 2755 (1987).
87. B. Manger, A. Weiss, C. Weyand, J. Goronzy, and J. D. Stobo, T cell activation: Differences in the signals required for IL 2 production by nonactivated and activated T cells, *J. Immunol.*, *135*: 3669 (1985).
88. B. Margolis, S. G. Rhee, S. Felder, M. Mervic, R. Lyall, A. Levitzki, A. Ullrich, A. Zilberstein, and J. Schlessinger, EGF induces tyrosine phosphorylation of phospholipase C-II: A potential mechanism for EGF receptor signaling, *Cell*, *57*: 1101 (1989).

89. J. Meisenhelder, P.-G. Suh, S. G. Rhee, and T. Hunter, Phospholipase C-gamma is a substrate for the PDGF and EGF receptor protein-tyrosine kinases in vivo and in vitro, *Cell*, *57*: 1109 (1989).
90. M. Mercep, J. S. Bonifacino, P. Garcia-Morales, L. E. Samelson, R. D. Klausner, and J. D. Ashwell, T cell CD3-ζη heterodimer expression and coupling to phosphoinositide hydrolysis, *Science*, *242*: 571 (1988).
91. M. Mercep, A. M. Weissman, S. J. Frank, R. D. Klausner, and J. D. Ashwell, Activation-driven programmed cell death and T cell receptor ζη expression, *Science*, *246*: 1162 (1989).
92. I. Merida, and G. N. Gaulton, Protein tyrosine phosphorylation associated with activation of the interleukin 2 receptor, *J. Biol. Chem.*, *265*: 5690 (1990).
93. I. Merida, J. C. Pratt, and G. N. Gaulton, Regulation of interleukin 2-dependent growth responses by glycosylphosphatidylinositol molecules, *Proc. Natl. Acad. Sci. USA*, *87*: 9421 (1990).
94. I. M. Merida, E. Diez, and G. N. Gaulton, Interleukin 2 binding activates a tyrosine phosphorylated phosphatidylinositol 3-kinase, *J. Immunol.*, *147*: 2202 (1991).
95. S. C. Meuer, K. A. Fitzgerald, R. E. Hussey, J. C. Hodgdon, S. F. Schlossman, and E. L. Reinherz, Clonotypic structures involved in antigen-specific human T cell function, *J. Exp. Med.*, *157*: 705 (1983).
96. S. C. Meuer, J. C. Hodgdon, R. E. Hussey, J. P. Protentis, S. F. Schlossman, and E. L. Reinherz, Antigen-like effects of monoclonal antibodies directed at receptors on human T cell clones, *J. Exp. Med.*, *158*: 988 (1983).
97. S. C. Meuer, R. E. Hussey, M. Fabbi, D. Fox, O. Acuto, A. Fitzgerald, J. C. Hodgdon, J. P. Protentis, S. F. Schlossman, and E. L. Reinherz, An alternative pathway of T-cell activation: A functional role for the 50 kd T11 sheep erythrocyte receptor protein, *Cell*, *36*: 897 (1984).
98. G. B. Mills, R. K. Cheung, S. Grinstein, and E. W. Gelfand, Increase in cytosolic free calcium concentration is an intracellular messenger for the production of interleukin 2 but not for expression of the interleukin 2 receptor, *J. Immunol.*, *134*: 1640 (1985).
99. G. B. Mills, R. K. Cheung, S. Grinstein, and E. W. Gelfand, Interleukin 2-induced lymphocyte proliferation is independent of increases in cytosolic free calcium concentrations, *J. Immunol.*, *134*: 2431 (1985b).
100. G. B. Mills, C. May, M. McGill, M. Fung, M. Baker, R. Sutherland, and W. C. Greene, Interleukin 2-induced tyrosine phosphorylation, *J. Biol. Chem.*, *265*: 3561 (1990).
101. G. M. Mills, D. J. Stewart, A. Mellors, and E. W. Gelfand, Interleukin 2 does not induce phosphatidylinositol hydrolysis in activated T cells, *J. Immunol.*, *136*: 3019 (1986).
102. A. R. Mire-Sluis, A. V. Hoffbrand, and R. G. Wickremasinghe, Evidence that guanine-nucleotide binding regulatory proteins couple cell-surface receptors to the breakdown of inositol-containing lipids during T-lymphocyte mitogenesis, *Biochem. Biophys. Res. Commun.*, *148*: 1223 (1987).
103. J. F. Modiano, R. Kolp, R. J. Lamb, and P. C. Nowell, Protein kinase C regulates both production and secretion of interleukin 2, *J. Biol. Chem.*, *266*: 10552 (1991).

104. K. Muegge, T. W. Williams, J. Kant, M. Karin, R. Chiu, A. Schmidt, U. Siebenlist, H. A. Young, and S. K. Durum, Interleukin-1 costimulatory activity on the interleukin-2 promoter via AP-1, *Science*, *246*: 249 (1989).
105. E. W. Mullner, B. Neupert, and L. C. Kuhn, A specific mRNA binding factor regulates the iron-dependent stability of cytoplasmic transferrin receptor mRNA, *Cell*, *58*: 373 (1989).
106. T. Mustelin, GTP dependence of the transduction of mitogenic signals through the T3 complex in T lymphocytes indicates the involvement of a G-protein, *FEBS Lett.*, *213*: 199 (1987).
107. T. Mustelin, K. M. Coggeshall, N. Isakov, and A. Altman, T cell antigen receptor-mediated activation of phospholipase C requires tyrosine phosphorylation, *Science*, *247*: 1584 (1990).
108. L. M. Neckers, S. Bauer, R. C. McGlennen, J. B. Trepel, K. Rao, and W. C. Greene, Diltiazem inhibits transferrin receptor expression and causes G_1 arrest in normal and neoplastic T cells, *Mol. Cell. Biol.*, *6*: 4244 (1986).
109. L. M. Neckers, and J. Cossman, Transferrin receptor induction in mitogen stimulated human T lymphocytes is required for DNA synthesis and cell division and is regulated by interleukin 2, *Proc. Natl. Acad. Sci. USA*, *80*: 3493 (1983).
110. J. J. O'Shea, K. B. Urdahl, H. T. Luong, T. M. Chused, L. E. Samelson, and R. D. Klausner, Aluminum fluoride induces phosphatidylinositol turnover, elevation of cytoplasmic free calcium, and phosphorylation of the T cell antigen receptor in murine T cells, *J. Immunol.*, *139*: 3463 (1987).
111. T. Ohmura, A. Sakata, and K. Onoue, A 68-kD GTP-binding protein associated with the T cell receptor complex, *J. Exp. Med.*, *176*: 887 (1992).
112. S. Ohta, A. Matsui, Y. Nazawa, and Y. Kagawa, Complete cDNA encoding a putative phospholipase C from transformed human lymphocytes, *FEBS Lett.*, *242*: 31 (1988).
113. N. E. Olashaw, S. G. Rhee, and W. J. Pledger, Cyclic AMP agonists induce the phosphorylation of phospholipase C-γ and of a 76 kDa protein coprecipitated by anti-(phospholipase C-γ) monoclonal antibodies in BALB/c-3T3 cells, *Biochem. J.*, *272*: 297 (1990).
114. H. L. Ostergaard, D. A. Shackelford, T. R. Hurley, P. Johnson, R. Hyman, B. M. Sefton, and I. S. Trowbridge, Expression of CD45 alters phosphorylation of the *lck*-encoded tyrosine protein kinase in murine lymphoma T-cell lines, *Proc. Natl. Acad. Sci. USA*, *86*: 8959 (1989).
115. J. T. Pingel, and M. L. Thomas, Evidence that the leukocyte-common antigen is required for antigen-induced T lymphocyte proliferation, *Cell*, *58*: 1055 (1989).
116. J. L. Pomerantz, F. Mauxion, M. Yoshida, W. C. Green, and R. Sen, A second sequence element located 3′ to the NF-κB-binding site regulates IL-2 receptor-α gene induction, *J. Immunol.*, *143*: 4275 (1989).
117. J. C. Pratt, and G. N. Gaulton, Purification and characterization of PI-PLC-γ from human T lymphocytes, (submitted, 1993).
118. J. C. Reed, J. D. Alpers, and P. C. Nowell, Sequential expression of protooncogenes during lectin-stimulated mitogenesis of normal human lymphocytes, *Proc. Natl. Acad. Sci. USA*, *83*: 3982 (1986).
119. J. C. Reed, J. D. Alpers, P. A. Scherle, R. G. Hoover, P. C. Nowell, and M. B.

119. J. C. Reed, J. D. Alpers, P. A. Scherle, R. G. Hoover, P. C. Nowell, and M. B. Prystowsky, Proto-oncogene expression in cloned T lymphocytes: Mitogens and growth factors induce different patterns of expression, *Oncogene*, *1*: 223 (1987).
120. H. Reiser, H. Oettgen, E. T. H. Yeh, C. Terhorst, M. G. Low, B. Benacerraf, and K. L. Rock, Structural characterization of the TAP molecule: A phosphatidylinositol-linked glycoprotein distinct from the T cell receptor/T3 complex and Thy-1, *Cell*, *47*: 365 (1986).
121. R. J. Robb, W. C. Greene, and C. M. Rusk, Low and high affinity cellular receptors for interleukin 2. Implications for the level of Tac antigen, *J. Exp. Med.*, *160*: 1126 (1984).
122. R. J. Robb, R. M. Kutny, M. Panico, H. R. Morris, and V. Chowdhry, Amino-acid sequence and post-translational modifications of human interleukin 2, *Proc. Natl. Acad. Sci. USA*, *81*: 6486 (1984).
123. R. J. Robb, A. Munck, and K. A. Smith, T-cell growth factor receptors: quantitation, specificity, and biological relevance, *J. Exp. Med.*, *154*: 1455 (1981).
124. K. L. Rock, E. T. H. Yeh, C. F. Gramm, S. I. Haber, H. Reiser, and B. Benacerraf, TAP, a novel T-cell-activating protein involved in the stimulation of MHC-restricted T lymphocytes, *J. Exp. Med.*, *163*: 315 (1986).
125. J. L. Rossio, F. W. Ruscetti, and W. L. Farrar, Ligand-specific calcium mobilization in IL2 and IL3 dependent cell lines, *Lymphokine Res.*, *5*: 163 (1986).
126. D. E. Sabath, H. E. Broome, and M. B. Prystowsky, Glyceraldehyde-3-phosphate dehydrogenase mRNA is a major interleukin 2-induced transcript in a cloned T-helper lymphocyte, *Gene*, *91*: 185 (1990).
127. D. E. Sabath, P. L. Podolin, P. G. Comber, and M. B. Prystowsky, cDNA cloning and characterization of interleukin 2-induced genes in a cloned T helper lymphocyte, *J. Biol. Chem.*, *265*: 12671 (1990a).
128. E. M. Saltzman, R. R. Thorn, and J. E. Casnellie, Activation of a tyrosine protein kinase is an early event in the stimulation of T lymphocytes by interleukin-2, *J. Biol. Chem.*, *263*: 6956 (1988).
129. L. E. Samelson, A. F. Phillips, E. T. Luong, and R. D. Klausner, Association of the *fyn* protein-tyrosine kinase with the T-cell antigen receptor, *Proc. Natl. Acad. Sci. USA*, *87*: 4358 (1990).
130. T. Sasaki, and H. Hasegawa-Sasaki, Activation of polyphosphoinositide phospholipase C by guanosine 5′-*O*-(3-thio)triphosphate and fluoroaluminate in membranes prepared from a human T cell leukemia line, JURKAT, *FEBS Lett.*, *218*: 87 (1987).
131. J. P. Secrist, L. Karnitz, and R. T. Abraham, T-cell antigen receptor ligation induces tyrosine phosphorylation of phospholipase C-γ1, *J. Biol. Chem.*, *266*: 12135 (1991).
132. D. A. Shackelford, and I. S. Trowbridge, Induction of expression and phosphorylation of the human interleukin 2 receptor by phorbol diester, *J. Biol. Chem.*, *259*: 11706 (1984).
133. M. Sharon, R. D. Klausner, B. R. Cullen, R. Chizzonite, and W. J. Leonard, Novel interleukin-2 receptor subunit detected by cross-linking under high-affinity conditions, *Science*, *234*: 859 (1987).

134. A. S. Shaw, K. E. Amrein, C. Hammond, D. F. Stern, B. M. Sefton, and J. K. Rose, The *lck* tyrosine protein kinase interacts with the cytoplasmic tail of the CD4 glycoprotein through its unique amino-terminal domain, *Cell*, *59*: 627 (1989).
135. P. M. Shipman, D. E. Sabath, A. H. Fischer, P. G. Comber, K. Sullivan, E. M. Tan, and M. B. Prystowsky, Cyclin mRNA and protein expression in recombinant interleukin 2-stimulated cloned murine T lymphocytes, *J. Cell. Biol.*, *38*: 189 (1988).
136. P. Shipman-Appasamy, K. S. Cohen, and M. B. Prystowsky, Interleukin 2-induced expression of proliferating cell nuclear antigen is regulated by transcriptional and post-translational mechanisms, *J. Biol. Chem.*, *265*: 19180 (1990).
137. J. P. Siegel, M. Sharon, P. L. Smith, and W. J. Leonard, The IL-2 receptor β chain (p70): Role in mediating signals for LAK, NK, and proliferative activities, *Science*, *238*: 75 (1987).
138. B. P. Sleckman, Y. Rosenstein, V. E. Igras, J. L. Greenstein, and S. J. Burakoff, Glycolipid-anchored form of CD4 increases intracellular adhesion but is unable to enhance T cell activation, *J. Immunol.*, *147*: 428 (1991).
139. K. A. Smith, Interleukin-2: Inception, impact and implications, *Science*, *240*: 1169 (1988).
140. K. A. Smith, Interleukin-2, *Sci. Am.*, *262*: 50 (1990).
141. I. Stefanova, V. Horejsi, I. J. Ansotegui, W. Knapp, and H. Stockinger, GPI-anchored cell-surface molecules complexed to protein tyrosine kinases, *Science*, *254*: 1016 (1991).
142. P. L. Stein, H.-M. Lee, S. Rich, and P. Soriano, $pp59^{fyn}$ mutant mice display differential signaling in thymocytes and peripheral T cells, *Cell*, *70*: 741 (1992).
143. J. B. Stern, and K. A. Smith, Interleukin-2 induction of T-cell G_1 progression and c-*myb* expression, *Science*, *233*: 203 (1986).
144. D. B. Strauss, and A. Weiss, Genetic evidence for the involvement of the *lck* tyrosine kinase in signal transduction through the T cell antigen receptor, *Cell*, *70*: 585 (1992).
145. G. P. Studzinski, Z. S. Brelvi, S. C. Feldman, and R. A. Watt, Participation of c-*myc* protein in DNA synthesis of human cells, *Science*, *234*: 467 (1986).
146. J. J. Sussman, M. Mercep, T. Saito, R. N. Germain, E. Bonvini, and J. D. Ashwell, Dissociation of phosphoinositide hydrolysis and Ca^{2+} fluxes from the biological responses of a T-cell hybridoma, *Nature*, *334*: 625 (1988).
147. T. Takeshita, H. Asao, K. Ohtani, N. Ishii, S. Kumaki, N. Tanaka, H. Munakata, M. Nakamura, and K. Sugamura, Cloning of the γ chain of the human IL-2 receptor, *Science*, *257*: 379 (1992).
148. T. Taniguchi, H. Matsui, T. Fujita, M. Hatakeyama, N. Kashima, A. Fuse, J. Hamuro, C. Nishi-Takaoka, and J. Yamada, Molecular analysis of the interleukin-2 system, *Immunol. Rev.*, *92*: 121 (1986).
149. T. Taniguchi, H. Matsui, T. Fujita, C. Takaoka, N. Kashima, R. Yoshimoto, and J. Hamuro, Structure and expression of a cloned cDNA for human interleukin-2, *Nature*, *302*: 305 (1983).
150. S. J. Taylor, H. Z. Chae, S. G. Rhee, and J. H. Exton, Activation of the β1 isozyme of phospholipase C by α subunits of the Gq class of G proteins, *Nature*, *350*: 516 (1991).

151. C. B. Thompson, P. G. Challoner, P. E. Neiman, and M. Groudine, Expression of the c-*myb* proto-oncogene during cellular proliferation, *Nature*, *319*: 374 (1986).
152. C. B. Thompson, T. Lindsten, J. A. Ledbetter, S. L. Kunkel, H. A. Young, S. G. Emerson, J. M. Leiden, and C. H. June, CD28 activation pathway regulates the production of multiple T cell-derived lymphokines/cytokines, *Proc. Natl. Acad. Sci. USA*, *86*: 1333 (1989).
153. M. A. Tigges, L. S. Casey, and M. E. Koshland, Mechanism of interleukin-2 signalling: Mediation of different outcomes by a single receptor and transduction pathway, *Science*, *243*: 781 (1989).
154. T. Torigoe, H. U. Saragovi, and J. C. Reed, Interleukin 2 regulates the activity of the *lyn* protein-tyrosine kinase in a B cell line, *Proc. Natl. Acad. Sci. USA*, *89*: 2674 (1992).
155. A. Truneh, F. Albert, P. Goldstein, and A.-M. Schmitt-Verhulst, Early steps of lymphocyte activation bypassed by synergy between calcium ionophores and phorbol ester, *Nature*, *313*: 318 (1985).
156. B. Turner, U. Rapp, H. App, M. Greene, K. Dobashi, and J. Reed, Interleukin-2 induces tyrosine phosphorylation and activation of p72-74 Raf-1 kinase in a T cell line, *Proc. Natl. Acad. Sci. USA*, *88*: 1227 (1991).
157. T. Uchiyama, S. Broder, and T. A. Waldmann, A monoclonal antibody (anti-Tac) reactive with activated and functionally mature T cells, *J. Immunol.*, *126*: 1393 (1981).
158. P. Vandenberghe, G. J. Freeman, L. M. Nadler, M. C. Fletcher, M. Kamoun, L. A. Turka, J. A. Ledbetter, C. B. Thompson, and C. H. June, Antibody and B7/BB1-mediated ligation of the CD28 receptor induces tyrosine phosphorylation in human T cells, *J. Exp. Med.*, *175*: 951 (1992).
159. A. Veillette, J. B. Bolen, and M. B. Bookman, Alterations in tyrosine protein phosphorylation induced by antibody-mediated cross-linking of the CD4 receptor of T lymphocytes, *Mol. Cell. Biol.*, *9*: 4441 (1989).
160. A. Veillette, M. A. Bookman, E. M. Horak, and J. B. Bolen, The CD4 and CD8 T cell surface antigens are associated with the internal membrane tyrosine-protein kinase $p56^{lck}$, *Cell*, *55*: 301 (1988).
161. A. Veillette, M. A. Bookman, E. M. Horak, L. E. Samelson, and J. B. Bolen, Signal transduction through the CD4 receptor involves the activation of the internal membrane tyrosine-protein kinase $p56^{lck}$, *Nature*, *338*: 257 (1989).
162. U. S. Vogel, R. A. F. Dixon, M. D. Schaber, R. E. Diehl, M. S. Marshall, E. M. Scolnick, I. S. Sigal, and J. B. Gibbs, Cloning of bovine GAP and its interaction with oncogenic *ras* p21, *Nature*, *335*: 90 (1988).
163. S. Volarevic, C. M. Burns, J. J. Sussman, and J. D. Ashwell, Intimate association of Thy-1 and the T-cell antigen receptor with the CD45 tyrosine phosphatase, *Proc. Natl. Acad. Sci. USA*, *87*: 7085 (1990).
164. M. I. Wahl, T. O. Daniel, and G. Carpenter, Antiphosphotyrosine recovery of phospholipase C activity after EGF treatment of A-431 cells, *Science*, *241*: 968 (1988).
165. M. I. Wahl, S. Nishibe, J. W. Kim, H. Kim, S. G. Rhee, and G. Carpenter, Identification of two epidermal growth factor-sensitive tyrosine phosphorylation sites of phospholipase C-γ in intact HSC-1 cells, *J. Biol. Chem.*, *265*: 3944 (1990).

166. T. A. Waldmann, The structure, function, and expression of interleukin-2 receptors on normal and malignant lymphocytes, *Science*, *232*: 727 (1986).
167. H. M. Wang, and K. A. Smith, The interleukin 2 receptor: Functional consequences of its bimolecular structure, *J. Exp. Med.*, *166*: 1055 (1987).
168. S. G. Ward, and D. A. Cantrell, Heterogeneity of the regulation of phospholipase C by phorbol esters in T lymphocytes, *J. Immunol.*, *144*: 3523 (1990).
169. V. M. Wasenius, M. Saraste, P. Salven, M. Eramaa, L. Holm, and V. P. Lehto, Primary structure of the brain α-spectrin, *J. Cell. Biol.*, *108*: 79 (1989).
170. A. Weiss, J. Imboden, D. Shoback, and J. Stobo, Role of T3 surface molecules in human T-cell activation: T3-dependent activation results in an increase in cytoplasmic free calcium, *Proc. Natl. Acad. Sci. USA*, *81*: 4169 (1984).
171. A. Weiss, G. Koretzky, R. C. Schatzman, and T. Kadlecek, Functional activation of the T-cell antigen receptor induces tyrosine phosphorylation of phospholipase C-γ1, *Proc. Natl. Acad. Sci. USA*, *88*: 5484 (1991).
172. A. Weiss, R. Shields, M. Newton, B. Manger, and J. Imboden, Ligand–receptor interactions required for commitment to the activation of the interleukin 2 gene, *J. Immunol.*, *138*: 2169 (1987).
173. A. Weiss, R. L. Wiskocil, and J. D. Stobo, The role of T3 surface molecules in the activation of human T cells: A two-stimulus requirement for IL2 production reflects events occurring at a pretranslational level, *J. Immunol.*, *133*: 123 (1984).

7

Molecular Aspects of Antigen Processing and Presentation

Vivian Lam Braciale
University of Virginia Health Sciences Center, Charlottesville, Virginia

I. HISTORICAL BACKGROUND

It has been known for quite some time that T lymphocytes recognize both native and denatured forms of antigen (1), whereas antibodies generally recognize conformational determinants present in native antigen, but absent in denatured antigen. The observation in the mid-1970s that T lymphocytes recognize foreign antigen in a major histocompatibility complex (MHC)-restricted fashion (2–5) led to studies seeking the nature of the T-cell receptor (TCR) as well as the nature of the foreign antigenic moiety recognized on antigen-presenting cells (APC) for helper T lymphocytes, or on target cells for cytolytic T lymphocytes (CTL).

Studies in the early 1980s (6,7) revealed that antigen (Ag) was likely taken up by the APC, here, the macrophage, and processed in some manner that was inhibitable by agents capable of causing an increase in the pH in lysosomal and endosomal compartments. In addition, it was demonstrated that fixed (paraformaldehyde-treated) APCs were unable to present protein antigens, but could do so if the antigen was taken up by the APC, and a few hours incubation at 37°C was permitted before fixation. These processing events (most likely proteolytic cleavage of antigen) could be bypassed by incubating CNBr or trypsin cleavage products (cell-free antigen processing) with the fixed APC, here the B cell (8). These observations have held true for processing of various soluble protein antigens (e.g., cytochrome c, lysozyme, ovalbumin, myoglobin, ribonuclease, and insulin).

The determinant selection hypothesis, first advanced by Rosenthal et al. (9), was extended in a model proposed by Heber-Katz et al. (10) to explain Ir gene responses and MHC-restricted recognition by T lymphocytes. The model defined areas of contact between the TCR, the antigenic fragment, and the MHC restriction molecule. A site on the TCR, termed the restitope, interacts with a site on the MHC molecule, the histiotope. Another site on the TCR, termed the paratope, interacts with a site on the antigenic fragment, the epitope. Finally, a site on the antigenic fragment, termed the agretope, interacts with a site on the MHC restriction molecule, the desetope. A trimolecular complex of TCR, class II MHC molecule, and the antigenic peptide fragment would thus be formed. A T-cell epitope,

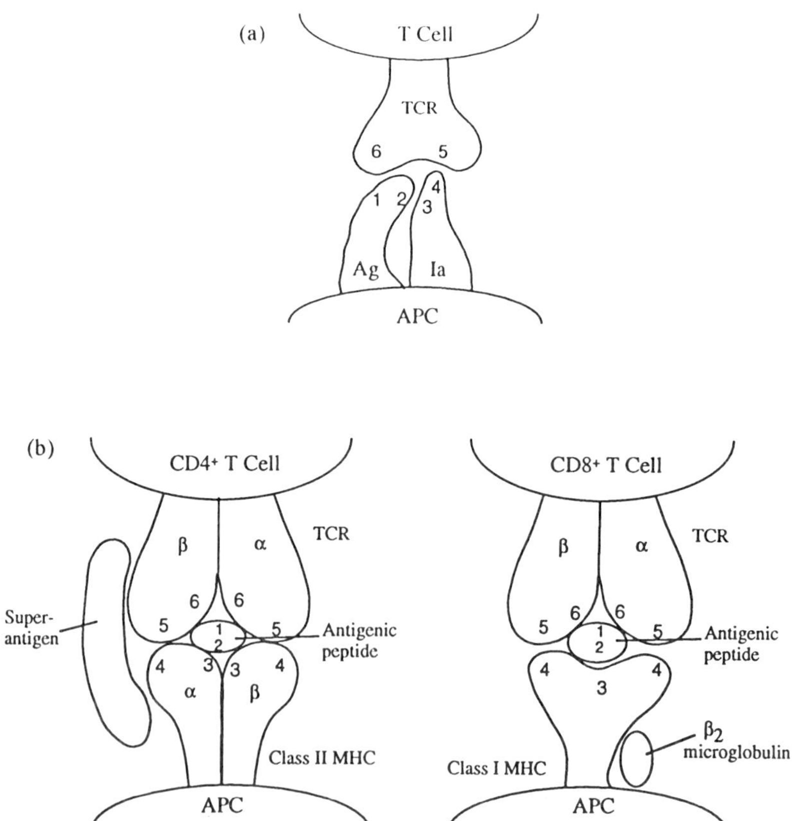

Figure 1 (a) 1 is the epitope, 2 is the agretope, 3 is the desetope, 4 is the histotope, 5 is the restitope, 6 is the paratope. (b) Numbers 1–6 correspond with those in (a). (Adapted from Ref. 134.)

therefore, would need to be able to bind MHC as well as be recognized by the TCR (Fig. 1a). The nomenclature in Figure 1a is rarely used, but does define important sites of MHC–Ag–TCR contact as depicted in Figure 1b.

An observation that supported the determinant selection hypothesis and represents a major breakthrough in our understanding of antigen processing, Ir gene responses, and MHC-restricted recognition came when Unanue and co-workers (11) demonstrated that isolated class II MHC molecules could bind antigenic peptide in an haplotype-specific manner. Also, with a few exceptions, the antigenic peptide could bind only the class II MHC, which serves as the restriction element in T-cell recognition of that peptide. That peptides restricted by the same MHC molecule can compete with each other for binding to the MHC, indicated that there was one peptide-binding site.

Another advance that supported the notion that MHC molecules can bind peptides directly came from the described structure of a human class I MHC molecule, which emerged from the data on crystals of HLA-A2 (12,13). There was apparent electron-dense material occupying the groove that was postulated to be peptide-bound by class I MHC molecules. It had, by this time, been demonstrated that synthetic peptides could also mimic T-cell antigenic determinants for class I MHC-restricted recognition. Thus, by analogy, both class I and class II MHC molecules can bind and present antigenic peptides to T cells. Direct binding of peptide to class I MHC was not easily demonstrable at that time. In addition, the crystal structure of class II MHC has not been determined and is presently modeled on the class I MHC structure. Figure 1b is a modification of Figure 1a in light of current information for class I and class II antigen presentation.

II. DISTINCT PATHWAYS FOR ANTIGEN PROCESSING AND PRESENTATION FOR CLASS I AND CLASS II MAJOR HISTOCOMPATIBILITY COMPLEX MOLECULES

To understand the nature of the antigen–MHC moiety recognized by T lymphocytes, several approaches were pursued to characterize which parts of a given foreign protein might serve as a T-lymphocyte determinant. In addition to defining such T-cell epitopes, it was of interest to understand the mechanism(s) by which antigen processing and presentation occur. There were several ways to mimic the processing events. This was first characterized for antigen presentation in the context of class II MHC molecules. Since the products of proteolytic cleavage of a given protein could sensitize APC for T-lymphocyte recognition, the logical path was to sequence the particular peptide fragment that could sensitize for T-cell recognition. The development of clonal T-cell populations was invaluable for dissecting the recognition of a given determinant within the protein. Antigen-specific, MHC-restricted T-cell clonal recognition of Ag–MHC was quantified in proliferation assays, whereas the Ag–MHC recognition by T-cell hybridomas correlated with the levels of interleukin (IL)-2 produced (measured in a bioassay) in the culture supernatants. The minimal peptide sequence that could mimic the T-cell epitope was determined by use of synthetic peptides that corresponded to various NH_2-terminal or COOH-terminal truncations of the original peptide fragment sequence. The part of a given protein that could serve as a T-cell determinant differed according to, and was dependent on, the mouse strain haplotype. Given the observation that there is demonstrable haplotype-specific binding for synthetic peptides, it was concluded that a prerequisite for a given peptide to be a T-cell determinant in a particular haplotype was the property that it bind an MHC molecule of that haplotype. In addition to mimicking the putative T-cell determinant, synthetic peptides could be used to elicit T-cell responses in place of the whole protein. If the synthetic peptide corresponds to a site generated by APC processing of whole protein, then T cells elicited from immunization with such a peptide will recognize APC-processed whole protein. Some synthetic peptides correspond to a sequence in the protein that fulfills the criterion that it bind the particular class II MHC molecule, but immunization with this peptide elicits T cells that cannot recognize APC-processed whole protein, yet are able to recognize APC pulsed with the immunizing peptide (14). This type of peptide might represent regions (termed cryptic epitopes) of the protein capable of binding the restriction element, but that may not be a T determinant because (1) it is destroyed during APC processing; (2) if produced, it is generated in too low a concentration; or (3) it is of too low an MHC-binding affinity to be successfully presented to T cells. In addition to using peptide-pulsed APC to characterize T-cell recognition, there have been studies employing isolated class II MHC molecules incorporated into planar membranes that could then be peptide pulsed and used to stimulate IL-2 production by T hybridomas (15).

In an effort to dissect the processing and presentation events that result in generation of the Ag–MHC complexes, investigators employed inhibitors of intracellular compartments as well as particular enzyme inhibitors. Chloroquine and ammonium chloride inhibit antigen processing and raise the pH of lysosomal and endosomal compartments. Since this change in pH might cause resident acid proteases to be inactive, inhibitors of various cellular proteases were screened for the ability to inhibit T-cell epitope generation. Berzofsky and colleagues showed that for sperm whale myoglobin, leupeptin and antipain (inhibitors of serine and thiol proteases), and E-64 (inhibitor of thiol proteases) were inhibitory, whereas pepstatin, (chymostatin, *N*-tosyl-L-phenylalanylchloromethyl ketone (TPCK), and tosyl lysine chloromethyl ketone (TLCK) were largely unable to inhibit processing (16). These results suggested a role for cathepsin B, a thiol protease present in endosomes, as a processing enzyme for myoglobin. Another candidate endosomal protease (17) that may process some, but not all, antigens is cathepsin D, which is inhibited by pepstatin.

Townsend and co-workers demonstrated that synthetic peptides could mimic T-cell epitopes for class I-restricted T lymphocytes (18), which implied that antigen was processed and presented in the context of the class I MHC. Unlike the peptides and class II MHC-restricted T-cell responses, it has been difficult, except for rare examples, to elicit class I MHC-restricted T-cell responses in vivo using peptides. In vitro restimulation of class I MHC-restricted T lymphocytes can be achieved with the antigenic synthetic peptide, but in vitro primary stimulation with peptide has been difficult to achieve in a uniform manner. In instances when peptide has elicited in vitro primary responses, the responder T cells were not able to recognize endogenously processed and presented antigen (19).

Studies of processing and presentation of influenza hemagglutinin (HA) to class I and class II MHC-restricted cytolytic T lymphocytes (CTL) led to the finding that the antigen-processing and presentation pathways for the two classes of MHC appeared to be distinct (20,21). In these studies, the parameters of antigen presentation were compared by using several class I and class II MHC-restricted CTL clones (22). The same cell line (a murine B lymphoma cell, A20), expressing both class I and class II MHC molecules, was used as the target for two CTL subsets in the standard ^{51}Cr release cytotoxicity assay. The results, summarized in Table 1, were that inactivated virus could provide

Table 1 Parameters of Antigen Processing and Presentation to Class I and Class II MHC-Restricted CTL Clones

^{51}Cr-labeled A20 B-lymphoma target cell sensitization	$CD8^+$ CTL clones	$CD4^+$ CTL clones
Uninfected	−	−
Infectious influenza virus[a]	+	−
UV-inactivated influenza virus[a]	−	+
Infectious influenza virus + emetine[b]	−	+
Infectious influenza virus + chloroquine[c]	+	−
UV-inactivated A/JAPAN/57 virus + chloroquine[c]	−	−
Soluble bromelain-cleaved hemagglutinin	−	+
Infectious recombinant vaccina virus encoding influenza HA	+	−
Synthetic peptide mimics of HA epitopes	+	+

[a]A/JAPAN/57 strain.
[b]Inhibitor of protein synthesis.
[c]Lysosomotropic agent.
Source: Adapted from Ref. 20.

a source of antigen for processing and presentation in the context of class II, but not of class I, MHC molecules. This has been confirmed by numerous reports of the difficulty of eliciting class I MHC-restricted responses with exogenous soluble protein antigen. In addition, it was demonstrated that protein synthesis inhibitors could inhibit infectious virus sensitization of target cells for recognition by class I MHC-restricted CTL, but not by class II MHC-restricted CTL. Finally, a recombinant vaccinia virus vector encoding the influenza hemagglutinin could sensitive targets for class I (newly synthesized hemagglutinin), but not class II, MHC-restricted CTL recognition (there was no source of hemagglutinin in the input virions). By this direct comparison of the two CTL subsets, using the same assay and the same target cell line, as well as numerous other findings from other laboratories, in separate studies of class I MHC-restricted CTL or class II MHC-restricted T-helper cells, it was suggested that there is an “endogenous” pathway (for class I MHC antigen processing and presentation) and an “exogenous” pathway (for class II MHC) of antigen processing and presentation (23,24).

To determine whether the lack of overlap of the class I endogenous and the class II exogenous pathways could be due to processing differences inherent in the proteolytic enzymes in the two pathways, Sweetser et al. used an expressed minigene product to sensitize target cells for recognition by class I and class II MHC-restricted CTL clones. In these studies, the minigene encoded a product corresponding to a synthetic peptide shown to be able to mimic an epitope for an H-2K^k-restricted influenza hemagglutinin epitope and also an epitope for an I-A^d-restricted influenza hemagglutinin epitope. When this minigene was expressed in the appropriate target cell expressing both H-2K^k and I-A^d, the targets were recognized by the class I-restricted CTL clone, but not by the class II-restricted CTL clone, implying that the peptide product expressed in the cytosol was not able to access the class II MHC-charging compartment (presumably an endosome), but was able to access the class I MHC-charging compartment (presumably the endoplasmic reticulum; ER) (25).

III. CLASS I MAJOR HISTOCOMPATIBILITY COMPLEX PROCESSING AND PRESENTATION PATHWAY

Both class I and class II MHC-restricted T cells can recognize nonnative fragments of antigen, but pathways of actual antigen processing and presentation for each appear distinct. The pathway for antigen processing and presentation in the context of class I MHC has been termed the *endogenous pathway,* because the antigenic peptides presented by class I MHC are generated intracellularly. Newly synthesized viral proteins can be processed and presented by this pathway, regardless of whether the protein is destined to be cytosolic, nuclear, a cell surface glycoprotein, or a secreted soluble protein. Townsend showed that cytosolic nucleocapsid of influenza could be recognized by $CD8^+$ CTL (26), thereby implicating the cytoplasm as a possible site of processing antigens for class I MHC-restricted recognition. Many $CD8^+$ CTL recognize viral membrane glycoproteins that contain a signal sequence important in targeting proteins to the endoplasmic reticulum (ER) and biosynthesis through the secretory pathway. Studies of a recombinant vaccinia vector expressing mutant hemagglutinin (HA) protein lacking the signal sequence (signal minus HA), where the mutant HA is not cotranslationally translocated into the ER, indicated that signal minus HA can be processed and presented in the context of class I MHC molecules (27). This raises the possibility that, even for wild-type full-length HA, some HA may be synthesized in the cytoplasm under conditions in which signal recognition protein (SRP) may be limiting. Hemagglutinin assembles as a trimer in the ER, but is likely to be monomeric in the cytosol, where it may be unstable and, therefore, readily degraded in the cytosol. Protein antigen introduced in the cytosol bypasses the requirement for new protein synthesis. Certain viruses have fusion capacity with the cell plasma membrane at neutral pH and, consequently, can inject virion proteins into the cytosol. This was demonstrated for a particular strain of influenza virus (28) and may be true for Sendai virus. Bevan and co-workers introduced soluble protein into the cell cytosol, by using a method of osmotic lysis of vesicles containing pinocytosed soluble protein, which was then processed and presented in the context of class I MHC molecules (29). Such findings suggest that the cytosol may contain proteases important in the processing of antigen for the endogenous pathway. It is also possible that some viral glycoprotein antigens, such as HA, can be degraded in the ER. There is as yet little known of ER resident proteases.

Studies using brefeldin A, an inhibitor of transport out of the ER, demonstrated that newly synthesized class I MHC was required for presentation of foreign antigen and that processed viral antigen may interact with newly synthesized class I MHC molecules in a pre-Golgi compartment (30,31). The demonstration that adenovirus E3/19K could bind class I MHC molecules and, thereby, retain them in the ER also supports the notion that newly synthesized class I MHC molecules are necessary for antigen presentation (32–34).

That newly synthesized class I MHC is necessary has implications for viral antigen presentation. Many viruses shut off host protein synthesis in the infected cells. This suggests that early viral proteins are more likely to be processed and presented before class I MHC synthesis is fully terminated. This is true for several viruses including cytomegalovirus. Other viruses, which might not shut off host protein synthesis, employ strategies such as binding of class I MHC molecules in the ER and, in effect, suppress antigen presentation by inhibiting egress of class I MHC molecules from the ER to the cell surface (35).

IV. CLASS II MAJOR HISTOCOMPATIBILITY COMPLEX PROCESSING AND PRESENTATION PATHWAY

The class II MHC antigen presentation pathway functions to process and present exogenous antigen that can enter the antigen-presenting cells by phagocytosis, receptor-mediated endocytosis, and fluid-phase pinocytosis. These different modes of antigen uptake and the type of cell acting as the APC can influence the processing and presentation of certain antigens. Many viruses have cell surface receptors which, on binding virions, undergo receptor-mediated endocytosis. Some antigens may be bound by B lymphocytes that express the particular cell surface immunoglobulin (sIg) receptor with specificity for that antigen. This type of receptor-mediated uptake can be 1000-fold more efficient than the nonspecific uptake of antigen by fluid-phase pinocytosis (36). It has been suggested that there may be an effect of the specificity of the cell surface immunoglobulin (Ig) receptor of B lymphocytes on the determination of the T epitopes of the antigen it specifically binds, internalizes, processes, and presents (37). In addition, the antigen bound to surface Ig receptor may be directed to the appropriate endosomal compartment and may be aided in the charging of class II MHC molecules. That there may be cell surface entities that can help class II MHC bind antigenic peptides, possibly by acting as a reservoir, has been suggested by Pierce and co-workers for a cell surface protein (38) that is a member of the HSP70 heat-shock protein family (39) and by Rock and colleagues for the cell membrane itself (40). Phagocytosis of particulate antigens by macrophages may be quite efficient or of intermediate efficiency.

A. Exceptions to Separate Pathways for Class I and Class II Major Histocompatibility Complex Antigen Processing and Presentation

Reports of exceptions to the rule of the exogenous pathway for class II MHC antigen processing and presentation were those demonstrating that endogenously synthesized antigen can be presented in the context of class II MHC molecules. Some are studies of APC that are permanent transfectants expressing the foreign antigen endogenously, such as the measles virus matrix and nucleocapsid proteins (41). In studies of permanent transfectants producing variant forms of immunoglobulin λ light chain, Weiss and Bogen demonstrated that all forms could be processed and presented, except if the mutant protein remains cytosolic or is transported to the nucleus (42). In these permanent transfectants, the antigen may be present in the cell culture medium, taken up by the cells from the exogenous milieu, and presented in the context of class II MHC. Other studies include those in which viral antigen is expressed briefly in virally infected target cells (43). These have mostly used human CTL effectors and Epstein–Barr virus (EBV)-transformed B-lymphoblastoid cell line (B-LCL) targets. A common thread in these studies is that the antigen, although endogenously synthesized, must somehow access the lysosomal–endosomal compartment to be processed or to bind the class II MHC molecules.

Presentation of endogenous proteins in the apparent absence of intersection with the endosomal–lysosomal pathways have been reported by Jin et al. for the human T-cell response to the surface antigen of hepatitis B virus (HBsAg) (44) and by Nuchtern et al. for recognition of influenza virus Λ matrix protein for class II MHC-restricted human T cells (43). In studies using B-lymphoma cells transfected with genes encoding soluble hen egg lysozyme (sHEL), an ER localized form of HEL (ER-HEL), or membrane-anchored HEL (mHEL), it was shown that each form of HEL was presented, with different efficiencies, and apparently not through uptake of antigen from the culture medium (45). More studies of this nature in the mouse as well as the human, utilizing different APCs, and proliferation or lymphokine production assays as well as cytotoxicity assays should reveal the details of an endogenous antigen presentation pathway for class II MHC.

V. RECENT ADVANCES: ROLE OF ANTIGEN PROCESSING AND PRESENTATION IN CLASS I AND CLASS II MAJOR HISTOCOMPATIBILITY COMPLEX BIOSYNTHESIS

In probing the steps in antigen processing and presentation for both class I and class II MHC molecules, it has become clear that although the binding of antigenic peptides may be the functional end purpose of these molecules, in addition, in the course of biosynthesis of these molecules, there may be a requirement for binding of peptide that is necessary and critical for the formation of functional MHC molecules. Information pertaining to this comes from studies of mutant cell lines defective in either class I or class II antigen processing and presentation.

A. Class I Major Histocompatibility Complex Biosynthesis

Recent advances in our understanding of steps in the class I MHC antigen-processing and antigen-presentation pathway (endogenous pathway) have come from initial studies on mutant cell lines—the murine cell line RMA-S (46) and the human cell line T2 (47)—that express low cell surface levels of class I MHC molecules and are defective in presenting viral protein antigens, but yet they could present synthetic antigenic peptides. Incubation of these mutant cells with the appropriate synthetic peptides could serve to increase the cell surface levels of class I MHC molecules. Although it was first suggested that the exogenous peptides were able to access the ER from the extracellular milieu, by some as yet unidentified pathway, it was subsequently determined that the peptides more than likely are bound to cell surface class I MHC molecules which, in this cell line, are unstable at 37°C (48). Therefore, the increase in cell surface expression was likely the result of peptide stabilization of cell surface class I MHC molecules (49), rather than from increased transit of class I MHC molecules to the cell surface (50). Additionally, it was suggested, from the differential recognition by monoclonal antibodies of the class I MHC molecules in these defective cells, that these antibodies may distinguish conformational differences between peptide-bound class I MHC and "empty" class I MHC molecules.

As previously described, there is indirect evidence that the cytosol may be the cellular compartment responsible for processing or fragmentation of proteins to antigenic fragments. Which protease(s) may be responsible for cytosolic antigen processing? A recently reported candidate for this is the high molecular weight cytosolic protease complex, named the proteasome by Goldberg (51,52). This proteolytic complex, which has a ringlike appearance in electron micrographs, appears to exist in several forms composed of heterosubunits. Some of the subunits have recently been mapped to the MHC locus. Genes in the class II region that are deleted in certain antigen presentation-defective mutant cells include those encoding for two subunits of a large molecular weight proteolytic complex (53–55). These proteolytic complexes are related to the multicatalytic proteinase complexes in the mouse (56) and humans (57).

What is the mechanism of peptide transport from the compartment in which they are generated to the compartment where charging of class I MHC takes place (likely to be the ER)? The current view of the endogenous pathway involves possible cytoplasmic proteolysis coupled to "transporters" of proteolytic fragments to the ER, where a molecule may bind and retain the peptide in the vicinity of class I MHC heavy chain (Fig. 2). Formation of a stable complex between the class I MHC heavy chain (45 kDa) and β_2-microglobulin in the endoplasmic reticulum and its subsequent transport to the cell surface, may require binding of peptide in the peptide-binding groove (46). Upon peptide binding by the heavy chain and β_2-microglobulin association, the appropriately conformed trimolecular complex egresses the ER or cis-Golgi.

There may be other routes for endogenous peptide to charge MHC. For instance, in minigene expression during which the sequence encoding the T-cell determinant is expressed in the cell, a peptide product that requires no processing may be produced. This peptide might access the ER compartment by the aforementioned peptide transporters, directly through the membrane, or through

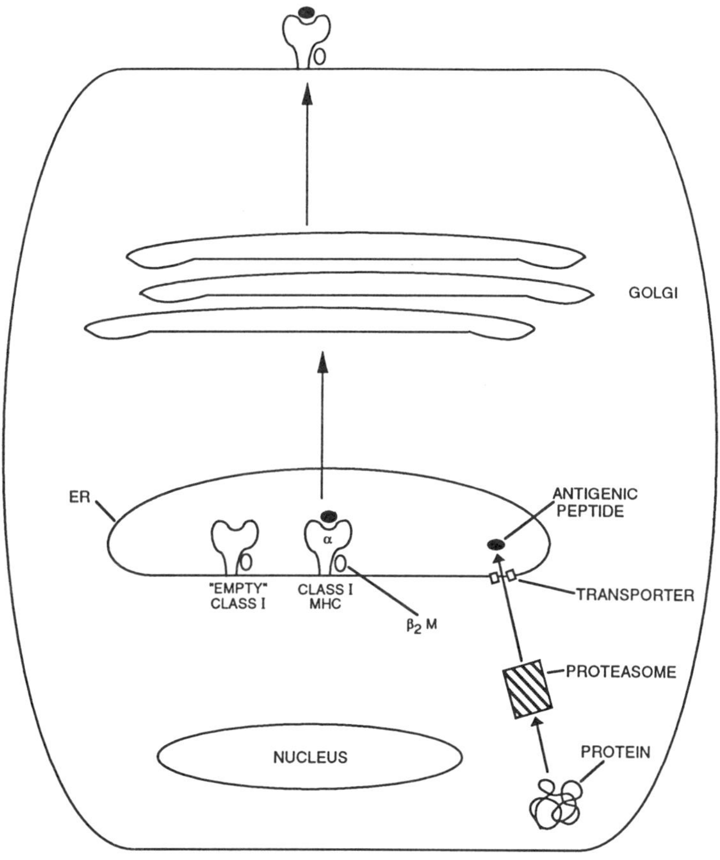

CLASS I PATHWAY

Figure 2 Class I MHC antigen-processing and presentation pathway. Protein is processed, undergoes proteolytic cleavage, in the cytosol. A putative transporter directs antigenic fragments to the ER. Newly synthesized class I MHC heavy chain (α) forms a trimolecular complex with β_2-microglobulin (β_2M) and antigenic peptide in the ER. This complex leaves the ER, traverses the Golgi, and is expressed on the cell surface. A complex between the class I heavy chain and β_2M (empty class I MHC) may egress to the cell surface where it is unstable at 37°C.

some other as yet undefined pore. Alternatively, an expressed minigene containing a signal sequence directly upstream from the minigene sequence can access the ER by the normal signal recognition protein (SRP)-dependent mechanism, whereby the minigene peptide product is then released in the ER following the signal peptide cleavage event (58).

The murine RMA-S (46), and the human cell lines LBL 721.174 and T2 (47,59) are mutant cell lines defective in presentation of viral antigens to $CD8^+$ CTL. Class I MHC heavy chain is apparently transcribed and translated at normal levels, and not mutated as far as sequence data indicate. The proposed defects in these cells are either transport of peptides to the ER for class I MHC charging, the inability of the MHC to bind peptide in this compartment, or the inability to process antigen appropriately. Recent evidence suggests that the defect may reside in a mutated or deleted transporter, mapping to the class II region of the MHC locus, which bears homology to the mammalian multidrug-resistance gene products and certain bacterial peptide transporters. These putative transporter genes contain regions homologous to ATP-binding domains, which may imply an ATP-dependent step in the transport of antigenic peptides to class I MHC in the ER. Two genes that belong to the superfamily of genes the products of which are involved in transport across membranes are named *Ham-1* and *Ham-2* in the mouse (60); *Cim* in the rat (61), and *Y3, PSF-1* (62), or *RING4* (63) in humans are homologues of *Ham-1,* whereas the human homologue of *Ham-2* is *Y1* or *RING11*.

Antigen presentation can be restored in certain antigen presentation-defective cell lines by a transfected gene encoding a transporter from a different or the same species (64). Expression of a cloned copy of the *Ham-2* gene in RMA-S cells restores ability to present class I MHC-restricted antigens and partial recovery of normal levels of cell surface class I MHC expression (65). The restoration of wild-type levels of class I MHC cell surface expression in a mutant human cell line, 724.134, was restored by expression of the human homologue of *Ham-1, PSF-1,* also called *Y3-1* or *RING4* (66). In addition, a minigene product that contains the epitope following a signal sequence can bypass the putative transporter defect in these cells and be presented (58).

Also, based on studies of mutant cell lines, as well as requirements for peptide sensitization of target cells for CTL killing, it has been suggested that association of β_2-microglobulin is critical to the proper class I MHC folding and stability. In vitro exposure of immature class I heavy chain showed that exogenously added human β_2-microglobulin could stabilize the binding of peptide to murine class I MHC molecules. The addition of peptide in vitro could also stabilize the binding of β_2-microglobulin to the class I heavy chain. These studies could not distinguish whether, in the course of class I MHC biosynthesis, the heavy chain first binds peptide, undergoes appropriate folding events, and then associates with β_2-microglobulin, or whether the heavy chain first associates with β_2-microglobulin, and then undergoes a conformational change when it binds peptide. Neither is it known whether a subset of empty class I MHC molecules egress to the cell surface, nor whether so-called empty class I MHC are the result of release of peptides that have a low-binding affinity for class I MHC. Some class I MHC molecules (e.g., L^d) weakly associate with murine β_2-microglobulin and are found at the cell surface in the absence of β_2-microglobulin. Other class I MHC molecules have a weaker affinity for murine β_2-microglobulin and are found with bovine β_2-microglobulin derived from fetal bovine serum in the culture medium. Two groups have reported that, for certain T-cell epitopes and the class I MHC molecule K^b, there is an apparent requirement for serum (as a source of exogenous β_2-microglobulin) or for purified human β_2-microglobulin for exogenously added synthetic peptides to associate with class I MHC on the cell surface (67, 68). These findings have not been generalizable to all class I MHC molecules, nor for all epitopes bound and presented by class I MHC (69).

B. Class II Major Histocompatibility Complex Biosynthesis

Recently described human mutant B-cell lines appear to be defective in class II antigen presentation (70). Genes affecting class II MHC antigen processing and presentation also map to the class II MHC region. These cells were mutated under the selective pressure of an anti-DR3 monoclonal antibody (MAb) plus complement. The cell surface class II MHC molecules appear to be less stable than class II MHC molecules from normal cells. Although these cells fail to present whole-protein antigen, they can present peptides. The defect in these cells could be mimicked in normal cells by long-term treatment with chloroquine. Therefore, it was suggested that there is somehow a deficit in the peptide pool available to newly synthesized class II MHC molecules in these cells that results in cell surface expression of a less stable, putatively empty form of class II MHC. These less stable class II MHC $\alpha\beta$ dimers dissociate into monomers in the presence of the detergent sodium dodecyl sulfate (SDS) unless they bind peptide, in which case, they are stable in SDS.

C. Role of Invariant Chain in Class II Major Histocompatibility Complex Antigen Presentation

There are several forms of invariant chain, the major of which is p33(31–33 kDa). It is a type II membrane glycoprotein in which the NH_2-terminal domain resides in the cytoplasm and the COOH-terminal domain is luminal. Other invariant chain species include p41, which results from an alternative splice with insert. Class II MHC surface molecules consist of heterodimers ($\alpha\beta$). The structure of class II MHC has been modeled on that of class I (as determined by x-ray crsytallographic data) by sequence homology, yielding a structure with a putative peptide-binding groove (71). Newly synthesized α and β chains of class II MHC associate with invariant (Ii) chain in the ER (72), and the class II MHC $\alpha\beta$–Ii complexes subsequently leave the ER (Fig. 3). The invariant chain appears to undergo proteolytic cleavage events and then dissociate from $\alpha\beta$ in an

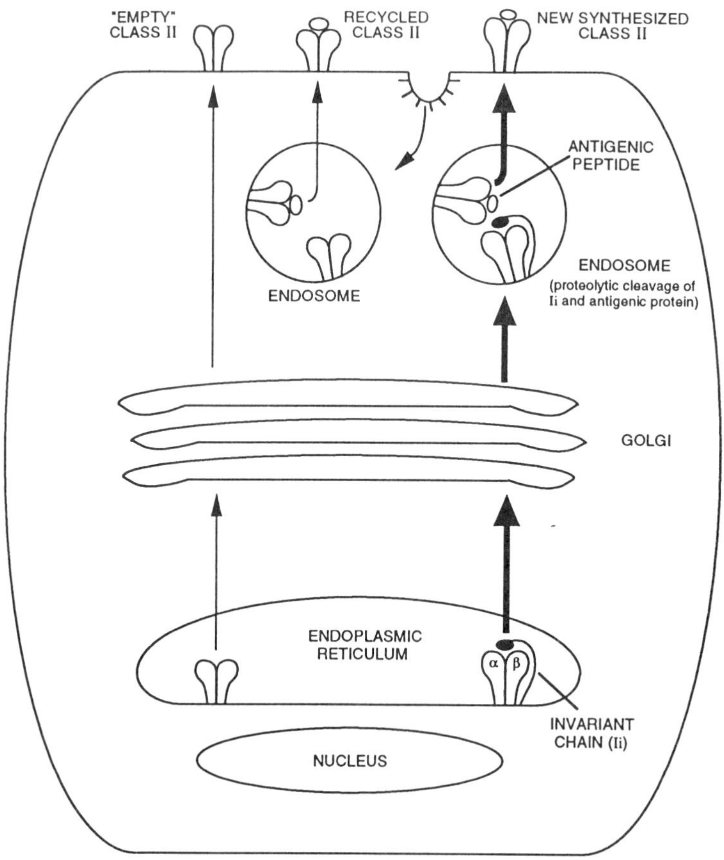

CLASS II PATHWAY

Figure 3 Class II MHC antigen-processing and presentation pathway. Protein is endocytosed through coated pits or through pinocytosis and is processed by acid proteases in a low-pH vesicular compartment (endosome). Newly synthesized class II MHC α and β chains complex with invariant chain (Ii) in the ER, which leaves the ER and traverses the Golgi. This complex is unable to bind antigenic peptide. The invariant chain targets the class II αβ–Ii to a low-pH vesicular compartment. Here the invariant chain undergoes proteolytic cleavage and, in so doing, reveals the peptide-binding groove. Here, also, newly synthesized class II αβ intersects with endocytosed processed protein. Class II αβ binds antigenic peptide and egresses to the cell surface. Class II αβ can recycle from the cell surface and may be reexpressed on the cell surface, binding a new peptide. In certain mutant cell lines or in the absence of invariant chain, class II αβ may egress by the "default" (bypassing a trip to endosomes) pathway and be expressed at the cell surface in an empty form.

endosomal compartment after transit through the Golgi. There appears to be a lag of up to 2 h during which class II MHC is thought to reside in a post-Golgi compartment before cell surface expression of class II αβ. The biosynthetic route of class II MHC molecules intersects with the endosomal compartment, as shown by studies demonstrating that newly synthesized class II in transit to the surface of human B-lymphoblastoid cell lines (B-LCL) interacts with transferrin–neuraminidase conjugates internalized by means of receptor-mediated endocytosis (73). Colocalization of class II, invariant chain, immunoglobulin, and proteolytic enzymes in an early endocytic compartment was visualized by immunoelectron microscopy (74–76). Endocytic vesicles contain class II MHC and cathepsins B and D. Leupeptin treatment of B cells results in accumulation of HLA-DR complexes associated with a fragment of invariant chain (77).

Two putative roles for invariant chain that have received some experimental support are (1) that it targets the class II MHC αβ to endosomal compartments, where it meets the proteolytic products of antigen (78,79); and (2) that the invariant chain, which contains a region with homology to antigenic peptides, can bind class II αβ, thereby blocking the class II MHC-binding site until the complex reaches an endosomal compartment where proteolytic cleavage of the invariant chain may result in exposure of the binding site. At this point, invariant chain no longer associates with αβ and the αβ trafficks to the cell surface (see Fig. 3). It has been suggested that the NH_2-terminal or cytoplasmic region of the invariant chain is capable of targeting the class II αβ dimers to the endosomal compartment (80).

These two roles for invariant chain in class II antigen processing and presentation are most likely not mutually exclusive and could account for differences between the endogenous and exogenous antigen-presentation pathways. In investigating the role of invariant chain in class II antigen presentation, the role of invariant chain in the biosynthesis of class II MHC molecules has become better understood. It was suggested early on, in work by Koch and co-workers, that invariant chain is required for the proper targeting of class II MHC molecules to the endosomes, where charging with peptide could occur. In invariant chain-negative class II MHC-transfected L cells, viral antigens were not presented to T cells. Others have shown that newly synthesized class II MHCs do not appear to accumulate in the endosomes in invariant chain-negative cells [for murine cells (81) and human cells (82)], but go to the cell surface, possibly by the default biosynthetic pathway (see Fig. 3). In certain invariant chain-negative mutant cells, class II MHC αβ form dimers that are expressed on the cell surface in a conformation that is not detected by certain anti-class II MHC monoclonal antibodies, which led to speculation that prior association with invariant chain is required for proper folding of class II MHC (83). A downward shift in the peptide dose required to sensitize invariant chain-negative cells supports the notion that empty class II molecules are expressed in greater proportion in such mutant cells, and are more readily available to bind exogenous peptide. It is not yet known whether peptides sensitize cell surface class II by exchange with self-peptides bound in the groove, or whether peptides sensitize newly synthesized class II in an endosomal compartment. Some studies suggest that cell surface class II MHC needs to recycle to be sensitized by exogenous peptide, most likely in an endosomal compartment. Adorini et al. reported that exogenous peptides can compete with the presentation of endogenous antigens on class II MHC in hen egg lysozyme (HEL) transfectants (84). It has been observed that for some peptides, class II may bind them better in a low-pH compartment. Although these observations support the importance of the class II MHC molecule transiting an endosomal compartment, there was yet another feature of the class II-presentation pathway that needed to be reconciled, which was the apparent inability of class II MHC to bind processed peptide in the ER, in a manner similar to that seen for class I MHC.

The second proposed role for invariant chain is that it covers the peptide-binding groove of the $\alpha\beta$ subunits in the ER (85) and through the biosynthetic pathway until it is proteolytically cleaved in an endosomal compartment and leaves the $\alpha\beta$ subunits. Evidence to support this concept came from studies in which purified class II MHC $\alpha\beta$ dimers were shown to be able to bind antigenic peptides, whereas purified class II $\alpha\beta$–Ii complexes were unable to do so (86). The class II $\alpha\beta$–peptide complexes are stable to SDS at room temperature, whereas the class II $\alpha\beta$–Ii complexes are unstable in the presence of SDS (86,87).

VI. T-CELL EPITOPES

With the increasing number of T-cell determinants that were characterized and mimicked by synthetic peptides, it was of interest to determine rules governing the makeup of an epitope, with the hope that the information would be of predictive value in finding epitopes for a variety of viral, bacterial, and parasitic, as well as self- and tumor antigens.

Several groups developed algorithms that would be predictive of putative T-cell antigenic determinants. These include the predictive algorithm of Berzofsky and co-workers that T-cell epitopes are most likely amphipathic helices (88). Rothbard and Taylor suggested a T-epitope motif, as well as an MHC-binding motif (89). Analyses of substituted peptide analogues were used to assess the role of individual residues in binding to MHC versus the TCR. One type of analysis was the 'Ala scan,' which involved the substitution of an alanine residue for each residue of a synthetic peptide. In aligning several peptides known to bind the same MHC molecule, and with the knowledge that there is likely one peptide-binding region in the MHC, it was hoped that information about the MHC-binding motifs might fall out. These types of studies allowed the description of apparent-binding motifs for several MHC molecules. On a cautionary note, it was reported that the role of a particular residue in these types of studies might be determined by the type of readout assay used as a gauge of T-cell recognition. Evavold and Allen reported that a single amino acid-substituted synthetic peptide analogue was capable of stimulating lymphokine (interleukin-4) production by a T-cell clone, but failed to stimulate proliferation (90).

Also of interest, was the quantitation of the number of peptide–MHC complexes that may be required for triggering a T-cell recognition event. Studies using quantitative immunoprecipitation of purified class II MHC from APC, given stimulatory doses of radiolabeled immunogenic peptide, yielded a minimal estimate of 200–300 complexes (91). Some studies indicated that, although antigen may be processed and presented by B cells and macrophages with different efficiencies, the estimate of the minimal number of class II–peptide complexes was the same (92). On the other hand, class II MHC-transfected L cells appeared to require more complexes, which may reflect a requirement for accessory adhesion molecules. Studies of peptide binding to class I MHC on living cells were used to quantitate the number of Ag–class I MHC complexes required for recognition and lysis by class I-restricted CTL (93). The absolute amounts of naturally processed viral peptides contained in infected cells were determined to be between 200 and 500 copies in studies by Falk et al. (94).

The true naturally processed and presented moieties for class I and class II MHC molecules have yet to be determined unambiguously. Synthetic peptides have been used to mimic antigenic epitopes for both class I and class II MHC-restricted T cells, but how close or true are they? Are endogenously expressed minigenes good mimics of T cell epitopes in the context of class I MHC? Recent reports on acid-eluted and high-performance liquid chromatography (HPLC)-purified peptides from purified class I MHC molecules indicate that nonamers are the most common size for naturally processed self-peptides, as well as for peptides derived from processed viral proteins (95,96). Peptides of this length appear to bind class I MHC molecules better (97,98) and, thereby, can more efficiently sensitize APC cells for recognition by T cells. Studies of the crystal structure of HLA-B27 revealed self-peptide nonamers bound in the groove in an extended conformation (99). Acid elution from immunoaffinity-purified class I MHC molecules of bound peptide followed by HPLC separation and sequencing yielded information concerning particular motifs dicernable for a given MHC (95,96, 100–102). In particular, the binding motif for the H-2K^d molecule appears to be - Y ------ L; ---- F -- L for K^b; ---- N --- M for D^b; - L ------ V for HLA-A2.1; and - R------- for HLA-B27. As previously mentioned, the class I and class II MHC peptide-binding grooves appear similar as seen by comparing the patterns of conserved and polymorphic residues of several class I and class II amino acid sequences and modeling on the class I MHC crystal structure (71). Evidence from peptides acid-eluted from class II MHC suggest self-peptides of variable lengths, 14 mers or greater, and of little sequence heterogeneity, occupy the class II-binding groove (103). In these studies, however, functionality of the eluted peptides was not tested. Further investigation along these lines will be useful in the elucidation of alloreactive epitopes, self-peptides responsible for autoimmune recognition, as well as possible common foreign epitopes for disparate MHC molecules that may be useful in vaccine design.

Viral expression vectors have been used to express T-cell epitopes, minigenes, by the endogenous route. Minigenes consist of the nucleotide sequence encoding an amino acid sequence thought to be the T-cell epitope. As with synthetic peptides, the minigenes have been used to mimic the processed T-cell determinant. Since the parameters of processing endogenous cytosolic antigen are unknown, it is not known whether minigenes truly mimic the processed moiety. The role of flanking sequences in the processing and presentation of epitiopes in the class I pathway has been investigated, and conflicting results have been reported. As to whether flanking sequences are important in epitope generation, several reports indicate that an epitope is appropriately generated when inserted into a protein other than the one from which the sequence is derived, or when it is relocated within the protein from which it was derived.

Maryanski and co-workers demonstrated that when synthetic oligonucleotides encoding region 170–182 of HLA.CW3 were inserted into the influenza nucleoprotein (NP) gene, cells transfected with the NP–oligorecombinant could present the H-2K^d-restricted epitope from HLA.CW3 (104). Hahn et al. relocated an influenza hemagglutinin epitope (actually two overlapping epitopes) in several different locations in the hemagglutinin molecule and noted no negative effect of different flanking sequences on the processing and presentation of the epitope (both overlapping sites were efficiently recognized) (105). There are many reports of minigene constructs that successfully mimic T-cell epitopes when expressed endogenously (25,106–108), implying that the lack of flanking sequences does not affect processing and presentation of the expressed preprocessed epitope. There are, however, recent reports in which flanking sequences could influence epitope presentation and reports in which particular minigene sequences were unable to mimic processed antigen (109,110). These studies indicate that minigene constructs do not always mimic a preprocessed antigenic epitope. Indeed, depending on the choice of the exact sequence thought to represent the antigenic epitope, the expressed product may or may not require processing. In addition, if the expressed product is incapable of acting as a substrate for the endogenous proteolytic-processing enzymes that would normally act on the full-length gene product, then there may be NH_2-terminal or COOH-terminal residues present on the expressed minigene product that are inhibitory of transport into the ER. The resolution of these conflicting reports will require studies of a similar nature for several T-cell epitopes to determine which is the generalizable and prevalent finding (i.e., to determine the role, if any, of flanking sequences for minigene or full-length gene products) and which might be true for a particular MHC or a particular T-cell epitope in processing events, transport, MHC association, and T-cell recognition. Indeed, expressed minigenes of varying lengths can be used to probe requirements to access or bypass the proteolytic machinery and to discover rules governing the nature of peptides transportable to the ER.

VII. NONCLASSIC CLASS I MAJOR HISTOCOMPATIBILITY COMPLEX ANTIGEN PRESENTATION

A minor histocompatibility antigen that appeared to be presented in an MHC-unrestricted manner has been characterized for the true restriction molecule, which is not a classic class I nor class II MHC molecule. Also, the identity of the self-antigen moiety presented has been determined. The maternally transmitted antigen (Mta) in the mouse is a complex of the maternally transmitted factor (MTF), β_2-Microglobulin, and the *Hmt* locus product. The MTF is derived from the NH_2-terminus of ND1, a mitochondrially encoded subunit for NADH-dehydrogenase (111). The formyl group on the NH_2-terminal methionine of the MTF peptide that defines this epitope is essential for binding of peptide to MTH, a nonclassic class I MHC molecule (112). Recently, *H-2M3* has been shown to be *Hmt,* which encodes the MHC class I-like molecule that presents the maternally transmitted antigen (113). It has been suggested that this neoclassic class I MHC may, in particular, present antigenic moieties derived from bacterial proteins in which the initial NH_2-terminal residue is a formylated methionine.

VIII. SUPERANTIGENS

In the past few years, studies of superantigens have captured the attention of scientists engaged in areas of research from antigen processing and presentation, to T-cell activation, and to autoimmunity. Although it had been reported well over a decade ago that certain bacterial enterotoxins could serve as polyclonal mitogens for T lymphocytes (114,115), it was not until the critical observation by White et al. (116) of the ability of staphylococcal enterotoxins (SEA, SEB) to stimulate a subset, rather than all T lymphocytes, that the interest in these antigens exploded. In addition, the reactive T-cell subset is correlated with the use of particular Vβ chains in the TCR. Kappler and Marrack's group was also instrumental in defining the longtime enigmatic minor lymphocyte-stimulating (Mls) loci determinants as being encoded by retroviral sequences present in certain strains of mice. The term *superantigen* originally coined by this group, describes antigens such as the bacterial toxins and Mls determinants that are capable of eliciting the reactivity of a large subset of T lymphocytes in an

MHC-unrestricted manner, but that are dependent on class II MHC molecules, and correlate with the use of particular Vβs in the TCR. Although SEB and SEA have been the most studied, other reported bacterial superantigens are streptococcal M protein (117), toxic shock syndrome toxin (TSST) (118), *Mycoplasma arthridites* (119), and streptococcal toxins A and B (120).

Initial studies in which the presence of class II MHC molecules was an absolute requirement for stimulation of T lymphocytes were interpreted as suggesting that the superantigens were processed and presented as peptides in the peptide-binding groove of class II MHC molecules, but that the MHC restriction was degenerate (116). Janeway and co-workers showed, in studies in which T cells recognized APC that were fixed before the addition of staphylococcal enterotoxin, that these superantigens may not require processing and suggested that the staphylococcal enterotoxin directly binds both the class II MHC molecule and the TCR receptor in the variable domain of the β-chain of the TCR (see Fig. 1b) (121). This latter model has been supported by reports that bacterial toxins can bind class II MHC molecules directly (in the mouse and human) (118, 122). Also, there appears to be no cross-blocking of antigenic peptide and superantigen for binding to class II MHC (123). Reduction and alkylation of staphylococcal enterotoxins results in loss of T-cell mitogenic activity, yet capacity to bind class II MHC molecules is retained (124). The region(s) of the class II MHC bound by the superantigens have been localized to distinct regions for various different superantigens such as the toxic shock syndrome toxin 1 (TSST-1) for HLA-DR to the α_1 domain (125) and for the staphylococcal toxins to the β_1 domain (125,126).

In work by Kappler and co-workers, the likely region of superantigen binding to the TCR β-chain variable region has been determined [for *Staphylococcus aureus* toxin (127) and for Mls-1a reactivity (128)]. Also, Gascoigne and Ames have demonstrated that a soluble truncated form of TCR β-chain interacts directly with superantigens when they are bound by class II MHC molecules (129). Furthermore, a series of mutated staphylococcal enterotoxin molecules have been used to map certain regions of the molecule for class II MHC-binding and TCR-binding regions, and some regions appear to influence binding to both class II MHC and TCR (130). There is a hierarchy of binding of class II MHC molecules in the human and the mouse. There is also a hierarchy of bacterial enterotoxin binding to the variable domain of certain TCR β chains. Recent studies also indicate that for the same TCR β chain use, there can be an influence of the paired α-chain on the affinity of superantigen binding (131).

The Mls loci determinants are sometimes referred to as self-superantigens, whereas the bacterial toxins were referred to as foreign superantigens. In light of the identity of Mls determinants as arising from retroviral-encoded products (132), it may be more accurate to refer to the Mls loci determinants as foreign endogenous superantigen [Kappler and Marrack's group have also shown that the product of exogenously introduced MMTV can provide the appropriate superantigen (133)]. Also, it was formerly thought the I-E molecule was a superantigen, but it is now known that it, too, binds and presents the products of retroviral sequences. Woodland et al. showed that Etc-1, which is presented by I-E molecules, is also an MMTV gene product [134]. Another finding is that, in an acquired immunodeficiency disease model in mice (MAIDS), a defective murine leukemia virus may encode a superantigen that stimulates Vβ5-bearing T cells (135). Therefore, true self-superantigens arising from self-protein not of viral origin have yet to be demonstrated.

Superantigens have been implicated as having a role in some autoimmune diseases. It is postulated that the ability of superantigens to activate large numbers of T lymphocytes, without consideration for antigen specificity, may lead to activation of or enhanced autoimmune responses. One way to look for superantigen reactivity in autoimmune disease is to look for enhanced or deleted subsets of T cells bearing certain Vβs in various disease states or in experimental autoimmune disease models.

IX. SUMMARY

Clearly, a great deal of progress in our understanding of the molecular aspects of antigen processing and presentation has taken place in the last few years. As the details of the processing and presentation pathways are revealed, our understanding of intracellular comparmentalization and transport increases. Furthermore, we are gaining insight into the symbiotic relations between antigenic peptides and the biosynthesis and function of class I and class II MHC molecules. However, it is not yet clear whether the requirement for peptide–MHC molecule binding during steps in the biosynthesis of functional MHC molecules is also necessary for the presentation of self-peptides such that positive- and negative-selection can occur with MHC molecules in an "occupied" conformation. Further studies should elucidate processing and presentation differences for different APC. In addition, differences in the processing and presentation of low-versus high-abundance self-proteins should emerge. Of importance will be methods to track degradative and minor pathways, as well as the normal biosynthetic pathway of putative antigenic proteins, which will help clarify issues of processing compartments and peptide transport or accessibility to compartment(s) where peptide binding to the MHC molecules might occur. The effect of infectious agents on the up-regulation of processing and presentation pathways for different APC needs to be investigated. Knowledge of these molecular details should permit the development of novel interventions in the triggering of allergic responses, the prevention of cessation of autoimmune responses, the enhancement of, or generation of, responses to infectious agents through vaccines that bypass or take advantage of the normal-processing and presentation pathways. The phenomenological observations of MHC-restricted recognition and Ir gene effects can now be explained at the level of antigen presentation and TCR specificity. What remains is the elucidation of the manner in which antigen presentation affects T-cell development and, therefore, T-cell repertoire and effector function.

REFERENCES

1. P. G. Gell, and B. Benacerraf, Studies on hypersensitivity. II. Delayed hypersensitivity to denatured proteins in guinea pigs, *Immunology, 2:* 64 (1959).
2. B. Kindred, and D. Shreffler, H-2 dependence of co-operation between T and B cells in vivo, *J. Immunol., 109:* 940 (1972).
3. A. S. Rosenthal, and E. M. Shevach, Function of macrophages in antigen recognition by guinea pig T lymphocytes. I. Requirement for histocompatible macrophages and lymphocytes, *J. Exp. Med., 138:* 1194 (1973).
4. R. M. Zinkernagel, and P. C. Doherty, Restriction of in vitro T cell-mediated cytotoxicity in lymphocytic choriomeningitis within a syngeneic or semiallogeneic system, *Nature, 248:* 701 (1974).
5. G. M. Shearer, T. G. Rhen, and A. M. Schmitt-Verhulst, Role of murine major histocompatibility complex in the specificity of in vitro T cell mediated lympholysis against chemically modified autologous lymphocytes, *Transplant. Rev., 29:* 222 (1976).
6. K. Ziegler, and E. R. Unanue, Identification of a macrophage antigen-processing event required for I-region-restricted antigen presentation to T lymphocytes, *J. Immunol, 127:* 1869 (1981).
7. E. R. Unanue, and P. M. Allen, The basis for the immunoregulatory role of macrophages and other accessory cells, *Science, 236:*551 (1987).
8. R. Shimonkevitz, J. Kappler, P. Marrack, and H. Grey, Antigen recognition by H-2- restricted T cells. I. Cell-free antigen processing, *J. Exp. Med., 158:* 303 (1983).
9. A. S. Rosenthal, M. A. Barcinski, and J. T. Blake, Determinant selection is a macrophage dependent immune response gene function, *Nature, 267:* 156 (1977).
10. E. Heber-Katz, D. Hansburg, and R. H. Schwartz, The Ia molecule of the antigen presenting cell plays a critical role in immune response gene regulation of T cell activation, *J. Mol. Cell. Immunol., 1:* 3 (1983).
11. B. P. Babbitt, P. M. Allen, G. Matsueda, E. Haber, and E. R. Unanue, Binding of immunogenic peptides to Ia histocompatibility molecules, *Nature, 317:* 359 (1985).
12. P. J. Bjorkman, M. A. Saper, B. Samraoui, W. S. Bennett, J. L. Strominger, and D. C. Wiley, Structure of the human class I histocompatibility antigen, HLA-A2, *Nature, 329:* 506 (1987).
13. P. J. Bjorkman, M. A. Saper, B. Samraoui, W. S. Bennett, J. L. Strominger, and D. C. Wiley, The foreign antigen binding site and T cell recognition regions of class I histocompatibility antigens, *Nature, 329:* 512 (1987).
14. A. Ametani, R. Apple, V. Bhardwaj, G. Gammon, A. Miller, and E. Sercarz, Examining the crypticity of antigenic determinants, *Cold Spring Harbor Symp. Quant. Biol., 1:* 505 (1989).

15. T. H. Watts, A. A. Brian, J. W. Kappler, P. Marrack, and H. M. McConnell, Antigen presentation by supported planar membranes containing affinity-purified I-A^d, *Proc. Natl. Acad. Sci. USA, 81:* 7564 (1984).
16. H. Takahashi, K. B. Cease, and J. A. Berzofsky, Identification of proteases that process distinct epitopes on the same protein, *J. Immunol., 142:* 2221 (1989).
17. S. Diment, and P. Stahl, Macrophage endosomes contain proteases which degrade endocytosed protein ligands, *J. Biol. Chem., 260:* 15311 (1985).
18. A. R. Townsend, J. Rothbard, F. M. Gotch, G. Bahadur, D. Wraith, and A. J. McMichael, The epitopes of influenza nucleoprotein recognized by cytotoxic T lymphocytes can be defined with short synthetic peptides, *Cell, 44:* 959 (1986).
19. F. R. Carbone, M. W. Moore, J. M. Sheil, and M. J. Bevan, Induction of cytotoxic T lymphocytes by primary in vitro stimulation with peptides, *J. Exp. Med. 167:* 1767 (1988).
20. L. A. Morrison, A. E. Lukacher, V. L. Braciale, D. P. Fan, and T. J. Braciale, Differences in antigen presentation to MHC class I- and II-restricted influenza virus-specific cytolytic T lymphocyte clones, *J. Exp. Med., 163:* 903 (1986).
21. T. J. Braciale, L. A. Morrison, M. T. Sweetser, J. Sambrook, M. J. Gething, and V. L. Braciale, Antigen presentation pathways to class I and class II MHC-restricted T lymphocytes, *Immunol. Rev., 98:* 95 (1987).
22. A. E. Lukacher, L. A. Morrison, V. L. Braciale, and T. J. Braciale, T lymphocyte function in recovery from experimental viral infection: The influenza model, *Mechanisms of Host Resistance to Infectious Agents, Tumors, and Allografts: A Conference in Recognition of the Trudeau Institute Centennial* (R. M. Steinman, and R. J. North, eds.), Rockefeller University Press, New York, p. 233 (1986).
23. R. N. Germain, The ins and outs of antigen processing and presentation, *Nature, 322:* 687 (1986).
24. M. J. Bevan. Antigen recognition. Class discrimination in the world of immunology, *Nature, 325:* 192 (1987).
25. M. T. Sweetser, L. A. Morrison, V. L. Braciale, and T. J. Braciale, Recognition of pre-processed endogenous antigen by class I but not class II MHC-restricted T cells, *Nature, 342:* 180 (1989).
26. A. R. Townsend, F. M. Gotch, and J. Davey, Cytotoxic T cells recognize fragments of the influenza nucleoprotein, *Cell, 42:* 457 (1985).
27. A. R. Townsend, J. Bastin, K. Gould, and G. G. Brownlee, Cytotoxic T lymphocytes recognize influenza haemagglutinin that lacks a signal sequence, *Nature, 324:* 575 (1986).
28. J. W. Yewdell, J. R. Bennink, and Y. Hosaka, Cells process exogenous proteins for recognition by cytotoxic T lymphocytes, *Science, 239:* 637 (1988).
29. M. W. Moore, F. R. Carbone, and M. J. Bevan, Introduction of soluble protein into the class I pathway of antigen processing and presentation, *Cell, 54:* 777 (1988).
30. J. G. Nuchtern, J. S. Bonifacino, W. E. Biddison, and R. D. Klausner, Brefeldin A implicates egress from endoplasmic reticulum in class I restricted antigen presentation, *Nature, 339:* 223 (1989).
31. J. W. Yewdell, and J. R. Bennink, Brefeldin A specifically inhibits presentation of protein antigens to cytotoxic T lymphocytes, *Science, 244:* 1072 (1989).

32. M. Andersson, S. Paabo, T. Nilsson, and P. A. Peterson, Impaired intracellular transport of class I MHC antigens as a possible means for adenoviruses to evade immune surveillance, *Cell, 43:* 215 (1985).
33. J. H. Cox, J. W. Yewdell, L. C. Eisenlohr, P. R. Johnson, and J. R. Bennink, Antigen presentation requires transport of MHC class I molecules from the endoplasmic reticulum, *Science, 247:* 715 (1990).
34. J. H. Cox, J. R. Bennink, and J. W. Yewdell, Retention of adenovirus E19 glycoprotein in the endoplasmic reticulum is essential to its ability to block antigen presentation, *J. Exp. Med., 174:* 1629 (1991).
35. J. W Yewdell, J. R. Bennink, K. B. Eager, and R. P. Ricciardi, CTL recognition of adenovirus-transformed cells infected with influenza virus: Lysis by anti-influenza CTL parallels adenovirus-12-induced suppression of class I MHC molecules, *Virology, 162:* 236 (1988).
36. A. Lanzavecchia, and S. Bove, Specific B lymphocytes efficiently pick up, process and present antigen to T cells, *Behring. Inst. Mitt.,* (1985).
37. H. W. Davidson, and C. Watts, Epitope-directed processing of specific antigen by B-lymphocytes, *J. Cell Biol., 109:* 85 (1989).
38. E. K. Lakey, E. Margoliash, and S. K. Pierce, Identification of a peptide binding protein that plays a role in antigen presentation, *Proc. Natl. Acad. Sci. USA, 84:* 1659 (1987).
39. A. Vanbuskirk, B. L. Crump, E. Margoliash, and S. K. Pierce, A peptide binding protein having a role in antigen presentation is a member of the HSP70 heat shock family, *J. Exp. Med., 170:* 1799 (1989).
40. L. D. Falo, B. Benacerraf, and K. L. Rock, Phospholipase treatment of accessory cells that have been exposed to antigen selectively inhibits antigen specific Ia-restricted, but not allospecific, stimulation of T lymphocytes, *Proc. Natl. Acad. Sci. USA, 83:* 6994 (1986).
41. S. Jacobson, R. P. Sekaly, C. L. Jacobson, H. F. McFarland, and E. O. Long, HLA class II-restricted presentation of cytoplasmic measles virus antigens to cytotoxic T cells, *J. Virol., 63:* 1756 (1989).
42. S. Weiss, and B. Bogen, MHC class II-restricted presentation of intracellular antigen, *Cell, 64:* 767 (1991).
43. J. G. Nuchtern, W. E. Biddison, and R. D. Klausner, Class II MHC molecules can use the endogenous pathway of antigen presentation, *Nature, 343:* 74 (1990).
44. Y. Jin, W. K. Shih, and I. Berkower, Human T cell response to the surface antigen of hepatitis B virus (HBsAg). Endosomal and nonendosomal processing pathways are accessible to both endogenous and exogenous antigen, *J. Exp. Med., 168:* 293 (1988).
45. A. Brooks, S. Hartley, N. L. Kjer, J. Perera, C. C. Goodnow, A. Basten, and J. McCluskey, Class II-restricted presentation of an endogenously derived immunodominant T-cell determinant of hen egg lysozyme, *Proc. Natl. Acad. Sci. USA, 88:* 3290 (1991).
46. A Townsend, C. Ohlen, J. Bastin, H. G. Ljunggren, L. Foster, and K. Karre, Association of class I major histocompatibility heavy and light chains induced by viral peptides, *Nature, 340:* 443 (1989).

47. N. A. Hosken, and M. J. Bevan, Defective presentation of endogenous antigen by a cell line expressing class I molecules, *Science, 248:* 367 (1990).
48. H. G. Ljunggren, N. J. Stam, C. Ohlen, et al., Empty MHC class I molecules come out in the cold, *Nature, 346:* 476 (1990).
49. T. N. Schumacher, M. T. Heemels, J. J. Neefjes, W. M. Kast, C. J. Melief, and H. L. Ploegh, Direct binding of peptide to empty MHC class I molecules on intact cells and in vitro, *Cell, 62:* 563 (1990).
50. A. Townsend, T. Elliott, V. Cerundolo, L. Foster, B. Barber, and A. Tse, Assembly of MHC class I molecules analyzed in vitro, *Cell, 62:* 285 (1990).
51. A. P. Arrigo, K. Tanaka, A. L. Goldberg, and W. J. Welch, Identity of the 19S "prosome" particle with the large multifunctional protease complex of mammalian cells (the proteasome), *Nature, 331:* 192 (1988).
52. L. R. Dick, C. R. Moomaw, G. N. DeMartino, and C. A. Slaughter, Degradation of oxidized insulin B chain by the multiproteinase complex macropain (proteasome), *Biochemistry, 30:* 2725 (1991).
53. C. K. Martinez, and J. J. Monaco, Homology of proteasome subunits to a major histocompatibility complex-linked LMP gene, *Nature, 353:* 664 (1991).
54. A. Kelly, S. H. Powis, R. Glynne, E. Radley, S. Beck, and J. Trowsdale, Second proteasome-related gene in the human MHC class II region, *Nature, 353:* 667 (1991).
55. V. Ortiz-Navarrete, A. Seelig, M. Gernold, S. Frentzel, P. M. Kloetzel, and G. J. Hammerling, Subunit of the "20S" proteasome (multicatalytic proteinase) encoded by the major histocompatibility complex, *Nature, 353:* 662 (1991).
56. M. G. Brown, J. Driscoll, and J. J. Monaco, Structural and serological similarity of MHC-linked LMP and proteasome (multicatalytic proteinase) complexes, *Nature, 353:* 355 (1991).
57. R. Glynne, S. H. Powis, S. Beck, A. Kelly, L. A. Kerr, and J. Trowsdale, A proteasome-related gene between the two ABC transporter loci in the class II region of the human MHC, *Nature, 353:* 357 (1991).
58. K. Anderson, P. Cresswell, M. Gammon, J. Hermes, A. Williamson, and H. Zweerink, Endogenously synthesized peptide with an endoplasmic reticulum signal sequence sensitizes antigen processing mutant cells to class I-restricted cell-mediated lysis, *J. Exp. Med., 174:* 489 (1991).
59. V. Cerundolo, J. Alexander, K. Anderson, C. Lamb, P. Cresswell, A. McMichael, F. Gotch, and A. Townsend, Presentation of viral antigen controlled by a gene in the major histocompatibility complex, *Nature, 345:* 449 (1990).
60. J. J. Monaco, S. Cho, and M. Attaya, Transport protein genes in the murine MHC: Possible implications for antigen processing. *Science, 250:*1723 (1990).
61. E. V. Deverson, I. R. Gow, W. J. Coadwell, J. J. Monaco, G. W. Butcher, and J. C. Howard, MHC class II region encoding proteins related to the multidrug resistance family of transmembrane transporters [see comments], *Nature, 348:* 738 (1990).
62. T. Spies, M. Bresnahan, S. Bahram, D. Arnold, G. Blanck, E. Mellins, D. Pious, and R. DeMars, A gene in the human major histocompatibility complex class II region controlling the class I antigen presentation pathway, *Nature, 348:* 744 (1990).
63. J. Trowsdale, I. Hanson, I. Mockridge, S. Beck, A. Townsend, and A. Kelly,

Sequences encoded in the class II region of the MHC related to the "ABC" superfamily of transporters, *Nature, 348:* 741 (1990).
64. S. J. Powis, A. R. Townsend, E. V. Deverson, J. Bastin, G. W. Butcher, and J. C. Howard, Restoration of antigen presentation to the mutant cell line RMA-S by an MHC-linked transporter, *Nature, 354:* 528 (1991).
65. M. Attaya, S. Jameson, C. K. Martinez, E. Hermel, C. Aldrich, J. Forman, K. F. Lindahl, M. J. Bevan, and J. J. Monaco, *Ham-2* corrects the class I antigen-processing defect in RMA-S cells, *Nature, 355:* 647 (1992).
66. T. Spies, and R. DeMars, Restored expression of major histocompatibility class I molecules by gene transfer of a putative peptide transporter, *Nature, 351:* 323 (1991).
67. K. L. Rock, L. E. Rothstein, S. R. Gamble, and B. Benacerraf, Reassociation with beta$_2$-microglobulin is necessary for Kb class I major histocompatibility complex binding of exogenous peptides, *Proc. Natl. Acad. Sci. USA, 87:* 7517 (1990).
68. A. Vitiello, T. A. Potter, and L. A. Sherman, The role of β_2-microglobulin in peptide binding by class I molecules, *Science, 250:* 1423 (1990).
69. L. A. Sherman, T. A. Burke, and J. A. Biggs, Extracellular processing of peptide antigens that bind class I major histocompatibility molecules, *J. Exp. Med., 175:* 1221 (1992).
70. E. Mellins, L. Smith, B. Arp, T. Cotner, E. Celis, and D. Pious, Defective processing and presentation of exogenous antigens in mutants with normal HLA class II genes, *Nature, 343:* 71 (1990).
71. J. H. Brown, T. Jardetzky, M. A. Saper, B. Samraoui, P. J. Bjorkman, and D. C. Wiley, A hypothetical model of the foreign antigen binding site of class II histocompatibility molecules, *Nature, 332:* 845 (1988).
72. S. Kvist, K. Wiman, L. Claesson, P. A. Peterson, and B. Dobberstein, Membrane insertion and oligomeric assembly of HLA-DR histocompatibility antigens, *Cell, 29:* 61 (1982).
73. P. Cresswell, Intracellular class II HLA antigens are accessible to transferrin–neuraminidase cojugates internalized by receptor-mediated endocytosis, *Proc. Natl. Acad. Sci. USA, 82:* 8188 (1985).
74. F. M. Brodsky, B. Koppelman, J. S. Blum, M. S. Marks, P. Cresswell, and L. Guagliardi, Intracellular colocalization of molecules involved in antigen processing and presentation by B cells, *Cold Spring Harbor Symp. Quant. Biol., 1:* 319 (1989).
75. L. E. Guagliardi, B. Koppelman, J. S. Blum, M. S. Marks, P. Cresswell, and F. M. Brodsky, Co-localization of molecules involved in antigen processing and presentation in an early endocytic compartment, *Nature, 343:* 133 (1990).
76. J. J. Neefjes, V. Stollorz, P. J. Peters, H. J. Geuze, and H. L. Ploegh, The biosynthetic pathway of MHC class II but not class I molecules intersects the endocytic route, *Cell, 61:* 171 (1990).
77. J. S. Blum, and P. Cresswell, Role for intracellular proteases in the processing and transport of class II HLA antigens, *Proc. Natl. Acad. Sci. USA, 85:* 3975 (1988).
78. L. Teyton, D. O'Sullivan, P. W. Dickson, V. Lotteau, A. Sette, P. Fink, and P.

A. Peterson, Invariant chain distinguishes between the exogenous and endogenous antigen presentation pathways, *Nature, 348:* 39 (1990).

79. V. Lotteau, L. Teyton, A. Peleraux, T. Nilsson, L. Karlsson, S. L. Schmid, V. Quaranta, and P. A. Peterson, Intracellular transport of class II MHC molecules directed by invariant chain, *Nature, 348:* 600 (1990).
80. O. Bakke, and B. Dobberstein, MHC class II-associated invariant chain contains a sorting signal for endosomal compartments, *Cell, 63:* 707 (1990).
81. J. Miller, and R. N. Germain, Efficient cell surface expression of class II MHC molecules in the absence of associated invariant chain, *J. Exp. Med., 164:* 1478 (1986).
82. R. P. Sekaly, C. Tonnelle, M. Strubin, B. Mach, and E. O. Long, Cell surface expression of class II histocompatibility antigens occurs in the absence of the invariant chain, *J. Exp. Med., 164:* 1490 (1986).
83. M. Peterson, and J. Miller, Invariant chain influences the immunological recognition of MHC class II molecules, *Nature, 345:* 172 (1990).
84. L. Adorini, J. Moreno, F. Momburg, G. J. Hammerling, J. C. Guery, A. Valli, and S. Fuchs, Exogenous peptides compete for the presentation of endogenous antigens to major histocompatibility complex class II-restricted T cells, *J. Exp. Med., 174:* 945 (1991).
85. W. L. Elliott, C. J. Stille, L. J. Thomas, and R. E. Humphreys, An hypothesis on the binding of an amphipathic, a helical sequence in I_i to the desetope of class II antigens, *J. Immunol., 138:* 2949 (1987).
86. P. A. Roche, and P. Cresswell, Invariant chain association with HLA-DR molecules inhibits immunogenic peptide binding. *Nature, 345:* 615 (1990).
87. R. N. Germain, and L. R. Hendrix, MHC class II structure, occupancy and surface expression determined by post-endoplasmic reticulum antigen binding, *Nature, 353:* 134 (1991).
88. H. Margalit, J. Spouge, J. Cornette, K. Cease, C. DeLisi, and J. Berzofsky, Prediction of immunodominant helper T cell antigenic sites from the primary sequence, *J. Immunol., 138:* 2213 (1987).
89. J. B. Rothbard, and W. R. Taylor, A sequence pattern common to T cell epitopes, *EMBO J., 7:* 93 (1988).
90. B. Evavold, and P. Allen, Separation of IL-4 production from Th cell proliferation by an altered T cell receptor ligand, *Science, 252:* 1308 (1991).
91. C. V. Harding, and E. R. Unanue, Quantitation of antigen-presenting cell MHC class II/peptide complexes necessary for T-cell stimulation, *Nature, 346:* 574 (1990).
92. S. Demotz, H. M. Grey, and A. Sette, The minimal number of class II MHC–antigen complexes needed for T cell activation, *Science, 249:* 1028 (1990).
93. E. Christinck, M. Luscher, B. Barber, and D. Williams, Peptide binding to class I MHC on living cells and quantitation of complexes required for CTL lysis, *Nature, 352:* 67 (1991).
94. K. Falk, O. Rotzschke, K. Deres, J. Metzger, G. Jung, and H. G. Rammensee, Identification of naturally processed viral nonapeptides allows their quantification in infected cells and suggests an allele-specific T cell epitope forecast, *J. Exp. Med., 174:* 425 (1991).

95. G. M. van Bleek, and S. G. Nathenson, Isolation of an endogenously processed immunodominant viral peptide from the class I H-2K^b molecule, *Nature, 348:* 213 (1990).
96. O. Rotzschke, K. Falk, K. Deres, H. Schild, M. Norda, J. Metzger, G. Jung, and H. Rammensee, Isolation and analysis of naturally processed viral peptides as recognized by cytotoxic T cells, *Nature, 343:* 252 (1990).
97. T. N. Schumacher, B. M. De, L. N. Vernie, W. M. Kast, C. J. Melief, J. J. Neefjes, and H. L. Ploegh. Peptide selection by MHC class I molecules, *Nature, 350:* 703 (1991).
98. V. Cerundolo, T. Elliott, J. Elvin, J. Bastin, H. G. Rammensee, and A. Townsend, The binding affinity and dissociation rates of peptides for class I major histocompatibility complex molecules, *Eur. J. Immunol., 21:* 2069 (1991).
99. D. Madden, J. Gorga, J. Strominger, and D. Wiley, The structure of HLA-B27 reveals nonamer self-peptides bound in an extended conformation, *Nature, 353:* 321 (1991).
100. K. Falk, O. Rotzschke, S. Stevanovic, G. Jung, and H.-G. Rammensee, Allele-specific motifs revealed by sequencing of self-peptides eluted from MHC molecules, *Nature, 351:* 290 (1991).
101. T. Jardetzky, W. Lane, R. Robinson, D. Madden, and D. Wiley, Identification of self-peptides bound to purified HLA-B27, *Nature, 353:* 326 (1991).
102. O. Rotzschke, and K. Falk, Naturally-occurring peptide antigens derived from the MHC class-I-restricted processing pathway. *Immunol. Today, 12:* 447 (1991).
103. A. Y. Rudensky, P. Preston-Hurlburt, S. C. Hong, A. Barlow, and C. A. J. Janeway, Sequence analysis of peptides bound to MHC class II molecules, *Nature, 353:* 622 (1991).
104. G. Chimini, P. Pala, J. Sire, B. R. Jordan, and J. L. Maryanski, Recognition of oligonucleotide-encoded T cell epitopes introduced into a gene unrelated to the original antigen, *J. Exp. Med. 169:* 297 (1989).
105. Y. S. Hahn, V. L. Braciale, and T. J. Braciale, Presentation of viral antigen to class I major histocompatibility complex-restricted cytotoxic T lymphocyte. Recognition of an immunodominant influenza hemagglutinin site by cytotoxic T lymphocyte is independent of the position of the site in the hemagglutinin translation product, *J. Exp. Med., 174:* 733 (1991).
106. J. L. Whitton, and M. B. Oldstone, Class I MHC can present an endogenous peptide to cytotoxic T lymphocytes, *J. Exp. Med., 170:* 1033 (1989).
107. M. T. Sweetser, V. L. Braciale, and T. J. Braciale, Class I MHC-restricted recognition of cells expressing a gene encoding a 41 amino acid product of the influenza hemagglutinin, *J. Immunol., 141:* 3324 (1988).
108. G. N. Milligan, L. A. Morrison, J. Gorka, V. L. Braciale, and T. J. Braciale, The recognition of a viral antigenic moiety by class I MHC-restricted cytolytic T lymphocytes is limited by the availability of the endogenously processed antigen. *J. Immunol., 145:* 3188 (1990).
109. V. M. Del, H. J. Schlicht, T. Ruppert, M. J. Reddehase, and U. H. Koszinowski, Efficient processing of an antigenic sequence for presentation by MHC class I molecules depends on its neighboring residues in the protein, *Cell, 66:* 1145 (1991).
110. L. C. Eisenlohr, J. W. Yewdell, and J. R. Bennink, Flanking sequences influence

the presentation of an endogenously synthesized peptide to cytotoxic T lymphocytes, *J. Exp. Med., 175:* 481 (1992).
111. B. E. Loveland, C.-R. Wang, E. Hermel, H. Yonekawa, and K. Fischer Lindahl, Maternally transmitted mouse histocompatibility antigen: A hydrophobic peptide from a mitochondrially encoded protein, *Cell, 60:* 971 (1990).
112. S. Shawar, J. Vyas, J. Rodgers, R. Cook, and R. Rich, Specialized functions of major histocompatibility complex class I molecules. II. Hmt binds *N*-formylated peptides of mitochondrial and prokaryotic origin, *J. Exp. Med., 174:* 941 (1991).
113. C.-R. Wang, B. E. Loveland, and K. Fischer Lindahl, *H-2M3* encodes the MHC class I molecule presenting the maternally transmitted antigen of the mouse, *Cell, 62:* 335 (1991).
114. D. L. Peavy, W. H. Adler, and R. T. Smith, The mitogenic effects of endotoxin and staphylococcal enterotoxin B on mouse spleen cells and human peripheral blood lymphocytes, *J. Immunol., 105:* 1453 (1970).
115. B. Fleischer, and H. Schrezenmeier, T cell stimulation by staphylococcal enterotoxins, *J. Exp. Med., 167:* 1697 (1988).
116. J. White, A. Herman, A. M. Pullen, R. Kubo, J. W. Kappler, and P. Marrack, The V beta-specific superantigen staphylococcal enterotoxin B: Stimulation of mature T cells and clonal deletion in neonatal mice, *Cell, 56:* 27 (1989).
117. M. Tomai, M. Kotb, G. Majumdar, and E. Beachey, Superantigenicity of streptococcal M protein, *J. Exp. Med., 172:* 359 (1990).
118. P. Scholl, A. Diez, and R. Geha, Staphylococcal enterotoxin B and toxic shock syndrome toxin-1 bind to distinct sites on HLA-DR and HLA-DQ molecules, *J. Immunol., 143:* 2583 (1989).
119. G. Cannon, B. Cole, J. Ward, J. Smith, and E. Eichwald, Arthritogenic effects of *Mycoplasma arthriditis* T cell mitogen in rats. *J. Rheumatol., 15:* 735 (1988).
120. J. Abe, J. Forrester, T. Nakahara, J. Lafferty, B. Kotzin, and D. Leung, Selective stimulation of human T cells with streptococcal erythrogenic toxins A and B, *J. Immunol., 146:* 3747 (1991).
121. C. A. J. Janeway, J. Yagi, P. J. Conrad, M. E. Katz, B. Jones, S. Vroegop, and S. Buxser, T-cell responses to Mls and to bacterial proteins that mimic its behavior, *Immunol. Rev., 107:* 61 (1989).
122. J. D. Fraser, High-affinity binding of staphylococcal enterotoxins A and B to HLA-DR, *Nature, 339:* 221 (1989).
123. P. Dellabona, J. Peccoud, J. Kappler, P. Marrack, C. Benoist, and D. Mathis, Superantigens interact with MHC class II molecules outside of the antigen groove, *Cell, 62:* 1115 (1990).
124. D. Grossman, R. Cook, J. Sparrow, J. Mollick, and R. Rich, Dissociation of the stimulatory activity of staphyloccal enterotoxins for T cells and monocytes, *J. Exp. Med., 172:* 1831 (1990).
125. D. Karp, C. Teletski, P. Scholl, R. Geha, and E. Long, The alpha I domain of the HLA-DR molecule is essential for high-affinity binding of the toxic shock syndrome toxin-1, *Nature, 346:* 474 (1990).
126. C. Pontzer, J. Russell, and H. Johnson, Structural basis for differential binding of staphylococcal enterotoxin A and toxic shock syndrome toxin 1 to class II major histocompatibility molecules, *Proc. Natl. Acad. Sci. USA, 88:* 125 (1991).

127. Y. W. Choi, A. Herman, D. DiGiusto, T. Wade, P. Marrack, and J. Kappler, Residues of the variable region of the T-cell-receptor beta-chain that interact with *S. aureus* toxin superantigens, *Nature, 346:* 471 (1990).
128. A. M. Pullen, T. Wade, P. Marrack, and J. W. Kappler, Identification of the region of T cell receptor beta chain that interacts with the self-superantigen MIs 1a, *Cell, 61:* 1365 (1990).
129. N. Gascoigne, and K. Ames, Direct binding of secreted T-cell receptor beta chain to superantigen associated with class II major histocompatibility complex protein, *Proc. Natl. Acad. Sci. USA, 88:* 613 (1991).
130. J. W. Kappler, A. Herman, J. Clements, and P. Marrack, Mutations defining functional regions of the superantigen staphylococcal enterotoxin B, *J. Exp. Med., 175:* 387 (1992).
131. M. Blackman, J. Kappler, and P. Marrack, The role of the T cell receptor in positive and negative selection of developing T cells, *Science, 248:* 1335 (1990).
132. W. N. Frankel, C. Rudy, J. M. Coffin, and B. T. Huber, Linkage of Mls genes to endogenous mammary tumor viruses of inbred mice. *Nature, 349:* 526 (1991).
133. P. Marrack, E. Kushnir, and J. W. Kappler, A maternally inherited superantigen encoded by mammary tumor virus, *Nature, 349:* 524 (1991).
134. D. Woodland, M. P. Happ, J. Bill, and E. Palmer, Requirement for cotolerogenic gene products in the clonal deletion of I-E reactive T cells, *Science, 247:* 964 (1989).
135. A. W. Hugin, M. S. Vacchio, and H. C. Morse, A virus-encoded "superantigen" in a retrovirus-induced immunodeficiency syndrome of mice, *Science, 252:* 424 (1991).
136. R. H. Schwartz, T-lymphocyte recognition of antigen in association with gene products of the major histocompatibility complex, *Annu. Rev. Immunol., 3:* 237 (1985).

II

Features of the Allergic Immune Response

8

The High-Affinity Receptor for Immunoglobulin E

Carole L. Jelsema
National Institute of Arthritis, Musculoskeletal and Skin Diseases, National Institutes of Health, Bethesda, Maryland

Jean-Pierre Kinet
National Institute of Allergy and Infectious Diseases, National Institutes of Health, Bethesda, Maryland

I. INTRODUCTION

Over 70 years ago, Prausnitz and Kustner discovered that sensitization to specific allergens could be transferred passively to a nonallergic individual by subcutaneous injection of a factor found in the serum of an allergic individual (1). These experiments established that, in addition to specific allergens, the allergic reaction was composed of two major components: a specific component or "reagin" found in the serum of allergic individuals and a nonspecific tissue component found in both allergic and nonallergic individuals. Over 40 years later, the reagin was identified as a specific class of antibody, immunoglobulin E (IgE), which is capable of recognizing specific antigens or allergens (2). The tissue component was shown to be mast cells and basophils (see Chap. 11).

The molecules and mechanisms responsible for the allergic reaction are now known in considerable detail. The B cells of allergic individuals secrete IgE specific for a variety of antigens. The IgE then binds to at least two types of cell surface receptors, a high-affinity receptor (FcϵRI) (3–5) and a low-affinity receptor (FcϵRII or CD23) (6–8). Although the low-affinity receptors are found on many cells of the immune system, the high-affinity receptor is found on relatively few cells: mast cells and basophils (3–5), some nongranulated non-B, non-T mononuclear cells (9), and more recently, on certain dermal dendritic cells (Langerhans cells) located predominantly among the dermal microvasculature in normal human skin (10,11). By coupling the presence of allergens with the cellular response, the FcεRI is the principal molecule through which IgE exerts its characteristic biological function (i.e., the release of inflammatory mediators and cytokines that mediate the allergic response).

The binding of IgE to FcεRI per se, however, appears to be a passive event that, of itself, is not sufficient to generate intracellular signals. Cell activation requires that the receptor-bound IgE be subsequently challenged with a multivalent antigen or allergen, leading to aggregation of the IgE–receptor complexes on the cell surface (3,12,13). It is this aggregation event that is the critical step ultimately leading to the explosive cellular degranulation and the accompanying synthesis or release of the mediators of allergic diseases (3,14). These mediators include histamine and arachidonate metabolites, such as leukotrienes and prostaglandins, which function as inflammatory agents, chemoattractants, and vasoactive compounds (15).

In addition, in the mast cell, the activation of FcεRI also causes the later synthesis and, in some cases, the secretion of many cytokines such as granulocyte–monocyte colony-stimulating factor (GM-CSF); interleukin (IL)-1α, IL-3, Il-4, IL-5, IL-6; tumor necrosis factor-alpha (TNF-α); and interferon gamma (IFN-γ) (16–18). This latter role for FcεRI is particularly important (see later) in that is the first example of a physiological, rather than a pathological, role for FcεRI and mast cells in general. Until now, FcεRI has only been linked to IgE-induced allergic diseases. Even the FcεRI present on epidermal Langerhans cells is thought to be linked to the pathobiology of atopic dermatitis (10).

II. STRUCTURE OF FcεRI

A. FcεRI: A Multimeric $\alpha\beta\gamma_2$ Complex

Initial attempts to purify the FcεRI receptor by classic affinity chromatography techniques led to the identification of a single IgE-binding polypeptide. This glycoprotein appeared as a broad band, with an apparent molecular mass (M_r) of 50–60 kDa on sodium dodecyl sulfate–polyacrylamide gel electrophoresis (SDS–PAGE) (19–21) and was later called the α chain. Several studies indicated a stoichiometry of one α molecule per bound IgE (22,23). At the same time, a variety of other studies began to suggest that the receptor might be composed of more than a single polypeptide. The most convincing evidence came from experiments using chemical cross-linking reagents, which demonstrated that the IgE-binding glycoprotein, the α subunit, was associated with other polypeptides (24,25). The high-affinity IgE receptor (FcεRI) is now known to be a tetrameric complex, composed of one α chain, one β chain, and a dimer of identical disulfide-linked γ chains. A schematic representation of the FcεRI receptor (Fig. 1) demonstrates the principal structural characteristics of the receptor (4). The α

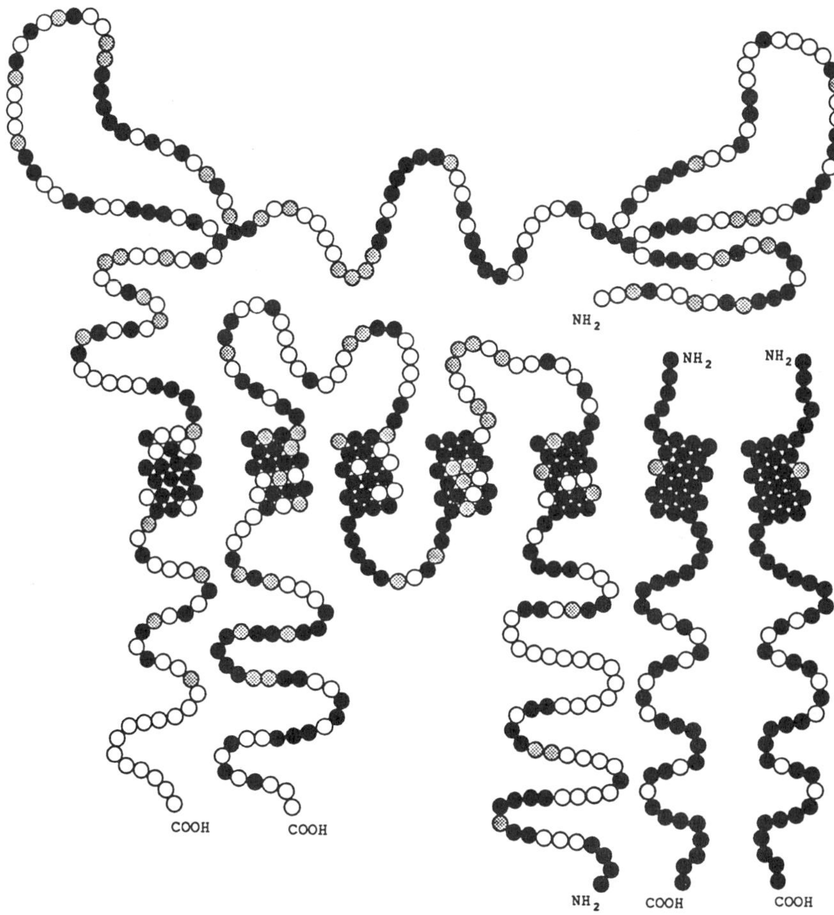

Figure 1 Model of FcεRI. The α subunit is shown at the top and the remainder of the chain on the left. To the right of the α subunit is the β subunit and its four transmembrane domains, and to the right of the β subunit is the dimer of γ chains. Each amino acid residue is represented by a circle. The polypeptides are shown in what is believed to be their fully processed form. The amino acid residues that are identical in rat, mouse, and human are indicated by filled circles. The residues that are homologous, but not identical, are indicated in shaded circles.

chain contains a long extracellular segment with two immunoglobulin-related domains, a single transmembrane domain, and a relatively short cytoplasmic tail. The β chain, in contrast, crosses the plasma membrane four times and has both its COOH- and NH_2-termini exposed to the cytoplasm. The γ chains form a disulfide-linked dimer. Each γ chain has a very short extracellular segment, a single transmembrane domain, and a relatively long cytoplasmic tail. The entire receptor, therefore, has seven transmembrane domains, five extracellular segments (one long and four short), and six cytoplasmic domains.

The initial failure to detect the associated β and γ subunits of the FcεRI receptor relates to the ease with which the tetrameric complex dissociates in mild detergent (25–27). Noncovalent hydrophobic and electrostatic interactions are responsible for maintaining the association of the FcεRI subunits in the plasma membrane, and these interactions are disrupted by detergent micelles, thereby dissociating the multimeric complex (26). The techniques classically employed for solubilizing receptors from the plasma membrane thus result not only in receptor solubilization, but dissociation of the tetrameric complex as well. These subunit interactions are, however, protected when solutions of submicellar detergent are used for receptor solubilization or when the deter- gent micelles are mixed with specific phospholipids (26–29). The finding of the specific conditions required to keep the complex associated proved to be crucial to the subsequent purification of the individual β and γ subunits. Microsequencing of purified tryptic peptides allowed the cloning of complementary DNAs for each of the subunits and the isolation of the corresponding genes.

In the membrane bilayer, the phospholipids may contribute to stabilizing the tetrameric receptor complex. The observation that the multimeric FcεRI complex may be stabilized by adding specific phospholipids to the detergent micelles used to solubilize the receptor supports this concept (27,28). Such a role for the membrane phospholipids further suggests that the transmembrane domains are the major regions for subunit interaction.

B. Transcripts and Polypeptide Structure of the α, β, and γ Subunits

1. *The* α *Subunit*

Complementary DNAs for the α subunit have been isolated from rat (30,31), mouse (32), and human (31,33), with a deduced amino acid sequence of 245, 250, and 260 residues, respectively. This difference in length among the species reflects a difference primarily in the length of the cytoplasmic tails, with 20, 25, and 31 residues reported for the rat, mouse, and human α, respectively.

The sequence of the unprocessed α subunit contains a characteristic NH_2-terminal signal peptide that is found in all three species (30–33). The basic characteristics of the processed α peptide are depicted in Figure 1. A hydropathicity plot of the sequence reveals a single transmembrane domain. The extracellular domain, which carries the high-affinity-binding site for IgE (discussed later), comprises about 180 residues and contains two immunoglobulin-related domains of 40 and 42 residues each. These domains are formed by disulfide bridges generated from two pairs of cysteine residues.

The heterogeneity in relative molecular mass observed on SDS–PAGE for the mature α chain (34–36) can be explained by differences in the glycosylation levels. There are seven (rat and human) or six (mouse) classic *N*-linked glycosylation sites in the extracellular domains. However, the position of only one of these sites is conserved among the three species. *O*-Linked glycosylation has also been reported (34), although the sites remain to be identified.

Conservation of the DNA sequence is substantial between the rodent and human α chains (68% homology). However, only 38% identical amino acid residues are found in the consensus sequence for the three species, and the level of conservation varies among the different domains. In the extracellular domain, there is 42% identity, whereas there is only 16% identity in the cytoplasmic tail. In contrast, in the transmembrane domain, 62% of the residues are identical, nine of which are consecutive residues.

One major mRNA species is found in both rodent (1.3 kb) (30) and in humans (1.1 kb) (37). As expected, the corresponding transcripts were found in cells of mast cell or basophil lineage (30,37) and have also been detected by polymerase chain reaction (PCR) in Langerhans cells (10,11), but not in other cells tested thus far. Three additional transcripts were identified by cDNA cloning (38). However, the genomic structure of the α gene indicates that two of these variant transcripts are most likely generated by aberrant splicing, whereas the third represents either an unspliced precursor or is generated from an alternative initiation site in intron 1. These variants are probably irrelevant, since no corresponding protein products have been isolated using an anti-α antibody (38).

2. *The β Subunit*

Complementary DNAs for rat (39), mouse (32), and human β (40,41) have also been isolated. The deduced amino acid sequences comprise 243, 235, and 244 residues, respectively. The open-reading frame in the rat sequence (729 bp) is slightly longer than in the mouse (705 bp) owing to a 24-bp in-frame insertion in the rat sequence. The level of homology between the rat and mouse β is considerable, regardless of whether DNA sequences (87%) or amino acid sequences (83% identical residues) are compared. The human β is very homologous to rodent β, with about 69% of the amino acid residues identical between human, mouse, and rat.

Hydropathicity plots of the three species strongly suggest that β has four transmembrane domains, with no leader peptide (39,40). Studies with two monoclonal antibodies indicate that both the COOH- and NH_2-termini of the β protrude into the cytoplasm (39) (see Fig. 1). Biosynthetic incorporation with labeled sugar indicates that, unlike the α subunit, the β chain does not contain any carbohydrate (25,39,42) even though its sequence contains a few potential *N*-linked glycosylation sites (the exact number differs among the species). In addition, within the β chain, both tyrosine and serine residues are phosphorylated (43,44) after receptor engagement (discussed later).

Two major species of mRNA (2.7 kb and 1.75 kb) have been observed in rodent mast cells, but not in any of the other cell lines tested. These two transcripts are the result of alternate polyadenylation (32,39). Other cDNA clones with an early termination are likely to be the result of aberrant splicing. The human β gene encodes two transcripts, which are detected as a doublet of about 3.9 kb (40). These transcripts are also present in cells of basophil or mast cell lineage (32,39,40) and have been detected by PCR in Langerhans cells (10,11), but, as yet, not in other cells tested. In mast cells, the β subunit has been reported to also associate with the FcγRIIIA receptor (45).

3. *The γ Subunits*

Complementary DNAs for rat (46), mouse (32), and human γ (47) have now been reported. The M_r of the unprocessed γ, which is 86 residues in length, is about 9.78 kDa. A hydropathicity plot of the sequence shows a typical leader peptide of 18-amino acid residues and a single transmembrane domain. The processed γ, therefore, is a membrane-attached peptide of 68 residues in length, with an M_r of about 7.8 kDa. This is in good agreement with previous estimates using SDS–urea gels for γ purified from the cell surface (48). There is no *N*-linked glycosylation site in the γ sequence. This is consistent with biosynthetic studies that indicate no incorporation of carbohydrate into the γ subunit (39). Despite a unique histidine residue in the COOH-terminus of γ, the subunit could not be biosynthetically labeled with [^{3}H]histidine (46). This led to the speculation that a posttranslational processing of the COOH-terminus would remove the unique histidine. However, recent studies with antipeptide antibodies directed against the unique histidine-containing COOH-terminal peptide indicates that this processing is not occurring (49).

Each γ chain consists of only five extracellular amino acid residues followed by a 21-amino acid membrane-spanning segment, and a cytoplasmic region of 43 residues (32,46,47) (see Fig. 1). The transmembrane domain contains one aspartic acid residue and is flanked by two cysteines that are 19 residues apart, one being located at the external and the other at the internal border of the plasma membrane. In the plasma membrane, the γ chain is found as a covalantly cross-linked dimer in association with the α and β subunits (25). On SDS–PAGE the dimeric γ is reduced to the monomeric form by the use of reducing agents. This suggests that a disulfide bond stabilizes the γ dimer, and recent data using mutated γ chains indicate that this disulfide is formed from the external or NH_2-terminal cysteines [residues 7 in the processed chains (see Fig. 1) and not those (residues 26) at the cytoplasmic aspect of the transmembrane domain] (50). Cysteine-7 of the γ chain is, therefore, homologous to the single cysteine in the ζ chain that forms the disulfide bridge linking the ζ dimers to each other or to the η chain in the T-cell receptor (TCR) (51,52). Although the internal cysteine may react covalently with fatty acyl chains (49), its role in determining the structure or function of the γ chain is now unclear.

The topology of the γ sequence is shown in Figure 1. In addition to forming disulfide-linked homodimers similar to the ζ chain of the T-cell receptor, the topology of the γ subunit is also similar to that of the ζ chain (51). This topology is based on two types of studies: surface iodination of intact cells and in vivo phosphorylation of the chains. The surface iodination did not label γ chains on intact cells, although iodination was observed using inverted vesicles (53). Furthermore, in vivo studies indicate that there are both tyrosine and threonine residues within γ that are phosphorylated (43). Since tyrosine and threonine residues are found only in the presumptive cytoplasmic tail of γ, it is likely that the short segment of five residues preceding the transmembrane domain is extracellular.

The γ subunits from rat, mouse, and human are highly conserved, with the consensus sequence for the three species containing 86% identical amino acid residues (32,46,47). This conservation through evolution reflects the functional importance of the γ chains (see later). A single transcript of 0.85 kb has been detected in the three species. The calculated total length of the mRNA (591 bp) is in agreement with results obtained by Northern blots (750 bp), if one accounts for the poly(A) tail. Unlike the α and β transcripts, the transcript for the γ subunit has been identified in macrophages (54), natural killer (NK) cells (49,55,56), and T cells (57–59) where they associate with the low-affinity receptor for IgG (FcγRIII) or with the T-cell antigen receptor.

C. Genes and Chromosomal Localization

1. *Genomic Organization of the α, β, and γ Genes*

The FcεRI α genes share features common to the Ig supergene family (13). This gene contains five exons and four introns and spans approximately 6.6 kb (60). The overall intron–exon organization of the FcεRIα gene is similar to that of the FcγRs, with the leader peptide encoded by two exons (exons 1 and 2) and each of the immunoglobulin-related domains encoded on a separate exon (exons 3 and 4) (60,61). Finally, the transmembrane and cytoplasmic domains and the 3′-untranslated region are encoded by the last exon, exon 5.

The coding sequence of the human β gene spans approximately 10 kb and is encoded by seven exons (40). There is a single transcription initiation site preceded by a TATA box. The first exon codes for the 5′-untranslated region and a portion of the NH_2-terminal cytoplasmic tail. Transmembrane (TM) segment 1 is encoded in exon 2 and 3, TM 2 in exon 3 and 4, TM 3 in exon 5 and TM 4 in exon 6. The seventh and final exon encodes the end of the COOH-terminal cytoplasmic tail and the 3′-untranslated sequence. The human β gene appears to be a single-copy gene (40).

The human γ gene spans 4.2 kb and contains five exons (47). The first exon contains the entire 5′-untranslated region and most of the leader peptide. This is followed by a large intron of 2.6 kb, which represents 67% of the total gene and contains Alu sequences. The second exon encodes the last five bp of the leader peptide, the short extracellular domain, the hydrophobic transmembrane region, and the beginning of the cytoplasmic tail. Three short exons encode the remainder of the polypeptide as well as the 3′-untranslated flanking sequence. The gene for the γ subunit belongs to the same family as the gene for the T-cell antigen receptor ζ chain (47). In addition to being homologous to the ζ chain of the T-cell receptor at the protein level, with both subunits forming disulfide-linked dimers and having similar topology (see foregoing), both genes exhibit a similar exon organization. A high level of homology between the γ and ζ genes is found in three of their respective exons, and the splice sites are identical (47). This homology and the fact that both genes are on the same chromosome (see next section) suggest that the two genes arose by gene duplication (62,63).

2. *Chromosomal Localization*

The genes for the α and γ subunits are close together at the distal end of mouse chromosome 1 (62–64) and on band 1q23 of human chromosome 1 (65). Both are linked to the gene encoding one of the FcRs for IgG (FcγRII or Ly-17) in a region that also harbors the genes for other FcγRs that are homologous to the FcεRI α chain. It is surprising that the genes for two associated peptides, such as the α and γ, are also associated on the same chromosome, especially when their sequences are completely unrelated. It is possible that selective pressures may have brought and kept these two genes together, possibly for transcriptional regulation. The relative distance between these two genes, when established, should indicate whether this is a possibility.

The mouse β gene is localized on chromosome 19, linked to the Ly-1 (Ly-12) locus (62). This region also contains the loci for Ly-10 and Ly-44 (66). Since the human analogues of Ly-44 and Ly-1, CD20 and CD5, respectively, have been localized as a syntenic unit on human chromosome 11, one can speculate that the human β gene will also be found on that chromosome. Interestingly, *Ly-44* and its human counterpart *CD20*, encode peptides that have clusters of homology with β (see later). This fact, combined with the information that the mouse *FcεRI*β and *Ly44* genes are closely associated on chromosome 19 (62), suggests that these genes have evolved by duplication.

III. RECEPTOR ASSEMBLY AND EXPRESSION

A. Ontogeny, Biosynthesis, and Assembly

Receptors are expressed very early during the differentiation of mouse mast cells and human basophils, at a time when the cells are still undifferentiated and intracellular granules are as yet not detectable (37). Although these cells most likely represent committed progenitors, these nongranulated cells are capable of receptor-mediated lymphokine production (9). Transcripts for the subunits of FcϵRI have also been detected in mature peripheral basophils (37). Therefore, it is likely that these cells have retained their ability to synthesize receptors de novo.

The relative abundance of the mRNA for each of the subunits is not equivalent (3). Specifically, the β transcripts are substantially less abundant than the α and γ transcripts. If the relative abundance of the mRNA has any correlation with the amount of protein synthesized, then the subunits are probably not synthesized in equal amounts. This could indicate that the subunits are not coordinately synthesized, assuming that the rate of translation of these mRNAs is similar. It is possible that the cell regulates the assembly and, consequently, the surface expression of the receptor by controlling the synthesis of the subunit that is present in limiting quantities. This has been observed for other multimeric complexes such as the T-cell receptor (67).

It is not yet clear how the tetrameric receptor assembles. Biosynthetic-labeling studies with RBL cells (Rivera and H. Metzger, unpublished results) suggest that α associates first with β and only later with γ. There is also good evidence that the subunits assemble before their final processing and before leaving the pre-Golgi compartment. However, the surface receptor appears to recycle and be degraded as a unit (68). Furthermore, the incubation of cells with monomeric IgE for prolonged periods (12–24 h) results in an increased number of cell surface receptors. Thus, the receptor up-regulation observed upon binding of monomeric IgE appears to result, at least in part, from decreased receptor degradation, as if IgE binding protected the receptor from enzymatic degradation during receptor recycling (68,69).

Subunit interactions probably occur through the transmembrane domains, although extracellular and intracellular interactions are possible. A model of the organization of the transmembrane domains has been previously proposed (70) based on the evidence for a direct interaction between β and γ, between α and γ, and between α and β. The detection of $\beta(\gamma)_2$ complexes support the existence of β–γ interactions. This complex is generated after immunoprecipitation of the receptor (71) and during the first stages of receptor dissociation, following exposure to detergent (72). Evidence for α–γ interactions is provided by transfection experiments in which the human α chain associates with the γ chain on the cell surface after cotransfection of the corresponding genes (73).

B. Expression of FcεRI by Gene Transfer

Transient transfection of Cos-7 cells with the rat α gene did not lead to any detectable surface IgE binding (30,31). Although surface expression of the rat, mouse, or human α was found to occur, the efficiency of expression was so low that detection of IgE binding was difficult (32). Subsequent studies revealed that efficient receptor expression required the simultaneous cotransfection of one or more of the other subunits. Surprisingly, the requirements for efficient receptor expression were found to be different for the rodent and human receptors. For the rodent receptor, simultaneous cotransfection of the three genes for α, β, and γ is necessary to induce surface expression (32,46). In contrast, cotransfection of human α and γ is sufficient for the human receptor expression, as if the β was not essential (73). This proved to be true whether the α human is transfected with rat, mouse, or human γ (32,47,73). Furthermore, human αγ and αβγ complexes are expressed with comparable efficiency, despite the species specificity of the α–β interaction (40), and the surface-expressed $\alpha(\gamma)_n$ receptor has the same binding characteristics as the complete $\alpha\beta(\gamma)_2$ receptor. These results leave open the possibility that human αγ receptor complexes occur in vivo. The recent cloning of the human β cDNA allowed the screening of a variety of cell types on allergic and nonallergic individuals (40). It is now clear that β is present in every individual and in mast cells, basophils, and the Langerhans cells (10,11,40). Nevertheless, the possibility that a fraction of the surface receptors may exist as αγ complexes cannot be excluded.

As described earlier, each of the receptor subunits is composed of at least one cytoplasmic extension and one transmembrane domain. On the basis of transfection studies using mutated receptor subunit cDNAs transfected into Cos-7 cells, the cytoplasmic extensions do not appear to be critical for the assembly of the receptor (50). In contrast, relatively modest changes in the transmembrane domains results in a substantial decrease in the efficiency of receptor expression (50). These results are consistent with studies of the isolated receptors, which implicated hydrophobic forces as the ones primarily responsible for stabilizing the subunit interactions (26) and lend support to the concept that the transmembrane domains are the likeliest sites for subunit interactions (70).

IV. FUNCTION OF FcεRI

A. Binding of Immunoglobulin E

1. *Stoichiometry, Affinity, and Specificity of Immunoglobulin E Binding*

Monomeric IgE binds to FcεRI with an affinity of about 10^{10} M^{-1} (74,75) and with a stoichiometry of one IgE bound per receptor (76). The high affinity of

IgE binding is principally due to a slow dissociation rate ($\leq 10^{-5}$ s^{-1}). The interaction of IgE with FcεRI cannot be inhibited by overwhelming excesses of IgG or other immunoglobulins; hence, the isotype specificity of FcεRI is absolute. This is in contrast with the FcγRII and FcγRIII receptors that have recently been shown in mice to bind IgE as well as IgG (see later). The species specificity of FcεRI, however, is not complete. Whereas rodent receptors bind only rodent IgE, human receptors bind human IgE as well as rat and mouse IgE (74).

2. *Sites of Interaction on Immunoglobulin E*

The binding of IgE to mast cells was shown to occur through the Fc region of the immunoglobulin which, in this instance, consists of six domains $(C_\epsilon 2 C_\epsilon 3 C_\epsilon 4)_2$. Studies with recombinant peptide fragments corresponding to a single 76-amino acid region of the ε chain of human IgE (residues 301–376) implicated a region centered around the $C_\epsilon 2{:}C_\epsilon 3$ interdomain interface as the site for IgE binding (77,78). The peptide, which was produced in bacteria, contains no carbohydrate, confirming earlier evidence of the lack of importance of the carbohydrate for IgE binding (79). Furthermore, since the active peptide appeared to remain as a monomer in solution, it suggested that only one of the ε chains may be involved in the binding to the receptor. There is now good evidence from studies using chimeric constructs that it is the $C_\epsilon 3$ domain of IgE that alone contains the site for interaction with FcεRI (80,81). Interestingly, overlapping but distinct sequences were critical for binding of IgE to low-affinity FcεRII, and the dimeric peptide was required (82).

A search for conformational changes in IgE following its binding to the receptor have failed to reveal any such alterations (83). Although there is a decreased susceptibility of IgE to proteases after receptor binding (84) as well as an inhibition of binding to certain anti-IgE antibodies (85), both of these results can be explained on the basis of steric factors alone.

3. *The Immunoglobulin E-Binding Site on the Receptor*

Several lines of evidence demonstrate that the α chain of FcεRI alone is sufficient for binding IgE. The first indication came from experiments in which the transfection of human α and γ subunits was found to allow IgE binding (32,47,73), thereby suggesting that the β chain was not essential for the binding reaction. Since the ectodomain of the γ chain is minimal (see foregoing), it thus appeared that the α chain alone was sufficient for IgE binding. Recently, this has been confirmed using both chimeric constructs and soluble α peptides. A chimeric construct containing only the ectodomains of the human α chain engrafted onto the transmembrane and cytoplasmic domains of the p55 IL-2 receptor protein, *Tac*, clearly localized the high-affinity binding site for IgE to the extracellular portion of the α chain of FcεRI (86). Similarly, a construct containing the ectodomain of α linked to a glycosylphosphatidylinositol group that anchors it to the plasma membrane, binds IgE normally (87). Recently, soluble peptides, corresponding to the extracellular domain of the α chain, have been produced and secreted from stably transfected cells (88). These peptides bind IgE with an affinity characteristic of membrane-bound receptors, providing further evidence that the extracellular domain is sufficient for binding IgE. Thus, the β and γ subunits are neither required for, nor do they appear to be involved in, the binding reaction. In addition, the IgE-binding ability of these soluble peptides clearly demonstrate that the membrane attachment of the receptor is also not a requisite for IgE binding. Furthermore, glycosylation of α is not required for high-affinity binding (88).

Studies with monoclonal antibodies against the α chain further indicate that at least two regions of the extracellular domain of α are involved in forming the pocket responsible for binding (89). The evidence to date suggests that both the Ig-like regions may contribute to the binding, but that the one closest to the membrane is the most important (89).

B. Signal Transduction

1. Role of Aggregation

Receptor aggregation has long been recognized as critical for the initiation of cell activation by FcεRI. Anti-α antibodies, but not their Fab fragments, trigger exocytosis (90,91). Similarly, chemically cross-linked IgE activates mast cells, whereas binding of monomeric IgE does not result in any apparent intracellular signaling (92). Thus, aggregation of receptor-bound IgEs by a multivalent antigen is required to trigger cell activation. The minimum size of an "active" aggregate appears to be a dimer (92,93), and a few thousands of receptor aggregates are sufficient (94,95). However, recent studies with fluorescent probes suggest that small receptor aggregates tend to form larger patches spontaneously (96). Therefore, the minimal size of an active aggregate is difficult to assess.

The issue of whether physical contact between receptor molecules is an additional prerequisite for signaling has also been investigated (97). The distance between two epitopes of a rigid bivalent hapten was varied to ascertain the maximal distance between receptors aggregated by IgEs specific for this hapten. When the distance between the two epitopes of the hapten was apparently large enough to prevent physical contact between aggregated receptors, the bivalent haptens were still capable of triggering. Therefore, contact between adjacent receptors is not required for cell activation (97). It may be that a specific redistribution of receptors with particular orientational constraints is the critical factor in "aggregation" (98). If so, this may explain why some anti-β antibodies, although apparently recognizing a surface epitope, fail to trigger cell activation.

The old observation that exocytosis and signaling induced by a multivalent hapten can be interrupted by an excess of monovalent hapten has, therefore, been given a new interpretation. The role of the monovalent hapten apparently is not so much to disaggregate previously aggregated receptors, as to prevent further receptor aggregation. Since signaling is halted after addition of monovalent hapten, despite the presence of aggregated receptors on the cell surface (99), these aggregated receptors must be inactive, and it is proposed that the receptors have been desensitized. Thus, aggregation of receptors, although required to initiate signaling, is not sufficient to sustain it. Continuous aggregation is required to sustain the signal. Consequently, the analysis of possible structural changes of active receptors, such as phosphorylation, should be performed very early after stimulation.

2. *The Amino Acid-Signaling Motifs of the Receptor*

The γ Subunit. An amino acid-signaling motif containing 18-amino acid residues has been found in the cytoplasmic tails of the γ chains of FcεRI (100–103). This activator motif contains six consensus amino acids, with two phylogenetically conserved tyrosine residues that are absolutely critical for activity (101,102). When fused to the cytoplasmic tail of other transmembrane proteins, this motif appears to be sufficient to confer receptorlike activity (101,108). Transfection of human T-cell line Jurkat cells with a chimeric construct containing the γ cytoplasmic tail linked to the ectodomain of CD4 led to a calcium response following activation of the transfected cells (101). Similarly, a chimera construct consisting of the γ cytoplasmic tail together with the ectodomain of the α chain of the IL-2 receptor resulted in serotonin release following activation of the transfected receptor in RBL cells (102). The truncation of the γ tails abolished the capacity for the chimeras to induce cell activation.

Similar signaling motifs are also found in the ε, δ, ζ, and η chains of the T-cell antigen receptor (but not the T-cell antigen receptor α or β chains), the MB1 component of the IgM antigen receptor on B cells and the envelope protein, gp30, of bovine leukemia virus (BLV) (103–107). These motifs, when introduced into cells by transfection of chimeras, generate signals resulting ultimately in proliferation and differentiation of B and T cells and histamine release in mast cells.

The ζ, η, and γ chains each contain 3, 2, and 1 such motifs, respectively (103). Consequently, the ζζ, ζη, ηη, ζγ, ηγ, and γγ-dimers contain 6, 5, 4, 3, and 2 such motifs. Interestingly, the ζ chain appears to be more potent than the γ chain in activating either T cells or RBL cells (102). It is possible that the number of these motifs per se may confer on each dimer its potency. Since it appears that the signaling capacity of the ζζ, ζη, and ηη dimers are the same after transfection in ζ-deficient T cells (108), it is conceivable that the different homo- and heterodimers initiate signals of different potency, rather than initiating different signals. The variable expression of these dimers would then give the cell a means of modulating the level of its activation. Alternatively, slight modifications in the signaling motifs may lead to activation of different signals. For example, a 22-amino acid motif within the CD3 ϵ chain cytoplasmic tail, which is homologous, but not identical, with the γ and ζ motifs, also results in cell activation when transfected into T cells (106). The level of tyrosine phosphorylation of different substrates, however, differs from that seen following transfection of the ζ cytoplasmic tail. It is possible that such differences in the phosphorylation levels of specific substrates may ultimately result in the activation of different pathways.

The β *Subunits.* Although the cytoplasmic domain of the FcϵRI β subunit similarly contains this activator motif (100), a chimera constructed with the COOH-terminal cytoplasmic domain of FcϵRI β is not sufficient to activate RBL cells (O. Letourneur, J. P. Kinet, unpublished results). However, a role for the cytoplasmic tail in the initiation of activation is suggested by the finding that truncation of the COOH-terminal cytoplasmic domain of FcϵRI β abolished the receptor activation in a reconstituted system (110). Truncation of either the NH_2-terminal cytoplasmic domain of FcϵRI β or the cytoplasmic tail of α, neither of which contain this motif, had no apparent effect on receptor activation (110). Thus, the FcϵRI β chain appears to play some role in receptor signaling. In addition, in P815 cells transfected with a mutated β subunit that is missing the COOH-terminal cytoplasmic domain, FcϵRI activation resulted in abnormal aggregation (111), suggesting a role for the β subunit in the aggregation event as well.

3. *Activation of Kinases and of Phosphatases*

Phosphorylation of receptors and other proteins is a common feature of cell activation in a variety of systems. Some receptors, as exemplified by the growth factor receptors, have an intrinsic kinase activity (112). The FcϵRI receptor, however, belongs to a class of receptors that lack intrinsic kinase activity, but are able to mediate protein phosphorylation by activating nonreceptor tyrosine and serine/threonine kinases. Among the members of this latter class are the antigen receptors on T and B cells and the Fc receptors for IgG (low-affinity FcγRII and FcγRIII). Interestingly, some of these same receptors not only share similarities in their structural elements and signaling mechanisms, but actually share common subunits. For example, FcϵRI, FcγRIII, and some forms of the T-cell antigen receptor have a common subunit identified as FcϵRI-γ (46,49,54–59) (discussed later). This subunit and the homologous T-cell antigen receptor ζ and η subunits contain amino acid motifs that function as transducing elements. The sharing of a common subunit or common signal-transducing motifs within specific subunits may explain the similarities in the signals transduced by these receptors (e.g., rise in intracellular calcium and phosphoinositide turnover).

Following engagement of the receptor, members of this latter class of receptors physically associate with and activate a variety of nonreceptor kinases. Such an interaction has been demonstrated for the T-cell antigen receptor and $p53^{fyn}$ kinase (113), and for the B-cell antigen receptor and the $p53^{fyn}$, $p56^{lyn}$, and $p55^{blk}$ kinases (114, 115). A tyrosine kinase activity, which was at the time considered to be a contaminant, coimmunoprecipitated with the FcεRI receptor in RBL cells (116). Recently, $p56^{lyn}$ kinase of the $p60^{c\text{-}src}$-related tyrosine kinase family, has been found to associate with the FcεRI receptor in RBL cells (117) and is activated following receptor aggregation. There is also evidence of increased activity of $p60^{c\text{-}src}$ following receptor aggregation (117), although this kinase does not associate with the receptor. Surprisingly, in another murine mast cell line, PT-18, another $p60^{c\text{-}src}$-related tyrosine kinase $p62^{c\text{-}yes}$ is associated with the FcεRI receptor (117). Thus, the FcεRI exhibits the capacity for association with some, but not all, $p60^{c\text{-}src}$-related tyrosine kinases, suggesting that molecular motifs critical for their interaction with FcεRI may be shared by different tyrosine kinases. Since a variety of cellular proteins (inclusive of specific receptor subunits) are tyrosine phosphorylated subsequent to receptor engagement, these tyrosine kinases may constitute a direct link between receptor aggregation and the subsequent biochemical events involved in transducing the signals that ultimately lead to degranulation and cytokine production.

There is also evidence of a role for phosphatases in the receptor-activated cascade. Tyrosine phosphatases have, for example, been demonstrated to have an obligatory role in the activation of other receptors coupled in nonreceptor tyrosine kinases (e.g., the antigen receptor on T and B cells) (118–120). In these cells, action of a protein tyrosine phosphatase, leukocyte common antigen (CD45), is required for cell activation. A role for a CD45-like phosphatase in the cell activation mediated by FcεRI is suggested by the observation that a monoclonal antibody against the tyrosine phosphatase CD45 modulated the FcεRI-mediated stimulation of basophils, but was without effect on the stimulation of these cells by other secretagogues, such as calcium ionophores (121). Although it is not yet known whether the effects of the anti-CD45 are related to activation or inactivation, the findings demonstrate a singular functional relation between FcεRI and CD45 in human basophils. Further support for a role for phosphatases derives from recent experiments examining the effect on mast cell signals of phenylarsine oxide (PAO), a potent thiol-reactive inhibitor of tyrosine-specific phosphatases such as CD45. In these studies, PAO uncoupled the receptor from later events in cell activation, inclusive of increased levels of intracellular Ca^{+2}, phosphoinositide turnover, and secretion (122). The PAO-sensitive step occurs at the level of phospholipase C (PLC) activation, by preventing the tyrosine phosphorylation [and, thereby, the activation (123)] of the γ1 isoform of PLC (122). This, in turn, prevents inositol phosphate formation, which is essential for the rise in intracellular Ca^{+2} and subsequently, secretion (discussed later). A PAO-sensitive phosphatase thus appears to play a major role in PLCγ1 activation specifically and FcεRI-mediated cell activation in general.

4. *Receptor Phosphorylation and Dephosphorylation*

Initial analysis of the phosphorylation state of the FcεRI complex revealed a dephosphorylation of the γ chain and a slight phosphorylation of the β chain following FcεRI activation (124). In these same studies, aggregated and nonaggregated receptors from the same cell exhibited no differences in their phosphorylation levels, leading to the conclusion that phosphorylation was not a primary event in the signaling cascade activated by FcεRI. However, a recent reanalysis of the phosphorylation status of the individual receptor subunits reveals that the resting receptor has some basal level of phosphorylation and that, after receptor engagement, there is immediate phosphorylation of both the β chain (tyrosine and serine) and the γ chain (tyrosine and threonine) (43). This phosphorylation signal is restricted to the activated receptors and, therefore, does not affect the nonaggregated receptors. It is indeed one of the earliest events observed after cell activation, occurring within 5 s of receptor engagement (43). Furthermore, receptor phosphorylation does not require extracellular Ca^{+2} or the activation of protein kinase C (43). Therefore, its activation is also independent of the other major FcεRI-induced–signaling pathways in the RBL cells.

A unique feature of the FcεRI on mast cells is that monomeric hapten can be used to interrupt the stimulation by polymeric antigen. This leads to immediate dephosphorylation of the receptor, thereby allowing direct visualization of the receptor-specific phosphatase activity (43). The tyrosine phosphatase involved in the hapten-induced dephosphorylation of the receptor subunits is PAO-insensitive (122). This is in contrast with the PAO sensitivity of PLCγ1 activation and subsequent events involved in cell activation (see foregoing section). These findings suggest that there are at least two different tyrosine phosphorylation pathways activated in mast cells following FcεRI engagement, one that is PAO-sensitive and leads to PLCγ1 activation, inositol phosphate release, enhanced levels of intracellular Ca^{+2} and secretion, and a second PAO-insensitive pathway responsible for phosphorylation of the activated receptor and the hapten-induced dephosphorylation of the activated receptor. However, there is also evidence of a PAO-sensitive phosphatase involved in constitutive dephosphorylation of the receptor, since PAO enhances receptor phosphorylation (tyrosine and serine/threonine) in unstimulated cells (122).

It is particularly striking that the activator motifs of γ and β contain critical tyrosine residues and that it is the γ and β chains containing these motifs that become phosphorylated on these tyrosines after triggering through FcεRI. This is analogous to the phosphorylation of the tyrosine residues in the ζ chain following engagement of FcγRIII (125,126). As to whether these specific phosphorylations activate or desensitize the receptor, two distinct models suggest themselves at present. In the first model, nonphosphorylated tyrosines are required for cell activation, and receptor phosphorylation uncouples the receptor from critical elements of the machinery. In the second model, phosphorylated tyrosines are required to couple the activated receptor to intracellular-signaling elements. Of the two models, more of the existing data are consistent with the latter. Phosphorylation of the β (serine and tyrosine) and γ (threonine and tyrosine) subunits of the receptors is specific to activated receptors and is virtually immediate. Furthermore, addition of monomeric hapten, which interrupts the stimulation by polymeric antigen, leads to disengagement of the receptor and immediate dephosphorylation of the receptor and other cellular proteins (43). Therefore, the specificity, immediacy, and reversibility of the receptor phosphorylation following disengagement of the receptor suggest that the receptor is active when phosphorylated. Moreover, FcεRI also associates physically with phosphorylated substrates following receptor activation and phosphorylation (43; and R. Pachini, R. Numerof, and J.-P. Kinet, submitted). Thus it seems likely that receptor phosphorylation promotes the coupling of the receptor to other elements. Phosphotyrosine residues are essential for the association of certain classes of proteins (e.g., SH2-containing proteins). Whether this coupling is required for activation remains to be demonstrated.

5. *Phosphorylation of Other Substrates*

The receptor-mediated activation of specific nonreceptor tyrosine kinases appears to serve as a direct link to subsequent receptor-mediated biochemical events. This is indicated by the fact that the kinases phosphorylate not only the receptor subunits upon receptor activation, but several other cellular substrates—as yet unidentified proteins of 36, 40, 55, 57, 60, 72, 77, 85, 91, 105, and 130 kDa (43,44,127,128)—as well as the γ1 isoform of phosphatidylinositol-specific phospholipase C (PLCγ1) (44,129) and a pp110-kDa protein that has been identified as the *vav* gene product (130).

The mechanisms whereby receptor phosphorylation or receptor-associated tyrosine kinases lead to the generation of intracellular signals remains to be identified. However, phosphorylation of specific tyrosine residues is critical for their subsequent association with proteins that contain SH2 domains (*src*-homology 2 domains) (131). These SH2 domains have a phosphotyrosine-recognizing function that allows the association of the SH2-containing protein with a variety of phosphotyrosine-containing proteins, including $p60^{v\text{-}src}$, but not with tyrosine phosphate-free forms of the proteins. One function this serves is to bring the proteins into physical proximity with tyrosine kinases such that the proteins can themselves be phosphorylated and activated by the kinases. This is the mechanism for activation of PLCγ1 (123). That tyrosine phosphorylation may prove essential for coupling receptor activation to the generation of subsequent signals in RBL cells is supported by the fact that tyrosine phosphorylation of PLCγ1 is key to the activation of PLC in the RBL cells. Similarly, phosphorylation of a p72 or p110 or both, has been implicated in the production of cytokines and histamine secretion (132–134). Yet another potential mechanism for linking tyrosine phosphorylation to cellular events is the observation that certain serine/threonine kinases are able to be activated by tyrosine phosphorylation (135), leading to the subsequent activation or inactivation of other enzymes.

6. *Intracellular Signals: Calcium GTP-Binding Proteins, Phospholipid Hydrolysis, and Protein Kinase C Activation*

Calcium was, for many years, thought of as "the" second-messenger in the FcεRI-mediated activation of secretion in mast cells. That calcium enters the cell under physiological conditions of triggering is clear, but studies with dialyzed mast cells indicate that calcium is neither necessary nor sufficient for exocytosis (136). In addition, permeabilized cells can be made to degranulate in the total absence of calcium (137). Nevertheless, it is likely that, under physiological conditions, calcium acts synergistically with other second-messenger systems to ensure fast and efficient secretion (138).

Both an initial rise in cytoplasmic calcium resulting from a release of internal stores and a subsequent, more massive, entry of external calcium have been implicated in the mast cell response to receptor activation (139–141). The mechanism for the initial release of calcium from internal stores involves a product of phosphoinositide (PI) hydrolysis, inositol 1,4,5-triphosphate (IP_3). This second-messenger induces release of calcium from the endoplasmic reticulum (141,142). The subsequent calcium influx, which is associated with a whole-cell current of 1–2 pA or less, with no channel activity detectable in whole-cell, patch-clamp recording (141), has recently been demonstrated to result from a depletion of intracellular Ca^{+2} alone, independent of inositol phosphate production (143).

Cell activation also results in the opening of large conductance (50 pS) cation channels. These channels represent yet another pathway for calcium influx which can be activated by GTPγS and inhibited by GDPβS (141). These results are consistent with other reports of a cholera toxin-sensitive GTP-binding protein (G-protein) controlling calcium influx (144) by a pathway independent of PI hydrolysis. Further verification of this observation was provided by the ability of mycophenolic acid, which depletes cells of GTP, to inhibit antigen-stimulated Ca^{+2} influx, without impairing antigen-stimulated PI turnover (145). These results are consistent with a role for G proteins in ion channel activity, but not phospholipase C (PLC) activation.

One group of investigators claims to have purified a putative calcium channel, the so-called cromolyn-binding protein (146), which coaggregates with the receptor in reconstituted bilayers, resulting in opening of the calcium channel. These observations are still controversial, and it is likely that the question will not be resolved without cloning these putative channels and achieving functional reconstitution by gene transfer.

Calcium-independent and calcium-dependent PI hydrolysis have both been implicated in FcεRI cell activation. The calcium-independent hydrolysis of PI appears to reflect activation of PLCγ1 by tyrosine kinase-mediated phosphorylation, since this phosphorylation (129) as well as the receptor-stimulated PI turnover is not only calcium-independent but is blocked by tyrosine kinase inhibitors such as genistein (129,134,147).

Studies with GTPγS have implicated a G-protein in the antigen-induced activation of PI hydrolysis (148,149). This mechanism appears to predominate in permeabilized cells (150) in which PLCγ1 is found to be tyrosine-phosphorylated basally (i.e., without receptor activation). This phosphorylation occurs to such an extent that there is no detectable difference following receptor engagement. Since the PLC isozyme known to be coupled by G-proteins, PLCβ1, is not detected in RBL cells (129), the observed GTPγS effect on PLC is unlikely to occur by direct activation of this enzyme. Even though it is possible that another G-protein–responsive PLC isozyme other than PLCβ1 may be present in these cells, it is more likely that the G-protein activation of PLC occurs secondary to G-protein activation of other signals.

Receptor activation triggers the action of other phospholipases as well. Phospholipase A_2, which generates arachidonic acid and, subsequently, prostaglandins and leukotrienes, is activated following engagement of the FcεRI receptor (151). This activation also appears to occur through a GTP-binding protein, but a requirement for Ca^{+2} ions is unclear (152). Phospholipase D is also activated, yielding phosphatidic acid, which is then converted to diacylglycerol (DAG). In fact, this may be the major pathway for DAG formation (153). In turn, DAG, whether arising from PI hydrolysis or action of phospholipase D, activates protein kinase C when Ca^{+2} is present. Arachidonic acid as well as some of its metabolites have also been implicated in the activation of protein kinase C (154). The role for protein kinase C in mast cell degranulation is unclear; although receptor engagement induces its membrane translocation, this may not be essential for exocytosis, since depletion of PKC does not abolish degranulation (155).

7. *Immobilization and Internalization of the Receptor*

Receptor aggregation leads to virtually instantaneous immobilization of FcεRI (111,156). This immobilization appears to be indicative of the extent to which the receptors are interacting with cytoskeletal elements, since upon aggregation, the receptors become progressively less soluble (111,157–159) and more slowly extractable (159). Other indications of an interaction of the activated mast cell receptors with the cytoskeleton include actin polymerization, membrane ruffling, and changes in the morphologic appearance of the cells from a microvillar to a lamellar architecture (159,160). The interaction of the receptors with the cytoskeleton ultimately leads to internalization of the receptor aggregates, but this occurs considerably more slowly than the initial signals generated by the receptor (161), suggesting that the cytoskeletal association of the receptor does not play a role in the signaling pathways leading to degranulation. This is further supported by the observation that the degree of aggregation required to detect cytoskeletal association is considerably greater than that required for the receptors to initiate a variety of biochemical perturbations and secretion (162).

Like many other membrane proteins, FcεRI are internalized following aggregation (163). In RBL cells, electron microcopic studies have demonstrated that the aggregation of the receptor leads to their association with coated pits and that the latter appear to mediate the internalization (160). As expected, dimerized IgE is sufficient to initiate this process, with the rate and extent of internalization proportional to the amount of receptor aggregation. In RBL cells, internalization precedes degranulation, whereas in peritoneal mast cells the opposite is true (161). This observation, as well as the finding that the sensitivity of internalization to various metabolic inhibitors does not parallel the effect of such compounds on secretion, demonstrates that receptor internalization is also not a prerequisite for secretion.

Specific sites of interactions between the receptor and other cellular components have been ascribed to the cytoplasmic domains by genetic engineering techniques. In transfection experiments with either Cos cells or cells of the mastocytoma line P815, in which wild-type subunits were added in conjunction with a specific mutated subunit, the role of the cytoplasmic domains of each subunit has been explored. The diffusion of such receptors before and after aggregation, the internalization of the receptors, and their ability to be solubilized before and after aggregation were assessed. Surprisingly, none of the cytoplasmic domains of any of the subunits was uniquely required to mediate these processes (111,159).

8. *Cytokine Production*

One of the more interesting recent developments in the field of mast cells is the discovery that following prolonged aggregation of the IgE receptor, mast cells are capable of synthesizing and secreting cytokines (16–18) that are involved in both the differentiation of the mast cell (e.g., IL-3) and in the differentiation of antibody-producing cells toward the synthesis of IgE (e.g., IL-4.) The pathway between the aggregated receptor and nuclear events will be interesting to explore. These changes, while requiring aggregated receptors be present, are manifested hours after the initiation of receptor aggregation and involve activation of genes coding for interleukins. The recent identification of the pp110 protein as the protooncogene *Vav* (130) may prove particularly relevant since this hematopoietic cell protein has motifs consistent with a role in the initiation of transcription. Specifically, the *Vav* gene product contains helix–loop–helix, leucine-zipper and zinc-finger motifs as well as nuclear localization signals, an SH2 domain for association with phosphotyrosine residues, and two SH3 domains for association with cytoskeletal elements (164). This protein appears to belong to a class of tyrosine kinase substrates for which tyrosine phosphorylation may provide a mechanism for direct signal transduction, linking receptor activation at the cell surface to transcriptional control.

V. COMPARISON BETWEEN FcεRI AND OTHER MEMBRANE PROTEINS

A. The α Subunit and Fcγ Receptors

The α subunit of FcεRI is homologous to most FcγRs (165). Of all the FcγRs, the α subunit is most closely related to the FcγRIII (CD16) receptors. In three different species (rat, mouse, human), there are nine consecutive amino acids that are conserved in the transmembrane domain of FcεRIα. Eight of these nine amino acids are also identical in the FcγRIII receptors. Since these receptors share an identical γ subunit, and the γ subunits are necessary for surface expression of both receptors, these conserved residues may be involved in the interaction of the α subunit with the γ subunits. In addition, the two immunoglobulin-related domains in the extracellular portion of FcεRIα are homologous both in size and sequence to the corresponding domains in the α subunit of the FcγRIII receptor. Of the 95 amino acids of FcεRIα that are identical in the three species, 61% of these are also in both human and mouse FcγRIII. This may explain why mouse FcγRIII also binds IgE.

B. The β Subunit and CD20

The chromosomal localization of the gene for β on mouse chromosome 19 led to the identification of striking similarities between β and Ly44 as well as its human counterpart, CD20 (66). Similar to β, the Ly44 and CD20 peptides contain four transmembrane domains and have both their NH_2- and COOH-termini protruding into the cytoplasm (66). Furthermore, β and CD20 have clusters of homologous residues in two of the four transmembrane domains (62), and 29% identical amino acid residues are found between mouse β (residues 45 to 141) and mouse Ly44 (residues 37 to 129). The human Ly44 counterpart, CD20, is a surface molecule specifically expressed in B cells. Although its role in B cells in not yet known, it will be interesting to see whether the limited homologies observed between the β and CD20 prove to be functionally relevant. For example, it has been suggested that CD20 regulates transmembrane calcium ion flux in B cells (166), and the aggregated FcεRI is known to activate calcium ion flux.

C. The γ Subunit and Other Receptors

Although initially identified as part of FcϵRI, γ chains identical with those of FcϵRI are part of other multimeric complexes. The γ homodimer also associates with mouse FcγRIII (54) and human FcγRIIIA in macrophages and NK cells (49,55,56). These same homodimers are also part of the TCR complex (57–59), and the ζ chain homodimers of the TCR clearly belong to the same family as the γ chain (47). Their genes are localized on the same mouse and human chromosome 1 (62,65,167,168), and their exons are organized analogously (47,169). A high level of homology (about 50%) is found in three of their respective exons at both the DNA and protein levels; therefore, it is likely that the genes are the product of gene duplication.

The resemblance is not limited to homologies: γ and ζ polypeptides each form disulfide-linked homodimers, and both are essential for the assembly and surface expression of their respective receptors (46,54). Furthermore, γ chains interact with the homologous ζ and η chains of the TCR to form various types of dimers. In some T cells, as many as six different types of dimers have been identified as part of the TCR: the homodimers ηη, ζζ, and γγ, as well as the heterodimers ηζ, ηγ, and ζγ (57–58). Three of these dimers, ζζ, ζγ, and predominantly γγ associate with human FcγRIIIA in human NK cells (49,58). Thus, the same coupling structures, including γγ homodimers, are part of different receptor complexes: γ chains are part of at least three different receptor complexes (FcϵRI, FcγRIII, and TCR), whereas the TCR ζ chains are a component of at least two receptors (FcγRIII and TCR). These structures contain an 18-amino acid signaling motif and seem to mediate similar functions (170 and see foregoing).

VI. FcγRs: ARE THEY ALSO RECEPTORS FOR IMMUNOGLOBULIN E?

More than a decade ago, two molecules with different M_r were isolated from extracts of rat basophilic leukemia cells (RBL cells) and rat mast cells by affinity chromatography on IgE–Sepharose (171,172). Subsequently, RBL cells were shown to express IgG receptors capable of binding IgE as well (173). In fact, these receptors seemed to have a greater affinity for rat IgE than rat IgG (173,174). Other reports indicated that rat IgE could inhibit the binding of IgG to rat lymphocytes (175) or to rat macrophages (176). However, this inhibition was characterized as unidirectional and was thought to be unique for rat cells (177). Recently, however, the low-affinity IgE receptors on mouse mast cells and macrophages have also been characterized and were neither Mac-2 nor CD23, the calcium- and thiol-dependent class of low-affinity IgE-binding structures, respectively. Rather, the low-affinity IgE receptors appear to be FcγRII and FcγRIII (178). Thus, the so-called IgG receptors FcγRII and FcγRIII are not isotype specific; they bind IgE and are thus functional Fcε-receptors as well as Fcγ-receptors.

VII. FcεRI AS A POTENTIAL TARGET FOR ALLERGY THERAPY

In theory, the inhibition of IgE binding to FcεRI, if successful in vivo, should help to prevent IgE-related allergic diseases. The feasibility of such an approach has recently been demonstrated in mice (179). It is possible that peptides such as the 76-amino acid peptide mentioned earlier will be used as a basis for the design of therapeutically useful drugs. Another approach would be to search randomly for compounds that are naturally capable of preventing high-affinity IgE binding. This strategy is made more feasible with the establishment of stably transfected cells that express millions of human FcεRI on their surface (73). Ultimately, once the receptor-binding site has been analyzed crystallographically, one can envision the use of molecular design to generate inhibitory molecules.

Interfering with receptor aggregation or with the intrinsic functions of the receptor also could be useful. One could, for example, interfere specifically with the region of the receptor that is involved in the interaction with kinases. Presumably, this could cure, rather than prevent, allergic reactions.

The recent discovery of cytokine production in mast cells in response to FcεRI activation and their role in regulating the IgE receptor-mediated response, may yet become another avenue for therapeutic manipulation. Interferon-γ, for example, which is produced in response to FcεRI activation, has suppressed IgE secretion in mast cells. Needless to say, efforts to provide allergy sufferers with potent and safe drugs should not be limited to one particular receptor or strategy, and innovative approaches should continue to be actively sought.

REFERENCES

1. C. Prauznitz, and H. Kustner, Studien uber die Uberentfindlichkeit, *Zentrabl. Bakteriol.*, *86*: 160 (1921).
2. K. Ishizaka, T. Ishizaka, and M. M. Hornbrook, Physiochemical properties of reaginic antibody. V. Correlation of reaginic activity with gamma-E-globulin antibody, *J. Immunol.*, *97*: 840 (1966).
3. J.-P. Kinet, The high affinity receptor for immunoglobulin E: Structure, function and role in allergy, *Inflammation: Basic Principles and Clinical Correlates* (J. I. Gallin, I. M. Goldstein, and R. Snyderman, eds.), Raven Press, New York (in press).
4. J.-P. Kinet, The high affinity receptor for immunoglobulin E, *Curr. Opin. Immunol.*, *2*: 499 (1990).
5. J.-P. Kinet, and H. Metzger, Genes, structures and action of the high affinity receptor for immunoglobulin E, *Fc Receptors and the Action of Antibodies* (H.

Metzger, ed.), American Society for Microbiology, Washington, D.C., p. 239 (1990).
6. D. H. Conrad, FcεRII/CD23: The low affinity receptor for IgE, *Annu. Rev. Immunol.*, *8*: 623 (1990).
7. M. L. Richards, and D. H. Katz, Biology and chemistry of the low affinity IgE receptor (FcεRII/CD23), *CRC Crit. Rev. Immunol.*, *11*: 65 (1991).
8. H. L. Spiegelberg, CD23. A novel multifunctional regulator of the immune system that binds IgE. FcεR2/CD23: Its discovery and possible functions, *Monogr. Allergy*, *29*: 1 (1991).
9. R. A. Seder, M. Plaut, S. Barbieri, J. Urban, Jr., F. D. Finkelman, and W. E. Paul, Purified Fc epsilon R^+ bone marrow and splenic non-B, non-T cells are highly enriched in the capacity to produce IL-4 in response to immobilized IgE, IgG2a, or ionomycin, *J. Immunol.*, *147*: 903 (1991).
10. B. Want, A. Rieger, O. Kilgus, K. Ochiai, D. Mauter, D. Fodinger, J.-P. Kinet, and G. Stingl, Epidermal Langerhans cells from normal human skin bind monomeric IgE via FcεRI, *J. Exp. Med.*, *175*: 1353 (1992).
11. T. Bieber, H. de la Salle, A. Wollenberg, J. Hakimi, R. Chizzonite, J. Ring, D. Hanau, and C. de la Salle, Human epidermal Langerhans cells express the high affinity receptor for immunoglobulin E (FcεRI), *J. Exp. Med.*, *175*: 1285 (1992).
12. H. Metzger, G. Alcaraz, R. Hohman, J.-P. Kinet, V. Pribluda, and R. Quarto, The receptor with high affinity for immunoglobulin E, *Annu. Rev. Immunol.*, *4*: 419 (1986).
13. J. V. Ravetch, and J.-P. Kinet, Fc receptors, *Annu. Rev. Immunol.*, *9*: 457 (1991).
14. T. Ishizaka, and K. Ishizaka, Activation of mast cells for mediator release through IgE receptors, *Prog. Allergy*, *34*: 188 (1984).
15. C. W. Parker, Lipid mediators produced through the lipoxygenase pathway, *Annu. Rev. Immunol.*, *5*: 65 (1987).
16. M. Plaut, J. H. Pierce, C. J. Watson, J. Hanley-Hyde, R. P. Nordan, and W. E. Paul, Mast cell lines produce lymphokines in response to cross-linkage of FcεRI or to calcium ionophores, *Nature*, *339*: 64 (1989).
17. A. Wodnar-Filipowicz, C. H. Heusser, and C. Moroni, Production of the haemopoietic growth factors GM-CSF and interleukin-3 by mast cells in response to IgE receptor-mediated activation, *Nature*, *339*: 150 (1989).
18. P. R. Burd, H. W. Rogers, J. R. Gordon, C. A. Martin, S. Jayaraman, S. D. Wilson, A. M. Dvorak, S. J. Galli, and M. E. Dirf, Interleukin 3-dependent and -independent mast cells stimulated with IgE and antigen express multiple cytokines, *J. Exp. Med.*, *170*: 245 (1989).
19. D. H. Conrad, and A. Froese, A characterization of the target cell receptor for IgE. II-D polyacrylamide gel analysis of the surface IgE receptor from normal rat mast cells and from rat basophilic leukemia cells. *J. Immunol.*, *116*: 319 (1976).
20. A. Kulczycki, Jr., T. A. McNearney, and C. W. Parker, The rat basophilic leukemia cell receptor for IgE. I. Characterization as a glycoprotein, *J. Immunol.*, *117*: 661 (1976).
21. J. Kanellopoulos, G. Rossi, and H. Metzger, Preparative isolation of the cell receptor for immunoglobulin E, *J. Biol. Chem.*, *254*: 7691 (1979).

22. C. Fewtrell, E. Kempner, G. Poy, and H. Metzger, Unexpected findings from target analysis of immunoglobulin E and its receptor, *Biochemistry*, *20*: 6589 (1981).
23. J. M. Kanellopoulos, T. Y. Liu, G. Poy, and H. Metzger, Composition and subunit structure of the cell receptor for immunoglobulin E, *J. Biol. Chem.*, *255*: 9060 (1980).
24. D. Holowka, H. Hartmann, J. Kanellopoulos, and H. Metzger, Association of the receptor for immunoglobulin E with an endogenous polypeptide on rat basophilic leukemia cells, *J. Receptor Res.*, *1*: 41 (1980).
25. R. Perez-Montfort, J.-P. Kinet, and H. Metzger, A previously unrecognized subunit of the receptor for immunoglobulin E, *Biochemistry*, *22*: 5722 (1983).
26. J.-P. Kinet, G. Alcaraz, A. Leonard, S. A. Wank, and H. Metzger, Dissociation of the receptor for immunoglobulin E in mild detergents, *Biochemistry*, *24*: 4117 (1985).
27. B. Rivnay, S. A. Wank, G. Poy, and H. Metzger, Phospholipids stabilize the interaction between the alpha and beta subunits of the solubilized receptor for immunoglobulin E, *Biochemistry*, *21*: 6922 (1982).
28. J.-P. Kinet, R. Quarto, R. Perez-Montfort, and H. Metzger, Noncovalently and covalently bound lipid on the receptor for immunoglobulin E, *Biochemistry*, *24*: 7342 (1985).
29. J.-P. Kinet, and B. Grasberger, Reconstitution of the high affinity receptor for immunoglobulin E, *Subcellular Biochemistry 14* (J. Harris, A. H. Etemadi, eds.), Plenum Press, New York, p. 321 (1989).
30. J.-P. Kinet, H. Metzger, J. Hakimi, and J. A. Kochan, A cDNA presumptively coding for the alpha subunit of the receptor with high affinity for immunoglobulin E, *Biochemistry*, *26*: 4605 (1987).
31. A. Shimizu, I. Tepler, P. N. Benfey, E. H. Berenstein, R. P. Siraganian, and P. Leder, Human and rat mast cell high-affinity immunoglobulin E receptors: Characterization of putative alpha-chain gene products, *Proc. Natl. Acad. Sci. USA*, *85*: 1907 (1988).
32. C. Ra, M.-H. E. Jouvin, and J.-P. Kinet, Complete structure of the mouse mast cell receptor for IgE (FcεRI) and surface expression of chimeric receptors (rat–mouse–human) on transfected cells, *J. Biol. Chem.*, *264*: 15323 (1989).
33. J. Kochan, L. F. Pettine, J. Hakimi, K. Kishi, and J.-P. Kinet, Isolation of the gene coding for the alpha subunit of the human high affinity IgE receptor, *Nucleic Acids Res.*, *16*: 3584 (1988).
34. J. M. Kanellopoulos, T. Y. Liu, G. Poy, and H. Metzger, Composition and subunit structure of the cell receptor for immunoglobulin E, *J. Biol. Chem.*, *255*: 9060 (1980).
35. B. L. Hempstead, C. W. Parker, and A. Kulczycki, Jr., The cell surface receptor for immunoglobulin E. Effect of tunicamycin on molecular properties of receptor from rat basophilic leukemia cells, *J. Biol. Chem.*, *256*: 10717 (1981).
36. A. Kulczycki, Jr., T. A. McNearney, and W. Parker, The rat basophilic leukemia cell receptors for IgE. I. Characterization as a glycoprotein, *J. Immunol.*, *117*: 661 (1978).
37. H. L. Thompson, D. D. Metcalfe, and J.-P. Kinet, Early expression of high-af-

finity receptor for immunoglobulin E (FcεRI) during differentiation of mouse mast cells and human basophils, *J. Clin. Invest.*, *85*: 1227 (1990).

38. F. T. Liu, K. Albrandt, and M. W. Robertson, cDNA heterogeneity suggests structural variants related to the high-affinity IgE receptor, *Proc. Natl. Acad. Sci. USA*, *85*: 5639 (1988).
39. J.-P. Kinet, U. Blank, C. Ra, K. White, H. Metzger, and J. Kochan, Isolation and characterization of cDNAs coding for the beta subunit of the high-affinity receptor for immunoglobulin E, *Proc. Natl. Acad. Sci. USA*, *85*: 6483 (1988).
40. H. Kuster, L. Zhang, A. T. Brini, D. W. J. MacGlashan and J.-P. Kinet, The gene and cDNA for the human high affinity immunoglobulin E receptor β chain and expression of the complete human receptor, *J. Biol. Chem.*, *267*: 12782 (1992).
41. K. Maekawa, N. Imagawa, Y. Tanaka, and S. Harada, Determination of the sequence coding for the β subunit of the human high-affinity IgE receptor, *FEBS Lett.*, *302*: 161 (1992).
42. D. Holowka, and H. Metzger, Further characterization of the β-component of the receptor for immunoglobulin E, *Mol. Immunol.*, *19*: 219 (1982).
43. R. Paolini, M.-H. E. Jouvin, and J.-P. Kinet, Phosphorylation and dephosphorylation of the high-affinity receptor for immunoglobulin E immediately after receptor engagement and disengagement, *Nature*, *353*: 855 (1991).
44. W. Li, G. Deanin, B. Margolis, J. C. Seagrave, J. Schlessinger, and J. M. Oliver, FcεRI-mediated tyrosine phosphorylation of multiple proteins, including phospholipase Cγ1 and the receptor βγ2 complex, in RBL-2H3 rat basophilic leukemia cell, *Mol. Cell. Biol.*, *12*: 3176 (1992).
45. T. Kurosaki, I. Gander, U. Wirthmeller, and J. V. Ravetch, The β subunit of FcεRI is associated with the FcγRIII on mast cells, *J. Exp. Med.*, *175*: 447 (1992).
46. U. Blank, C. Ra, L. Miller, K. White, H. Metzger, and J.-P. Kinet, Complete structure and expression in transfected cells of high affinity IgE receptor, *Nature*, *337*: 187 (1989).
47. H. Kuster, H. Thompson, and J.-P. Kinet, Characterization of the gene for the human Fc receptor γ subunit. Definition of a new gene family, *J. Biol. Chem.*, *265*: 6448 (1990).
48. G. Alcaraz, J.-P. Kinet, T. Y. Liu, and H. Metzger, Further characterization of the subunits of the receptor with high affinity for immunoglobulin E, *Biochemistry*, *26*: 2569 (1987).
49. O. Letourneur, I. C. S. Kennedy, A. T. Brini, J. R. Ortaldo, J. J. O'Shea, and J.-P. Kinet, Characterization of the family of dimers associated with Fc receptors (FcεRI and FcγRIII), *J. Immunol.*, *147*: 2652 (1991).
50. N. Varin-Blank, and H. Metzger, Surface expression of mutated subunits of the high affinity mast cell receptor for IgE, *J. Biol. Chem.*, *265*: 15685 (1990).
51. A. M. Weissman, M. Baniyash, D. Hou, L. E. Samelson, W. H. Burgess, and R. D. Klausner, Molecular cloning of the zeta chain of the T cell antigen receptor, *Science*, *239*: 1018 (1988).
52. Y. J. Jin, L. K. Clayton, F. D. Howard, S. Koyasu, M. Sieh, R. Steinbrich, G. E. Tarr, and E. L. Reinherz, Molecular cloning of the CD3η subunit identifies a CD3ζ-related product in thymus-derived cells, *Proc. Natl. Acad. Sci. USA*, *87*: 3319 (1990).

53. D. Holowka, and B. Baird, Lactoperoxidase-catalyzed iodination of the receptor for immunoglobulin E at the cytoplasmic side of the plasma membrane, *J. Biol. Chem.*, *259*: 3720 (1984).
54. C. Ra, M.-H. E. Jouvin, U. Blank, and J.-P. Kinet, A macrophage Fcγ receptor and the mast cell receptor for IgE share an identical subunit, *Nature*, *341*: 752 (1989).
55. M. L. Hibbs, P. Selvaraj, O. Carpen, T. A. Springer, J. Kuster, M.-H. E. Jouvin, and J.-P. Kinet, Mechanisms for regulating expression containing membrane isoforms of FcγRIII (CD16), *Science*, *246*: 1608 (1989).
56. T. Kurosaki, and J. V. Ravetch, A single amino acid in the glycosylphosphatidylinositol attachment domain determines the membrane topology of FcγRIII, *Nature*, *342*: 85 (1989).
57. D. G. Orloff, C. Ra, S. J. Frank, R. D. Klausner, and J.-P. Kinet, The zeta and eta chains of the T cell receptor and the gamma chain of Fc receptors form a family of disulfide-linked dimers, *Nature*, *347*: 189 (1990).
58. E. Vivier, N. Rochet, J. P. Kochan, D. R. Dresky, S. F. Schlossman, and P. Anderson, Structural similarity between Fc receptors and T cell receptors. Expression of the gamma subunit of Fc epsilon RI in human T cells, natural killer cells and thymocytes, *J. Immunol.*, *147*: 4263 (1991).
59. S. Koyasu, L. D'Adamio, A. R. Arulanandam, S. Abraham, L. K. Clayton, and E. L. Reinherz, T cell receptor complexes containing Fc epsilon RI gamma homodimers in lieu of CD3 zeta and CD3 eta components: A novel isoform expressed on large granular lymphocytes, *J. Exp. Med.*, *175*: 203 (1992).
60. I. Tepler, A. Shimizu, and P. Leder, The gene for the rat mast cell high affinity IgE receptor α chain. Structure and alternative mRNA splicing patterns, *J. Biol. Chem.*, *264*: 5912 (1989).
61. W. Q. Qiu, D. de Bruin, B. H. Brownstein, R. Pearse, and J. V. Ravetch, Organization of the human and mouse low-affinity Fc R genes: Evidence for duplication and recombination, *Science*, *248*: 732 (1990).
62. K. Huppi, D. Siwarski, B. A. Mock, and J.-P. Kinet, Gene mapping of the three subunits of the high affinity FcR for IgE on mouse chromosome 1 and 19, *J. Immunol.*, *143*: 3787 (1989).
63. M. F. Seldin, H. C. Morse, R. C. LeBoeuf, and A. D. Steinberg, Establishment of a molecular genetic map of distal mouse chromosome 1: Further definition of a conserved linkage group syntenic with human chromosome 1q, *Genomics*, *2*: 48 (1988).
64. K. Huppi, B. A. Mock, J. Hilgers, J. Kochan, and J.-P. Kinet, Receptors for Fcϵ and Fcγ are linked on mouse chromosome 1, *J. Immunol.*, *141*: 2807 (1988).
65. M. Le Coniat, J.-P. Kinet, and R. Berger, The human genes for the alpha and gamma subunits of the mast cell receptor for immunoglobulin E are located on human chromosome band 1q23, *Immunogenetics*, *32*: 183 (1990).
66. T. F. Tedder, G. Klejman, C. M. Disteche, D. A. Adler, S. F. Schlossman, and H. Saito, Cloning of a complementary DNA encoding a new mouse β lymphocyte differentiation antigen, homologous to the human (B1) CD20 antigen, and localization of the gene to chromosome 19, *J. Immunol.*, *141*: 4388 (1988).
67. J. J. Sussman, J. S. Bonifacino, J. Lippincott-Schwartz, A. M. Weissman, T.

Saito, R. D. Klausner, and J. D. Ashwell, Failure to synthesize the T cell CD3-zeta chain: Structure and function of a partial T cell receptor complex, *Cell*, *52*: 85 (1988).

68. R. Quarto, J.-P. Kinet, and H. Metzger, Coordinate synthesis and degradation of the alpha-, beta- and gamma-subunits of the receptor for immunoglobulin E, *Mol. Immunol.*, *22*: 1045 (1985).
69. K. Furuichi, J. Rivera, and C. Isersky, The receptor for immunoglobulin E on rat basophilic leukemia cells: Effect of ligand binding on receptor expression, *Proc. Natl. Acad. Sci. USA*, *82*: 1522 (1985).
70. J.-P. Kinet, H. Metzger, G. Alber, U. Blank, K. Huppi, M.-H. E. Jouvin, H. Kuster, L. Miller, C. Ra, J. Rivera, and N. Varin-Blank, The high affinity receptor for immunoglobulin E: New developments, *Prog. Immunol.*, *7*: 716 (1989).
71. J.-P. Kinet, R. Perez-Montfort, and H. Metzger, Covalent cross-linking of subunits of the receptor for immunoglobulin E induced by immunoprecipitation, *Biochemistry*, *22*: 5729 (1983).
72. J. Rivera, J.-P. Kinet, J. Kim, C. Pucillo, and H. Metzger, Studies with a monoclonal antibody to the beta subunit of the receptor with high affinity for immunoglobulin E, *Mol. Immunol.*, *25*: 334 (1988).
73. L. Miller, U. Blank, H. Metzger, and J.-P. Kinet, Expression of high-affinity binding of human immunoglobulin E by transfected cells, *Science*, *244*: 334 (1989).
74. D. H. Conrad, J. R. Wingard, and T. Ishizaka, The interaction of human and rodent IgE with the human basophil IgE receptor, *J. Immunol.*, *130*: 327 (1983).
75. S. A. Wank, C. Delisi, and H. Metzger, Analysis of the rate-limiting step in ligand-receptor interaction: The immunoglobulin E system, *Biochemistry*, *22*: 954 (1983).
76. G. R. Mendoza, and H. Metzger, Distribution and valency of receptor for IgE on rodent mast cells and related tumor cells, *Nature*, *264*: 548 (1976).
77. B. Helm, P. Marsh, D. Vercelli, E. Padian, H. Gould, and R. Geha, The mast cell binding site on human immunoglobulin E, *Nature*, *331*: 180 (1988).
78. B. Helm, D. Kebo, D. Vercelli, M. M. Glovsky, H. Gould, K. Ishizaka, R. Geha, and T. Ishizaka, Blocking of passive sensitization of human mast cells and basophil granulocytes with IgE antibodies by a recombinant human ϵ-chain fragment of 76 amino acids, *Proc. Natl. Acad. Sci. USA*, *86*: 9465 (1989).
79. A. Kulczycki, Jr., and V. L. Vallina, Specific binding of non-glycosylated IgE to Fc epsilon receptor, *Mol. Immunol.*, *18*: 723 (1981).
80. A. Nissim, M.-H. Jouvin, and Z. Eshhar, Mapping of the high affinity Fc epsilon receptor binding site to the third constant region domain of IgE, *EMBO J.*, *10*: 101 (1991).
81. M. Weetall, B. Shopes, D. Holowka, and B. Baird, Mapping the site of interaction between murine IgE and its high affinity receptor with chimeric Ig, *J. Immunol.*, *145*: 3849 (1990).
82. D. Vercelli, B. Helm, P. Marsh, E. Padian, R. S. Geha, and H. Gould, The B-cell binding site on human immunoglobulin E, *Nature*, *338*: 649 (1989).
83. Y. Zheng, B. Shopes, D. Holowka, and B. Baird, Conformations of IgE bound to its receptor $Fc_{\epsilon}RI$ and in solution, *Biochemistry*, *30*: 9125 (1991).

84. R. Perez-Montfort, and H. Metzger, Proteolysis of soluble IgE-receptor complexes: Localization of sites on IgE which interact with the Fc receptor, *Mol. Immunol.*, *19*: 1113 (1982).
85. H. Metzger, Molecular aspects of receptors and binding factors for IgE, *Adv. Immunol.*, *43*: 277 (1988).
86. J. Hakimi, C. Saito, J. A. Kondas, L. Pettine, W. Danho, and J. Kochan, The α subunit of the human IgE receptor (FcεRI) is sufficient for high affinity IgE binding, *J. Biol. Chem.*, *265*: 22079 (1990).
87. G. Alber, and H. Metzger, The high affinity IgE receptor [book chapter] (in press).
88. U. Blank, C. Ra, and J.-P. Kinet, Characterization of truncated α chain products from human, rat and mouse high affinity receptor for immunoglobulin E, *J. Biol. Chem.*, *266*: 2639 (1991).
89. F. Riske, J. Hakimi, M. Mallamaci, M. Griffin, B. Pilson, N. Tobkes, P. Lin, W. Danho, J. Kochan, and R. Chizzonite, High affinity human IgE receptor (FcεRI). Analysis of functional domains of the α-subunit with monoclonal antibodies, *J. Biol. Chem.*, *266*: 11245 (1991).
90. T. Ishizaka, and K. Ishizaka, Triggering of histamine release from rat mast cells by divalent antibodies against IgE-receptors, *J. Immunol.*, *120*: 800 (1978).
91. C. Isersky, J. D. Taurog, G. Poy, and H. Metzger, Triggering of cultured neoplastic mast cells by antibodies to the receptor for IgE, *J. Immunol.*, *121*: 549 (1978).
92. R. P. Siraganian, W. A. Hook, and B. B. Levine, Specific in vitro histamine release form basophils by bivalent haptens: Evidence for activation by simple bridging of membrane bound antibodies, *Immunochemistry*, *12*: 149 (1975).
93. D. M. Segal, J. D. Taurog, and H. Metzger, Dimeric immunoglobulin E serves as a unit signal for mast cell degranulation, *Proc. Natl. Acad. Sci. USA*, *74*: 2993 (1979).
94. C. DeLisi, and R. P. Siraganian, Receptor cross-linking and histamine release. I. The quantitative dependence of basophil degranulation on the number of receptor doublets, *J. Immunol.*, *122*: 22286 (1979).
95. C. Fewtrell, and H. Metzger, Larger oligomers of IgE are more effective than dimers in stimulating rat basophilic leukemia cells, *J. Immunol.*, *125*: 701 (1980).
96. A. K. Menon, D. Holowka, and B. Baird, Small oligomers of immunoglobulin E (IgE) cause large-scale clustering of IgE receptors on the surface of rat basophilic leukemia cells, *J. Cell Biol.*, *98*: 577 (1984).
97. P. M. Kane, D. Holowka, and B. Baird, Cross-linking of IgE-receptor complexes by rigid bivalent antigens greater than 200 A in length triggers cellular degranulation, *J. Cell Biol.*, *107*: 969 (1988).
98. E. Ortega, R. Schweitzer-Stenner, and I. Pecht, Possible orientational constraints determine secretory signals by aggregation of IgE receptors on mast cells, *EMBO J.*, *20*: 4101 (1988).
99. J. C. Seagrave, G. G. Deanin, J. C. Martin, B. H. Davis, and J. M. Oliver, DNP-phycobiliproteins, fluorescent antigens to study dynamic properties of antigen–IgE–receptor complexes on RBL-2H3 rat mast cells, *Cytometry*, *8*: 287 (1987).

100. M. Reth, Antigen receptor tail clue, *Nature*, *338*: 383 (1989).
101. C. Romeo, and B. Seed, Cellular immunity to HIV activated CD4 fused to T cell or Fc receptor polypeptides, *Cell*, *64*: 1037 (1991).
102. F. Letourneur, and R. D. Klausner, T cell and basophil activation through the cytoplasmic tail of T-cell receptor ζ family proteins, *Proc. Natl. Acad. Sci. USA*, *88*: 8905 (1991).
103. J.-P. Kinet, The γ–ζ dimers of Fc receptors as connectors to signal transduction, *Curr. Opin. Immunol.*, *4*: 43 (1992).
104. B. A. Irving, and A. Weiss, The cytoplasmic domain of the T cell receptor ζ chain is sufficient to couple to receptor-associated signal transduction pathways, *Cell*, *64*: 891 (1991).
105. C. Romeo, M. Amiot, and B. Seed, Sequence requirements for induction of cytolysis by the T cell antigen/Fc receptor ζ chain, *Cell*, *68*: 889 (1992).
106. F. Letourneur, and R. D. Klausner, Activation of T cells by a tyrosine kinase activation domain in the cytoplasmic tail of CD3ε, *Science*, *255*: 79 (1992).
107. A.-M. K. Wegener, F. Letourneur, A. Hoeveler, T. Brocker, F. Luton, and B. Malissen, The T cell receptor complex is composed of at least two autonomous transduction molecules, *Cell*, *68*: 83 (1992).
108. A. Bauer, B. J. McConckey, F. D. Howard, L. K. Clayton, V. Novick, S. Koyasu, and E. L. Reinherz, Differential signal transduction via T-cell receptor CD3ζ2, CD3ζ–η and CD3η2 isoforms, *Proc. Natl. Acad. Sci. USA*, *88*: 3842 (1991).
109. S. Koyasu, L. D'Adamio, L. K. Clayton, and E. L. Reinherz, T cell receptor isoforms and signal transduction, *Curr. Opin. Immunol.*, *3*: 32 (1991).
110. G. Alber, L. Miller, C. L. Jelsema, N. Varin-Blank, and H. Metzger, Structure–function relationships in the mast cell high affinity receptor for IgE. Role of the cytoplasmic domains and of the β subunit, *J. Biol. Chem.*, *206*: 22613 (1991).
111. S.-Y. Mao, N. Varin-Blank, M. Edidin, and H. Metzger, Immobilization and internalization of mutated IgE receptors in transfected cells, *J. Immunol.*, *146*: 958 (1991).
112. Y. Yardin, and A. Ullrich, Growth factor receptor tyrosine kinases, *Annu. Rev. Biochem*, *57*: 443 (1988).
113. L. E. Samelson, A. F. Phillips, E. T. Luong, and R. D. Klausner, Association of the *fyn* protein-tyrosine kinase with the T cell antigen receptor, *Proc. Natl. Acad. Sci. USA*, *87*: 4358 (1990).
114. A. L. Burkhardt, M. Brunswick, J. B. Bolen, and J. L. Mond, Anti-immunoglobulin stimulation of B lymphocytes activates *src*-related protein tyrosine kinases, *Proc. Natl. Acad. Sci. USA*, *88*: 7410 (1991).
115. Y. Yamanashi, T. Kakiuchi, J. Mizugushi, T. Yamamoto, and K. Toyoshima, Association of B cell antigen receptor with protein tyrosine kinase *lyn*, *Science*, *251*: 192 (1991).
116. R. Quarto, and H. Metzger, The receptor for immunoglobulin E: Examination for kinase activity and as a substrate for kinases, *Mol. Immunol.*, *23*: 1215 (1986).
117. E. Eiseman, and J. B. Bolen, Engagement of the high-affinity IgE receptor activates *src*-protein-related tyrosine kinases, *Nature*, *355*: 78 (1992).
118. G. A. Koretzky, J. Picus, M. L. Thomas, and A. Weiss, Tyrosine phosphatase

CD45 is essential for coupling T-cell antigen receptor to the phosphatidyl inositol pathway, *Nature*, *346*: 66 (1990).
119. G. A. Koretsky, J. Picus, T. A. Schultz, and A. Weiss, Tyrosine phosphatase CD45 is required for T-cell antigen receptor and CD2-mediated activation of a protein tyrosine kinase and interleukin 2 production, *Proc. Natl. Acad. Sci. USA*, *88*: 2037 (1991).
120. L. B. Justement, K. S. Campbell, N. C. Chien, and J. C. Cambier, Regulation of B cell antigen receptor signal transduction and phosphorylation by CD45, *Science*, *252*: 1839 (1991).
121. W. A. Hook, E. H. Berenstein, F. U. Zinsser, C. Fischler, and R. P. Siraganian, Monoclonal antibodies to the leukocyte common antigen (CD45) inhibit IgE-mediated histamine release from human basophils, *J. Immunol.*, *147*: 2670 (1991).
122. M. Adamczewski, R. Paolini, and J.-P. Kinet, Evidence for two distinct phosphorylation pathways activated by receptors coupled to non-receptor-kinases. The high affinity IgE receptor as a model, *J. Biol. Chem.*, (submitted).
123. S. G. Rhee, and K. O. Choi, Regulation of inositol phospholipid-specific phospholipase C isozymes, *J. Biol. Chem.*, *267*: 12393 (1992).
124. R. Perez-Montfort, C. Fewtrell, and H. Metzger, Changes in the receptor for immunoglobulin E coincident with receptor-mediated stimulation of basophilic leukemia cell, *Biochemistry*, *22*: 5733 (1983).
125. J. J. O'Shea, A. M. Weissman, I. C. Kennedy, and J. R. Ortaldo, Engagement of the natural killer cell IgG Fc receptor results in tyrosine phosphorylation of the zeta chain, *Proc. Natl. Acad. Sci. USA*, *88*: 350 (1991).
126. E. Vivier, P. Morin, C. O'Brien, B. Druker, S. F. Schlossman and P. Anderson, Tyrosine phosphorylation of the FcγRIII (CD16):ζ complex in human natural killer cells. Induction by antibody-dependent cytotoxicity but not by natural killing, *J. Immunol.*, *146*: 206 (1991).
127. M. Benhamou, J. S. Gutkind, K. C. Robbins, and R. P. Siraganian, Tyrosine phosphorylation coupled to IgE receptor-mediated signal transduction and histamine release, *Proc. Natl. Acad. Sci. USA*, *87*: 53327 (1990).
128. T. Kawakami, N. Inagaki, M. Takei, H. Fukamachi, K. M. Coggeshall, K. Ishizaka, and T. Ishizaka, Tyrosine phosphorylation is required for mast cell activation by FcϵRI cross-linking, *J. Immunol.*, *148*: 3513 (1992).
129. D. J. Park, H. Y. Min, and S. G. Rhee, IgE-induced tyrosine phosphorylation of phospholipase C-γ1 in rat basophilic leukemia cells, *J. Biol. Chem.*, *266*: 24237 (1991).
130. B. Margolis, P. Hu, S. Katzav, W. Li, J. M. Oliver, A. Ullrich, A. Weiss, and J. Schlessinger, Tyrosine phosphorylation of *vav* proto-oncogene product containing SH2 domain and transcription factor motifs, *Nature*, *356*: 71 (1992).
131. M. Matsuda, B. J. Mayer, and H. Hanafusa, Identification of domains of the v-*crk* oncogene product sufficient for association with phosphotyrosine-containing proteins, *Mol. Cell. Biol.*, *11*: 1607 (1991).
132. A. D. Keegan, J. H. Pierce, J. Atrip, M. Plaut, and W. E. Paul, Ligand stimulation of transfected and endogenous growth factor receptors enhances cytokine production by mast cells, *EMBO J.*, *10*: 3675 (1991).
133. K.-T. Yu, R. Lyall, N. Jariwala, A. Zilberstein, and H. Haimovich, Antigen and

ionophore-induced signal transduction in rat basophilic leukemia cells involve protein tyrosine phosphorylation, *J. Biol. Chem.*, *266*: 22564 (1991).

134. V. Stephan, M. Benhamou, J. S. Gutkind, K. C. Robbins, and R. P. Siraganian, FcϵRI-induced protein tyrosine phosphorylation of pp72 in rat basophilic leukemic cells (RBL-2H3). Evidence for a novel signal transduction pathway unrelated to G protein activation and phosphatidylinositol hydrolysis, *J. Biol. Chem.*, *267*: 5434 (1992).
135. D. K. Morrison, D. R. Kaplan, J. A. Escobedo, U. R. Rapp, T. M. Roberts, and L. T. Williams, Direct activation of the serine/threonine kinase activity of *raf-1* through tyrosine phosphorylation by the PDGF-β receptor, *Cell*, *58*: 649 (1989).
136. E. Neher, The influence of intracellular calcium concentration on degranulation of dialyzed mast cells from rat peritoneum, *J. Physiol.*, *895*: 193 (1988).
137. R. J. Hohman, Jr., Aggregation of IgE receptors induces degranulation in RBL cells permeabilized with α-toxin, *Proc. Natl. Acad. Sci. USA*, *85*: 1624 (1988).
138. R. Penner, and E. Neher, Secretory responses of rat peritoneal mast cells to high intracellular calcium, *FEBS Lett.*, *226*: 307 (1988).
139. C. Fewtrell, F. C. Mohr, T. A. Ryan, and P. J. Millard, Calcium: An important second messenger in mast cells, *Ciba Found. Symp.*, *147*: 114 (1989).
140. P. J. Millard, T. A. Ryand, W. W. Webb, and C. Fewtrell, Immunoglobulin E receptor crosslinking induces oscillations in intracellular free ionized calcium in individual tumor mast cells, *J. Biol. Chem.*, *264*: 19730 (1989).
141. G. Matthews, E. Neher, and R. Penner, Second-messenger-activated calcium influx in rat peritoneal mast cells, *J. Physiol.*, *418*: 105 (1989).
142. T. Meyer, D. Holowka, and L. Stryer, Highly cooperative opening of calcium channels by inositol 1,4,5-trisphosphate, *Science*, *240*: 652 (1988).
143. M. Hoth, and R. Penner, Depletion of intracellular calcium stores activates a calcium current in mast cells, *Nature*, *355*: 353 (1992).
144. V. Narasimhan, D. Holowka, C. Fewtrell, and B. Baird, Cholera toxin increases the rate of antigen-stimulated calcium influx in rat basophilic leukemia cells, *J. Biol. Chem.*, *263*: 19626 (1988).
145. B. S. Wilson, G. G. Deanin, J. C. Standefer, D. Vanderjaat, and J. M. Oliver, Depletion of guanine nucleotides with mycophenolic acid suppresses IgE receptor-mediated degranulation in rat basophilic leukemia cells, *J. Immunol.*, *143*: 259 (1989).
146. S. Hammerich, and I. Pecht, Isolation and purification of an Fc epsilon receptor-activated ion channel from the rat mast cell line RBL-2H3, *Biochemistry*, *27*: 7488 (1988).
147. T. Akiyama, J. Ishida, S. Nakagawa, H. Ogawara, S. Watanabe, N. Itoh, M. Shibuya, and Y. Fukami, Genestein, a specific inhibitor of tyrosine specific protein kinases, *J. Biol. Chem.*, *262*: 5592 (1987).
148. S. Cockcroft, and B. D. Gomperts, Role of guanine nucleotide binding protein in the activation of polyphosphoinositide phosphodiesterase, *Nature*, *314*: 534 (1985).
149. H. Ali, J. R. Cunha-Melo, and M. A. Beaven, Receptor-mediated release of IP_3 and IP_2 in RBL-2H3 cells permeabilized with streptolysin O, *Biochim. Biophys. Acta*, *1010*: (1989).

150. K. Yamada, C. L. Jelsema, and M. A. Beaven, Certain inhibitors of protein serine/threonine kinases also inhibit tyrosine-phosphorylation of phospholipase Cγ1 and other proteins and reveal distinct roles for tyrosine kinase(s) and protein kinase C in stimulated, rat basophilic RBL-2H3 cells, *J. Immunol.*, *149*: (in press).
151. C. W. Parker, Lipid mediators produced through the lipoxygenase pathway, *Annu. Rev. Immunol.*, *5*: 65 (1987).
152. V. Narasimhan, D. Holowka, and B. Baird, A guanine nucleotide-binding protein participates in IgE receptor-mediated activation of endogenous and reconstituted phospholipase A_2 in a permeabilized cell system, *J. Biol. Chem.*, *265*: 1459 (1990).
153. R. S. Gruchella, T. T. Dinh, and D. A. Kennerly, An indirect pathway of receptor-mediated 1,2-diacylglycerol formation in mast cells. I. IgE receptor-mediated activation of phospholipase D, *J. Immunol.*, *144*: 2334 (1990).
154. J. Axelrod, R. M. Burch, and C. L. Jelsema, Receptor-mediated activation of phospholipase A_2 via GTP-binding proteins: Arachidonic acid and its metabolites as second messengers, *Trends Neurosci.*, *11*: 117 (1988).
155. K. N. White, and H. Metzger, Translocation of protein kinase C in rat basophilic leukemia cells induced by phorbol ester or by aggregation of IgE receptors, *J. Immunol.*, *141*: 942 (1988).
156. J. Schlessinger, W. W. Webb, E. L. Elson, and H. Metzger, Lateral motion and valence of Fc receptors on rat peritoneal mast cells, *Nature*, *264*: 550 (1976).
157. D. Robertson, D. Holowka, and B. Baird, Cross-linking of immunoglobulin E–receptor complexes induces their interaction with the cytoskeleton of rat basophilic leukemia cells, *J. Immunol.*, *136*: 4565 (1986).
158. J. R. Apgar, Antigen-induced cross-linking of the IgE receptor leads to an association of the detergent-insoluble membrane skeleton of rat basophilic leukemia (RBL-2H3) cells, *J. Immunol.*, *145*: 3814 (1990).
159. S.-Y. Mao, G. Alber, J. Rivera, J. Kochan, and H. Metzger, Interaction of aggregated native and mutant IgE receptors with the cellular skeleton, *Proc. Natl. Acad. Sci. USA*, *89*: 222 (1992).
160. J. R. Pfeiffer, J. C. Seagrave, B. H. Davis, G. G. Deanin, and J. M. Oliver, Membrane and cytoskeletal changes associated with IgE-mediated serotonin release from rat basophilic leukemia cells, *J. Cell Biol.*, *101*: 2145 (1985).
161. K. Furuichi, J. Rivera, and H. Metzger, The fate of IgE bound to rat basophilic leukemia cells. III. Relationship between antigen-induced endocytosis and serotonin release, *J. Immunol.*, *133*: 1513 (1984).
162. J. R. Apgar, Association of the crosslinked IgE receptor with the membrane skeleton is independent of the known signalling mechanisms in rat basophilic leukemia cells, *Cell Regul.*, *2*: 181 (1991).
163. C. Isersky, J. Rivera, D. M. Segal, and T. Triche, The fate of IgE bound to rat basophilic leukemia cells. II. Endocytosis of IgE oligomers and effect on receptor turnover, *J. Immunol.*, *131*: 388 (1983).
164. X. R. Bustelo, J. A. Ledbetter, and M. Barbacid, Product of *vav* proto-oncogene defines a new class of tyrosine protein kinase substrates, *Nature*, *356*: 68 (1992).
165. J.-P. Kinet, Antibody cell interactions: Fc receptors, *Cell*, *57*: 351 (1989).
166. T. F. Tedder, Seventh International Congress of Immunology, abstr. 6440 (1989).
167. F. Letourner, G. Mattei, and B. Malissen, The mouse CD3 γ, δ and ϵ genes

reside within 50 kb on chromosome 9, whereas ζ maps to chromosome 1, band H, *Immunogenetics*, *29*: 265 (1989).

168. A. M. Weissman, D. Hou, D. G. Orloff, W. S. Modi, H. Seuanez, S. J. O'Brien, and R. D. Klausner, Molecular cloning and chromosomal localization of the human T-cell receptor ζ chain: Distinction from the molecular CD3 complex, *Proc. Natl. Acad. Sci. USA*, *85*: 9709 (1988).
169. M. Baniyash, V. W. Hsu, M. F. Seldin, and R. D. Klausner. The isolation and characterization of the murine T-cell receptor ζ chain gene, *J. Biol. Chem.*, *264*: 13252 (1989).
170. J.-P. Kinet, The γ–ζ dimers of Fc receptors as connectors to signal transduction, *Curr. Opin. Immunol.*, *4*: 43 (1992).
171. D. H. Conrad, and A. Froese, Characterization of the target cell receptor for IgE. III. Properties of the receptor isolated from rat basophilic leukemia cells by affinity chromatography, *J. Immunol.*, *120*: 429 (1978).
172. A. Froese, The presence of the two kinds of receptors for IgE on rat mast cells, *J. Immunol.*, *125*: 981 (1980).
173. D. M. Segal, S. O. Sharrow, J. F. Jones, and R. P. Siraganian, Fc(IgG) receptors on rat basophilic leukemia cells, *J. Immunol.*, *126*: 138 (1981).
174. M. R., Kepron, H. Bazin, and A. Froese, The interaction of IgG subclasses with solubilized Fc receptors of rat basophilic leukemia cells, *Mol. Immunol.*, *25*: 599 (1988).
175. J. Yodoi, and K. Ishizaka, Lymphocytes bearing receptors for IgE. III. Transition of FcγR(+) cells to FcεR(+) cells by IgE, *J. Immunol.*, *123*: 2004 (1979).
176. G. Boltz-Nitulescu, H. Bazin, and H. L. Spiegelberg, Specificity of Fc receptors for IgE on lymphocytes, monocytes, and macrophages, *J. Exp. Med.*, *154*: 374 (1981).
177. H. Spiegelberg, Structure and function of Fc receptors for IgE on lymphocytes, monocytes, and macrophages, *Adv. Immunol.*, *35*: 61 (1984).
178. F. Takizawa, M. Adamczewski, and J.-P. Kinet, Identification of the low affinity receptor for IgE on mouse mast cells and macrophages as FcγRII and FcγRIII, *J. Exp. Med.* (in press).
179. S. Kitani, D. Kroft, C. Fischler, S. E. Mergenhagen, and R. Siraganian, Inhibition of allergic reactions with monoclonal antibodies to the high affinity IgE receptor, *J. Immunol.*, *140*: 2585 (1988).

9

Immunoglobulin E FcєRII: Molecular Biology and Role in Immunoglobulin-Mediated Inflammation

Jean-Yves Bonnefoy
Glaxo Institute for Molecular Biology, Geneva, Switzerland

I. INTRODUCTION

This review summarizes recent data on CD23, a low-affinity receptor for IgE (FcєRII). CD23 is the only FcR that does not belong to the immunoglobulin gene superfamily. The CD23 molecule was discovered independently as an IgE receptor on human lymphoblastoid B cells (1), as a cell surface marker expressed on Epstein–Barr virus-transformed B cells (EBVCS) (2), and as a B-cell activation antigen (Blast 2) (3). CD23 and the low-affinity receptor for IgE were shown to be the same molecule (4,5).

Similar to most FcR, soluble forms of CD23 (sCD23) are released into extracellular fluids. The soluble fragments formed by proteolytic cleavage of surface CD23 are not only capable of binding IgE (IgE-binding factors), but also exhibit multiple functions that are not IgE-related. These observations, together with the finding that CD23 displays significant homology with Ca^{2+}-dependent (C-type) animal lectins, suggested the existence of natural ligands other than IgE. The recent finding that CD23 interacts with membrane CD21 (6) indicates that CD23 should be viewed not only as a low-affinity IgE receptor, but also as an adhesion molecule involved in cell–cell interaction.

After a brief overview of the molecular structure, cellular distribution, and regulation of CD23 expression reviewed in detail recently (7–9), there follows a discussion of the biological activities ascribed to CD23.

II. MOLECULAR STRUCTURE

CD23 exists as a single-copy gene (10) located on chromosome 19 (11). Two transcripts have been identified that differ in their 5′-untranslated sequences and in their intracytoplasmic region (12). CD23 is a type II membrane glycoprotein, with a long COOH-terminal extracellular region (277 residues), a single transmembrane domain (20 residues), and a short NH_2-terminal intracytoplasmic tail (23 residues). The region homologous to C-type lectins spans from cysteine-163 to cysteine-282 and contains four highly conserved (positions 191, 259, 273, 282) and two partially conserved cysteines (positions 163 and 174) (13–15). The lectin domain comprises the IgE-binding site (16) and more precisely the region delimited by Cys-163 and Cys-282 encodes all the structural information required to form the IgE-binding site (17). The cytokine effects of CD23 are mediated by an epitope distinct from the IgE-binding site (18). The positions of the two disulfide bonds in the IgE-binding region have been mapped: residue 259 is joined to residue 273 and residue 191 to residue 282 (19) (Fig. 1). A family of proteins displaying this lectin motif binds specific carbohydrates in a Ca^{2+}-dependent fashion (20). This is the case for the selectins, a family of adhesion molecules comprising ELAM-1, CD62, and Mel-14 (21). The binding of IgE to CD23 is calcium-dependent (22) as is the interaction between CD23 and CD21 (23).

The specific sugar(s) recognized by CD23 on its ligands has yet to be characterized. Fucose-1-phosphate inhibits binding of IgE to CD23 (9) and CD23–CD21 interaction (23). Protein–protein interaction may also occur with IgE, since nonglycosylated IgE has been reported to bind CD23-bearing B cells (24).

CD23 bears a triplet of amino (Asp, Gly, Arg) at the COOH-terminus, which in reverse orientation is a common recognition site for the integrin receptors. However, the role of this motif in CD23 function(s) remains to be determined. This motif is absent from mouse CD23 (25,26).

Three short consensus repeats of leucine (or isoleucine), forming a "leucine zipper" motif, are present in the CD23 molecule. This structure might be involved in the formation of CD23 oligomers at the cell surface (27). CD23 also associates with other cell surface proteins, such as HLA-DR, on the surface of B cells (28). Surface CD23 is cleaved into soluble fragments, ranging in size from 37 to 16 kDa. Letellier et al. (29) have suggested an autocatalytic process, although CD23 exhibits no homology with any known proteases. The rate of cleavage is reduced by IgE and anti-CD23 antibodies and increased by agents that prevent glycosylation of CD23 (30).

In addition to the extracellular cleavage, a fraction of CD23 is internalized and subsequently processed intracellularly to yield a 16-kDa fragment. This endocytosis is enhanced by anti-CD23 monoclonal antibodies (31).

In comparison with the human, murine CD23 has two instead of one *N*-linked glycosylation site(s), no DGR motif and four instead of three consensus repeats

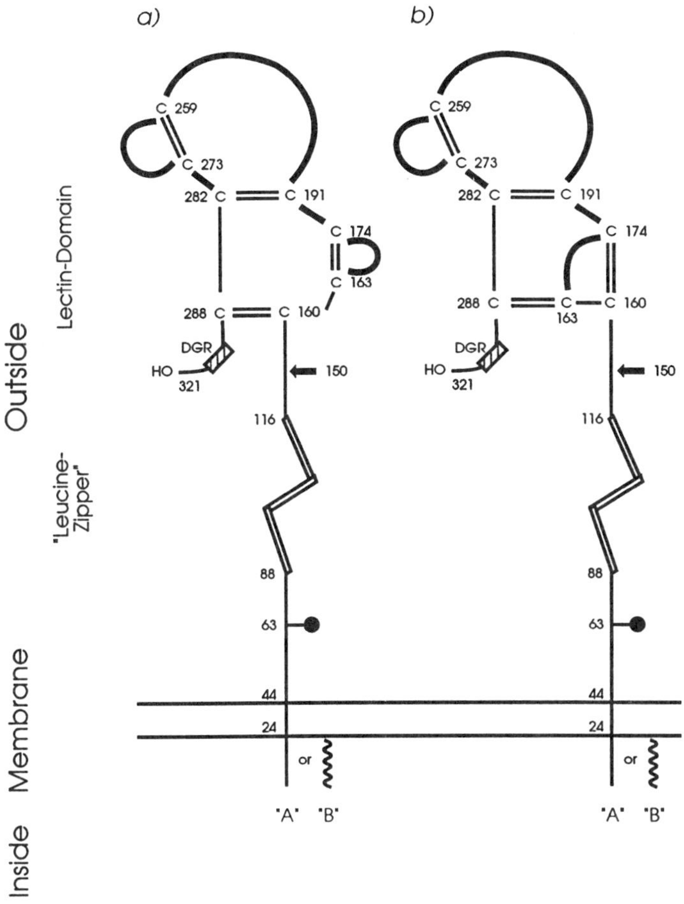

Figure 1 (a) and (b) represent alternative models based on disulfide bonding. A and B represent the two different forms of CD23 (12). ---•, *N*-glycosylation site;□, repeat sequences; ←, main cleavage site; —, IgE-binding region (17). See text for discussion. (Adapted from Refs. 9 and 19.)

(25,26). Two CD23 isoforms have been reported in the mouse, differing only by a few amino acids at their intracytoplasmic NH_2-terminus domain (32).

III. CELLULAR DISTRIBUTION

The CD23 molecule is expressed on a wide variety of cells, including B cells, T cells, a subset of thymic epithelial cells, follicular dendritic cells, Langerhans cells, monocytes, eosinophils, platelets, and EBV-containing nasopharyngeal carcinoma cells (9,33). The expression of type A CD23 is restricted to B cells only, whereas the type B isoform is found in all other cell types capable of expressing CD23. In the mouse, CD23 is exclusively found on B cells (7) and perhaps also on a subset of $CD8^+$ T cells (34,35). On B cells, CD23 is a differentiation marker that is selectively expressed on sIgM–sIgD double-bearing cells. CD23 is lost after immunoglobulin isotype switching (36) as well as during the differentiation of B cells into immunoglobulin-secreting cells, regardless of isotype switching (37).

IV. REGULATION OF CD23 EXPRESSION

The expression of CD23 is tightly regulated in a tissue-specific manner. Interleukin (IL)-4 is the most potent inducer of CD23 expression, acting on all cell types that are capable of expressing this molecule. Freshly isolated normal B cells express a small amount of CD23 of the type A isoform. Upon incubation at 37°C, in the absence of stimuli, cells rapidly lose both CD23 protein and mRNA, indicating that the expression of CD23 on normal resting cells is not constitutive (30). When cultured in the presence of IL-4, resting B cells express substantial amount of CD23 types A and B. This enhancing effect of IL-4 on B cells (38) is inhibited by interferon (IFN)-γ, IFN-α, transforming growth factor (TGF)-β, and glucocorticoids (9). Interleukin-2 was reported to up-regulate CD23 on activated, but not on resting, B cells; the effect of IL-2 is much less pronounced than that of IL-4 (39). Several B-cell activation signals markedly increase the IL-4-induced expression of CD23; these signals may be delivered by contact-dependent interactions with T cells (40), or by the engagement of either sIg, CD40, or CD72 (41–44). Autacoids, such as leukotriene (LTB)-4 or platelet-activating factor (PAF), can enhance the IL-4-induced expression of CD23 on normal B cells, as well as on monocytes (45–47).

The foregoing observations, together with the recent report that engagement of CD40 may override the suppressive effects of IFNs and TGF-β on the IL-4-induced CD23 expression (43), clearly indicate that the regulation of CD23 expression on the B cell is complex and tightly controlled. It is noteworthy that, there is as yet no evidence documenting a differential regulation of CD23 type A and B in normal B cells. The regulation of CD23 expression on monocytes is different from that on B cells. For example, freshly isolated normal monocytes do not express CD23 and, in contrast with B cells, IFN-γ enhances CD23 expression on monocytes. As in B cells, glucocorticoids and TGF-β inhibit CD23 on monocytes (9). Interestingly, 1,25-dihydroxyvitamin D_3 inhibits CD23 expression on IL-4-stimulated monocytes, but does not affect expression of either the type A or B CD23 isoforms on highly purified B cells (48). Finally, the regulation of CD23 on T cells appears to be similar to that on B cells (49,50).

CD23 on T cells is preferentially induced by stimulation with antigens that induce an IgE response (50). On Langerhans cells, eosinophils, and platelets, CD23 seems to be regulated in the same way as on monocytes (51, 52).

Immunoglobulin E alone is a sufficient stimulus for CD23 upregulation (49,53–55) by protection of the CD23 from degradation into sCD23. Epstein–Barr virus (EBV) induces high levels of CD23 expression in B cells through the *EBNA-2* gene (56).

V. FUNCTIONS OF THE CD23 MOLECULE

This section is divided into two main parts: The first deals with CD23 function as an IgE receptor, and the second, with the function of CD23 as a ligand for other receptors.

A. CD23 as a Receptor for Immunoglobulin E

The receptor function seems to depend on the cell expressing the CD23. Thus, CD23 on macrophages, eosinophils, and platelets mediates IgE-dependent cytotoxicity and promotes phagocytosis of IgE-coated particles (57); CD23 acts in synergy with CD11b (58). Immunoglobulin E binding to CD23 on monocytes induces the release of IL-1 and TNF-α (59) and the generation of superoxides (60). Given the cellular distribution of the forms of CD23, all these functions must be ascribed to the B form. This is supported by transfection experiments showing that type B CD23 mediates the phagocytosis of IgE-coated particles, whereas type A mediates the endocytosis of soluble IgE complexes (61). Therefore, type B CD23 plays a role in protective immunity against parasites and in IgE-mediated inflammatory responses.

CD23 expressed on B cells regulates the presentation of antigen to T cells and B-cell differentiation into IgE-producing cells. The role of CD23 in antigen presentation has been related to its physical association with HLA-DR molecules on the B-cell surface (28). Evidence for this was provided by Flores-Romo et al. (62), who showed that anti-CD23 monoclonal antibodies and IgE inhibit presentation of alloantigen in the mixed lymphocyte reaction. One putative role for CD23 in antigen presentation could be as an adhesion molecule enhancing the T-cell–antigen-presenting cell interaction.

Normal murine B cells and EBV-transformed human B cells are very efficient at IgE-dependent antigen presentation to T cells (63,64). Specific IgE antibodies and CD23 may amplify the T-cell response to the corresponding antigen by expanding the pool of antigen-specific T-cell clones.

Cross-linking of CD23 on B cells by IgE–immune complexes (IC) prevents the activation and differentiation of these cells (65). IgE could down-regulate its own production by preventing the recruitment of new IgE-secreting cells from precursor B cells (9).

Interestingly, the inhibition of B-cell activation by IgE–IC is observed only when the B cells have been exposed to IL-4. Because IL-4 not only increases the density of CD23 on the B-cell surface, but also induces CD23 type B, it is possible that the inhibitory signal is secondary to the engagement of this CD23 isoform. However, other studies suggested that signaling by CD23 may deliver growth-promoting signal to human B cells (66). Some anti-CD23 monoclonal antibodies (MAbs) induce the proliferation of phorbol ester-preactivated human B lymphocytes. CD23 is also capable of delivering a growth signal to mouse B cells, as shown by their proliferative response to the cross-linking of CD23 with their surface immunoglobulin (67,68). Taken together, the foregoing studies indicate that CD23 may indeed regulate B-cell proliferation; however, depending on the experimental conditions, engagement of CD23 may either enhance or inhibit the progression of B cells from G_1 to S phase of the cell cycle. Further support for the role of CD23 in B-cell activation and proliferation is found in the ability of this molecule to signal the B cell through the phosphoinositide pathway (69). Interestingly, engagement of CD23 on the surface of monocytic cells fails to trigger phosphatidylinositol hydrolysis and the associated intracellular $[Ca^{2+}]$ increase (70). Hence, it is possible that CD23 A and B isoforms are connected to different-signal pathways. The involvement of CD23 in the regulation of B-cell growth is underlined by its abnormal expression and regulation in two malignant diseases resulting from the clonal expansion of sIgM–sIgD mature B cells [i.e., chronic lymphocytic leukemia (CLL) and small lymphocytic lymphoma] (71). CD23 may also play a role in CLL; since it is abnormally regulated (72), it has prognostic value (73), and engagement of membrane CD23 inhibits their in vitro proliferative response to several stimuli (9).

B. CD23 as a Ligand for CD21

Interleukin-4, which induces CD23 expression, also induces IgE production by normal B cells by causing IgE isotype switching (74). Interleukin-4-induced IgE synthesis was blocked by IFN-γ, as well as by IFN-γ and IFN-α, and prostaglandin E_2 (PGE_2) (75), factors that also inhibit the induction by IL-4 of CD23 on B cells (38).

Soluble CD23 augments ongoing spontaneous IgE production by B cells derived from atopic individuals (76) and acts synergistically with a suboptimal concentration of IL-4 to induce IgE production by normal human B cells (75). The importance of CD23 or its soluble fragments in the synthesis of human IgE in vitro was highlighted by the demonstration that a specific subset of anti-CD23 monoclonal antibodies blocked IL-4-induced IgE production by normal B cells in vitro (75,77,78) and in vivo (78a), as well as the spontaneous production of IgE by B cells from atopic patients (77).

Physical interaction between T cell and B cells is a process known to be required for IgE production (79). Therefore, one possible role for CD23 in IgE regulation would be in enhancing the T-cell–B-cell interaction (80). In support of such a role, we have found that certain anti-CD23 monoclonal antibodies inhibit T–B-cell conjugate formation in vitro (81) and that this inhibition is restricted mainly to CD4-positive T cells that form conjugates with B cells (Fig. 2). In addition, we have recently found that CD23 specifically interacts with another cell surface molecule, in addition to IgE, that is most probably involved in T-cell–B-cell interaction. Recombinant full-length CD23 reconstituted into fluorescent liposomes binds to a subset of T cells and B cells (23). Studies using inhibitory anti-CD21 antibodies, as well as the binding of the CD23–liposomes to recombinant CD21-transfected cells, revealed that CD23 binds to a subtype of CD21 (6), a molecule that had been previously identified as the receptor for EBV and as the complement receptor-2. CD23 acts as a Ca^{2+}-dependent lectin, probably recognizing a sugar moiety on CD21

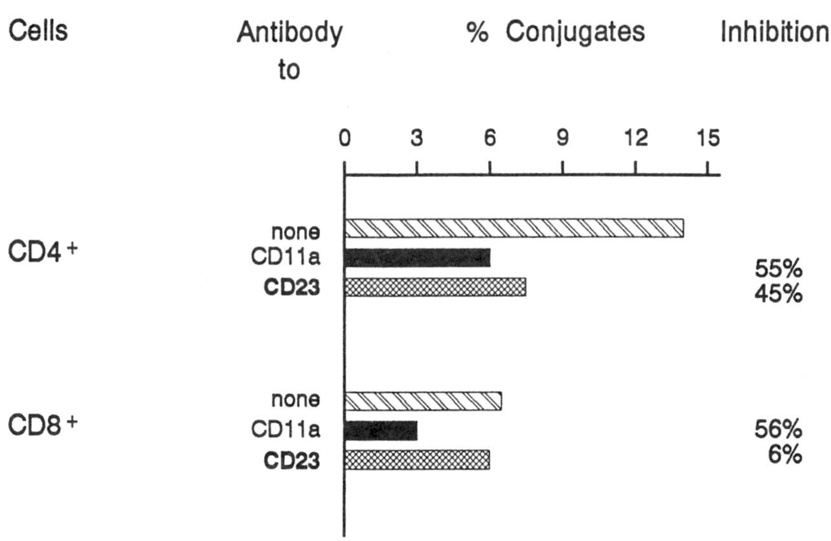

Figure 2 Peripheral blood T cells were separated into $CD4^+$ and $CD8^+$ subsets by negative depletion using the appropriate monoclonal antibodies and immunomagnetic beads. T-cell subpopulations were labeled with BCECF-AM and then used to form conjugates (81) with indo-1-AM labeled RPMI 8866 cells in the presence or absence of anti-CD23 (EBVCS#4) or anti-CD11a monoclonal antibodies.

(23). CD21 is expressed on B cells, follicular dendritic cells, and on T cells. Anti-CD21 antibodies, similar to anti-CD23 antibodies, decreased T–B conjugate formation (6), as would be expected if CD23–CD21 pairing has a role in the T-cell–B-cell interaction. Thus CD23 appears to be another molecule involved in the T-cell–B-cell interaction, along with LFA-1–ICAM-1, CD2–LFA-3, CD28–B7, and CD5–CD72. But the particular characteristic of the CD21–CD23 pair, when compared with the others, is that this pairing controls IgE synthesis in an isotype-specific manner.

Engagement of CD21 on B cells by some anti-CD21 MAbs increased IL-4-induced IgE production, as did treatment with recombinant sCD23 (6). Molecular analysis revealed that triggering of CD21 increased the IL-4-induced germline ϵ transcription levels and had a synergistic effect on the expression of the ϵ transcript induced by T cells (81a). On the basis of observations that triggering of CD21 by either anti-CD21 antibodies (81a) or by recombinant sCD23 enhances IgE production in both T-cell-dependent and T-cell-independent systems, and from the reported induction of CD23 on T cells by IL-4 and allergen (50), one can speculate that in allergic individuals, T cell-associated CD23 could interact with B-cell-associated CD21, leading to an increase in IgE production. Conversely, the T-cell–B-cell interaction leading to IgE production would not occur in normal persons, owing to the absence of CD23 expression on T cells.

The CD23–CD21 interaction may have other important effects, independent of IgE regulation. We recently demonstrated that some anti-CD21 antibodies promoted the survival of germinal center B cells and their differentiation toward a "plasmocyte-type" cell (82), thus mimicking the activity of recombinant sCD23 (83).

We also demonstrated that CD21 is expressed on basophilic cells, and that CD21 controls histamine production upon ligand-induced stimulation (CD23 or anti-CD21 MAb) (83a). The CD23–CD21 interaction, therefore, is a novel mechanism for basophil activation and increases histamine release. The CD23–CD21 pair of adhesion molecules is able to regulate IgE production and is also able to act at another stage of the allergic reaction by regulating effector cells such as basophils.

Interestingly, EBV–B-cell lines and nasopharyngeal carcinoma cells express both CD23 and CD21; therefore, the CD23–CD21 interaction may well contribute to autostimulatory growth of these transformed cells.

An increasing number of reports have demonstrated that CD23 also plays a role in various activities, such as B-cell growth (84,85), germinal center B-cell survival (83), prothymocyte maturation (86), myeloid precursor proliferation (87), and inhibition of macrophage migration (88). Whether all of these CD23-mediated activities occur through binding to CD21, or to yet another ligand, clearly remains to be determined.

VI. CONCLUSION

CD23 becomes a more and more interesting molecule. It is an Fc receptor for IgE and, as such, is involved in IgE-mediated inflammatory responses and in protective immunity against parasites. It is also an adhesion molecule that interacts with the CD21 antigen and, by this means, is involved in differentiation of B cells into IgE-secreting plasma cells, in survival of germinal center B cells, and in the release of histamine from basophils. Furthermore, CD23 is cleaved into soluble fragments acting as cytokines in vitro and, therefore, could be viewed as a membrane-bound cytokine, such as TNFα or CD40-ligand. Finally, because of its multiple roles on various target cells, it is quite likely that CD23 may well interact with yet other unidentified ligand(s).

ACKNOWLEDGMENTS

I would like to thank J.-P. Aubry, K. Bacon, L. Flores-Romo, J.-F. Gauchat, P. Graber, S. Henchoz, K. Jansen, P. Life, G. Mazzei, S. Pochon, and J. Shields for their past, current, and future invaluable scientific contribution and M. Thomasset for typing the manuscript.

REFERENCES

1. A. Gonzalez-Molina, and H. L. Spiegelberg, Binding of IgE myeloma proteins to human cultured lymphoblastoid cells, *J. Immunol.*, *117*: 1838 (1976).
2. C. Kintner, and B. Sugden, Identification of antigenic determinants unique to the surfaces of cells transformed by Epstein–Barr virus, *Nature*, *294*: 458 (1981).
3. D. A. Thorley-Lawson, L. M. Nadler, A. K. Bahn, and R. T. Schooley, BLAST-2 [EBVCS], an early cell surface marker of human B cell activation, is superinduced by Epstein-Barr virus, *J. Immunol.*, *143*: 3007 (1985).
4. K. Yukawa, H. Kikutani, H. Howaki, K. Yamasaki, A. Yokota, H. Makamura, E. L. Barsumian, R. R. Hardy, M. Suemura, and T. Kishimoto, A B cell-specific differentiation antigen CD23, is a receptor for IgE (FcεR) on lymphocytes, *J. Immunol.*, *138*: 2576 (1987).
5. J. Y. Bonnefoy, J. P. Aubry, C. Peronne, J. Wijdeness, and J. Banchereau, Production and characterization of a monoclonal antibody specific for the human lymphocyte low affinity receptor for IgE: CD23 is a low affinity receptor for IgE, *J. Immunol.*, *138*: 2970 (1987).
6. J. P. Aubry, S. Pochon, P. Graber, K. U. Jansen, and J. Y. Bonnefoy, CD21 is a ligand for CD23 and regulated IgE production, *Nature*, *358*: 505 (1992).
7. D. H. Conrad, FcεRII/CD23: The low affinity receptor for IgE, *Annu. Rev. Immunol.*, *8*: 623 (1990).
8. J. Gordon, CD23: A novel multifunctional regulator of the immune system that binds IgE, *Monographs in Allergy* (L. A. Hanson and F. Shakib, eds.), Karger, Basel, *29* (1991).

9. G. Delespesse, M. Sarfati, C. Y. Wu, S. Fournier, and M. Letellier, The low affinity receptor for IgE, *Immunol. Rev.*, *125*: 77 (1992).
10. U. Suter, R. Bastos, and H. Hofstetter, Molecular structure of the gene and the 5′-flanking region of the human lymphocyte immunoglobulin E receptor, *Nucleic Acids Res.*, *15*: 7295 (1987).
11. V. Wendel-Hansen, M. Riviere, I. Jansson, J. Szpirer, M. Q. Islam, G. Levan, G. Klein, J. Yodoi, and A. Rosen, The gene encoding CD23 leukocyte antigen (FcϵRII) is located on human chromosome 19, *Somat. Cell. Mol., Genet.*, *16*: 283 (1990).
12. A. Yokota, H. Kikutani, T. Tanaka, R. Sato, E. L. Barsumian, M. Suemura, and T. Kishimoto, Two species of human Fcϵ receptor II (FcϵRII/CD23): Tissue-specific and IL-4-specific regulation of gene expression, *Cell*, *55*: 611 (1988).
13. H. Kikutani, S. Inui, R. Sato, E. L. Barsumian, H. Owaki, K. Yamasaki, T. Kaisho, N. Uchibayashi, R. R. Hardy, T. Hirano, S. Tsumasawa, F. Sakiyama, M. Suemura, and T. Kishimoto, Molecular structure of human lymphocyte receptor for immunoglobulin E, *Cell*, *47*: 657 (1986).
14. C. Ludin, H. Hofstetter, M. Sarfati, C. A. Levy, U. Suter, D. Alaimo, E. Kilchherr, H. Frost, and G. Delespesse, Cloning and expression of the cDNA coding for a human lymphocyte IgE receptor, *EMBO J.*, *6*: 109 (1987).
15. K. Ikuta, M. Takami, L. W. Kim, T. Honjo, T. Miyoshi, Y. Tagaya, T. Kawabe, and J. Yodoi, Human lymphocyte Fc receptor for IgE: Sequence homology of its cloned cDNA with animal lectins, *Proc. Natl. Acad. Sci. USA*, *84*: 819 (1987).
16. B. Bettler, R. Maier, D. Ruegg, and H. Hofstetter, Binding site of IgE of the human low affinity Fcϵ receptor (FCϵRII/CD23) is confined to the domain homologous with animal lectins, *Proc. Natl. Acad. Sci. USA*, *86*: 718 (1989).
17. B. Bettler, G. Texido, S. Raggini, D. Rüegg, and H. Hofstetter, Immunoglobulin E-binding site in Fcϵ receptor (FcϵRII/CD23) identified by homolog-scanning mutagenesis, *J. Biol. Chem.*, *267*: 185 (1992).
18. M. D. Mossalayi, M. Arock, G. Delespesse, H. Hofstetter, B. Bettler, A. H. Dalloul, E. Kilchherr, F. Quaaz, P. Debré, and M. Sarfati, Cytokine effects of CD23 are mediated by an epitope distinct from the IgE binding site, *EMBO J.*, *11*: 4323 (1992).
19. K. Rose, G. Turcatti, P. Graber, S. Pochon, P. O. Regamey, K. U. Jansen, E. Magnenat, N. Aubonney, and J. Y. Bonnefoy, Partial characterization of natural and recombinant human soluble CD23, *J. Biochem.*, *286*: 819 (1992).
20. K. Drickamer, Two distinct classes of carbohydrate-recognition domains in animal lectins, *J. Biol. Chem.*, *263*: 9557 (1988).
21. L. M. Stoolman, Adhesion molecules controlling lymphocyte migration, *Cell*, *56*: 907 (1989).
22. M. L. Richards, and D. H. Katz, The binding of IgE to murine FcϵRII is calcium-dependent but not inhibited by carbohydrates, *J. Immunol.*, *144*: 2638 (1990).
23. S. Pochon, P. Graber, M. Yaeger, K. Jansen, A. R. Bernard, J. P. Aubry, and J. Y. Bonnefoy, Demonstration of a second ligand for the low affinity receptor for immunoglobulin E (CD23) using recombinant CD23 reconstituted into fluorescent liposomes, *J. Exp. Med.*, *176*: 389 (1992).

24. D. Vercelli, B. Helm, P. Marsh, E. Padlan, R. S. Geha, and H. Gould, The B-cell binding site on human immunoglobulin E, *Nature*, *338*: 649 (1989).
25. B. Bettler, H. Hofstetter, M. Rao, W. M. Yokoyama, F. Kilchherr, and D. H. Conrad, Molecular structure and expression of the murine lymphocyte low affinity receptor for IgE (FcϵRII), *Proc. Natl. Acad. Sci. USA*, *86*: 7566 (1989).
26. S. O. Gollnick, M. L. Trounstine, L. C. Yamashita, M. R. Kehry, and K. W. Moore, Isolation, characterization, and expression of cDNA clones encoding the mouse Fc receptor for IgE (FcϵRII), *J. Immunol.*, *144*: 1974 (1990).
27. A. Beavil, R. Edmeades, H. Gould, and B. Sutton, α-Helical coiled-coil stalks in the low-affinity receptor for IgE (FcϵRII/CD23) and related C-type lectins, *Proc. Natl. Acad. Sci. USA*, *89*: 753–757 (1992).
28. J. Y. Bonnefoy, O. Guillot, H. Spits, D. Blanchard, K. Ishizaka, and J. Banchereau, The low-affinity receptor for IgE (CD23) on B lymphocytes is spatially associated with HLA-DR antigens, *J. Exp. Med.*, *167*: 57 (1988).
29. M. Letellier, T. Nakajima, G. Pulido-Cejudo, H. Hofstetter, and G. Delespesse, Mechanism of formation of human IgE-binding factors (soluble CD23): III. Evidence for a receptor (FCϵRII)-associated proteolytic activity, *J. Exp. Med.*, *172*: 693 (1990).
30. G. Delespesse, M. Sarfati, and M. Hofstetter, Human IgE-binding factors, *Immunol. Today*, *10*: 159 (1989).
31. N. Grenier-Brossette, I. Bourget, C. Akoundi, J. Y. Bonnefoy, and J. L. Cousin, Spontaneous and ligand-induced endocytosis of CD23 (FCϵ receptor II) from the surface of B lymphocytes generates a 16-kDa intracellular fragment, *Eur. J. Immunol.*, *22*: 1573 (1992).
32. M. L. Richards, D. H. Katz, and F. T. Liu, Complete genomic sequence of the murine low affinity Fc receptor for IgE, *J. Immunol.*, *147*: 1067 (1991).
33. J. Gordon, L. Flores-Romo, J. A. Cairns, M. J. Millsum, P. J. Lane, G. D. Johnson, and I. C. MacLennan, CD23: A multi-functional receptor/lymphokine? *Immunol. Today*, *10*: 153 (1989).
34. A. Mathur, S. Maekawa, Z. Ovary, and R. G. Lynch, Increased T-epsilon cells in BALB/c mice with an IgE-secreting hybridoma, *Mol. Immunol.*, *23*: 1193 (1986).
35. A. Mathur, D. H. Conrad, and R. G. Lynch, Characterization of the murine T cell receptor for IgE (Fc epsilon RII). Demonstration of shared and unshared epitopes with the B cell Fc epsilon RII, *J. Immunol.*, *141*: 2661 (1988).
36. H. Kikutani, M. Suemura, H. Howaki, H. Nakamura, R. Sato, K. Yamasaki, E. L. Barsumian, R. R. Hardy, and T. Kishimoto, Fcϵ receptor, a specific differentiation marker transiently expressed on mature B cell prior to isotype switching, *J. Exp. Med.*, *164* 1455 (1986).
37. C. M. Snapper, J. J. Hooley, Jr., F. G. Urban, and F. D. Finkelman, Lack of FcϵRII expression by murine B cells after in vivo immunization is directly associated with Ig secretion and not Ig Isotype switching, *J. Immunol.*, *146*: 2161 (1991).
38. T. Defrance, J. P. Aubry, F. Rousset, B. Vanbervliet, J. Y. Bonnefoy, N. Arai, Y. Takebe, T. Yokota, F. Lee, K. Arai, J. de Vries, and J. Banchereau, Human recombinant interleukin 4 induces Fcϵ receptors (CD23) on normal B lymphocytes, *J. Exp. Med.*, *165*: 1459 (1987).
39. C. Hivroz, A. Valle, J. C. Brouet, J. Banchereau, and C. Grillot-Courvalin,

Regulation by interleukin 2 of CD23 expression of leukemic and normal B cells: Comparison with interleukin 4, *Eur. J. Immunol., 19*: 1025 (1989).

40. M. K. Crow, B. Kushner, J. A. Jovner, S. M. Friedman, S. E. Mechanic, and W. Stohl, Human peripheral blood T helper cell-induced B cell activation results in B cell surface expression of the CD23 (BLAST-2) antigen, *Cell. Immunol., 121*: 99 (1989).
41. D. H. Conrad, A. D. Keegan, K. R. Kalli, R. Van Dusen, M. Rao, and A. D. Levine, Superinduction of low affinity IgE receptors on murine B lymphocytes by LPS and Interleukin-4, *J. Immunol., 141*: 1091 (1988).
42. A. D. Keegan, C. M. Snapper, R. Van Dusen, W. E. Paul, and D. H. Conrad, Superinduction of the murine B cell FcεRII by T helper cell clones. Role of interleukin-4, *J. Immunol., 142*: 3868 (1989).
43. J. Gordon, A. Katira, A. J. Strain, and S. Gillis, Inhibition of interleukin 4-promoted CD23 production in human B lymphocytes by transforming growth factor-β, interferons or anti-CD19 antibodies overriden on engaging CD40, *Eur. J. Immunol., 21*: 1917 (1991).
44. W. C. Fanslow, D. Anderson, K. H. Grabstein, E. A. Clark, D. Cosman, and R. J. Armitage, Soluble forms of CD40 inhibit biological responses of human B cells, *J. Immunol., 149*: 655 (1992).
45. K. A. Yamaoka, H. E. Claesson, and A. Rosen, Leukotriene B_4 enhances activation, proliferation and differentiation of human B lymphocytes, *J. Immunol., 143*: 1996 (1989).
46. N. Paul-Eugene, B. Dugas, S. Picquot, V. Lagente, J. M. Mencia-Huerta, and P. Braquet, Influence of interleukin-4 and platelet-activating factor on the Fc-epsilon-RII/CD23 expression on human monocytes, *J. Lipid Med., 2*: 95 (1990).
47. B. Dugas, N. Paul-Eugene, J. Cairns, J. Gordon, A. Calenda, J. M. Mencia-Huerta, and P. Braquet, Leukotriene B_4 potentiates the expression and release of FcεRII/CD23, and proliferation and differentiation of human B lymphocytes induced by IL-4, *J. Immunol., 145*: 3406 (1990).
48. C. Fargeas, C. Wu, H. Luo, M. Sarfati, G. Delespesse, and J. Wu, 1,25-$(OH)_2$ vitamin D_3 inhibits the CD23 gene expression by human peripheral blood monocytes, *J. Immunol., 145*: 4053 (1990).
49. T. Kawabe, M. Takami, M. Hosoda, Y. Maeda, S. Sato, M. Mayumi, H. Mikawa, K. Arai, and J. Yodoi, Regulation of FcεR2/CD23 gene expression by cytokines and specific ligands (IgE and anti-FcεR2 monoclonal antibody). Variable regulation depending on the cell types, *J. Immunol., 141*: 1376 (1988).
50. J. C. Prinz, X. Bauer, G. Mazur, and E. P. Rieber, Allergen-directed expression of Fc receptors for IgE (CD23) on human T lymphocytes is modulated by interleukin-4 and interferon-α, *Eur. J. Immunol., 20*: 1259 (1990).
51. T. Bieber, A. Rieger, C. Neuchrist, J. C. Prinz, E. P. Rieber, G. Boltz-Nitulescu, O. Scheiner, D. Kraft, J. Ring, and G. Stingl, Induction of FcεR2/CD23 on human epidermal Langerhan's cells by recombinant interleukin-4 and γ-interferon, *J. Exp. Med., 170*: 309 (1989).
52. V. Pancre, M. Joseph, A. Capron, J. Wietzerbin, J. P. Kusnierz, H. Vorng, and C. Auriault, Recombinant human interferon induces increased IgE receptor expression on human platelets, *Eur. J. Immunol., 18*: 829 (1988).

53. J. Yodoï, T. Ishizaka, and K. Ishizaka, Lymphocytes bearing Fc receptors for IgE. II. Induction of Fcϵ-receptor bearing rat lymphocytes by IgE, *J. Immunol.*, *123*: 455 (1979).
54. M. Daeron, and K. Ishizaka, Induction of Fc epsilon receptors on mouse macrophages and lymphocytes by homologous IgE, *J. Immunol.*, *136*: 1612 (1986).
55. M. Hosoda, S. Makino, T. Kawabe, Y. Maeda, S. Satoh, M. Takami, M. Mayumi, K. Arai, H. Saitoh, and J. Yodoi, Differential regulation of the low affinity Fc receptor for IgE (FcϵR2/CD23) and the IL-2 receptor (Tac/p55) on eosinophilic leukemia cell line (Eol-1 and Eol-3), *J. Immunol.*, *143*: 147 (1989).
56. F. Wang, C. D. Gregory, M. Rowe, A. B. Rickinson, D. Wang, M. Birkenbach, H. Kikutani, T. Kishimoto, and E. Kieff, Epstein–Barr virus nuclear antigen 2 specifically induces expression on the B-cell activation antigen CD23, *Proc. Natl. Acad. Sci. USA*, *84*: 3452 (1987).
57. A. Capron, J. P. Dessaint, M. Capron, M. Joseph, J. C. Ameisen, and A. B. Tonnel, From parasites to allergy: A second receptor for IgE, *Immunol. Today*, *7*: 15 (1986).
58. M. Capron, M. D. Kazatchkine, E. Fischer, M. Joseph, A. E. Butterworth, J. P. Kusnierz, L. Prin, J. P. Papin, and A. Capron, Functional role of the α-chain of complement receptor type 3 in human eosinophil-dependent antibody-mediated cytotoxicity against schistosomes, *J. Immunol.*, *139*: 2059 (1987).
59. L. Borish, J. J. Mascali, and L. J. Rosenwasser, IgE-dependent cytokine production by human peripheral blood mononuclear phagocytes, *J. Immunol.*, *246*: 63 (1991).
60. H. Kikutani, A. Yokota, N. Uchibayashi, K. Yakawa, T. Tanaka, K. Sugiyama, E. L. Barsumian, M. Suemura, and T. Kishimoto, Structure and function of Fc epsilon receptor II (Fc epsilon RII/CD23): A point of contact between the effector phase of allergy and B cell differentiation, *Ciba Found. Symp.*, *147*: 23 (1989).
61. A. Yokota, K. Yukawa, A. Yamamoto, K. Sugiyama, M. Suemura, Y. Tashiro, T. Kishimoto, and H. Kikutani, Two forms of the low-affinity Fc receptor for IgE differentially mediate endocytosis and phagocytosis: Identification of the critical cytoplasmic domains, *Proc. Natl. Acad. Sci. USA*, *89*: 5030 (1992).
62. L. Flores-Romo, G. D. Johnson, A. Veronesi, A. Ghaderi, D. R. Stanworth, and J. Gordon, Functional implication for the topographical association between MHC class II and the low affinity IgE receptor: Occupancy of CD23 blocks the ability of B cells to present alloantigen to T cells, *Eur. J. Immunol.*, *20*: 2465 (1990).
63. M. R. Kehry, and L. C. Yamashita, Fcϵ receptor II (CD23) function on mouse B cells: Role in IgE dependent antigen focusing, *Proc. Natl. Acad. Sci. USA*, *86*: 7566 (1989).
64. U. Pirron, T. Schlunch, J. C. Prinz, and E. P. Rieber, IgE-dependent antigen focusing by human B lymphocytes is mediated by the low-affinity receptor for IgE, *Eur. J. Immunol.*, *20*: 1547 (1990).
65. H. Luo, H. Hofstetter, J. Banchereau, and G. Delespesse, Cross-linking of CD23 antigen by its natural ligand (IgE) or by anti-CD23 antibody prevents B lymphocyte proliferation and differentiation, *J. Immunol.*, *146*: 2122 (1991).
66. J. Gordon, M. Rowe, L. Walker, and G. Guy, Ligation of the CD23 p45 (BLAST-2, EBVCS) antigen triggers the cell-cycle progression of activated B lymphocytes, *Eur. J. Immunol.*, *16*: 1975 (1986).

67. K. A. Campbell, A. Lees, F. D. Finkelman, and D. H. Conrad, Crosslinking FcεRII and surface IgD enhances anti-IgD mediated B cell activation, *FASEB J.*, *5*: 1384A (1991).
68. T. J. Waldschmidt, and L. Tygrett, Crosslinking surface Ig and the low affinity IgE Fc receptor induces B cells to enter cell cycle, *FASEB J.*, *5*: 1389A (1991).
69. J. P. Kolb, D. Renard, B. Dugas, E. Génot, E. Petit-Koskas, M. Sarfati, G. Delespesse, and J. Poggioli, Monoclonal anti-CD23 antibodies induce a rise in $[Ca^{2+}]_i$ and polyphosphoinositide hydrolysis in human activated B cells. Involvement on a Gp protein, *J. Immunol.*, *145*: 429 (1990).
70. J. P. Kolb, E. Genot, J. Poggioli, A. Abadie, M. Sarfati, G. Delespesse, and B. Dugas, Transduction through CD23. Different signalling pathways in human B cells and monocytes, *Monogr. Allergy*, *29*: 135 (1991).
71. M. Sarfati, D. Bron, L. Lagneaux, C. Fonteyn, H. Frost, and G. Delespesse, Elevation of IgE binding factors in serum of patients with B cell-derived chronic lymphocytic leukemia, *Blood*, *71*: 94 (1988).
72. M. Sarfati, S. Fournier, M. Christoffersen, and G. Biron, Expression of CD23 antigen and its regulation by IL-4 in chronic lymphocytic leukemia, *Leukemia Res.*, *14*: 47 (1990).
73. C. H. Geisler, J. K. Larsen, N. E. Hansen, M. M. Hansen, B. E. Christensen, B. Lund, H. Nielsen, T. Plesner, K. Thorling, E. Andersen, and P. K. Andersen, Prognostic importance of flow cytometric immunophenotyping of 540 consecutive patients with B-cell chronic lymphocytic leukemia, *Blood*, *78*: 1795 (1991).
74. D. A. Lebman, and R. L. Cofman, Interleukin 4 causes isotype switching to IgE in T cell stimulated clonal B cell cultures, *J. Exp. Med.*, *168*: 853 (1988).
75. J. Pene, F. Rousset, F. Briere, I. Chretien, J. Y. Bonnefoy, H. Spits, H. Yokota, N. Arai, K. I. Arai, and J. Banchereau, IgE production by human lymphocytes is induced by interleukin 4 and suppressed by interferon γ and α and prostaglandin E_2, *Proc. Natl. Acad. Sci. USA*, *85*: 6880 (1988).
76. M. Sarfati, E. Rector, M. Rubio-Trujillo, K. Wong, A. H. Sehon, and G. Delespesse, In vitro synthesis of IgE by human lymphocytes. II. Enhancement of the spontaneous IgE synthesis by IgE-binding factors secreted by RPMI 8866 lymphoblastoid B cells, *Immunology*, *53*: 197 (1984).
77. M. Sarfati, and G. Delespesse, Possible role of human lymphocyte receptor for IgE (CD23) or its soluble fragments in the in vitro synthesis of human IgE, *J. Immunol.*, *141*: 2195 (1988).
78. J. Y. Bonnefoy, J. G. Shields, and J. J. Mermod, Inhibition of human interleukin-4-induced IgE synthesis by a subset of anti-CD23/FcεRII monoclonal antibodies, *Eur. J. Immunol.*, *20*: 139 (1990).
78a. L. Flores-Romo, J. Shields, Y. Humbert, P. Graber, J-P. Aubry, J-F. Gauchat, G. Ayala, B. Allet, N. Chavez, H. Bazin, N. Capron, and J-Y. Bonnefoy, Inhibition of an in vivo antigen-specific IgE response by antibodies to CD23, *Science*, *261*: 1038 (1993).
79. D. H. Vercelli, H. Jabara, K. Arai, and R. S. Geha, Induction of human IgE synthesis requires interleukin 4 and T/B cell interactions involving the T cell receptor/CD23 complex and MHC class II antigen, *J. Exp. Med.*, *169*: 1295 (1989).
80. J. Y. Bonnefoy, S. Pochon, J. P. Aubry, P. Graber, J. F. Gauchat, K. Jansen, and

L. Flores-Romo, A new pair of surface molecules involved in human IgE regulation, *Immunol. Today*, *14*: 1 (1993).

81. J. P. Aubry, J. G. Shields, K. U. Jansen, and J. Y. Bonnefoy, A multiparameter flow cytometric method to study surface molecules in interactions between subpopulations of cells, *J. Immunol. Methods*, *159*: 161 (1993).

81a. S. Henchoz, J-F. Gauchat, J-P. Aubry, P. Graber, S. Pochon, and J-Y. Bonnefoy, Stimulation of human IgE production by a subset of anti-CD21 monoclonal antibodies: requirement of a CO-signal to modulate ϵ transcripts, *Immunology* (in press).

82. J. Y. Bonnefoy, S. Henchoz, M. J. Holder, D. Hardie, and J. Gordon, A subset of anti-CD21 antibodies promote the rescue of germinal center B cells from apoptosis, *Eur. J. Immunol.* *23*: 969 (1993).

83. Y. J. Liu, J. A. Cairns, M. Holder, S. D. Abbot, K. Jansen, J. Y. Bonnefoy, J. Gordon, and I. C. M. MacLennan, Recombinant 25 kDa CD23 and interleukin-1 promote the survival of germinal center B cells: Evidence for bifurcation in the development of centrocytes rescued from apoptosis, *Eur. J. Immunol.*, *21*: 1107 (1991).

83a. K. Bacon, J-F. Gauchat, J-P. Aubry, S. Pochon, P. Graber, S. Henchoz, and J-Y. Bonnefoy, CD21 expressed on basophilic cells is involved in histamine release triggered by CD23 and anti-CD21 antibodies, *Eur. J. Immunol.* (in press).

84. J. Cairns, and J. Gordon, Intact 45 kDa (membrane) form of CD23 is consistently mitogenic for normal and transformed B lymphoblasts, *Eur. J. Immunol.*, *20*: 539 (1990).

85. K. Jansen, J. Shields, J. Gordon, J. Cairns, P. Graber, and J. Y. Bonnefoy, Expression of human recombinant CD23 in insect cells, *J. Recept. Res.*, *11*: 507 (1991).

86. M. D. Mossalayi, J. C. Lecron, A. H. Dalloul, M. Sarfati, J. M. Bertho, H. Hofstetter, G. Delespesse, and P. Debre, Soluble CD23 (FCϵRII) and interleukin-1 synergistically induce early human thymocyte maturation, *J. Exp. Med.*, *171*: 959 (1990a).

87. M. D. Mossalayi, M. Arock, J. M. Bertho, C. Blanc, A. H. Dalloul, H. Hofstetter, M. Sarfati, G. Delespesse, and P. Debre, Proliferation of early human myeloid precursors induced by interleukin-1 and recombinant soluble CD23, *Blood*, *75*: 1924 (1990b).

88. L. Flores-Romo, J. A. Cairns, M. J. Millsum, and J. Gordon, Soluble fragments of the low-affinity IgE receptor (CD23) inhibit the spontaneous migration of U937 monocytic cells: Neutralization of MIF-activity by a CD23 antibody, *Immunology*, *67*: 547 (1989).

10

Cellular and Molecular Regulation of Immunoglobulin E Synthesis

Donata Vercelli and Raif S. Geha
Children's Hospital, Harvard Medical School, Boston, Massachusetts

I. INTRODUCTION

The role of immunoglobulin E (IgE) in allergic disease is by now well recognized (reviewed in 1,2). The study of the cellular basis of IgE regulation, therefore, has been actively pursued to gain insights into the pathogenesis of a disease that affects a considerable proportion of the population. More recently, the molecular events underlying IgE synthesis have become the focus of intense interest, inasmuch as they represent a suitable model to characterize the signals involved in isotype-specific regulation of immunoglobulin (Ig) synthesis in humans.

In the past 2–3 years, evidence from different laboratories has indicated that induction of IgE synthesis requires two signals: one is IgE isotype-specific, and is delivered by a cytokine, interleukin-4 (IL-4), the other is a B-cell-activating signal, which can be delivered through a variety of pathways (reviewed in 1) (Fig. 1). The challenge in IgE research is now to determine how the two signals required for IgE induction interact at the gene level. The cellular studies on IgE regulation have been extensively reviewed elsewhere (2,3). In this chapter, we will review what is currently known about the mechanisms underlying the synthesis of IgE in humans, as they are revealed through molecular biology approaches.

II. ISOTYPE SWITCHING

The antigen-binding specificity of an Ig molecule is determined by the NH_2-terminus of the Ig heavy and light chain, which is highly variable. By contrast, the

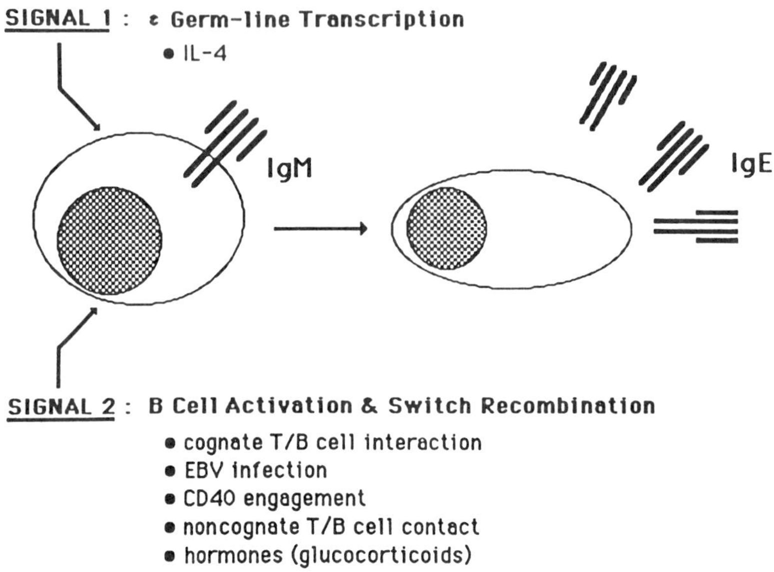

Figure 1 A two-signal model for the induction of IgE synthesis in humans.

COOH-terminus of the heavy chain has a constant amino acid sequence that determines the effector functions of the Ig molecule, such as binding to Fc receptors on different cell types and activation of secondary pathways (reviewed in 4). The variable region of Ig is encoded by multiple germline elements that are assembled into complete V(D)J variable regions during B-cell differentiation (reviewed in 5) by a common enzymatic activity (recombinase), which might be related to the products of the *RAG-1* and *RAG-2* genes (6,7).

During an immune response, a B lymphocyte can express different heavy chain isotypes sharing the same VDJ region. This phenomenon (heavy chain class switching) allows a single B-cell clone to produce antibodies that retain variable region specificity in association with a different C_H region gene (i.e., with a different effector function). Class switching is thought to result from a recombination event that juxtaposes a downstream C_H gene to the expressed *V(D)J* gene. According to the looping out-deletion model, intervening sequences, including the previously expressed C_H gene, are excised from the chromosome as a circular DNA (8). Detection of DNA circles (9–11) has provided direct evidence for this model of switch recombination.

Recombination involves characteristic repetitive sequences (switch regions: S regions) located 5′ of the C_μ gene and corresponding S regions located immediately 5′ of each C_H gene, except $C\delta$. S regions are 2–10 kb in length, and their sequence consists of short tandem repeats (e.g., GAGCT, GGGCT).

S region recombination sites are not flanked by obvious consensus elements, and joining has been observed at various locations within the S regions (reviewed in 12,13). Switch recombination has generally been thought to occur within one chromosome. However, recent studies in transgenic mice carrying an Ig μ heavy chain transgene show that switch recombination can occur between two different chromosomes (14).

Several studies have suggested that isotype switching may be a directed (nonrandom), rather than a stochastic, process. In particular, the observation that switch recombination in hybridomas and normal murine B cells is directed to the same S region on both the productive and the nonproductive C_H alleles (15–17) favors directed switching. The C_H rearrangements in human B cells appear to be more random (i.e., switching to the same C_H gene on both alleles is not uniformly observed; 18). However, the limited sample size may have influenced the pattern observed in these studies.

III. LYMPHOKINE-DEPENDENT REGULATION OF ISOTYPE SWITCHING: THE ROLE OF GERMLINE TRANSCRIPTS

For some time it has been known that different stimuli cause switching to different isotypes. In the mouse, lipopolysaccharide (LPS) induces switching to IgG3 and IgG2b, whereas a combination of interleukin (IL)-4 and LPS induce B cells to switch to IgG1 and IgE, and suppress switching to IgG2b and IgG3 (19–23). The molecular mechanisms of Ig class switching are largely unknown. No switch recombinase has yet been characterized; thus, it is not known whether there are selective recombinases, one for each S region, or whether there is one recombinase for all S regions. The latter seems more likely, because S regions are composed of highly repetitive sequences, many of which are shared among different S regions. It has been recently proposed that the specificity of switching is regulated by modulating the accessibility of S regions to a putative common switch recombinase, as has been suggested for *VDJ* rearrangements (12,24). The accessibility model is supported by several observations. Before switching, the chromatin structure of the relevant S regions undergoes changes, such as increased sensitivity to DNase I digestion and hypomethylation (20,25,26). Inhibition of methylation by 5-azacytidine during B-cell activation profoundly alters the direction and frequency of class switching (26).

Furthermore, and most importantly, in murine B cells that are predisposed to undergo class switching to a particular isotype (γ1, γ2b, or ϵ), the C_H locus to which switching is directed is transcribed, before switching, into a germline transcript (19–23,25,27). Germline transcripts initiate about 2 kb upstream of the S region involved in switch recombination and are processed to an mRNA in which a 100 to 500-bp–upstream exon (designated I exon) is spliced to the

C_H-coding exons. The resulting germline transcripts do not appear capable of initiating the synthesis of a mature protein, because the I exon contains stop codons in all three reading frames. Therefore, germline transcripts are also referred to as *sterile* transcripts. Alternatively, germline transcripts are referred to as *truncated* transcripts, because the *I* exon is usually 200–300 bp shorter than the *VDJ* exon present in mature transcripts.

Although the precise role of germline transcripts in isotype switching remains to be defined, several lines of evidence suggest that they may be important, or even critical, for switching. All germline transcripts yet characterized share a similar general structure: each consists of a 5′ exon derived from germline sequences upstream from the corresponding S region, that is spliced to the C_{H1} gene immediately downstream. Although the different transcripts contain no significant sequence homologies, these structural similarities suggest an analogous function. In addition, germline transcripts may themselves promote specific recombination through interaction with DNA in the S regions. In a recent study, in vitro transcription of a supercoiled plasmid containing the polypurine strand of the murine IgA switch region [$(AGGAG)_n$] resulted in the formation of stable, less supercoiled, more relaxed conformational isomers (conformers) of the plasmid (28,29). The conformers were stabilized by the nascent RNA transcripts, which remained base-paired with their DNA templates, forming DNA–RNA hybrids. Interestingly, the polypurine strand of the murine Sα region is the one transcribed in vivo to produce the α germline transcripts before switching (30). The formation of such RNA–DNA hybrids in transcriptionally active switch regions may play an important role in switch recombination, by altering the conformation of the relevant switch region.

Other possibilities, not necessarily mutually exclusive, should also be considered. Transcription may be a by-product of other events (e.g., binding of specific factors) involved in promoting recombinational accessibility. Alternatively, germline transcripts could conceivably be used in interchromosomal RNA transsplicing, whereby transcripts from rearranged *V(D)J* exons are joined to C_H germline transcripts (31). This process, which would result in heavy chain class switching in the absence of deletional RNA rearrangement in the heavy chain locus, has been recently identified as the mechanism underlying multiple Ig isotype expression in the TG.SA transgenic mouse model (32). Splenic B cells from TG.SA mice expressed the transgenic human IgM and endogenous murine IgG simultaneously after stimulation with LPS and IL-4. The B cells bearing double isotypes expressed several mRNA species, i.e., transgenic human *VDJ* properly spliced to the endogenous mouse $C\gamma$ genes (trans-mRNA), together with the transgenic human μ RNA and germline transcripts derived from the mouse $C\gamma$ gene. No rearrangement of the transgene was detected (33). It is conceivable that B cells expressing multiple Ig isotypes might be intermediates for class switching. Therefore, transsplicing involving germline transcripts might be a mechanism of general importance for class switching.

IV. INTERLEUKIN-4, THE FIRST SIGNAL FOR SWITCHING TO IMMUNOGLOBULIN E, IS SUFFICIENT FOR INDUCTION OF ϵ GERMLINE TRANSCRIPTS

Recent studies from various laboratories have shown that the cytokine IL-4 is required for the synthesis of IgE protein in both mice (34,35) and humans (36–38). At the molecular level, the effects on IgE isotype switching are thought to result from the ability of IL-4 to activate transcription through the ϵ locus.

In murine B cells, stimulation with IL-4 and LPS induces the appearance of 1.7 to 1.9-kb germline Cϵ transcripts that contain an *I*ϵ exon, located about 2 kb upstream from Sϵ, spliced to the *C*ϵ*1–C*ϵ*4* exons (23,27). The *I*ϵ exon has been reported to span about 245 bp in I29 B-lymphoma cells (39), and 60–100 bp in the Abelson murine leukemia virus-transformed pre-B-cell line A20 (40). Nuclear run-on transcription assays have recently shown that induction of ϵ germline transcripts by LPS and IL-4 in A20 cells occurs at the transcriptional level (41). Analysis of constructs transiently transfected into a murine B-cell lymphoma line demonstrated that a 170-bp segment of DNA spanning the *I*ϵ transcription initiation site confers LPS–IL-4-inducible expression to a heterologous reporter gene (41). Mutational analysis demonstrated that this expression depended on DNA sequences contained within a conserved region directly upstream from the *I*ϵ transcription initiation site. One nuclear protein that is constitutively expressed in normal B cells binds to the downstream end of the conserved sequence (Fig. 2). Two additional proteins, which are induced by IL-4, bind to the transcription initiation site of *I*ϵ (41). Interestingly, these

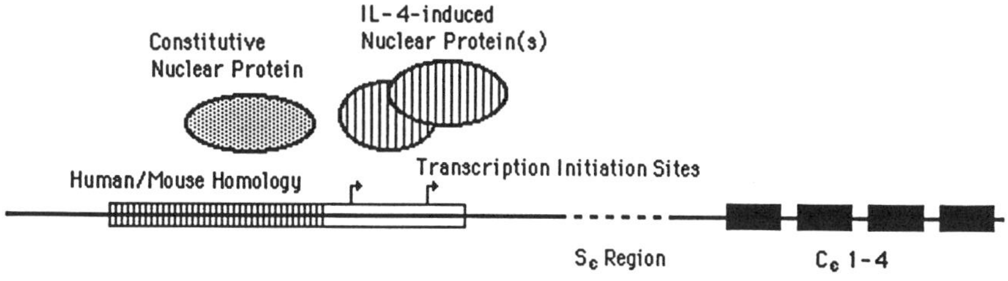

Figure 2 A schematic representation of the murine Iϵ exon. The region of high homology between human and mouse sequences is indicated, together with the putative binding regions for constitutive and IL-4-induced nuclear proteins. (Modified from Ref. 41.)

proteins are indistinguishable in binding assays from proteins known to bind an enhancer region of the murine class II major histocompatibility complex gene *A*α (42).

The structure of ϵ germline transcripts expressed by human B cells following stimulation with IL-4 was recently characterized (43) (Fig. 3). DNA complementary to the human *I*ϵ exon was synthesized by reverse transcription of poly(A)$^+$ RNA, tailed with dGTP, and amplified by anchored polymerase chain reaction (PCR) (44) using oligonucleotides complementary to Cϵ and to the dGTP tail as primers. The cloning of several amplified fragments revealed that the human ϵ germline transcript consists of approximately 134-bp *I*ϵ exon located about 3.5 kb upstream from *C*ϵ, in a region that is deleted by switch recombination in the expressed allele of the U266 IgE myeloma (45,46). Transcription initiation sites are scattered within a stretch of 90 bp. The *I*ϵ exon is directly spliced to *C*ϵ*1* by removal of the intervening sequences from the primary transcript (43). The germline transcript contains all four *C*ϵ exons, as well as the 3′-untranslated region. The human *I*ϵ exon does not contain an initiation codon in-frame with the *C*ϵ region, and stop codons were observed in all three reading frames (43), suggesting that no protein is encoded by these transcripts. Similar to what has been found in the mouse (39–41), the region upstream from the initiation sites lacks well-defined TATAA motifs (43).

Interestingly, ϵ germline transcripts were also induced in resting human B cells isolated by Percoll density gradients, indicating that no previous in vivo activation is required for expression of ϵ germline transcripts following IL-4 stimulation (47). Thus, IL-4 is, by itself, sufficient for the initiation of germline transcription through the human ϵ locus. However, additional stimuli are required for switch recombination and expression of mature ϵ mRNA.

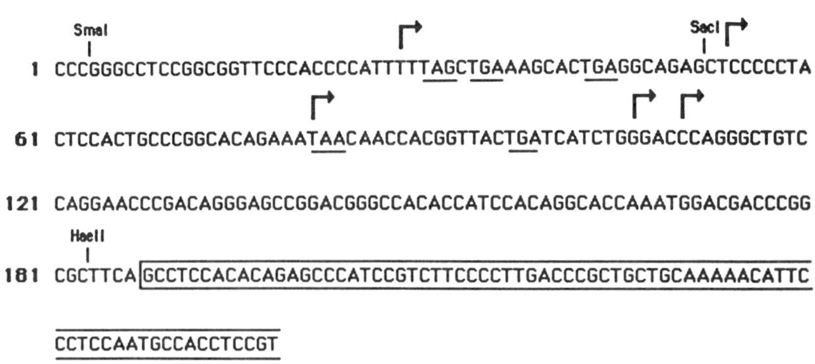

Figure 3 The human *I*ϵ exon. Transcription initiation sites are indicated by arrows. Stop codons are underlined. (Modified from Ref. 43.)

V. A SECOND B-CELL-ACTIVATION SIGNAL IS REQUIRED FOR SWITCH RECOMBINATION AND EXPRESSION OF MATURE Cε TRANSCRIPTS

Interleukin-4, although necessary, is not sufficient for the induction of IgE synthesis by B cells. Highly purified normal human B cells did not produce IgE following stimulation with IL-4 and mixtures of T-cell-derived lymphokines (48). At the molecular level, IL-4 induces germline ε transcripts, but not DNA switch recombination (discussed later) (47), nor expression of mature Cε transcripts (containing VDJ spliced to *Cε1–4*) (49). The latter events required the delivery of an additional signal, which led to B-cell activation and synergized with IL-4 in inducing isotype switching to IgE. In humans, a variety of second IgE-inducing signals have been described, both T-cell-dependent (cognate, as well as noncognate interactions between T and B cells) and T-cell-independent [transformation with Epstein–Barr virus (EBV), engagement of CD40, glucocorticoids] (reviewed in 2).

The T-cell-independent systems of IgE induction are suitable for dissection at the molecular level and are currently actively investigated using assay systems recently developed to follow the steps underlying IgE synthesis. Expression of germline or mature Cε RNA transcripts can be detected by Northern blot, using DNA probes hybridizing to the *I*ε exon (contained only in germline transcripts) or the *C*ε exons (contained in both germline and mature ε transcripts). The IgE switch fragments resulting from DNA recombination between the Sμ and Sε regions can be detected by PCR.

A. Analysis of Cε Transcription

Stimulation with IL-4 and EBV induces T-cell-independent IgE synthesis in human surface IgE-negative B cells (49,50). RNA from IgE-secreting B-cell lines obtained by EBV transformation in the presence of IL-4 was analyzed for the presence of germline as well as mature Cε transcripts using genomic DNA probes for the *I*ε exon and the *Cε1–Cε2* exons (49). The *Cε1–Cε2* probe hybridized to two mRNA species in IgE-secreting EBV B cells: a 2.0-kb band, and a more prominent 1.8-kb band. The 1.8-kb band, but not the 2.0-kb band, also hybridized with the *I*ε probe and, therefore, corresponded to germline Cε transcripts. No signal was detected in RNA from EBV B cells that were not treated with IL-4 and secreted Ig isotypes other than IgE. These results indicated that IgE-secreting B cells obtained by activation with EBV and IL-4 contain both mature and germline Cε transcripts. Interestingly, when IL-4 was added to IgM- or IgG-secreting EBV B-cell lines that had been maintained in culture for over 3 months, induction of germline, but not mature, Cε transcripts was observed. Thus, in established EBV B-cell lines stimulated with IL-4, there is a dissociation between germline Cϵ transcription and isotype switching to IgE expression.

It has been recently shown that anti-CD40 MAbs induce IgE synthesis by highly purified surface IgE-negative human B cells, in the presence of IL-4 (51–54). This is the first system in which the B-cell-activating signal required for human IgE synthesis is delivered by engagement of a discrete B-cell surface antigen.

The effect of CD40 engagement on Cϵ transcription was analyzed with the set of DNA probes used for the EBV–IL-4 system (47). The Cϵ germline transcripts, but not mature ϵ mRNA, were detected in B cells stimulated with IL-4 for 5 or 10 days. Anti-CD40 MAb by itself did not induce either germline transcripts or mature Cϵ transcription, but synergized with IL-4 to enhance ϵ-germline transcript accumulation. Densitometric analysis of ϵ germline mRNA bands versus actin RNA showed a sevenfold increase in ϵ germline transcripts following addition of anti-CD40 MAb at both time points. A 2.0-kb band, corresponding to mature ϵ mRNA, as indicated by comigration with the major band in RNA from the IgE-secreting plasmacytoma U266, was detected only in B cells stimulated with both IL-4 and anti-CD40 MAb for 10 days. These results indicated that the CD40 signal induced isotype switching and the production of mature ϵ mRNA. The effects of CD40 engagement were totally dependent on the presence of IL-4. These findings are compatible with the hypothesis that the induction of germline transcripts may result in increased accessibility of the downstream S region for deletional switch recombination with Sμ (47).

B. Analysis of DNA Recombination

The molecular mechanisms that underlie the production of mature ϵ mRNA have long been a matter of debate. Because of technical limitations, it was difficult to establish whether IgE synthesis is normal human B cells results from deletional switch recombination, or from alternative mechanisms, such as alternative splicing of a long mRNA transcript or transsplicing. The human IgE myeloma cell line, U266, has undergone switch recombination to Sϵ, at least on the productive chromosome (55,56). In contrast, a cloned EBV B-cell line obtained by limiting dilution, in the absence of IL-4, coexpresses IgE and IgM (57). The IgE production was not accompanied by DNA rearrangement, because the ϵ locus was maintained in its germline configuration on both alleles, with no deletional switch rearrangement (58). Expression of mature VDJCϵ mRNA was thought to occur by alternative processing of a long transcript, as previously proposed for the simultaneous expression of multiple isotypes, including IgE, in neoplastic murine B-cell lines, as well as in normal B cells (59–61). However, a long primary transcript was not detected, and the possibility of transsplicing was not explored.

The molecular mechanisms underlying IgE isotype switching have been more recently investigated taking advantage of T-cell-independent systems of IgE induction, and using "nested primer" PCR amplification of ϵ switch fragments, followed by cloning and DNA sequencing (46,47). Nested primer PCR is based on the use of two sets of primers, located one inside the other. The external set is used in the initial round of amplification, whereas the internal set primes the second amplification cycle. This approach, which greatly enhances both the sensitivity and the specificity of PCR amplification, was validated by showing the expected Sμ/Sε rearrangement, with deletion of intervening sequences in the IgE myeloma cell line U266 (46). Switch junction fragments were subsequently amplified from B cells stimulated with both IL-4 and EBV, or IL-4 and anti-CD40 MAb, or IL-4 and glucocorticoids (hydrocortisone), but not from B cells stimulated with EBV, anti-CD40 MAb, hydrocortisone, or IL-4, alone (46,47,62). The DNA sequence of these switch fragments revealed that deletional switch recombination had occurred with direct joining of Sμ to Sε. The recombination sites within Sμ were clustered within 900 bps at the 5′-end of the S region, suggesting that there are hot spots for recombination within Sμ. The Sε recombination sites were scattered throughout the Sε region (46) (Fig. 4). It should be noted, however, that nested primer PCR amplifies switch fragments from as little as 10 pg (one diploid genome) of DNA from the IgE plasmacytoma U266 (46). If all the B cells containing cytoplasmic IgE in the CD40–IL-4 system (≈10% of the total B cells) had undergone deletional switch recombination, switch fragments should have been detected in 100-pg DNA samples. In contrast, 100-fold more DNA (10 ng) was required to amplify "switch fragments." These data suggest either that only a fraction of the switch fragments could be amplified, possibly because PCR is biased against amplification of fragments longer than 1.6–2 kb, or that nondeletional recombination mechanisms for IgE switching (33,57,63) were also operative.

*C*ε gene rearrangement, with deletion of both *C*μ genes, has also been observed by Southern blot analysis in stable IgE-secreting B-cell lines obtained by stimulation with EBV and IL-4, followed by limiting-dilution cloning (64). Furthermore, analysis of Ig circular DNA excised in murine B cells stimulated with IL-4 and *Nippostrongylus brasiliensis* has recently provided evidence for successive deletional class switching from μ to ε through γ1 (65). These results altogether provide formal evidence for the occurrence of switch recombination in B cells that have been stimulated with IL-4 and an appropriate second activating signal.

VI. MODULATION OF INTERLEUKIN-4-DEPENDENT IMMUNOGLOBULIN E SYNTHESIS BY CYTOKINES OTHER THAN INTERLEUKIN-4 AND CD23-MEDIATED SIGNALS

T cells are crucial for IgE induction, not only because of their ability to deliver the second B-cell activation signal, but also as a source of lymphokines (66).

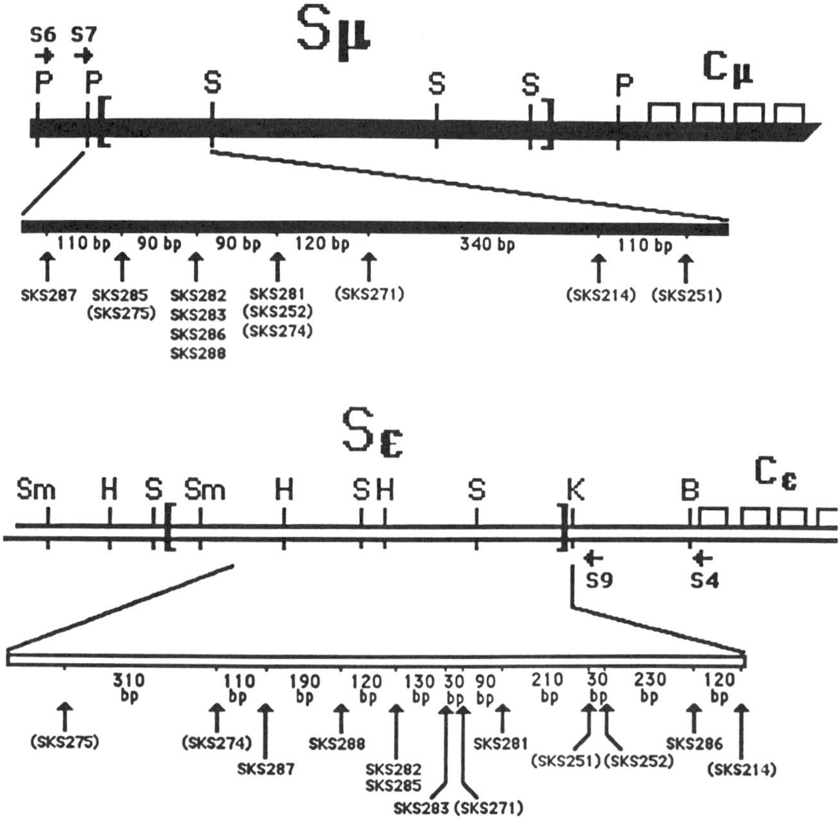

Figure 4 Switch recombination sites for Sμ/Sϵ "switch fragments." The positions of the PCR primers upstream from Sμ (S6, S7) and downstream from Sϵ (S4, S9) are shown. The first round of PCR used primers S6/S4, whereas the second or nested round of PCR used primers S7/S9. Each switch fragment comprises the DNA segment upstream from the recombination site in Sμ joined to the DNA segment downstream from the recombination site with the same designation in Sϵ. Sites of recombination for switch fragments isolated from IgE-secreting EBV-transformed B cells and from U266 are shown in parentheses. Approximate map distances between recombination sites are noted (rounded off to the nearest 10 bp): B, *Bam*HI; H, *Hin*dIII; K, *Kpn* 1; P, *Pst* 1; S, *Sac* 1; Sm, *Sma* 1.

The role of lymphokines—and more generally, of cytokines—in the regulation of IgE synthesis has been demonstrated by studies with T-cell clones obtained from patients with atopic disease and parasitic infections, as well as by the use of recombinant molecules.

The $CD4^{+}$T-cell clones established from atopic patients, and specific for

Dermatophagoides pteronyssinus and *Lolium perenne*, produced very high levels of IL-4 and IL-5, and very low levels of IFN-γ, compared with CD4$^+$T-cell clones from the same patients not specific for allergens, or CD4$^+$ clones from healthy persons. The IL-4-producing T-cell clones secreted low amounts of IFN-γ (67,68), and thus had the Th2 clone profile described in mice (69) (Table 1). Most importantly, the Th2-like allergen-specific clones derived from allergic subjects strongly supported the induction of IgE synthesis (70). Furthermore, the vast majority of helper T-cell clones isolated from patients with filariasis, toxocariasis, and ascaridiasis consistently produced high levels of IL-4, and low IL-2 and IFN-γ, indicating that in these conditions, as well as in allergic disorders, the cytokine profile of antigen-specific clones is skewed towards the Th2 phenotype (71,72,73). Because IL-4 production is known to require repeated antigen stimulation (74), these data seem to suggest that chronic stimulation by selected antigens (allergens) may select for allergen-specific T cells in persons whose T cells are intrinsically prone to secrete large amounts of IL-4 following activation.

In studies with recombinant lymphokines, IL-5 and IL-6 up-regulate the IgE response induced by IL-4 in unfractionated mononuclear cells (48,75). The IgE up-regulation by rIL-5 was more evident when suboptimal concentrations of rIL-4 were used, whereas rIL-6 could up-regulate IgE synthesis even in the presence of high concentrations of rIL-4. The effects of the combination of rIL-4, rIL-5, and rIL-6 were additive in some experiments, but were clearly synergistic in others. By contrast, IL-2, IL-1, and TNF-α had no amplifying effect. Most interestingly, endogenous IL-6 was found to be crucially involved in IL-4-driven IgE induction, both in T-cell-dependent and T-cell-independent systems, because an anti-IL-6 antibody strongly inhibited the production of IgE induced by IL-4 in PBMC (48) and in highly purified B cells by IL-4 and anti-CD40 MAb (51). Interestingly, both IL-4 and anti-CD40 MAb induce IL-6 production in B cells (76,77), whereas IL-6 increases the phosphorylation of CD40 in human B cells, thereby leading to a CD40–IL-6 loop (77). IL-6 acts at a late stage in B-cell differentiation, with no isotype preference (78). It induces the secretion of IgG1

Table 1 Cytokine Profiles of Th1 and Th2 Helper T-Cell Clones

Cytokine	Th1	Th2
IL-2	+	−
IFN-γ	+	−
IL-4	−	+
IL-5	−	+
IL-3	+	+
GM-CSF	+	+

by coordinated transcriptional activation, selective accumulation of mRNA for the secreted form of IgG, and possibly differential mRNA stabilization (79). IL-6 may amplify IgE secretion by similar mechanisms.

Interferon-γ has been reported to inhibit IL-4-dependent IgE synthesis in both mice (34,80,81) and humans (37,38). It also antagonizes the effects of IL-4 in other systems [e.g., it blocks the IL-4-dependent induction of FcεR2/CD23 on B cells (82), although not on monocytes (83)]. However, alloreactive T-cell clones secreting both IL-4 and IFN-γ in high amounts can induce IgE synthesis by normal B lymphocytes (66). Of particular interest is the finding that the expression of IFN-γ is down-regulated by IL-4, both at the protein and at the mRNA level (84,85). Induction of IgE synthesis by IL-4 may thus reflect not only the ability of this lymphokine to target the Sε region for switch recombination, but also the simultaneous suppression of a major functional IL-4 antagonist, IFN-γ.

The mechanisms by which IFN-γ inhibits IgE synthesis are not well defined. This cytokine profoundly suppressed the expression of ε germline transcripts in murine B cells stimulated with IL-4 and LPS (86). By contrast, no inhibition was observed when IFN-γ was added to highly purified human B cells stimulated with IL-4 (43) or PBMC (87), suggesting that activation of germline transcription and switch recombination may not be necessarily coupled. Unlike IFN-γ, TGF-β (88) and paf-acether (89) have been recently reported to suppress both ε germline transcription and IgE synthesis in human B cells stimulated with IL-4. Thus, TGF-β and paf-acether may inhibit human IgE production by acting on germline transcription, whereas IFN-γ seems to affect posttranscriptional events. These results suggest that the molecular events underlying IgE production may represent independent targets for specific regulatory stimuli.

Finally, the issue of the effect(s) on human IgE production of signals delivered by CD23, the low-affinity receptor for IgE (FcεR2), deserves a separate discussion. Several reports have described an inhibition, altogether modest, of IL-4-induced IgE synthesis by anti-CD23 MAb (37,90). Furthermore, engagement of FcεR2 by IgE immune complexes or anti-CD23 MAb partially inhibited ongoing spontaneous IgE production in AF10 cells, a mycoplasma-free subclone of the U266 IgE myeloma (91).

The AF10 system has been recently investigated at the molecular level, particularly focusing on the effect of CD23-mediated stimulation on the expression of mRNA for membrane-bound and secreted ε protein. Suppression of IgE production by FcεR2 was accompanied by a decrease in mRNA for secreted ε protein (corresponding to 2.1- and 3.0-kb bands on Northern blots), whereas levels of both membrane IgE protein and mRNA coding for membrane ε (corresponding to a 3.8-kb band on Northern blots) were unaltered (92). Because λ light chains were decreased at the RNA and protein level, as well, FcεR2 stimulation does not seem to selectively affect ε proteins. The role of posttranscriptional control mechanisms leading to increased turnover of secreted mRNA forms is presently under investigation.

VII. CONCLUSIONS

The data discussed in the foregoing clearly indicate that the second signal(s) required for IgE production can be delivered to B cells through different pathways. These second signals share the ability to induce DNA switch recombination and expression of mature Cε mRNA. Currently, no single explanatory mechanism can be proposed to unify all different observations. Little is known about signaling by MHC class II molecules during T–B-cell interactions; the role played by hydrocortisone is unclear, and the molecules involved in noncognate T–B-cell interactions are as yet undefined. The nature of the EBV-derived signal is presently unknown. Activation by EBV does not simply result from engagement of the EBV receptor–CD21–CR2. Monoclonal antibodies to different CD21 epitopes could not synergize with IL-4 in IgE induction. On the other hand, the EBV-induced proteins that are essential for IgE induction have not been identified. The observation that engagement of CD40 (93) and full expression of latent EBV proteins (94) share the ability to protect human B cells from programmed cell death (apoptosis) is certainly intriguing, but it is unclear whether this effect is in any way related to IgE induction. However, it seems plausible to speculate that the pathways mentioned earlier may share the ability to activate the switch recombination machinery in B cells in which the ε locus has been transcriptionally activated or rendered accessible by IL-4. In addition, the synthesis of membrane versus secretory Ig might also be a target for regulation (95). A strategy based on transfection of "switch plasmids" containing Sμ and Sε into IL-4-responsive B-cell lines should allow us to distinguish between these possibilities.

REFERENCES

1. D. Vercelli, and R. S. Geha, Regulation of IgE synthesis in man, *J. Clin. Immunol.*, *9*: 75 (1989).
2. D. Vercelli, and R. S. Geha, Regulation of IgE synthesis in humans: A tale of two signals, *J. Allergy Clin. Immunol.*, *88*: 285 (1991).
3. D. Vercelli, and R. S. Geha, Control of IgE synthesis, *Allergy: Principles and Practice* (Middleton et al., eds.), Mosby-Year Book, St. Louis (1993).
4. C. A. Hasemann, J. D. Capra, Immunoglobulins: Structure and function. *Fundamental Immunology* (W. E. Paul, ed.), Raven Press, New York (1989).
5. E. E. Max, Immunoglobulins: Molecular genetics. *Fundamental Immunology* (W. E. Paul, ed.), Raven Press, New York (1989).
6. D. G. Schatz, M. A. Oettinger, and D. Baltimore, The V(D)J recombination activating gene, *RAG-1*, *Cell*, *59*: 1035 (1989).

7. M. A. Oettinger, D. G. Schatz, C. Gorka, and D. Baltimore, *RAG-1* and *RAG-2* adjacent genes that synergistically activate V(D)J recombination, *Science*, *248*: 1517 (1990).
8. T. Honjo, and T. Kataoka, Organization of immunoglobulin heavy chain genes and allelic deletion model, *Proc. Natl. Acad. Sci. USA*, *75*: 2140 (1978).
9. U. von Schwedler, H.-M. Jäck, and M. Wabl, Circular DNA is a product of the immunoglobulin class switch rearrangement, *Nature*, *345*: 452 (1990).
10. M. Matsuoka, K. Yoshida, T. Maeda, S. Usuda, and H. Sakano, Switch circular DNA formed in cytokine-treated mouse splenocytes: Evidence for intramolecular DNA deletion in immunoglobulin class switching, *Cell*, *62*: 135 (1990).
11. T. Iwasato, A. Shimizu, T. Honjo, and H. Yamagishi, Circular DNA is excised by immunoglobulin class switch recombination, *Cell*, *62*: 143 (1990).
12. T. K. Blackwell, and F. W. Alt, Immunoglobulin genes, *Molecular Immunology* (Hames and Glover, eds.), IRL Press, Oxford (1988).
13. C. Esser, and A Radbruch, Immunoglobulin class switching: Molecular and cellular analysis, *Annu. Rev. Immunol.*, *8*: 717 (1990).
14. R. M. Gerstein, W. N. Frankel, C.-L. Hsieh, J. M. Durdik, S. Rath, J. M. Coffin, A. Nisonoff, and E. Selsing, Isotype switching of an immunoglobulin heavy chain transgene occurs by DNA recombination between different chromosomes, *Cell*, *63*: 537 (1990).
15. A. Radbruch, W. Müller, and K. Rajewsky, Class switch recombination is IgG1 specific on active and inactive IgH loci of IgG1-secreting B cell blasts, *Proc. Natl. Acad. Sci. USA*, *83*: 3954 (1986).
16. M. Hummel, J. K. Berry, and W. Dunnick, Switch region content of hybridomas: The two spleen cell *Igh* loci tend to rearrange to the same isotype. *J. Immunol.*, *138*: 3539 (1987).
17. M. R. Kepron, Y.-W. Chen, J. W. Uhr, and E. S. Vitetta, IL-4 induces the specific rearrangement of γ1 genes on the expressed and unexpressed chromosomes of lipopolysaccharide-activated normal murine B cells, *J. Immunol.*, *143*: 334 (1989).
18. G. V. Borzillo, M. D. Cooper, H. Kubagawa, A. Landay, and P. D. Burrows, Isotype switching in human B lymphocyte malignancies occurs by DNA deletion: Evidence for nonspecific switch recombination, *J. Immunol.*, *139*: 1326 (1987).
19. G. D. Yancopoulos, R. A. DePinho, K. A. Zimmerman, S. G. Lutzker, N. Rosenberg, and F. W. Alt, Secondary genomic rearrangement events in pre-B cells: V_HDJ_H replacement by a LINE-1 sequence and directed class switching, *EMBO J.*, *5*: 3259 (1986).
20. J. Stavnezer-Nordgren, J. and S. Sirlin, Specificity of immunoglobulin heavy chain switch correlates with activity of germline heavy chain genes prior to switching, *EMBO J.*, *5*: 95 (1986).
21. S. Lutzker, P. Rothman, R. Pollock, R. Coffman, and F. W. Alt, Mitogen- and IL-4-regulated expression of germ-like Ig γ2b transcripts: Evidence for directed heavy chain class switching, *Cell*, *53*: 177 (1988).
22. S. Lutzker, and F. W. Alt, Structure and expression of germ line immunoglobulin γ2b transcripts, *Mol. Cell. Biol.*, *8*: 1849 (1988).
23. J. Stavnezer, G. Radcliffe, Y.-C. Lin, J. Nietupski, L. Berggren, R. Sitia, and E. Severinson, Immunoglobulin heavy-chain switching may be directed by prior

induction of transcripts from constant-region genes, *Proc. Natl. Acad. Sci. USA*, *85*: 7704 (1988).

24. T. K. Blackwell, M. W. Moore, G. D. Yancopoulos, H. Suh, S. Lutzker, E. Selsing, and F. W. Alt, Recombination between immunoglobulin variable region gene segments is enhanced by transcription, *Nature*, *324*: 585 (1986).
25. M. T. Berton, J. W. Uhr, and E. S. Vitetta, Synthesis of germ-line γ1 immunoglobulin heavy-chain transcripts in resting B cells: Induction by interleukin 4 and inhibition by interferon γ, *Proc. Natl. Acad. Sci. USA*, *86*: 2829 (1989).
26. C. Burger, and A. Radbruch. Protective methylation of immunoglobulin and T cell receptor (TcR) gene loci prior to induction of class switch and TcR recombination, *Eur. J. Immunol.*, *20*: 2285 (1990).
27. P. Rothman, S. Lutzker, W. Cook, R. Coffman, and F. W. Alt, Mitogen plus interleukin 4 induction of Cε transcripts in B lymphoid cells, *J. Exp. Med.*, *168*: 2385 (1988).
28. D. A. Collier, J. A. Griffith, and R. D. Wells, Non-B right-handed DNA conformations of homopurine homopyrimidine sequences in the murine immunoglobulin Cα switch region, *J. Biol. Chem.*, *263*: 7397 (1988).
29. M. E. Reaban, and J. A. Griffin. Induction of RNA-stabilized DNA conformers by transcription of an immunoglobulin switch region, *Nature*, *348*: 342 (1990).
30. G. Radcliffe, Y.-C. Lin, M. Julius, K. B. Marcu, and J. Stavnezer, Structure of germline immunoglobulin α heavy chain RNA and its location on polysomes, *J. Mol. Cell. Biol.*, *10*: 382 (1990).
31. J. Durdik, R. M. Gerstein, S. Rath, P. F. Robbins, A. Nisonoff, and E. Selsing, Isotype switching by a microinjected μ immunoglobulin heavy chain gene in transgenic mice, *Proc. Natl. Acad. Sci. USA*, *86*: 2346 (1989).
32. A. Shimizu, M. Nussenzweig, T.-R. Mizuta, P. Leder, and T. Honjo, Immunoglobulin double-isotype expression by trans-mRNA in a human immunoglobulin transgenic mouse, *Proc. Natl. Acad. Sci. USA*, *86*: 8020 (1989).
33. A. Shimizu, M. C. Nussenzweig, H. Han, M. Sanchez, and T. Honjo, Transsplicing as a possible molecular mechanism for the multiple isotype expression of the immunoglobulin gene, *J. Exp. Med.*, *173*: 1385 (1991).
34. R. L. Coffman, and J. Carty, A T cell activity that enhances polyclonal IgE production and its inhibition by interferon-γ, *J. Immunol.*, *136*: 949 (1986).
35. R. L. Coffman, J. Ohara, M. W. Bond, J. Carty, A. Zlotnik, and W. E. Paul, B cell stimulatory factor-1 enhances the IgE response of lipopolysaccharide-activated B cells, *J. Immunol.*, *136*: 4538 (1986).
36. H. H. Jabara, S. J. Ackerman, D. Vercelli, T. Yokota, K. Arai, J. Abrams, A. M. Dvorak, M. C. Lavigne, J. Banchereau, J. deVries, D. Y. M. Leung, and R. S. Geha, Induction of interleukin-4 dependent IgE synthesis and interleukin-5 dependent eosinophil differentiation by supernatants of a human helper T cell clone, *J. Clin. Immunol.*, *8*: 437 (1988).
37. J. Pene, F. Rousset, F. Briere, I. Chretien, J. Y. Bonnefoy, H. Spits, T. Yokota, N. Arai, K. Arai, J. Banchereau, and J. deVries, IgE production by normal human lymphocytes is induced by interleukin 4 and suppressed by interferons γ and α and prostaglandin E_2, *Proc. Natl. Acad. Sci. USA*, *85*: 6880 (1988).
38. G. F. Del Prete, E. Maggi, P. Parronchi, I. Chretien, A. Tiri, D. Macchia, M.

Ricci, J. Banchereau, J. deVries, and S. Romagnani, IL-4 is an essential factor for the IgE synthesis induced in vitro by human T cell clones and their supernatants, *J. Immunol.*, *140*: 4193 (1988).

39. S. Gerondakis, Structure and expression of murine germ-line immunoglobulin ϵ heavy chain transcripts induced by interleukin 4, *Proc. Natl. Acad. Sci. USA*, *87*: 1581 (1990).
40. P. Rothman, Y.-Y. Chen, S. Lutzker, S. C. Li, V. Stewart, R. Coffman, and F. W. Alt, Structure and expression of germ line immunoglobulin heavy-chain ϵ transcripts: Interleukin-4 plus lipopolysaccharide-directed switching to Cϵ, *Mol. Cell Biol.*, *10*: 1672 (1990).
41. P. Rothman, S. C. Li, B. Gorham, L. Glimcher, F. Alt, and M. Boothby, Identification of a conserved lipopolysaccharide-plus-interleukin-4 responsive element located at the promoter of germ line ϵ transcripts, *Mol. Cell. Biol.*, *11*: 5551 (1991).
42. M. Boothby, E. Gravallese, H.-C. Liou, and L. H. Glimcher, A DNA binding protein regulated by IL-4 and by differentiation in B cells, *Science*, *242*: 1559 (1988).
43. J.-F. Gauchat, D. A. Lebman, R. L. Coffman, H. Gascan, and J. E. de Vries, Structure and expression of germline ϵ transcripts in human B cells induced by interleukin 4 to switch to IgE production, *J. Exp. Med.*, *172*: 463 (1990).
44. E. Y. Loh, J. F. Elliott, S. Cwirla, L. L. Lanier, and M. M. Davis, Polymerase chain reaction with single-sided specificity: Analysis of T cell receptor δ chain, *Science*, *243*: 217 (1989).
45. Y. Nishida, T. Miki, H. Hisajima, and T. Honjo, Cloning of human immunoglobulin ϵ chain genes: Evidence for multiple Cϵ genes, *Proc. Natl. Acad. Sci. USA*, *79*: 3833 (1982).
46. S. K. Shapira, H. H. Jabara, C. P. Thienes, D. J. Ahern, D. Vercelli, H. J. Gould, and R. S. Geha, Deletional switch recombination occurs in IL-4 induced isotype switching to IgE expression by human B cells, *Proc. Natl. Acad. Sci. USA*, *88*: 7528 (1991).
47. S. K. Shapira, D. Vercelli, H. H. Jabara, S. M. Fu, and R. S. Geha, Molecular analysis of the induction of IgE synthesis in human B cells by IL-4 and engagement of CD40 antigen, *J. Exp. Med.*, *175*: 289 (1992).
48. D. Vercelli, H. H. Jabara, K. Arai, T. Yokota, and R. S. Geha, Endogenous IL-6 plays an obligatory role in IL-4 induced human IgE synthesis, *Eur. J. Immunol.*, *19*: 1419 (1989).
49. H. H. Jabara, L. C. Schneider, S. K. Shapira, C. Alfieri, C. T. Moody, E. Kieff, R. S. Geha, and D. Vercelli, Induction of germ-line and mature Cϵ transcripts in human B cells stimulated with rIL-4 and EBV, *J. Immunol.*, *145*: 3468 (1990).
50. G. Thyphronitis, G. C. Tsokos, C. H. June, A. D. Levine, and F. D. Finkelman, IgE secretion by Epstein–Barr virus-infected purified human B lymphocytes is stimulated by interleukin 4 and suppressed by interferon γ, *Proc. Natl. Acad. Sci. USA*, *86*: 5580 (1989).
51. H. H. Jabara, S. M. Fu, R. S. Geha, and D. Vercelli, CD40 and IgE: Synergism between anti-CD40 MAb and IL-4 in the induction of IgE synthesis by highly purified human B cells, *J. Exp. Med.*, *172*: 1861 (1990).

52. F. Rousset, E. Garcia, and J. Banchereau, Cytokine-induced proliferation and immunoglobulin production of human B lymphocytes triggered through their CD40 antigen, *J. Exp. Med.*, *173*: 705 (1991).
53. K. Zhang, E. A. Clark, and A. Saxon, CD40 stimulation provides an IFN-γ-independent and IL-4-dependent differentiation signal directly to human B cells for IgE production, *J. Immunol.*, *146*: 1836 (1991).
54. H. Gascan, J.-F. Gauchat, G. Aversa, P. van Vlasselaer, and J. E. de Vries, Anti-CD40 monoclonal antibodies or $CD4^+$ T cell clones and IL-4 induce IgG4 and IgE switching in purified human B cells via different signaling pathways, *J. Immunol.*, *147*: 8 (1991).
55. M. Seno, T. Kurokawa, Y. Ono, H. Onda, R. Sasada, K. Igarashi, M. Kikuchi, Y. Sugino, Y. Nishida, and T. Honjo, Molecular cloning of human immunoglobulin epsilon chain cDNA, *Nucleic Acids Res.*, *11*: 719 (1983).
56. L. Hellman, S. Josephson, H. Jernberg, K. Nilsson, and U. Pettersson, Immunoglobulin synthesis in the human myeloma cell line U-266; expression of two immunoglobulin heavy chain isotypes (ϵ and α) after long-term cultivation *in vitro*, *Eur. J. Immunol.*, *18*: 905 (1988).
57. T. MacKenzie, and H. M. Dosch, Clonal and molecular characteristics of the human IgE-committed B cell subset, *J. Exp. Med.*, *169*: 407 (1989).
58. M. A. Chan, S. H. Benedict, H.-M. Dosch, M. E. Huy, and L. D. Stein, Expression of IgE from a nonrearranged ϵ locus in cloned B-lymphoblastoid cells that also express IgM, *J. Immunol.*, *144*: 3563 (1990).
59. Y. Yaoita, Y. Kumagai, K. Okumura, and T. Honjo, Expression of lymphocyte surface IgE does not require switch recombination, *Nature, 297*: 697 (1982).
60. A. P. Perlmutter, and W. Gilbert, Antibodies of the secondary response can be expressed without switch recombination in normal B cells, *Proc. Natl. Acad. Sci. USA*, *81*: 7189 (1984).
61. Y. Chen, C. Word, V. Dev, J. W. Uhr, E. S. Vitetta, and P. W. Tucker, Double isotype production by a neoplastic B cell line. II. Allelically excluded production of μ and γ1 heavy chains without C_H gene rearrangement, *J. Exp. Med.*, *164*: 562 (1986).
62. H. H. Jabara, R. Loh, N. Ramesh, D. Vercelli, and R. S. Geha, Sequential switching from μ to ϵ via γ4 in human B cells stimulated with IL-4 and hydrocortisone, *J. Immunol.*, *151*: 1 (1993).
63. C. M. Snapper, F. D. Finkelman, D. Stefany, D. H. Conrad, W. E. Paul, IL-4 induces co-expression of intrinsic membrane IgG1 and IgE by murine B cells stimulated with lipopolysaccharide, *J. Immunol.*, *141*: 489 (1988).
64. G. Thyphronitis, E. E. Max, and F. D. Finkelman, Generation and cloning of stable human IgE-secreting cells that have rearranged the Cϵ gene, *J. Immunol.*, *146*: 1496 (1991).
65. K. Yoshida, M. Matsuoka, S. Usuda, A. Mori, K. Ishizaka, and H. Sakano, Immunoglobulin switch circular DNA in the mouse infected with *Nippostrongylus brasiliensis*: Evidence for successive class switching from μ to ϵ via γ1, *Proc. Natl. Acad. Sci. USA*, *87*: 7829 (1990).
66. D. Vercelli, H. H. Jabara, K. Arai, and R. S. Geha, Induction of human IgE synthesis requires interleukin 4 and T/B cell interactions involving the T cell receptor/CD3 complex and MHC class II antigens, *J. Exp. Med.*, *169*: 1295 (1989).

67. E. A. Wierenga, M. Snoek, J. D. Bos, H. M. Jansen, and M. L. Kapsenberg, Comparison of diversity and function of house dust mite-specific T lymphocyte clones from atopic and non-atopic donors, *Eur. J. Immunol.*, *20*: 1519 (1990).
68. E. A. Wierenga, M. Snoek, C. de Groot, I. Chrétien, J. D. Bos, H. M. Jansen, and M. L. Kapsenberg, Evidence for compartmentalization of functional subsets of $CD4^+$ T lymphocytes in atopic patients, *J. Immunol.*, *144*: 4651 (1990).
69. T. R. Mosmann, H. Cherwinski, M. W. Bond, M. Giedlin, and R. L. Coffman, Two types of murine helper T cell clone. I. Definition according to profiles of lymphokine activities and secreted proteins, *J. Immunol.*, *136*: 2348 (1986).
70. P. Parronchi, D. Macchia, M.-P. Piccinni, P. Biswas, C. Simonelli, E. Maggi, M. Ricci, A. A. Ansari, and S. Romagnani, Allergen- and bacterial antigen-specific T-cell clones established from atopic donors show a different profile of cytokine production, *Proc. Natl. Acad. Sci. USA*, *88*: 4538 (1991).
71. S. Romagnani, Regulation and deregulation of human IgE synthesis, *Immunol. Today*, *11*: 316 (1990).
72. G. F. Del Prete, M. De Carli, C. Mastromauro, R. Biagiotti, D. Macchia, P. Falagiani, M. Ricci, and S. Romagnani, Purified protein derivative of *Mycobacterium tubercolosis* and excretory-secretory antigen(s) of *Toxocara canis* expand in vitro human T cells with stable and opposite (type 1 T helper or type 2 T helper) profile of cytokine production, *J. Clin. Invest.*, *88*: 346 (1991).
73. F. D. Finkelman, E. J. Pearce, J. F. Urban, and A. Sher, Regulation and biological function of helminth-induced cytokine responses, *Immunol. Today*, *12*: 62 (1991).
74. G. D. Powers, A. K. Abbas, and R. A. Miller, Frequencies of IL-2- and Il-4-secreting T cells in naive and antigen-stimulated lymphocyte populations, *J. Immunol.*, *140*: 3352 (1988).
75. J. Pene, F. Rousset, F. Briere, I. Chretien, J. Wideman, J. Y. Bonnefoy, and J. E. deVries, Interleukin 5 enhances interleukin-4-induced IgE production by normal human B cells. The role of soluble CD23 antigen, *Eur. J. Immunol.*, *18*: 929 (1988).
76. E. B. Smeland, H. K. Blomhoff, S. Funderud, R. R. Shalaby, and T. Espevik, Interleukin 4 induces selective production of interleukin 6 from normal human B lymphocytes, *J. Exp. Med.*, *170*: 1463 (1989).
77. E. A. Clark, and G. Shu, Association between IL-6 and CD40 signaling. IL-6 induces phosphorylation of CD40 receptors, *J. Immunol.*, *145*: 1400 (1990).
78. A. Muraguchi, T. Hirano, B. Tang, T. Matsuda, Y. Horii, K. Nakajima, and T. Kishimoto, The essential role of B cell stimulatory factor 2 (BSF-2/IL-6) for the terminal differentiation of B cells, *J. Exp. Med.*, *167*: 332 (1988).
79. M.-C. Raynal, Z. Liu, T. Hirano, L. Mayer, T. Kishimoto, and S. Chen-kiang, Interleukin 6 induces secretion of IgG1 by coordinated transcriptional activation and differential mRNA accumulation, *Proc. Natl. Acad. Sci. USA*, *86*: 8024 (1989).
80. E. M. Rabin, J. J. Mond, J. Ohara, and W. E. Paul, Interferon-γ inhibits the action of B cell stimulatory factor (BSF)-1 on resting B cells, *J. Immunol.*, *137*: 1573 (1986).
81. C. M. Snapper, F. D. Finkelman, and W. E. Paul, Differential regulation of IgG1 and IgE synthesis by interleukin 4, *J. Exp. Med.*, *167*: 183 (1988).
82. T. Defrance, J. Aubry, F. Rousset, B. Vanbervliet, J. Bonnefoy, N. Arai, Y. Takebe, T. Yokota, F. Lee, K. Arai, J. deVries, and J. Banchereau, Human

recombinant interleukin 4 induces Fcϵ receptors (CD23) on normal B lymphocytes, *J. Exp. Med.*, *165*: 1459 (1987).

83. D. Vercelli, H. H. Jabara, B. Lee, N. Woodland, R. S. Geha, and D. Y. M. Leung, Human recombinant interleukin 4 induces FcϵR2/CD23 on normal human monocytes, *J. Exp. Med.*, *167*: 1406 (1988).
84. R. Peleman, J. Wu, C. Fargeas, and G. Delespesse, Recombinant interleukin 4 suppresses the production of interferon γ by human mononuclear cells, *J. Exp. Med.*, *170*: 1751 (1989).
85. D. Vercelli, H. H. Jabara, R. P. Lauener, and R. S. Geha, Interleukin-4 inhibits the synthesis of interferon-γ and induces the synthesis of IgE in mixed lymphocyte cultures, *J. Immunol.*, *144*: 570 (1990).
86. E. Severinson, C. Fernandez, and J. Stavnezer. Induction of germ-line immunoglobulin heavy chain transcripts by mitogens and interleukins prior to switch recombination, *Eur. J. Immunol.*, *20*: 1079 (1990).
87. G. Qiu, J.-F. Gauchat, M. Vogel, M. Mandallaz, A. L. De Weck, and B. M. Stadler, Human IgE mRNA expression by peripheral blood lymphocytes stimulated with interleukin 4 and pokeweed mitogen, *Eur. J. Immunol.*, *20*: 2191 (1990).
88. J.-F. Gauchat, H. Gascan, R. de Waal Malefyt, and J. E. de Vries, Regulation of germ-line ϵ transcription and induction of ϵ switching in cloned EBV-transformed and malignant human B cell lines by cytokines and $CD4^+$ T cells, *J. Immunol.*, *148*: 2291 (1992).
89. S. Deryckx, R. de Waal Malefyt, J.-F. Gauchat, E. Vivier, Y. Thomas, and J. E. de Vries, Immunoregulatory functions of paf-acether. VIII. Inhibition of IL-4-induced human IgE synthesis in vitro, *J. Immunol.*, *148*: 1465 (1992).
90. M. Sarfati, and G. Delespesse, Possible role of human lymphocyte receptor for IgE (CD23) or its soluble fragments in the in vitro synthesis of human IgE, *J. Immunol.*, *141*: 2195 (1988).
91. E. Sherr, E. Macy, H. Kimata, and A. Saxon, Binding the low affinity Fc epsilon receptor on B calls suppresses ongoing human IgE production, *J. Immunol.*, *142*: 481 (1989).
92. A. Saxon, M. Kurbe-Leamer, K. Behle, E. E. Max, and K. Zhang, Inhibition of human IgE production via FcϵR-II stimulation results from a decrease in the mRNA for secreted but not membrane ϵ H chains, *J. Immunol.*, *147*: 4000 (1991).
93. Y.-J. Liu, D. E. Joshua, G. T. Williams, C. A. Smith, J. Gordon, and I. C. M. MacLennan, Mechanism of antigen-driven selection in germinal centres, *Nature*, *342*: 929 (1989).
94. C. D. Gregory, C. Dive, S. Henderson, C. A. Smith, G. T. Williams, J. Gordon, and A. B. Rickinson, Activation of Epstein–Barr virus latent genes protects human B cells from death by apoptosis, *Nature*, *349*: 612 (1991).
95. J. M. Purkerson, and P. C. Isakson, Isotype switching in anti-immunoglobulin-activated B lymphoblasts: Differential requirements for interleukin 4 and other lymphokines to elicit membrane vs. secreted IgG1, *Eur. J. Immunol.*, *21*: 707 (1991).

11

The Molecular and Cell Biology of Mast Cells and Basophils

Lawrence B. Schwartz
Virginia Commonwealth University, Richmond, Virginia

I. INTRODUCTION TO MAST CELL BIOLOGY

Mast cells and basophils are the major effector cells in allergic disease. In this chapter, mast cells and basophils are reviewed to cover cellular and molecular aspects of their biology, heterogeneity, and differentiation. The chapter will focus on human mast cells and on human basophils, because of their obvious relevance to human disease, but selected information from animal models potentially relevant to human mast cells will also be discussed. Traditionally, mast cells and basophils have been understood primarily in their role in pathogenesis of immediate hypersensitivity, but new evidence suggests that mast cells are also important cytokine-producing cells during nonallergic processes. In addition, mast cells have become an important model system for differentiation of hematopoietic cells and for the transcriptional regulation of cell-specific markers.

The first definitive description of mast cells is attributed to Paul Ehrlich (1), who observed the metachromatic staining of mast cell granules in 1877. Ehrlich applied aniline dyes to tissues and noticed that certain cells had prominent granules that stained not only in the natural color of the dye (blue), but also metachromatically (reddish-purple), and that the stain of the granule was insoluble in alcohol. Ehrlich also described the basophil separately as a metachromatic cell that circulated in the blood (2). Even today, metachromasia remains an important criterion for a mature cell to be called a mast cell or basophil. Historically, because mast cells and basophils are both metachromatic, both are major repositories of histamine (3), and both bind immunoglobulin E (IgE) with high affinity (4,5), older textbooks have reflected a long-standing, albeit incorrect, assumption that the two cells are intimately related in lineage. Although having these features in common, mast cells and basophils have other features that identify them as two distinctly different cells. These distinguishing features include nuclear morphology, location in vivo, mediator content, cell surface markers, and response to growth and differentiation cytokines and to chemical activating agents.

Mature human mast cells are usually 9–12 μm in diameter, and their appearance in tissues can be round, spindle-shaped, or spiderlike. They have numerous cytoplasmic granules and thin, elongated folds of their plasma membrane. Mast cell nuclei are rather large unsegmented ovals, elliptically placed, and rarely show mitotic figures in tissue. Human basophils are polymorphonuclear, contain larger and fewer granules, and exhibit a relatively smooth plasma membrane contour. At the ultrastructural level, there appear to be differences in granule morphologic configurations among different species: human mast cell granules have intricate and varied patterns, whereas rodent mast cell granules have an amorphous appearance. In the human, some differences in granule architecture correspond to different protease compositions in the two human mast cell subpopulations called MC_T and MC_{TC} (see later discussion). Only granules with chymase exhibit a grating or lattice substructure, whereas only regions lacking this protease exhibit complete scroll patterns. Neither pattern appears in basophils, which exhibit granules more heterogeneous in size and having amorphous and particulate features.

Mast cells are positioned in connective tissue or mucosa at potential portals of entry of noxious substances. Although basophils normally reside in the circulation, they can enter tissues at sites of inflammation, particularly during the late phase of allergic reactions and during the early phase of delayed-type hypersensitivity reactions. Mast cells are frequently found at the interface of the internal and external environment, such as skin, conjunctiva, gut, and respiratory mucosal surfaces, where allergic symptoms are best appreciated. In most species, including humans, bone and cartilage have few, if any, mast cells; parenchymatous organs such as liver, kidney, and adrenals generally have modest numbers of mast cells. In the human, mast cell concentrations are particularly high in the alveolar wall, bowel mucosa, dermis, nasal and conjunctival mucosa, and around blood vessels, consistent with the clinical alterations known to occur in allergic diseases at these sites.

Most considerations of mast cell and basophil function have centered on the pathological role of inflammatory mediators in extruded mast cell granules, whereas the normal physiological functions of mast cells and basophils are less well understood. Mast cells release their granules as the result of a variety of natural stimuli that perturb the plasma membrane, including the reaction of cell-bound IgE with appropriate antigen (4,5), binding of the anaphylatoxins C3a and C5a (6), and exposure to natural degranulating agents, such as mellitin in bee venom (7) or cell-derived histamine-releasing factors (8). The relative role of mast cells and basophils in various clinical conditions, as well as the duration, intensity, and tissue distribution of a particular response, depends on various characteristics of the agonist, immunological sensitivity of the host, the target tissue involved, and any underlying abnormalities.

Another level of complexity concerns the type of mast cell involved. Two types of mast cells have been observed in human and rodent tissues, based on mediator composition and apparent dependency on T-lymphocyte-derived cytokines. Mucosal (atypical) mast cells of rodents and MC_T cells of humans appear to be somewhat analogous, as are connective tissue (typical) mast cells of rodents and MC_{TC} cells of humans. Although differences in fixation and histochemical staining have been applied in rodents to distinguish different types of mast cells, these techniques lack precision in humans. The distinctive tissue distributions of each mast cell type suggest a purposeful presence for each and a higher level of complexity in pathological mechanisms and pharmacological control of mast cell-mediated reactions than previously appreciated.

II. MEDIATORS OF MAST CELLS AND BASOPHILS

A. Preformed Granule-Associated Mediators

1. *Biogenic Amines*

Histamine. Histamine (β-imidazolylethylamine) is the sole biogenic amine in human mast cells and basophils from normal subjects (Table 1). Histamine formation from histidine in mast cells and basophils is catalyzed by histidine decarboxylase (9) (Fig. 1). The cDNA for histidine decarboxylase has been cloned (10). Histamine is stored in secretory granules where the pH is acidic when the cell is unstimulated (11,12). With degranulation, histamine is externalized and dissociates from the proteoglycan–protein complex by cation exchange with extracellular sodium at neutral pH. Histamine is the only preformed mediator of human mast cells with direct potent vasoactive and smooth-muscle spasmogenic effects. Extracellular histamine is metabolized within minutes of release. The rapid metabolism of histamine suggests that it is destined to act locally near the site of its release.

Histamine levels of 1–3 pg per human mast cell (both MC_T and MC_{TC} types) and basophils have been reported and are similar to the levels in mouse and rat mast cells, with the exception of the rat peritoneal mast cell in which histamine levels of 10–30 pg/cell are found. Histamine concentrations of about 0.1 M are

Table 1 Preformed Mediators of Mast Cells and Basophils

	Biogenic amine	Neutral protease	Proteoglycan
Human			
MC_T (mucosal)	Histamine	Tryptase	Heparin, chondroitin sulfates A,E
MC_{TC} (connective tissue)	Histamine	Tryptase, chymase, carboxypeptidase, cathepsin G-like protease	Heparin, chondroitin sulfates A,E
Basophils	Histamine		Chondroitin sulfate A
Rat			
Mucosal (RBL-1)	Histamine	Chymase II	Chondroitin sulfate di-B
Connective tissue	Histamine serotonin	Chymase I, tryptase carboxypeptidase A	Heparin
Mouse			
Mucosal	Histamine	Chymases MMCP[a]:1,2,4	Chondroitin sulfate A,E
Connective tissue	Histamine, serotonin	Chymases MMCP:3,4,5, tryptase MMCP-6, carboxypeptidase A	Heparin
Bone marrow-derived IL-3-dependent	Histamine, serotonin	Chymase MMCP-5 mRNA, tryptase MMCP-6 mRNA, carboxypeptidase A mRNA	Chondroitin sulfate A,E
Dog	Histamine	Chymase, tryptase	Heparin, chondroitin di/mono-sulfates

[a]MMCP, murine mast cell protease.

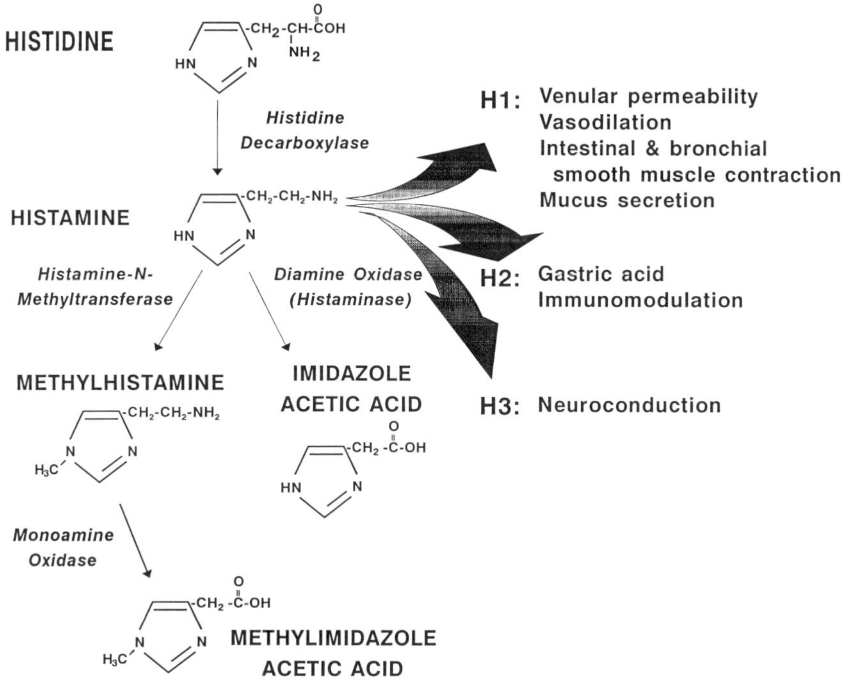

Figure 1 Histamine metabolism and actions.

estimated to exist inside secretory granules; whereas concentrations of about 2 nM exist in plasma.

The effects of histamine are mediated by its interactions with cell-surface H_1, H_2, and H_3 receptors. The H_1-mediated effects include enhanced vasopermeability between venular endothelial cells, vasodilation, contraction of bronchial and gastrointestinal smooth muscle, and increased mucous secretion at various sites. Subcutaneous injection produces the triple response: central erythema, followed by a pale central wheal and a surrounding erythematous flare. The combined effects of H_1 receptor- and H_2 receptor-mediated activities of histamine are required for the full expression of vasoactivity. The most notable H_2 receptor-mediated activity is stimulation of gastric acid secretion by parietal cells, but several immunomodulatory effects also have been described in vitro. These include inhibition of granule mediator secretion by cytotoxic lymphocytes, neutrophils, and basophils; suppression of tumor necrosis factor (TNF)-α expression (13); augmentation of suppression by T lymphocytes (14); and enhancement of epithelial permeability across human airways (15). Histamine activates endothelial cells through H_2 receptors to release a potent inhibitor of platelet aggregation, prostacyclin (PGI_2). The H_3 receptor-mediated effects of histamine that have been most clearly elucidated affect neurotransmitter release in the central and peripheral nervous systems, as well as histamine formation in these regions (16). Bronchial hyperreactivity in atopics with asthma to irritant stimuli may be mediated, in part, by histamine-mediated cholinergic hyperexcitability.

The sentinel location of mast cells near vessels and tissue surfaces is particularly well-suited for the enhancement of local permeability by released histamine, together with certain newly generated lipid mediators with vasoactive properties. This, in turn, will facilitate the tissue deposition of factors from plasma that may be important for tissue growth and repair. However, enhanced deposition of foreign material or immune complexes will result in tissue inflammation. Newly arriving material from the circulation also will be positioned to interact with the macromolecular complexes of neutral proteases and proteoglycans that are released in parallel with histamine from human mast cells.

2. *Neutral Proteases*

Neutral proteases are enzymes that catalyze the cleavage of peptide bonds and perform best near neutral pH. Such enzymes are the dominant protein components of secretory granules in human and rodent mast cells (17). In addition, such enzymes serve as selective markers that distinguish mast cells from other cell types, including basophils, and different mast cell subpopulations from one another. Whereas human and rodent mast cells are rich in serine proteases of the trypsin and chymotrypsin families, human basophils have relatively little esterase or protease activity. Because of the selective expression of these proteases in mast cells, one could hypothesize that these enzymes play an important role in the biology of mast cells, that such enzymes might serve as sensitive and specific markers of mast cells and of mast cell activation, and that regulation of the transcription of their corresponding genes is related to the regulation of the differentiation of this cell type. This section will concentrate on proteases of human mast cells.

Neutral Proteases of Human Mast Cells. *Tryptase* is the principal enzyme accounting for the trypsin-like activity first detected in human mast cells by histochemical techniques (18,19). Substantial amounts of tryptase are present in MC_{TC} cells derived from foreskin (35 pg/cell) and in MC_{T} cells derived from lung (10 pg/cell), where it is located in secretory granules and is released in parallel with histamine during degranulation (20). These levels appear to account for 20–50% of the entire protein in a mast cell. Negligible amounts have been measured in human basophils (0.04 pg/cell). Other cell types found in normal lung, skin, and bowel, as well as eosinophils, neutrophils, monocytes, and lymphocytes from peripheral blood have no detectable tryptase. Thus, the enzyme is a discriminating marker of human mast cells. A sensitive and specific immunoassay for tryptase was developed that distinguished tryptase from other trypsinlike enzymes and was used to quantify tryptase in complex biological fluids (21,22). This has enabled the measurement of tryptase release in vivo and, thereby, a precise assessment of mast cell activation.

Tryptase hydrolyzes proteins and peptides on the COOH-terminal side of the basic residues, lysine and arginine. However, such substrates also are cleaved by other trypsinlike enzymes and, therefore, are not specific. Although classified as a serine protease, the absence of inhibition by the classic inhibitors of serine esterases present in plasma, lung, and urine, as well as by lima bean, soy bean, and ovomucoid trypsin inhibitors, clearly distinguish tryptase from pancreatic trypsin and from most other serine esterases (23).

Tryptase is a tetrameric endoprotease of 134,000 Da, with subunits of 31,000–34,000 Da, each with an active enzymatic site (18) and common antigenic sites (24). Two tryptase cDNA molecules (α and β) have been cloned from a human lung mast cell cDNA library and sequenced (25,26), three (1, 2, and 3) from a human skin mast cell library (27). The α-tryptase cDNA codes for a 27,364 Da, 244-amino acid catalytic protein, with two putative carbohydrate-binding sites (25). The β-tryptase cDNA codes for a 245-amino acid catalytic protein that is 90% identical with α-tryptase and 98–99% identical with tryptases 1, 2, and 3. There are two *N*-linked carbohydrate-binding sites in the α, 1, and 3 types of tryptase, and one carbohydrate-binding site in the other tryptase sequences, suggesting an explanation for a portion of the electrophoretic heterogeneity of natural tryptase in sodium dodecylsulfate–polyacrylamide gel electrophoresis (SDS–PAGE). The amino acid sequence of α-tryptase contains 24-amino acid differences and 1-amino acid deletion compared with the β-tryptase sequence. Interestingly, the region of α-tryptase where the amino acid deletion occurs is identical with the sequence reported for dog tryptase (28).

The prepropeptide portion is always 30 amino acids long; these regions are identical in β, 1, 2, and 3 forms of tryptase, but differ by four amino acids from the α-tryptase prepropeptide. The activation peptide for tryptase has not been clearly defined, but its termination in glycine is unusual. The propeptide sequences for pancreatic serine proteases terminate in a basic residue; most leukocyte and mast cell granule proteases terminate in an acidic residue. Neither the prepropeptide nor the propeptide portions of tryptase have been detected in enzyme purified from human lung mast cells, consistent with tryptase being stored in an active form. The prepropeptide may be responsible for directing tryptase to secretory granules or for the assembly of tryptase as an active tetramer, and most likely are removed before or shortly after tryptase enters the secretory granules.

By applying polymerase chain reaction (PCR) techniques to DNA taken from human–hamster somatic cell hybrids, at least two tryptase sequences were localized to the normal human haploid genome on chromosome 16 (26). One sequence corresponds to α-tryptase and tryptase 3, the other to β-tryptase, tryptase 1, and tryptase 2. One tryptase gene has been sequenced that appears to match tryptase 1 cDNA. It contains six exons and five introns. The first intron falls between the 5′-noncoding and the ATG-methionine start site, which is rare, if not unique among proteases. Positions of the four introns within the coding region are similar to those for trypsin and glandular kallikrein, but different from those for proteases involved in blood coagulation and for mast cell chymases. The 5′ region has several consensus sites for regulating transcription, including TATA and CAAT boxes, and a region showing homology to the enhancer elements for the rat chymase II gene (29) and for protease genes expressed in the pancreas. If the latter proves to be a mast cell-specific enhancer region, it would be of interest to examine its regulation during the differentiation of human mast cells.

Tryptase is not inhibited by classic biological inhibitors of serine esterases present in plasma and tissues. Instead, the enzyme is uniquely stabilized in its active tetrameric form by heparin (30), to which it is ionically bound under physiological conditions. When free in solution, tryptase subunits rapidly and irreversibly dissociate from one another into inactive monomers, without any evidence for autodegradation. Chondroitin sulfate E and polyglutamic acid stabilize tryptase somewhat less well than heparin; and chondroitin monosulfates are little better than buffer at stabilizing tryptase under physiological conditions. Thus, negative-charge density, rather than carbohydrate structure, is the primary determinant for stabilizing tryptase activity (31). The physiological means of dissociating tryptase from heparin or chondroitin sulfate E in vivo is not totally clear, but the capacity of divalent cations (32) and of heparin-binding proteins, such as antithrombin III (23), to destabilize heparin-bound tryptase at physiological concentrations in vitro may be relevant to the situation in vivo.

Even though the biological role of tryptase has not been convincingly demonstrated, several activities of potential biological interest have been examined in vitro (Table 2). For example, tryptase slowly generates the anaphylatoxin C3a from C3 (33), but has no activity against C5. Locally produced C3a, by virtue of its capacity to contract smooth muscle and augment vasopermeability, could amplify activities directed by mast cell mediators. Tryptase does not appear to affect bradykinin generation (34); bradykinin or lysylbradykinin generated in vivo during immediate hypersensitivity events, therefore, is not due to tryptase activity. Tryptase rapidly cleaves and inactivates fibrinogen as a coagulable substrate for thrombin (35); the lack of fibrin deposition and rapid resolution of urticaria or angioedema reactions may, in part, reflect this same activity in vivo. Finally, tryptase activates latent collagenase derived from rheumatoid synovial cells, apparently by first activating prostromelysin (metalloproteinase III) (36) which, in turn, activates latent collagenase. The increased numbers of mast cells found in rheumatoid synovium (37) and in inflammatory cutaneous lesions of scleroderma (38),

Table 2 Potential Biological Properties of Tryptase and Chymase

Protease	Property
Tryptase	Fibrinogenolysis[a]
	High molecular weight kininogen destruction[a]
	Prostromelysin activation[a]
	C3a generation[a]
	Latent TGF-β activation[a]
	Fibroblast proliferation[b]
	Pulmonary smooth-muscle hyperreactivity to histamine[b]
	Neuropeptide degradation[a,b]
Chymase	Generation of angiotensin II[a]
	Basement membrane degradation[a]
	Glandular mucous secretion[b]
	Neuropeptide degradation[b]

[a]Human
[b]Dog

together with the increased tissue turnover that occurs in each condition, suggest a related role for tryptase.

Chymase is the principal enzyme accounting for the chymotrypsinlike activity present in human cutaneous mast cells. The enzyme was purified from human skin (39) and was selectively localized to mast cells in normal human lung, small bowel, and skin by immunohistochemical techniques (40) and, in bowel, by enzyme histochemical techniques (41). In contrast with tryptase, chymase is present only in a subpopulation of mast cells, those termed MC_{TC}, that are preferentially located in skin and bowel submucosa. Dispersed MC_{TC} cells obtained from foreskin contain 4.5 pg of chymase per mast cell by immunoassay. The MC_T cells lack immunoreactive chymase as well as an antigenically distinct form of a chymotrypsinlike enzyme.

The cDNA for human chymase codes for a catalytic portion with 226 amino acids and a relative molecular mass (M_r) of 25,000. Two consensus *N*-linked glycosylation sites are present. An 18-amino acid signal peptide and 2-amino acid activation peptide are predicted from the cDNA sequence. The corresponding 3-kb gene has been cloned and localized to human chromosome 14 (42,43). It contains five exons, with active site histidine, aspartic acid, and serine residues located on exons 2, 3, and 5, respectively. This motif is conserved in the chymase subfamily among different species, even though sequence homologies are only about 80% with dog chymase and 60% with rodent chymases. Southern blot analysis of human genomic DNA suggests that there is only one gene for chymase (43), unlike the situation in rodents for which multiple genes exist.

Human chymase is a monomer of 30,000 Da with endopeptidase activity that is measurable by cleavage of synthetic *p*-nitroanilide–peptide conjugates or of angiotensin I (44) and by immunoassay (45). Inhibition of chymase by diisopropylfluorophosphate and tosyl-L-phenyalanine chloromethylketone indicate the presence of serine and histidine in the active site. Similar to tryptase, chymase is stored fully active in mast cell secretory granules, presumably bound to heparin or to chondroitin sulfate E. Unlike tryptase, chymase stability is not substantially affected by heparin (46), and its activity is inhibited by classic biological inhibitors of serine proteinases. In plasma, chymase activity is inhibited by α_1-antichymotrypsin, α_1-proteinase inhibitor, and α_2-macrogobulin; however, these inhibitors are inefficient, serving as better substrates than inhibitors (47). Such inhibition of chymase enzymatic activity also attenuates its measurement by immunoassay, which currently, precludes its measurement in complex biological fluids. Human chymase is antigenically unrelated to cathepsin G from human neutrophils, bovine chymotrypsin, or rat mast cell chymase I (44). Whether additional chymase molecules reside in human mast cells, analogous to the situation in rodents, is not totally resolved. No chymotrypsinlike enzymatic activity has been detected in MC_T cells. However, a chymaselike enzyme was recently purified from cardiac tissue that, similar to skin chymase, was a potent activator of angiotensin (48). The enzyme appears to be identical to mast cell chymase, and is largely derived from mast cells in heart tissue.

Potential biological activities of chymase, like those of tryptase, are based on in vitro observations (see Table 2). An idealized substrate has been predicted to have the following motif: nXaa-Pro-[Phe, Tyr or Trp]-Yaa-Yaa, where $n \geq 6$, Xaa = any amino acid, and Yaa = any amino acid, but Pro (49). Of potential biological interest are findings that human chymase converts angiotensin I to II more efficiently than angiotensin-converting enzyme (44,49), inactivates bradykinin (50), and attacks the lamina lucida of the basement membrane at the dermal–epidermal junction of human skin (51). Blister formation, possibly related to chymase activity, occurs with systemic mastocytosis in infants, but is rare in adults and does not occur in classic mast cell-dependent urticarial reactions. Chymase purified from dog mast cells stimulates production of mucus by glandular cells in vitro (52) and cleaves substance P and vasoactive intestinal peptide (53). However, these latter neuropeptides are poor substrates for human chymase (48), suggesting important differences in the substrate specificities of chymases among different species. These differences, in part, may be due to residues corresponding to amino acids 21–29 in human chymase, which form a cleft near the substrate-binding pocket and vary considerably in sequence among different chymases (43).

Mast Cell Cathepsin G. This is a neutral protease found in neutrophils and monocytes. The presence of a cathepsin G-like protein in MC_{TC} cells has been demonstrated with immunohistochemical, Western blot, and enzymatic techniques (54). Western blot analysis showed a stained band at 30,000 Da. This comigrated with chymase and a minor species of cathepsin G present in neutrophils. The major form of neutrophil cathepsin G migrates at 28,000 Da. Analysis of the catalytic chymotryptic activity in extracts of cutaneous mast cells revealed that approximately 30% was inhibited by aprotinin and anticathepsin G antibody, both of which inhibit cathepsin G and not chymase. Thus, substantial levels of this cathepsin G-like protease are present in MC_{TC} cells and extend the compositional differences between MC_{TC} and MC_{T} types of human mast cells.

Mast cell carboxypeptidase was detected and localized to secretory granules in mast cells dispersed from skin and lung by the IgE-mediated release of an activity that cleaves the COOH-terminal His^{9}-Leu^{10} bond of angiotensin I and behaves like a zinc metalloexopeptidase (55). Human mast cell carboxypeptidase is a monomeric enzyme with a M_r of 34,500 Da by SDS–PAGE (56). Based on the cDNA-derived amino acid sequence of the catalytic portion of the enzyme, the human enzyme is 86% identical with mouse mast cell carboxypeptidase A and has a calculated M_r of 36,100 Da and a net positive charge of 16 (57). Although the total amino acid sequence and gene structure of the human enzyme is more homologous to human pancreatic carboxypeptidase B than A, the composition of the catalytic site is more homologous to human pancreatic carboxypeptidase A, as are its substrate specificities. Human mast cell carboxypeptidase removes COOH-terminal phenylalanine and leucine residues, but does not recognize COOH-terminal arginine nor lysine residues. Unlike bovine pancreatic carboxypeptidase A, the human mast cell enzyme does not cleave des-Arg^{9} bradykinin or substance P. The human enzyme has been called either human mast cell carboxypeptidase A or simply human mast cell carboxypeptidase (HMC-CP).

Murine monoclonal antibody prepared against human mast cell carboxypeptidase selectively stained the MC_{TC} type of human mast cell (58). The MC_{T} cells were not stained. This finding further extends the biochemical differences between MC_{T} and MC_{TC} types of human mast cells to include compositional differences in carboxypeptidase content as well as in the content of chymase and cathepsin G-like protease. Within mast cell granules, chymase and carboxypeptidase appear to reside in a complex with proteoglycan that is distinct and separable from the macromolecular complex with tryptase (59). Mast cells dispersed from neonatal foreskin, adult facial skin, and adult foreskin contain 0.5, 5, and 16 pg of carboxypeptidase per mast cell, respectively, as estimated by the enzymatic activity in mast cell extracts and the specific activity of the purified enzyme (56). These levels presumably reflect the percentage and maturity of the MC_{TC} cells in each preparation.

The cDNA for human mast cell carboxypeptidase codes for a 417-amino acid preproenzyme that contains a 15-amino acid signal peptide and a 94-amino acid activation peptide (57). The 318-amino acid mature enzyme has a predicted M_r of 36,058 and a net charge of +16. The human gene for mast cell carboxypeptidase localized to chromosome 3 and contains 11 exons over a 32-kb region (60). Although the substrate-binding pocket is more homologous to pancreatic carboxypeptidase A than B, the intron–exon organization and the overall nucleotide and amino acid sequences are more homologous to pancreatic carboxypeptidase B than A. Mast cell carboxypeptidase mRNA, like the protein, is selectively expressed in mast cells (60).

Neutral Proteases of Rat Mast Cells. In rats there are two chymotrypsinlike enzymes that serve to distinguish mast cells from other cell types and mast cells subtypes from one another. *Rat chymase I* (rat mast cell protease I; RMCP I) is found exclusively in the connective tissue type mast cell; *rat chymase II* (rat mast cell protease II; RMCP II) is found only in the mucosal-type mast cell (61). Chymase I accounts for 24–30 pg of protein per rat serosal mast cell, about 25% of the total cellular protein (62). Chymase II in cultured rat mucosal mast cells accounted for 26 pg of protein per cell (63). Although chymases I and II are derived from different genes (64), they appear to belong to a larger gene family. A mast cell-specific enhancer region upstream from the rat chymase II promoter and a corresponding DNA-binding protein in rat basophil leukemia cells have been reported (29). *Rat mast cell carboxypeptidase A* cleaves peptide and ester bonds on the NH_2-terminus of COOH-terminal amino acids with aromatic residues, similar to pancreatic bovine carboxypeptidase A. The enzyme is present in connective tissue rat mast cells (65), accounting for 20–25 pg/rat serosal mast cell (66). Complexes of chymase and carboxypeptidase A with heparin proteoglycan are released from rat connective tissue mast cells, but remain near the activated cell in extracellular labyrinths as an insoluble complex (66). In contrast, histamine, various acid hydrolases from all mast cells, and chymase II from mucosal mast cells are fully soluble after release and free to diffuse away from these sites. Ingestion of the protease–proteoglycan complex by tissue fibroblasts (67) or by eosinophils, neutrophils, and macrophages, that enter the tissue during the late phase of immediate hypersensitivity reactions may be a mechanism for their removal, or they may exert an effect on their new host cell. *Rat mast cell tryptase* has been detected in small amounts (approximately 0.5 pg/cell) in serosal mast cells, where reportedly it resides in association with an inhibitor of the enzyme called trypstatin (68). Unlike human tryptase, rat mast cell tryptase is inhibited by soy bean trypsin inhibitor, α_1-antitrypsin, antipain, and aprotinin, and does not bind to heparin in physiological buffer. The size of rat mast cell tryptase, as well as a similar and possibly identical enzyme purified from rat skin (69), has been reported as 160,000 Da, with four equally sized subunits of 40,000 Da.

Neutral Proteases in Murine Mast Cells. The production of Kirsten-*ras* immortalized mast cells (70) provided the material for production of cDNA libraries from which a series of murine mast cell proteases have been cloned, and the expression of several have been assessed in different mast cell types. The cloned proteases include six chymases **(MMCP-1,2,3,4,5′,5)** (71–73), mast cell **carboxypeptidase A** (70), and two **tryptases** (71,74). Expression at the protein level of each has not been verified.

Examination of mast cells for protease mRNA revealed that MMCP-4 is present in both mucosal (intestinal) and connective tissue (serosal) mast cell types, MMCP-1 and 2 are present in mucosal mast cells alone, and MMCP-3, 5, and 6, and mast cell carboxypeptidase A are present in the connective tissue mast cell alone. Surprisingly, IL-3-dependent bone marrow-derived mast cells contained a pattern of protease mRNA reflecting the connective tissue phenotype, even though they lack substantial quantities of heparin (75). This provides a new explanation for why bone marrow-derived cells develop a connective tissue phenotype after coculture with fibroblasts (i.e., that fibroblasts provide a maturation signal, not a signal to switch to a distinct developmental pathway).

Neutral Proteases of Canine Mast Cells. **Tryptase** (76) and **chymase** (77) have been purified from dog mastocytoma cells and cloned (42). Dog tryptase has an M_r of 132,000–144,000 Da and is a tetramer composed of two closely related subunits of 30,000–35,000 DA each (76). Similar to human tryptase, dog tryptase is insensitive to inhibition by α_1-antitrypsin and plasma and is stabilized by heparin. Chymase, a monomeric protein, produced several bands after SDS–PAGE in the size range of 25,000–30,000 Da. Recent evidence that dog chymase potentiates secretion of mucus by bovine tracheal gland serous cells and that dog tryptase potentiates histamine-mediated contraction of guinea pig pulmonary smooth muscle (78) and proliferation of fibroblasts (79) suggest the possibility of similar properties for human proteases.

3. Proteoglycans

The presence of highly sulfated proteoglycans in secretory granules of mast cells and basophils results in metachromasia when these cells are stained with basic dyes. This has been the characteristic histological criterion for identification of mast cells and basophils by light microscopy since their initial detection by Ehrlich in 1878 (1). Proteoglycans are composed of glycosaminoglycan side chains (repeating unbranched disaccharide units of a uronic acid and hexosamine moieties that are variably sulfated), covalently linked to a single-chain protein core by a specific trisaccharide–protein linkage region consisting of -Gal-Gal-Xyl-Ser. Although proteoglycans are important constituents of extracellular matrix material, where they are most abundant, this review is concerned only with the intracellular proteoglycans of mast cells and basophils. Two proteoglycans, **heparin** and **chondroitin sulfate E**, have been associated with purified human mast cells (80,81), whereas **chondroitin sulfate A** is the preponderant proteoglycan found in human basophils (82). Heparin is selectively concentrated only in mast cells; chondroitin sulfate E also has been found in human monocytic cells (83), and chondroitin sulfate A is present in eosinophils (84), natural killer cells (85), and probably most other cell types with cytoplasmic granules.

The characteristic disaccharide structures of heparin, chondroitin sulfate E, and chondroitin sulfate A are shown in Figure 2. The average number of sulfate residues per respective disaccharide is 2.5, 1.5, and 1.0. Heparin consists of either glucuronic acid without a sulfate residue or iduronic acid with a 2-*O*-sulfate that is linked to a glucosamine containing a 4-*O*-sulfate and a 2-*N*-sulfate. Occasionally a 2-*N*-acetyl rather than *N*-sulfate is found. The characteristic susceptibility of heparin to nitrous acid is due to attack at the *N*-sulfate residue; chondroitin sulfates lack this residue and are resistant to nitrous acid. Chondroitin sulfates A and E each consist of galactosamine with a 4-*O*-sulfate linked to iduronic acid. Type E, in addition, has a 6-*O*-sulfate on galactosamine. The recent development of an accurate, quantitative and sensitive technique for identifying characteristic disaccharides obtained after enzymatic degradation of the glycosaminoglycans by high-performance liquid chromatography (HPLC) has facilitated the identification of proteoglycans in mast cells and basophils (86).

Synthesis of proteoglycans proceeds by the sequential addition of monosac-

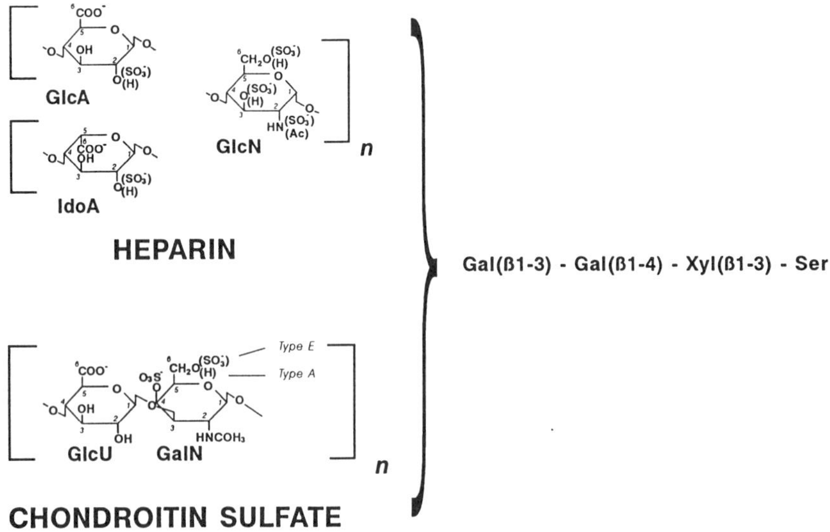

Figure 2 Proteoglycan structure.

charides conjugated to uridine diphosphate (UDP-glucuronic acid and UDP-*N*-acetyglucosamine for heparin) to the nonreducing ends of glycosaminoglycans attached to the protein core. The newly formed proteoglycans are then processed to their mature forms. For example, sulfation occurs through a 5′-phosphoadenosyl-3′-phosphosulfate donor intermediate. Degradation of heparin may proceed by digestion with heparinases or by fragmentation when exposed to oxygen radicals (87).

The same peptide core is associated with heparin and chondroitin sulfate proteoglycans. In humans the core protein is 17,600 Da and contains a glycosaminoglycan attachment region of 18 amino acids (88). The attachment region is rich in alternating Ser-Gly residues and the gene is located on chromosome 10. Thus, the choice of glycosaminoglycan a cell will synthesize is dictated by the enzymes available to synthesize the glycosaminoglycan, rather than by the peptide core. In rats, the core protein has a greater number of glycosaminoglycan attachment sites and larger glycosaminoglycan chains (rat, 60,000 Da; human, 20,000 Da) that together account for the larger size of rat (about 10^6 Da) (89) compared with human (about 60,000 Da) (90,91) heparin proteoglycan.

Each mast cell type appears to have the capacity to synthesize more than one class of proteoglycan, although often one class is preponderant. Rat peritoneal mast cells primarily synthesize heparin, but in the presence of *p*-nitrophenyl-β-D-xyloside, a substitute for the protein core onto which glycosaminoglycans can be synthesized, these cells synthesize a chondroitin disulfate (92). They also contain preformed chondroitin sulfate in their secretory granules, which may have been synthesized at an earlier stage of differentiation. Rat mucosal mast cells (RBL-1) make preponderantly chondroitin sulfate di-B as well as smaller amounts of heparin (93,94). Mouse bone marrow-derived, IL-3-dependent mast cells synthesize chondroitin sulfate E of 100,000–300,000 Da as well as small amounts of heparin (75,92). In this mouse mast cell, both heparin and chondroitin sulfate E proteoglycans are synthesized on each core peptide (95).

Different types of human mast cells, unlike those in rodents, are not easily distinguished by histochemistry with basic dyes (96). An explanation for this observation was provided by analysis of human mast cells in situ at the ultrastructural level after staining heparin with antithrombin III–gold conjugates. All human mast cells contain heparin, in contrast with rodent mucosal mast cells that lack this proteoglycan.

The biological functions of endogenous mast cell proteoglycans are somewhat speculative. These proteoglycans bind to histamine, neutral proteases, and acid hydrolases at the acidic pH inside mast cell secretory granules and may facilitate uptake and packaging of these preformed mediators into the secretory granules. Under these conditions, mast cell proteoglycans may regulate the stability and activities of many of the enzymes present. After release from mast cells, the proteoglycans may continue to affect those substances that remain associated, particularly the neutral proteases. The stabilizing effect of heparin and, to a lesser degree, chondroitin sulfate E, on human tryptase activity (30,31) may be crucial for the full expression of mast cell-mediated events. Heparin and chondroitin sulfate E (to a lesser extent) may express anticoagulant, anticomplement, antikallikrein, and Hageman factor autoactivation activities. Heparin neutralizes the ability of eosinophil-derived major basic protein to kill schistosomula and enhances the binding of fibronectin to collagen. The migration of capillary endothelial cells in vivo is stimulated by heparin. Heparin also may sequester or protect growth factors, such as fibroblast growth factor in the extracellular matrix, and modulate the cell adhesion properties of matrix proteins, such as vitronectin, fibronectin, and laminin (97,98). The anticoagulant activity of purified human and rodent heparin is similar to commercially available porcine heparin (99) and may depend, in part, on a specific pentasaccharide sequence (100). However, when heparin is saturated with tryptase, as it is when released from mast cells, many of these activities may be attenuated.

B. Newly Generated Lipid Mediators Derived from Arachidonic Acid

Liberation of the 20-carbon, polyunsaturated fatty acid called arachidonic acid from cellular lipid stores occurs with mast cell activation, whereupon it is either reincorporated into lipids or oxidatively metabolized. Unstimulated human mast cells derived from lung incorporate exogenous arachidonic acid into neutral lipids and phospholipids, in a ratio of about 7:2 (101), and store these lipids in membranes and cytoplasmic lipid bodies (102). Hydrolysis of these lipids to yield free arachidonate occurs through enzymatic pathways involving phospholipase A_2, diacylglycerol lipase, after either phospholipase D with phosphatidic acid phosphatase, or phospholipase C, or neutral lipid lipases. Arachidonic acid metabolism then proceeds along either the cyclooxygenase pathway to prostaglandins (PGs) and thromboxanes (TXs), or the 5-, 12-, or 15-lipoxygenase pathway to monohydroxyl fatty acids, leukotrienes (LTs) that include both LTB_4 and the sulfidopeptides LTC_4, LTD_4, and LTE_4 *(slow-reacting substances of anaphylaxis; SRS-A)*, and lipoxins. Synthesis of selected products by different cell types is governed by the repertoire of metabolic enzymes present in each cell; the major eicosanoid products of human mast cells are PGD_2 and LTC_4.

Dispersed and purified preparations of human mast cells obtained from lung (101,103), skin (104,105), and intestine (106), upon activation, produce PGD_2 and LTC_4 in weight ratios of approximately 5:1 (Table 3 and Fig. 3) (107). Smaller amounts of LTB_4 isomers are also produced. There is now no evidence that lipid products can be used to distinguish MC_T from MC_{TC} cell activation. In contrast, peripheral blood basophils obtained from normal human subjects

Table 3 Newly Generated Metabolites of Arachidonic Acid (ng/10^6 cells) from Mast Cells and Basophils

	Lipoxygenase		Cyclooxygenase
Cell Type	LTC_4	LTB_4	PGD_2
Human			
MC_T cell	++	+	++
MC_{TC} cell	++	+	++
Basophil	++	++	–
Rat			
Bone marrow-derived IL-3-dependent mast cell	+		+
Connective tissue mast cell	+	+	++
Mouse			
Bone marrow-derived IL-3-dependent mast cell	+	+	–
Connective tissue mast cell	+		++

++, > 10 ng/10^6 cells; +, 1–10 ng/10^6 cells; –, none detected; MC_T, tryptase$^+$, chymase$^-$ mast cells; MC_{TC}, tryptase$^+$, chymase$^+$ mast cells.

synthesize LTC_4, but no PGD_2 (108). Thus, the absence of PGD_2 in the presence of LTC_4 and histamine implicates activation of basophils. One caution for this interpretation is that basophils obtained from subjects with chronic myelogenous leukemia apparently produce substantial amounts of PGD_2 (109).

Prostaglandin D_2 is generated by human mast cells, human alveolar macrophages (110), and platelets (111); substantial quantities of PGD_2 are also generated in the central nervous system (CNS) neurons, where it promotes sleep (112). Thus, when PGD_2 is detected in a complex biological milieu, there is ambiguity about its cell source. In fact, the major cell source of PGD synthetase appears to be antigen-processing cells (APC) in various rat (113) and possibly human tissues. Concentrations of PGD_2 in normal plasma, measured by mass spectrometry, range from 1 to 5 pg/mL, and its half-life in the circulation is similar to that of histamine, about 2 min (114). Elevated levels of PGD_2 have been detected in venous blood that drains sites of induced cutaneous urticarial reactions (115); in nasal lavage fluid after allergen challenge (116); and in bronchoalveolar lavage fluid after allergen challenge (117). In addition, 9α, 11β-PGF_2, a major metabolite of PGD_2, has bronchospastic activity when introduced into human airways (118). Another metabolite of PGD_2 (9α-hydroxy-11,15-diketo-2,3,18,19-tetranorprost-5-ene-1,20-dioic acid) has been detected in the urine of patients with active systemic mastocytosis, and has correlated with recurrent hypotensive episodes (119). Aerosol administration of PGD_2 causes bronchoconstriction in human subjects and potentiates airway responsiveness to

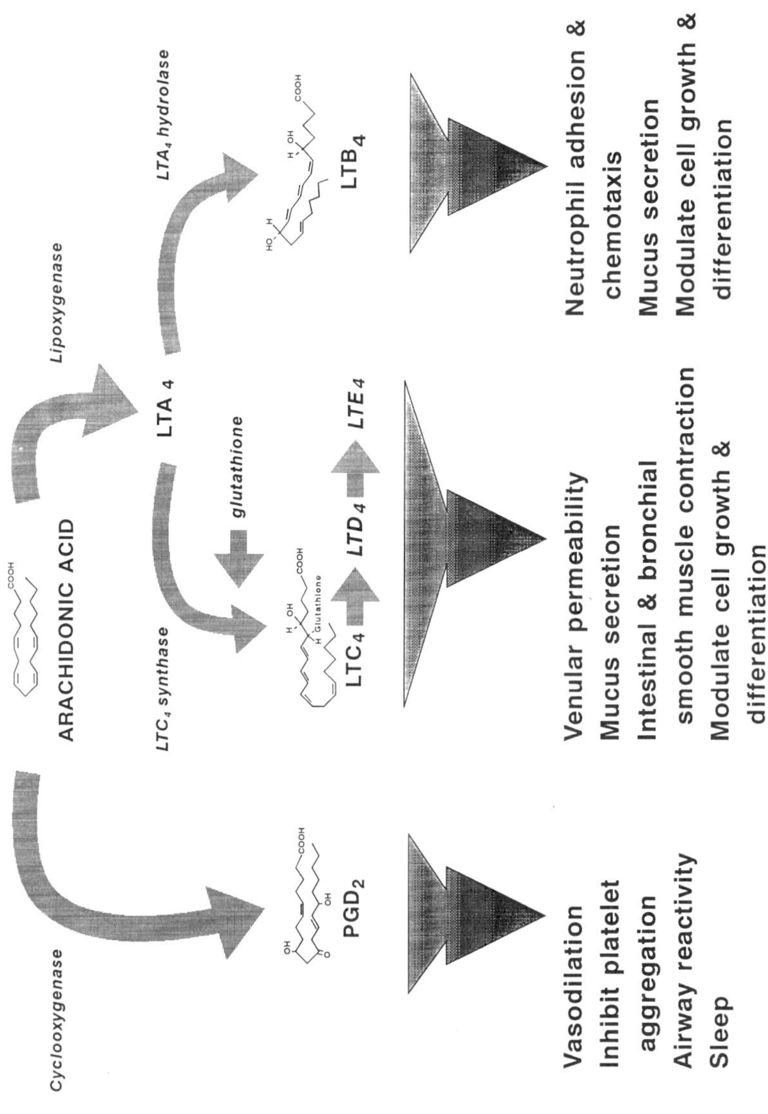

Figure 3 Metabolites of arachidonic acid.

histamine and methacholine (see Fig. 3) (120,121). Prostaglandin D_2 is chemokinetic for human neutrophils (122), and it augments the LTB_4-mediated accumulation of neutrophils in human skin (123). Similar to PGI_2, PGD_2 is a potent inhibitor of human platelet aggregation (124).

Leukotriene B_4 is the preponderant de novo lipoxygenase product of human neutrophils, monocytes, and macrophages, whereas LTC_4 is the preponderant product made by human eosinophils, mast cells, and basophils. Basophils release histamine, but not LTC_4, when activated by complement anaphylatoxins C3a and C5a. However, IL-3, IL-5, and granulocyte–monocyte colony-stimulating factor (GM-CSF) prime basophils, such that both mediators are generated in response to complement anaphylatoxins (125,126). The production of LTC_4 and PGD_2 are augmented after coculture of human lung-derived mast cells with mouse 3T3 fibroblasts (127). Leukotriene B_4 expresses potent neutrophil chemotactic and aggregating activity on human neutrophils in vitro at molar concentrations similar to those at which C5a attracts neutrophils (see Fig. 3) (128). In humans subcutaneous injection of LTB_4 elicits a transient wheal-and-flare, followed 4–6 h later by a tender indurated lesion that contains fibrin and is infiltrated with neutrophils (123). This activity is augmented by the coadministration of PGD_2. The LTB_4-mediated appearance of neutrophils in tissues in vivo is preceded by their margination (129), an observation most likely related to the in vitro finding of increased neutrophil adherence to endothelial cell monolayers in the presence of LTB_4 (130). Leukotriene B_4 is 100- to 1000-fold less potent as a spasmogen than the sulfidopeptide leukotrienes (131).

Sulfidopeptide leukotrienes enhance venular permeability; constrict arterial, arteriolar, intestinal, and bronchial smooth muscle; and enhance bronchial mucous secretion (see Fig. 3) (132). Decreased myocardial contractility and glomerular filtration are largely due to decreased tissue perfusion. An important role for these molecules in asthma has been postulated, based on their potent bronchospastic activities, resistance to inhibition by available drugs, secretion in vitro from allergen-challenged lung tissue obtained from asthmatics, and their mucosecretory activity. Synergistic decrements in airway caliber occur in the presence of histamine or PGD_2 with LTC_4 (133). Human lung contains few, if any, LTC_4 receptors; most of the LTC_4-mediated effects occur owing to conversion of LTC_4 to LTD_4 and LTE_4. Leukotriene antagonists with adequate specificity, safety, and potency for use in humans will permit a more accurate assessment of the role of these leukotrienes in airway pathophysiology.

C. Cytokines

Compelling evidence that rodent mast cells, when activated, produce cytokines (Table 4) emerged in 1987 when expression of IL-4 mRNA (134) and a cytolytic factor, appearing to be tumor necrosis factor (TNF) (135), were reported. Rodent mast cells in culture constitutively transcribe cytokines TNF-α and IL-6;

Table 4 Cytokines Produced by Mast Cells and Basophils

	Cytokines	
Cell type	mRNA	Protein
Human		
Mast cells	TNF-α	TNF-α
	IL-4	IL-4
Basophils		TNF-α
		IL-4
Rodent mast cells	TNF-α	TNF-α
	GM-CSF	
	TGF-β	
	SCF	
	IFN-γ	
	MIP-1α,β	
	MCAF	
	IL-1,2,3,4,5,6,10	IL-1,3,4,6

TNF, tumor necrosis factor; IL, interleukin; GM–CSF, granulocyte–macrophage colony-stimulating factor; TGF, transforming growth factor; IFN, interferon; MIP, macrophage inflammatory protein; MCAF, monocyte chemotactic and activating factor.

corresponding biological activities also are present. Activated mast cells in addition, transcribe macrophage inflammatory factor (MIP)-1α, MIP-1β, TCA3, IL-1, IL-2, IL-3, IL-4, IL-5, IL-10, GM-CSF, TGFβ and IFN-γ; biological activities have been detected for IL-1, IL-3, and IL-4. Preliminary evidence suggests that rodent mast cells also contain mRNA for stem cell factor (SCF) (T. F. Huff, et al., unpublished observation). Particularly interesting is the capacity of non-T, non-B murine spleen cells to generate IL-4 when stimulated by FcεRI. Both immature mast cells (c-*kit*$^+$) and immature basophils (c-*kit*$^-$) apparently are responsible for producing this IL-4. Preliminary reports suggest that an analogous situation holds true for human bone marrow cells. Also, induction of autocrine cytokine production through FcεRI may play a role in their differentiation to mast cells. Frequently, induction of cytokine synthesis occurs hours after mast cells are activated, in marked variance to the unloading of granule mediators within minutes and the generation of arachidonic acid metabolites beginning within minutes and lasting up to about 30 min. In humans, production of TNF-α has been demonstrated in human skin mast cells and in basophils (136,137), and IL-4 in human skin and lung mast cells (138). Possibly relevant to the late cellular phase of immediate hypersensitivity reactions is the finding that the increased levels of the adhesin ELAM-1 on endothelial cells is induced by TNF-α from stimulated cutaneous mast cells (139).

III. MAST CELL HETEROGENEITY

A. Different Types of Mast Cells

Mast cells of two major types have emerged as a consistent observation in human and rodent tissues. Recognition of this complexity is crucial for understanding mast cell biology and, potentially, for treating mast cell-associated disease (140). Several nomenclatures are in use for describing particular mast cell types. None is optimal, because they do not relate to functional differences. Mucosal and connective tissue are the terms currently in most common use, particularly for rodent mast cells. These terms refer to mast cells possessing certain phenotypic characteristics and should be used only to describe mast cells on the basis of phenotype, rather than on tissue localization, because each type of mast cell may be found both in mucosal and connective tissue sites.

In rodents, mast cell types have been discriminated from one another based on proteoglycan content, neutral protease expression, arachidonic acid metabolites, and surface antigenic determinants. Heparin accounts for most of the proteoglycan synthesized by rat serosal connective tissue mast cells (CTMC), but only for a small portion of the proteoglycan produced by rat intestinal mucosal mast cells (MMC) (141) and mouse bone marrow-derived mast cells (BMMC) (142), which produce mostly the oversulfated chondroitin sulfate proteoglycans, di-B in rats, E in mice (143). A single gene encodes the core peptide of both types of proteoglycan (144), suggesting that the class of glycosaminoglycan added onto the core may be regulated by signals independent of mast cell type. The neutral protease composition in the granules qualitatively distinguishes rat CTMC from rat MMC. Rat chymase I is found in CTMC of skin, tongue, intestinal serosa and submucosa, and lung parenchyma (145); whereas rat chymase II is found only in MMC of the rat intestine and bronchial epithelium. A carboxypeptidase A-like enzyme is selectively concentrated in the rat peritoneal CTMC (65,66). Discrimination of CTMC from MMC based on neutral protease expression also has been reported in the mouse (146,147). Upon activation, rat connective tissue mast cells from the serosal surface metabolize arachidonic acid largely to PGD_2 with little production of leukotrienes, whereas immunologically activated rat intestinal MMC generate LTC_4 and LTB_4 as well as PGD_2 (103, 148). Mouse BMMC, which are IL-3-dependent and appear to be immature, generate LTC_4 and LTB_4, with little production of PGD_2 (149). The pattern of lipid mediators is more likely to reflect local environmental effects, such as priming or stage of maturation, rather than differences in lineage. Also, mouse bone marrow-derived MMC and serosal CTMC display characteristic surface phenotypes, as defined by a panel of rat monoclonal antibodies (150,151).

Coculture experiments indicate that at least one of the glycolipid antigens is passively acquired from fibroblasts, indicating that environment, rather than lineage, may dictate the presence of certain surface markers (152).

In humans a nomenclature is used based on neutral protease composition, MC_T for mast cells with tryptase alone, and MC_{TC} for those with tryptase, chymase, carboxypeptidase, and a cathepsin G-like enzyme. The MC_T and MC_{TC} cells correspond, for the most part, to the mucosal and connective tissue mast cells, respectively, but this only applies to humans. Bone marrow-derived, IL-3-dependent mast cells in culture, are sometimes assumed to be a mucosal mast cell type, but in fact are relatively immature, and may contain mast cells committed to either type. This section will first concentrate on the characteristics that distinguish these different types of mast cells from one another and, then, analyze the information pertaining to the corresponding pathways of differentiation and maturation.

B. Histochemistry and Composition of MC_T and MC_{TC} Cells

The two distinct types of human mast cells are distinguished and named based on different protease compositions of the secretory granules. The MC_{TC} cells contain chymase, carboxypeptidase, cathepsin G-like protease, and tryptase, and MC_T cells contain only tryptase (Table 5) (153). Mast cells that lack tryptase,

Table 5 Characteristics of Human MC_{TC} and MC_T Cell Types

Characteristic	MC_{TC} cell	MC_T cell
Normal tissue distribution		
Skin	++	–
Small intestine		
Submucosa	++	+
Mucosa	+	++
Lung		
Alveolar wall	–	++
Bronchi/bronchioles	+	++
Nasal mucosa	++	++
Conjunctiva	++	+
Synovium	++	–
Protease content	Tryptase Chymase Cathepsin G-like protease Carboxypeptidase	Tryptase
T-lymphocyte dependency	No	Yes
Granule morphology	Grating/lattice Complete scroll-poor	Complete scroll-rich

but contain other mast cell proteases have been searched for, but never convincingly found. Analysis of histamine levels in dispersed preparations of MC_{TC} cells obtained from adult foreskin MC_{TC} (1.9 pg per mast cell) and newborn foreskin MC_{TC} (1.6 pg per mast cell) are similar to those in MC_T cells obtained from adult lung (1.5 pg per mast cell), showing that histamine content is not helpful in distinguishing among these different types of human mast cells. Thus, cell-specific proteases are the most discriminating markers of different types of human mast cells.

Human mast cell types cannot be precisely distinguished from one another by the histochemical staining (alcian blue/safranin) and fixation (formalin/Carnoy's fluid) techniques originally used to distinguish mature connective tissue (alcian blue$^+$, safranin$^+$ after formalin and Carnoy's fluid fixation) and mucosal (alcian blue$^+$, safranin$^-$ only after Carnoy's fluid fixation) mast cells from one another in rodents. Presumably, the presence of heparin as the preponderant proteoglycan in connective tissue mast cells and of chondroitin sulfates in mucosal mast cells account for these differences in rodents. However, heparin proteoglycan has been measured biochemically in mast cells derived from skin and lung (80,81,90,91, 154) and has been detected by antithrombin III–gold electron microscopy in all mast cells, abolishing the discriminating basis of these histochemical techniques in humans. Human basophils have negligible amounts of tryptase and no detectable chymase, carboxypeptidase, or cathepsin G. The preponderant proteoglycan in human basophils is chondroitin sulfate A, although human leukemic basophils (109) and cultured human peritoneal macrophages (83) also produce chondroitin sulfate E.

In histologically normal tissues, MC_T cells are the preponderant type found in the lung, particularly alveoli, and in the small-intestinal mucosa, whereas MC_{TC} cells are the preponderant type found in the skin and in the gastrointestinal submucosa (40). In normal synovium, MC_{TC} cells are preponderant. Table 5 summarizes the distribution of MC_T and MC_{TC} cells in tissue sections obtained from apparently normal human tissues. Although a particular mast cell type is often preponderant in a particular tissue, it is worthwhile to note that smaller amounts of the other mast cell type also are usually present, and that the relative abundance of MC_{TC} and MC_T cells may change with tissue inflammation or fibrosis, making it impossible to base a subtype designation on location alone.

C. Functional and Pharmacological Differences of MC_{TC} and MC_T Cells

Degranulation of human mast cells in response to various agents and inhibition of degranulation by antiallergic drugs have been studied in dispersed preparations of mast cells obtained from various anatomical sites. It is important to keep in mind that such preparations often contain a mixture of MC_T and MC_{TC} cells, depending on the site examined, the pathological condition of the donor organ, and the method of dispersion and purification. In addition, differences in the secretory response between mast cells isolated from different tissues may relate to microenvironmental influences, rather than to lineage. For example, rat mucosal mast cells, RBL-2H3, when cocultured with fibroblasts, develop the capacity to be activated by compound 48/80, but continue to synthesize only the mucosal mast cell protease, chymase II (155). When removed from the fibroblasts, the RBL-2H3 cells revert to the compound 48/80-resistant state more characteristic of rat mucosal mast cells.

In humans, the relative weight of lineage versus environmental factors on mast cell function is unresolved. Nevertheless, mast cells derived from foreskin of the MC_{TC} type release histamine in response to nonimmunological agents, such as compound 48/80, basic polypeptides (polylysine, polyarginine), morphine sulfate, fMet peptides, substance P, and the anaphylatoxin C5a, whereas mast cells isolated from lung (mostly MC_T cells) do not (156). Also, cromolyn exerts a modest inhibitory effect on IgE-mediated activation of lung-derived, but not skin-derived mast cells (157–160). Thus, different profiles of non-immunological activation and of pharmacological inhibition appear to exist among mast cells isolated from different tissue sites.

IV. GROWTH AND DIFFERENTIATION OF MAST CELLS

The cellular origin and lineage of mature mast cells subpopulations has been studied in both human (Fig. 4) and rodent species. The first definitive studies of mast cell differentiation were done in the rodent, but human mast cell differentiation may differ slightly and must be compared with basophil differentiation, so this topic is treated separately.

Historically, the hematopoietic origin of the basophil has always been understood, but the early descriptions of mast cells as connective tissue cells caused long-standing confusion over their origin. It is now clear that mast cells, similar to basophils and other blood cells, are derived from bone marrow stem cells; but unlike blood cells, undergo only part of their differentiation in marrow and complete their differentiation in peripheral mucosal or connective tissue microenvironments rich in fibroblasts or stromal cells.

A. Rodents

The bone marrow origin of mast cells was first demonstrated through a series of in vivo reconstitution studies (161–166). Injection of bone marrow cells from Beige mice, the mast cells of which are marked by giant granules, into dermis of normal mice resulted in tissue mast cells arranged in clonal patterns of either donor cells (giant granules) or host cells (normal granules), but never were cell types mixed within the same cluster (162). Injection of normal bone marrow

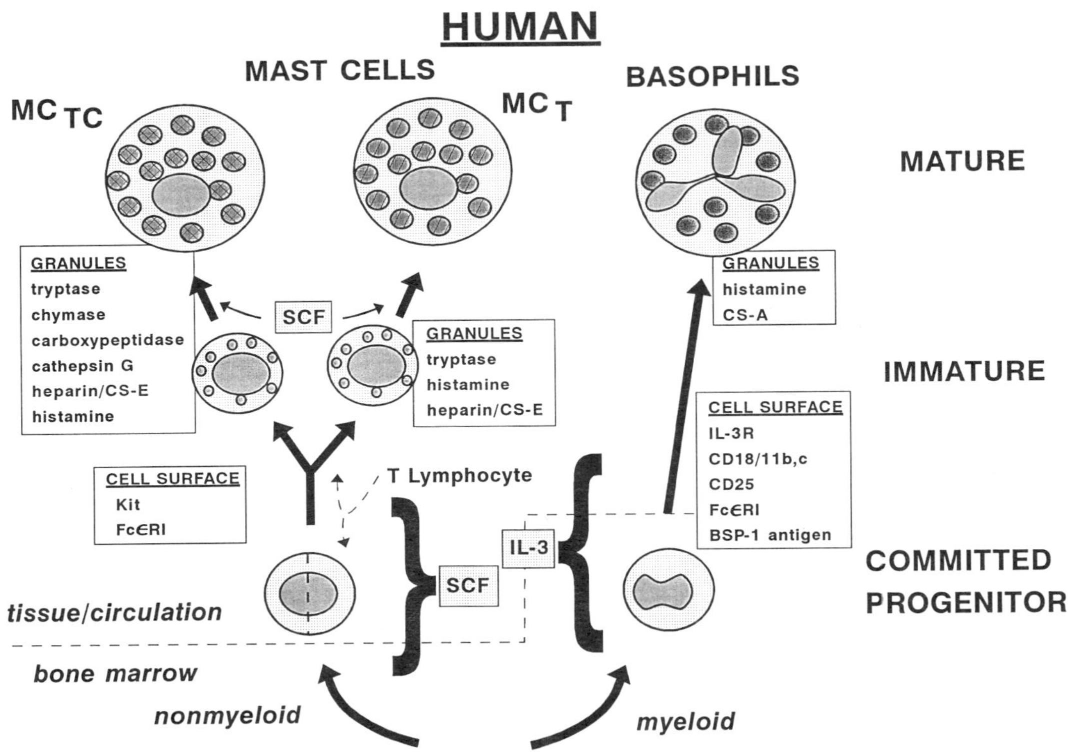

Figure 4 Developmental pathways for human mast cells and basophils, CS, chondroitin sulfate; SCF, stem cell factor.

cells into mice with homozygous mutations at the white spotting locus *(W)*, which causes a profound mast cell deficiency (167), reconstitutes mast cells in many tissues (168). Mice with homozygous mutations at the steel *(Sl)* locus are also profoundly mast cell-deficient, but cannot be cured by injection of normal bone marrow cells. These and other results suggested that mice with *Sl* mutations, unlike those with *W* mutations, have normal bone marrow progenitors for mast cells, but the local microenvironment where mast cells complete their differentiation is defective (166,169). The molecular explanation for these mast cell deficiencies was provided when the *W* and the *Sl* genes were cloned (170–177). The *W* gene encodes the c-*kit* protooncogene product, a receptor tyrosine kinase highly expressed on mast cell surfaces. The *Sl* gene encodes a hematopoietic growth factor (Table 6) called stem cell factor (SCF), kit ligand (KL) or mast cell growth factor (MCGF), which is the ligand for Kit. Stem cell factor is highly expressed on the surface of fibroblasts and stromal cells and also is detected in a cell-free soluble form (reviewed in 178). Thus, mast cell differentiation, unlike basophil differentiation, can be separated into a bone marrow phase and a peripheral phase (see Fig. 4), which may depend on contact with fibroblasts or stromal cells. Both phases are mediated by growth factors acting together or in sequence.

Within the bone marrow, the precise lineage of mast cells is still unknown, unlike basophils, which are known to be part of the granulocytic lineage. Recent experimental evidence suggests several possibilities for a mast cell lineage: the direct differentiation from pluripotent stem cells, or a branch pathway along the erythrocyte–megakaryocyte or the monocyte–macrophage lineages. Mast cells may be related to pluripotent stem cells as a default lineage, not functionally related to nor dependent on any other lineage, because in rodents both mast cells and stem cells highly express Kit and the IL-3 receptor. The ligands for these receptors, SCF and IL-3, respectively, were both originally defined as mast cell growth factors, but were later defined as growth factors capable of promoting self-renewal of stem cells. In fact, if naive bone marrow from rodents is cultured continuously with either SCF or IL-3, without other lineage-specific growth factors, the cultures yield more than 90% mast cells after a few weeks. No other lineage is so singularly stimulated by growth factors active on primitive stem cells. However, in humans, bone marrow cultures do not respond to IL-3 alone by producing mast cells, and mature human mast cells lack high-affinity IL-3 receptors (discussed later). Instead, human SCF is the major growth factor for human mast cells. The evidence that mast cells may be related in lineage to erythrocytes and megakaryocytes centers on the fact that all three lineages, but not other lineages, express the nuclear transcription factor GATA-1. GATA-1 is a *cis*acting DNA-binding protein that uses two novel zinc-finger domains to activate transcription message for cell-specific products, such as globin (179) and granule proteases (180). However, mast cells, unlike erythrocytes, also

Table 6 Cytokines Involved in Mouse Mast Cell Growth and Differentiation

Cytokine	Alternative name	Receptor on mast cell	Effect on mast cell progenitor	Effect on granulated mast cell or mast cell line
Stem cell factor (SCF)	Kit ligand (KL), Mast cell growth factor (MCGF), *Steel* factor (SLF)	Kit	Proliferation, differentiation and possibly homing of committed progenitors	Proliferation, priming, survival
Interleukin-3 (IL-3)	Multi-CSF, mast cell growth factor (MCGF)	IL-3R (α and β chain)	Proliferation and limited differentiation of committed progenitors	Proliferation, priming, survival
Interleukin-4 (IL-4)	B-cell stimulatory factor 1 (BSF-1)	IL-4R	Cofactor for proliferation; promotes connective tissue phenotype	Growth cofactor
Interleukin-9 (IL-9)	P-40; mast cell growth-enhancing activity (MEA)		Unknown	Growth cofactor; induction of IL-6 production
Interleukin-10 (IL-10)	Cytokine synthesis-inhibitory factor (CSIF)		Growth cofactor; induces expression of mucosal proteases	Growth cofactor
Nerve growth factor (NGF)		NGFR	Primes proliferation	Growth cofactor
IgE and IgE immune complexes		FcϵRI; FcγRII	Induces production of autocrine IL-3; progression from uncommitted to committed progenitor	Growth of connective tissue-type mast cells; induces production of cytokines
Transforming growth factor-β (TGF-β)			Inhibits proliferation	Inhibits proliferation
Interferon-γ (IFN-γ)		IFN-γR	Inhibits proliferation	Inhibits proliferation

employ GATA-2 and GATA-3 transcription factors. Finally, recent evidence of high levels of cytokine production by mast cells (reviewed in 181), and the presence of macrophage-associated surface markers on human mast cells, suggests that the mast cell lineage may have more in common with the macrophage, another cytokine-producing cell, than with myeloid cells such as basophils. In contrast, the finding of myeloid-associated surface markers on human basophils supports their distinct developmental pathway (182).

To begin the peripheral phase of the mast cell lineage, IL-3-dependent progenitors, derived from bone marrow, progress to the level of mast cell-committed progenitors, which no longer require exogenous IL-3 (183), but can respond to SCF alone. This committed state may be due to the induction of autocrine cytokine biosynthesis by IgE or IgG immune complexes known to be present under conditions during which committed progenitors are generated (184, 185). These committed progenitors migrate to and settle in fibroblast-rich connective tissues or mucosal sites by a process that may be mediated by the binding of their laminin receptors (186) with laminin in tissues or their c-*kit* receptors (187) with SCF on fibroblasts. At such sites, committed progenitors undergo limited clonal expansion and final differentiation into granulated, phenotypically distinct mast cells. Once differentiated, mature granulated mast cells, even apparently quiescent ones, can be stimulated to undergo limited proliferation by many of the same factors that act on the progenitors. For example, mature peritoneal mast cells can respond to SCF (188) or IgE immune complexes (189) to form small colonies of mast cells in semisolid medium.

Regulation of mast cell phenotype, mucosal versus connective tissue, has not been fully resolved. Although colonies from granulated peritoneal mast cells may have a connective tissue phenotype, the peritoneal cavity also contains a low-density, nongranulated mast cell progenitor that forms larger colonies of mast cells with a mucosal or immature phenotype (190). These and other data (see discussion on human mast cell differentiation) suggest the presence of distinct mast cell progenitors that are committed to develop along either mucosal or connective tissue lineages. Conversely, histochemical evidence that bone marrow-derived mast cells can mature as either mucosal or connective tissue mast cells in the proper environment in vivo has also been reported (191,192), suggesting that mast cells can transdifferentiate or are interconvertible (193). Many of the precise molecular and cellular events that control mast cell proliferation and differentiation, particularly those that involve fibroblasts, have been determined in tissue culture experiments.

In vitro systems for studying differentiation of mast cells under the influence of fibroblasts were first developed in landmark experiments by Ginsburg (194–196), who demonstrated that cultures of mouse thymocytes or hyperimmune lymph node cells plated on embryonic fibroblast monolayers developed mast cells after several weeks of culture. Later studies in the rat, by Ishizaka and her colleagues, confirmed these observations and demonstrated the presence of histamine and high-affinity IgE receptors in cultured mast cells (197,198). Later, other investigators used suspension cultures without fibroblast monolayers to grow mast cells from bone marrow in various cell-conditioned media containing a mast cell growth factor (199–205) that was shown to be identical with IL-3 (206). In mice, IL-3 induces the growth and differentiation of mast cells from bone marrow cells, but the cells, although granulated, tend to be immature. For cxamplc, these cells contain negligible amounts of the proteases that are present in great abundance in mast cells detected in vivo. However, mRNA for mast cell carboxypeptidase A and tryptase (MMCP-6), two markers of the connective tissue phenotype, are expressed in these IL-3-dependent mast cells, indicating progenitors of this type of mast cell. Consistent with this conclusion is the absence of mRNA by Northern blot analysis for a chymase (MMCP-2) that is expressed in mucosal but not connective tissue mast cells (71). The mRNA for another chymase, MMCP-4, is expressed late in both connective tissue and mucosal mast cells, but not in immature IL-3-driven bone marrow-cultured mast cells (72,73). When these cells are exposed to SCF in vitro, they mature and produce the proteases associated with CTMC. The factor(s) crucial for generation of MMC are less clear, although preliminary reports suggest that IL-10 induces expression of the mRNA molecules for MMCP-1 and MMCP-2 characteristic of that mast cell type.

The importance of fibroblasts and a connective tissue microenvironment in the development and maintenance of mast cell phenotype has been established in vitro (207–209) and in vivo (161,210,211) using mature mast cells or mast cell progenitors (212,213). Fibroblasts appear to play a more important role late in mast cell differentiation, after more primitive progenitors leave the bone marrow and become committed to a mast cell lineage. This is illustrated by the observation that mice with the *W* mutation, which lack c-*kit* function, are capable of producing IL-3-dependent mast cell progenitors in bone marrow, but cannot make a functional mast cell-committed progenitors, such as those isolated in the mesenteric lymph nodes of hookworm-infected mice (213) or in the peritoneal cavity (214). These mast cell-committed progenitors do not require exogenous IL-3 and, indeed, are not likely to encounter much IL-3 in a fibroblast-rich environment (209) such as dermis, but they can respond through their c-*kit* receptors (187) to SCF, which is highly expressed on the surface or is shed from fibroblasts. Other fibroblast-derived growth factors, such as nerve growth factor (215), can affect mast cell phenotype, whereas others such as interferons can inhibit mast cell proliferation (216). At sites of developing immune responses, mast cell hyperplasia can result from other T lymphocyte-derived cytokines such as IL-3, IL-4, IL-9, and IL-10 and from nerve growth factor (see Table 6).

B. Humans

In humans, the growth factors for mast cells and basophils appear to be somewhat different from those in rodents. For example, IL-3 is a growth factor for human basophils in vitro (217,218) and in vivo in primates (219), but not for human mast cells. Consistent with this observation is the presence of IL-3-binding sites on basophils (220), but not mast cells (221). Interleukin-4 lacks substantial activity on the growth of either mast cells or basophils. On the other hand, cocultures of cord blood mononuclear cells (222) and fetal liver cells (223) with murine 3T3 fibroblasts routinely yield mast cells in excess of basophils. A great portion of fibroblast-derived activity may be human SCF which, by itself, promotes the growth of mast cells from cord blood and fetal liver (A. A. Irani and L. B. Schwartz, unpublished data).

Although the question of distinct lineages for human MC_T and MC_{TC} cell types ultimately may be best answered in vitro, two experiments suggest that these mast cell types develop along distinct pathways. First, in humans with inherited combined immunodeficiency disease, as well as in those with acquired immunodeficiency syndrome, marked and selective decreases in MC_T cell concentrations occur in the bowel, whereas the concentration and distribution of MC_{TC} cells are unaffected (224). This suggests that the appearance in tissues of MC_T cells is dependent on functional T lymphocytes and that MC_{TC} cell development proceeds independently. Second, ultrastructural studies of immature human mast cells were also consistent with distinct pathways for development (225). Specifically, immature MC_T cells contained tryptase alone in all granules, and complete scrolls were the characteristic ultrastructural feature of these granules. Immature MC_{TC} had tryptase together with chymase in nearly all granules, but unlike mature MC_{TC}, the characteristic feature of these granules was one or more electron-dense core, surrounded by a less electron-dense matrix. In MC_{TC} cells of intermediate maturity, dense cores and grating or lattice features were observed in the same granules, as well as in different granules of the same cell, suggesting these immature cells were truly precursors of the mature MC_{TC} cell. These ultrastructural and compositional differences between the two types of immature mast cells indicate that distinct developmental pathways for these cells are present at or before the stage when granule formation begins.

A current working model for the lineage of human mast cells and basophils is shown in Figure 4. The MC_T and MC_{TC} cells arise as in rodents, from a common hematopoietic progenitor, perhaps derived directly from stem cells or from monocytic or megakaryotic–erythroid precursors; basophils likely arise from myeloid precursors. Mast cell progenitors are Kit^+, possibly $IL\text{-}3R^-$, whereas basophil progenitors appear to be Kit^-, $IL\text{-}3R^+$. These progenitor cells with commitment to a mast cell lineage are released from hematopoietic tissue into the circulation. Mast cells are the preponderant cells that develop from fetal liver cells in the presence of rhuSCF, whereas basophils along with other myelocytic cells develop in the presence of IL-3. At the time of commitment, or soon thereafter, they enter their tissue sites of residence and develop along MC_{TC} or MC_T pathways to the final mature forms. Basophils complete their maturation in the bone marrow and enter the circulation as mature granulated cells. The presence of receptors for laminin on the surface of rodent mast cells (226,227), although not yet shown in humans, may facilitate the tissue localization of human mast cells. The persistent presence of Kit on the mast cell surface during their differentiation and after their maturation is complete may also serve a homing and anchoring function because SCF, the Kit-ligand, is expressed on the surface of fibroblasts.

Whether the predominance of a particular type of mast cell in different tissues is due to local influences on the development of committed mast cell progenitors to a particular mast cell type, or is due to the selective recruitment of mast cells already directed toward a particular type, is unclear. T-lymphocyte-associated activity appears to be important in the development of MC_T cells, whereas fibroblast-derived factors, particularly SCF, appear to be important in the development of both mast cell types. However, the T-lymphocyte factors involved and the stage of maturation or recruitment that they affect is unknown. In contrast, IL-3 is critical for basophil maturation. The particular maturational stages at which certain capabilities are acquired or lost, such as proliferation, dedifferentiation, interconversion, and activation–secretion, also need to be clarified.

V. BIOLOGY AND PATHOBIOLOGY OF HUMAN MAST CELLS

The two major barriers to the assessment of mast cell involvement in clinical events have been the localization of this cell type in tissue, rather than the circulation, and the lack of a marker, detectable in biological fluids, with sufficient sensitivity and specificity to indicate activation of mast cells occurring in tissues.

A. Mast Cell Types and Human Disease

Several studies have begun to address the presence of distinct mast cell types in human diseases not typically associated with atopy. For example, mast cell hyperplasia has been noted in rheumatoid synovium near sites of cartilage erosion (228). The MC_{TC} cells appear to be associated with areas of dense fibrosis, whereas variable ratios of both MC_T and MC_{TC} cells were found at sites of active inflammation (229). Increased mast cell numbers also have been documented in subjects with scleroderma (38). A possible role for mast cells in the tissue destruction and fibrosis that occurs with rheumatoid arthritis and scleroderma has been suggested. The ability of purified tryptase to activate latent collagenase derived from rheumatoid synovium (36) may be relevant to these observations. Characterization of mast cell types by the presence of specific markers has provided a means with which to begin to delineate mast cell involvement in disease processes. These findings are likely to have important pathogenetic, diagnostic, and therapeutic outcomes.

B. Mast Cell Activation

Several mediators of mast cells have been considered as potentially useful indicators of mast cell or basophil activation. Lipid mediators, such as LTC_4 or PGD_2, or their associated metabolites, lack sufficient cell specificity and are rapidly metabolized. For example, eosinophils produce high levels of LTC_4, and antigen-processing cells, such as cutaneous Langerhans cells, exhibit high levels of PGD synthase (113). Histamine is present in basophils as well as mast cells (230), is rapidly metabolized in the circulation (231), may be derived from certain ingested foods, and reportedly is produced and secreted by other cell types (232). These factors impede its use as a practical marker for clinicians, in spite of its value in research. A suitable assay to evaluate the usefulness of heparin as a specific indicator of mast cell activation is not available. Certain neutral proteases have greater potential use than other mediators because of their abundant presence in, and selective localization to, mast cell secretory granules. Specific immunoassays for tryptase, which is present in secretory granules of all mast cells, were developed to quantify total amounts of the enzyme in complex biological fluids (21,22). In serum, tryptase levels in normal subjects are undetectable (<1 ng/mL), whereas elevated levels are detected in systemic mast cell disorders, such as anaphylaxis and mastocytosis. Thus, tryptase levels provide a more precise assessment of mast cell activation in vivo than was previously available.

Histamine, unlike tryptase, is present in mast cells and basophils at comparable amounts (1–3 pg/cell). Therefore, histamine, by itself, does not discriminate between mast cell and basophil involvement. Histamine levels, when elevated in the absence of detectable tryptase or PGD_2, suggests basophil involvement. However, to prove basophil involvement, the presence of basophils in the tissue also should be demonstrated. This approach has been used in the skin (233,234) and nose (235) to implicate mast cell activation during the immediate response and basophil activation during the late response to allergen challenge. The response in lung appears to be analogous. Unfortunately, no one mediator has yet been found that can serve as a specific indicator of basophil activation.

Tryptase quantities in serum are elevated in systemic anaphylaxis (236). During bee sting-induced anaphylaxis, circulating levels are maximal at 60–120 min, and then decline with a half-life of 1.5–2.5 h (231). Elevated levels of mast cell tryptase in postmortem sera reflect antemortem mast cell activation and suggest that tryptase levels also may be used as a practical indicator of fatal anaphylaxis (237).

Ongoing mast cell activation in asthma, detected by elevated levels of tryptase and PGD_2 in bronchoalveolar lavage fluid, higher spontaneous release of histamine by mast cells obtained from the bronchoalveolar lavage fluid of asthmatics, and ultrastructural analysis of mast cells in pulmonary tissue, appears to be a characteristic of the inflammation of asthma (238–241). The early and late response to endobronchial challenge of atopic subjects without asthma at different doses of allergen showed findings analogous to those in the skin, with eosinophil and perhaps basophil recruitment and activation occurring during the late-phase response (242). Mast cell activation appears to be involved in the immediate respiratory and vascular responses to oral aspirin challenge of sensitive subjects (243). On the other hand, mast cell activation may not be a feature of exercise-induced asthma (244).

A role for mast cells in resistance to certain parasitic organisms has been enticing because of the IgE-response elicited by such organisms. The T-lymphocyte-dependent hyperplasia of mucosal mast cells in response to intestinal parasite infestations has been well documented in rodents (245–248), less so in humans. In addition, elevated levels of chymase II from mucosal mast cells of rats during natural expulsion of intestinal nematode suggest a role for mast cells in parasite resistance (249). However, treatment of mice with anti-IL-4 and anti-IL-3 antibodies largely suppresses both the increase in IgE and mucosal mast cells to *Nippostrongylus brasiliensis* infection, without substantially altering the clinical course (250). Thus, the exact role for mast cells in parasite infestations is unclear.

REFERENCES

1. P. Ehrlich, Beitrage zur Theorie und Praxis der Histologischen Farbung, [Doctoral Thesis], University of Leipzig, Germany (1878).
2. P. Ehrlich, Uber die Specifischen Granulatoinen des Blutes, *Arch. Anat. Physiol. Phys. Abt.*, 571 (1879).
3. J. F. Riley, and G. B. West, The presence of histamine in tissue mast cells, *J. Physiol.*, *120*: 528 (1953).
4. K. Ishizaka, K. Tomioka, and T. Ishizaka, Mechanisms of passive sensitization. I. Presence of IgE and IgG molecules on human leukocytes, *J. Immunol.*, *105*: 1459 (1970).
5. K. Tomioka, and K. Ishizaka, Mechanisms of passive sensitization. II. Presence of receptors for IgE on monkey mast cells, *J. Immunol.*, *107*: 971 (1971).
6. A. Johnson, T. E. Hughli, and H. J. Muller-Eberhard, Release of histamine from rat mast cells by the complement peptides C3a and C5a, *Immunology*, *28*: 1067 (1975).

7. R. D. Higginbotham, and S. Karnella, The significance of the mast cell response to bee venom, *J. Immunol.*, *106*: 233 (1970).
8. A. P. Kaplan, M. Baeza, S. Reddigari, and P. Kuna, Histamine releasing factors, *Clin. Exp. Allergy*, *21*(Suppl. 1): 8 (1991).
9. R. W. Schayer, Histidine decarboxylase in mast cells, *Ann. N. Y. Acad. Sci.*, *103*: 164 (1963).
10. K. Yamauchi, R. Sato, Y. Tanno, Y. Ohkawara, K. Maeyama, T. Watanabe, K. Satoh, M. Yoshizawa, S. Shibahara, and T. Takishima, Nucleotide sequence of the cDNA encoding L-histidine decarboxylase derived from human basophilic leukemia cell line, KU-812-F, *Nucleic Acids Res.*, *18*: 5891 (1990).
11. D. Lagunoff, and A. Rickard, Evidence for control of mast cell granule protease in situ by low pH, *Exp. Cell Res.*, *144*: 353 (1983).
12. R. G. Johnson, S. E. Carty, B. J. Fingerhoff, and A. Scarpa, The internal pH of mast cell granules, *FEBS Lett.*, *120*: 75 (1980).
13. E. Vannier, L. C. Miller, and C. A. Dinarello, Histamine suppresses gene expression and synthesis of tumor necrosis factor alpha via histamine H_2 receptors, *J. Exp. Med.*, *174*: 281 (1991).
14. K. L. Melmon, R. E. Rocklin, and R. P. Rosenkranz, Autacoids as modulators of the inflammatory and immune response, *Am. J. Med.*, *71*: 100 (1981).
15. S. Braude, C. Coe, D. Royston, and P. J. Barnes, Histamine increases lung permeability by an H_2-receptor mechanism, *Lancet*, *2*: 372 (1984).
16. J. M. Arrang, B. Devaux, J. P. Chodkiewicz, and J. C. Schwartz, H_3-receptors control histamine release in human brain, *J. Neurochem.*, *51*: 105 (1988).
17. Neutral proteases of mast cells, *Monogr. Allergy*, *27*: 1 (1990).
18. L. B. Schwartz, R. A. Lewis, and K. F. Austen, Tryptase from human pulmonary mast cells. Purification and characterization, *J. Biol. Chem.*, *256*: 11939 (1981).
19. V. K. Hopsu, and G. G. Glenner, A histochemical enzyme kinetic system applied to the trypsin-like amidase and esterase activity in human mast cells, *J. Cell Biol.*, *17*: 503 (1963).
20. L. B. Schwartz, R. A. Lewis, D. Seldin, and K. F. Austen, Acid hydrolases and tryptase from secretory granules of dispersed human lung mast cells, *J. Immunol.*, *126*: 1290 (1981).
21. S. Wenzel, A. M. Irani, J. M. Sanders, T. R. Bradford, and L. B. Schwartz, Immunoassay of tryptase from human mast cells, *J. Immunol. Methods*, *86*: 139 (1986).
22. I. Enander, P. Matsson, J. Nystrand, A.-S. Andersson, E. Eklund, T. R. Bradford, and L. B. Schwartz, A new radioimmunoassay for human mast cell tryptase using monoclonal antibodies, *J. Immunol. Methods*, *138*: 39 (1991).
23. S. C. Alter, J. A. Kramps, A. Janoff, and L. B. Schwartz, Interactions of human mast cell tryptase with biological protease inhibitors, *Arch. Biochem. Biophys.*, *276*: 26 (1990).
24. L. B. Schwartz, Monoclonal antibodies against human mast cell tryptase demonstrate shared antigenic sites on subunits of tryptase and selective localization of the enzyme to mast cells, *J. Immunol.*, *134*: 526 (1985).
25. J. S. Miller, E. H. Westin, and L. B. Schwartz, Cloning and characterization of complementary DNA for human tryptase, *J. Clin. Invest.*, *84*: 1188 (1989).

26. J. S. Miller, G. Moxley, and L. B. Schwartz, Cloning and characterization of a second complementary DNA for human tryptase, *J. Clin. Invest.*, *86*: 864 (1990).
27. P. Vanderslice, S. M. Ballinger, E. K. Tam, S. M. Goldstein, C. S. Craik, and G. H. Caughey, Human mast cell tryptase: Multiple cDNAs and genes reveal a multigene serine protease family, *Proc. Natl. Acad. Sci. USA*, *87*: 3811 (1990).
28. P. Vanderslice, C. S. Craik, J. A. Nadel, and G. H. Caughey, Molecular cloning of dog mast cell tryptase and a related protease: Structural evidence of a unique mode of serine protease activation, *Biochemistry*, *28*: 4148 (1989).
29. J. Sarid, P. N. Benfey, and P. Leder, The mast cell-specific expression of a protease gene, RMCP II, is regulated by an enhancer element that binds specifically to mast cell trans-acting factors, *J. Biol. Chem.*, *264*: 1022 (1989).
30. L. B. Schwartz, and T. R. Bradford, Regulation of tryptase from human lung mast cells by heparin. Stabilization of the active tetramer, *J. Biol. Chem.*, *261*: 7372 (1986).
31. S. C. Alter, D. D. Metcalfe, T. R. Bradford, and L. B. Schwartz, Regulation of human mast cell tryptase. Effects of enzyme concentration, ionic strength and the structure and negative charge density of polysaccharides, *Biochem. J.*, *248*: 821 (1987).
32. S. C. Alter, and L. B. Schwartz, Effect of histamine and divalent cations on the activity and stability of tryptase from human mast cells, *Biochim. Biophys. Acta*, *991*: 426 (1989).
33. L. B. Schwartz, M. S. Kawahara, T. E. Hugli, D. Vik, D. T. Fearon, and K. F. Austen, Generation of C3a anaphylatoxin from human C3 by human mast cell tryptase, *J. Immunol.*, *130*: 1891 (1983).
34. L. B. Schwartz, Tryptase from human mast cells: Biochemistry, biology and clinical utility, *Monogr. Allergy*, *27*: 90 (1990).
35. L. B. Schwartz, T. R. Bradford, B. H. Littman, and B. U. Wintroub, The fibrinogenolytic activity of purified tryptase from human lung mast cells, *J. Immunol.*, *135*: 2762 (1985).
36. B. L. Gruber, M. J. Marchese, K. Suzuki, L. B. Schwartz, Y. Okada, H. Nagase, and N. S. Ramamurthy, Synovial procollagenase activation by human mast cell tryptase dependence upon matrix metalloproteinase 3 activation, *J. Clin. Invest.*, *84*: 1657 (1989).
37. D. G. Malone, R. L. Wilder, A. M. Saavedra-Delgado, and D. D. Metcalfe, Mast cell numbers in rheumatoid synovial tissues. Correlations with quantitative measures of lymphotic infiltration and modulation by anti-inflammatory therapy, *Arthritis Rheum.*, *30*: 130 (1987).
38. R. A. Hawkins, H. N. Claman, R. A. Clark, and J. C. Steigerwald, Increased dermal mast cell populations in progressive systemic sclerosis: A link in chronic fibrosis, *Ann. Intern. Med.*, *102*: 182 (1985).
39. N. M. Schechter, J. E. Fraki, J. C. Geesin, and G. S. Lazarus, Human skin chymotryptic protease. Isolation and relation to cathepsin G and rat mast cell proteinase, *J. Biol. Chem.*, *258*: 2973 (1983).
40. A. A. Irani, N. M. Schechter, S. S. Craig, G. DeBlois, and L. B. Schwartz, Two types of human mast cells that have distinct neutral protease compositions, *Proc. Natl. Acad. Sci. USA*, *83*: 4464 (1986).

41. I. A. R. Osman, J. R. Garrett, and R. E. Smith, Enzyme histochemical discrimination between tryptase and chymase in mast cells of human gut, *J. Histochem. Cytochem.*, *37*: 415 (1989).
42. G. H. Caughey, E. H. Zerweck, and P. Vanderslice, Structure, chromosomal assignment, and deduced amino acid sequence of a human gene for mast cell chymase, *J. Biol. Chem.*, *266*: 12956 (1991).
43. H. Urata, A. Kinoshita, D. M. Perez, K. S. Misono, F. M. Bumpus, R. M. Graham, and A. Husain, Cloning of the gene and cDNA for human heart chymase, *J. Biol. Chem.*, *266*: 17173 (1991).
44. B. U. Wintroub, N. B. Schechter, G. S. Lazarus, C. E. Kaempfer, and L. B. Schwartz, Angiotensin I conversion by human and rat chymotryptic proteinases, *J. Invest. Dermatol.*, *83*: 336 (1984).
45. N. M. Schechter, J. K. Choi, D. A. Slavin, D. T. Deresienski, S. Sayama, G. Dong, R. M. Lavker, D. Proud, and G. S. Lazarus, Identification of a chymotrypsin-like proteinase in human mast cells, *J. Immunol.*, *137*: 962 (1986).
46. S. Sayama, R. V. Iozzo, G. S. Lazarus, and N. M. Schechter, Human skin chymotrypsin-like proteinase chymase. Subcellular localization to mast cell granules and interaction with heparin and other glycosaminoglycans, *J. Biol. Chem.*, *262*: 6808 (1987).
47. N. M. Schechter, J. L. Sprows, O. L. Schoenberger, G. S. Lazarus, B. S. Cooperman, and H. Rubin, Reaction of human skin chymotrypsin-like proteinase chymase with plasma proteinase inhibitors, *J. Biol. Chem.*, *264*: 21308 (1989).
48. H. Urata, A. Kinoshita, K. S. Misono, F. M. Bumpus, and A. Husain, Identification of a highly specific chymase as the major angiotensin II-forming enzyme in the human heart, *J. Biol. Chem.*, *265*: 22348 (1990).
49. A. Kinoshita, H. Urata, F. M. Bumpus, and A. Husain, Multiple determinants for the high substrate specificity of an angiotensin II-forming chymase from the human heart, *J. Biol. Chem.*, *266*: 19192 (1991).
50. C. F. Reilly, D. A. Tewksbury, N. M. Schechter, and J. Travis, Rapid conversion of angiotensin I to angiotensin II by neutrophil and mast cell proteinases, *J. Biol. Chem.*, *257*: 8619 (1982).
51. R. A. Briggaman, N. M. Schechter, J. Fraki, and G. S. Lazarus, Degradation of the epidermal-dermal junction by a proteolytic enzyme from human skin and human polymorphonuclear leukocytes, *J. Exp. Med.*, *160*: 1027 (1984).
52. C. P. Sommerhoff, G. H. Caughey, W. E. Finkbeiner, S. C. Lazarus, C. B. Basbaum, and J. A. Nadel, Mast cell chymase. A potent secretagogue for airway gland serous cells, *J. Immunol.*, *142*: 2450 (1989).
53. G. H. Caughey, F. Leidig, N. F. Viro, and J. A. Nadel, Substance P and vasoactive intestinal peptide degradation by mast cell tryptase and chymase, *J. Pharmacol. Exp. Ther.*, *244*: 133 (1988).
54. N. M. Schechter, A.-M. A. Irani, J. L. Sprows, J. Abernethy, B. Wintroub, and L. B. Schwartz, Identification of a cathepsin G-like proteinase in the MC_{TC} type of human mast cell, *J. Immunol.*, *145*: 2652 (1990).
55. S. M. Goldstein, C. E. Kaempfer, D. Proud, L. B. Schwartz, A. M. Irani, and B. U. Wintroub, Detection and partial characterization of a human mast cell carboxypeptidase, *J. Immunol.*, *139*: 2724 (1987).

56. S. M. Goldstein, C. E. Kaempfer, J. T. Kealey, and B. U. Wintroub, Human mast cell carboxypeptidase: Purification and characterization, *J. Clin. Invest.*, *83*: 1630 (1989).
57. D. S. Reynolds, D. S. Gurley, R. L. Stevens, D. J. Sugarbaker, K. F. Austen, and W. E. Serafin, Cloning of cDNAs that encode human mast cell carboxypeptidase A, and comparison of the protein with mouse mast cell carboxypeptidase A and rat pancreatic carboxypeptidases, *Proc. Natl. Acad. Sci. USA*, *86*: 9480 (1989).
58. A.-M. A. Irani, S. M. Goldstein, B. U. Wintroub, T. Bradford, and L. B. Schwartz, Human mast cell carboxypeptidase: Selective localization to MC_{TC} cells, *J. Immunol.*, *147*: 247 (1991).
59. A. M. Dvorak, The fine structure of human basophils and mast cells, *Mast Cells, Mediators and Disease* (S. T. Holgate, ed.), Kluwer Academic Publishers, London, p.29. (1988).
60. D. S. Reynolds, D. S. Gurley, and K. F. Austen, Cloning and characterization of the novel gene for mast cell carboxypeptidase A, *J. Clin. Invest.*, *89*: 273 (1992).
61. H. R. P. Miller, J. F. Huntley, G. F. J. Newlands, and J. Irvine, Granule chymases and the characterization of mast cell phenotype in rat and mouse, *Monogr. Allergy*, *27*: 1 (1990).
62. L. B. Schwartz, C. Riedel, J. P. Caufield, S. I. Wasserman, and K. F. Austen, Cell association of complexes of chymase, heparin proteoglycan, and protein after degranulation by rat mast cells, *J. Immunol.*, *126*: 2017 (1981).
63. D. M. Haig, T. A. McKee, E. E. E. Jarrett, R. Woodbury, and H. R. P. Miller, Generation of mucosal mast cells is stimulated in vitro by factors derived from T cells of helminth-infected rats, *Nature*, *300*: 188 (1982).
64. P. N. Benfey, F. H. Yin, and P. Leder. Cloning of the mast cell protease, RMCP II. Evidence for cell-specific expression and a multi-gene family, *J. Biol. Chem.*, *262*: 5377 (1987).
65. M. T. Everitt, and H. Neurath, Rat peritoneal mast cell carboxypeptidase: Localization, purification and enzymatic properties, *FEBS Lett.*, *110*: 292 (1980).
66. L. B. Schwartz, C. Riedel, J. J. Schratz, and K. F. Austen, Localization of carboxypeptidase A to the macromolecular heparin proteoglycan–protein complex in secretory granules of rat serosal mast cells, *J. Immunol.*, *128*: 1128 (1982).
67. P. V. Subba Rao, M. M. Friedman, F. M. Atkins, and D. D. Metcalfe, Phagocytosis of mast cell granules by cultured fibroblasts, *J. Immunol.*, *130*: 341 (1983).
68. H. Kido, Y. Yokogoshi, and N. Katunuma. Kunitz-type protease inhibitor found in rat mast cells. Purification, properties, and amino acid sequence, *J. Biol. Chem.*, *263*: 18104 (1988).
69. V. J. Braganza, and W. H. Simmons, Tryptidase: A novel trypsin-like enzyme from rat skin [abstract], *Fed. Proc.*, *2*: A1006 (1988).
70. D. S. Reynolds, R. L. Stevens, and K. F. Austen, Isolation and characterization of a cDNA that encodes a mouse mast cell carboxypeptidase A (CPA)—a prominent secretory granule enzyme of connective tissue-type mast cells (CTMC), *Fed. Proc.*, *3*: A1276 (1989).
71. D. S. Reynolds, R. L. Stevens, W. S. Lane, M. H. Carr, K. F. Austen, and W.

E. Serafin, Different mouse mast cell populations express various combinations of at least six distinct mast cell serine proteases, *Proc. Natl. Acad. Sci. USA*, *87*: 3230 (1990).

72. W. E. Serafin, D. S. Reynolds, S. Rogelj, W. S. Lane, G. A. Conder, S. S. Johnson, K. F. Austen, and R. L. Stevens, Identification and molecular cloning of a novel mouse mucosal mast cell serine protease, *J. Biol. Chem.*, *265*: 423 (1990).
73. R. Huang, T. Blom, and L. Hellman, Cloning and structural analysis of MMCP-1, MMCP-4 and MMCP-5, three mouse mast cell-specific serine proteases, *Eur. J. Immunol.*, *21*: 1611 (1991).
74. W. Chu, D. A. Johnson, and P. R. Musich, Identification and molecular cloning of a mouse mast cell tryptase [abstract], *FASEB J.*, *4*: A2159 (1991).
75. E. Razin, R. L. Stevens, F. Akiyama, K. Schmid, and K. F. Austen, Culture from mouse bone marrow of a subclass of mast cells possessing a distinct chondroitin sulfate proteoglycan with glycosaminoglycans rich in *N*-acetylgalactosamine-4,6-disulfate, *J. Biol. Chem.*, *257*: 7229 (1982).
76. G. H. Caughey, N. F. Viro, J. Ramachondran, S. C. Lazarus, D. B. Borson, and J. A. Nadel, Dog mastocytoma tryptase: Affinity purification, characterization and amino-terminal sequence, *Arch. Biochem. Biophys.*, *258*: 555 (1987).
77. G. H. Caughey, N. F. Viro, S. C. Lazarus, and J. A. Nadel, Purification and characterization of dog mastocytoma chymase: Identification of an octapeptide conserved in chymotryptic leukocyte proteinases, *Biochim. Biophys. Acta*, *952*: 142 (1988).
78. K. Sekizawa, G. H. Caughey, S. C. Lazarus, W. M. Gold, and J. A. Nadel, Mast cell tryptase causes airway smooth muscle hyperresponsiveness in dogs, *J. Clin. Invest.*, *83*: 175 (1989).
79. S. J. Ruoss, T. Hartmann, and G. H. Caughey, Mast cell tryptase is a mitogen for cultured fibroblasts, *J. Clin. Invest.*, *88*: 493 (1991).
80. H. L. Thompson, E. S. Schulman, and D. D. Metcalfe, Identification of chondroitin sulfate E in human lung mast cells, *J. Immunol.*, *140*: 2708 (1988).
81. R. L. Stevens, C. C. Fox, L. M. Lichtenstein, and K. F. Austen, Identification of chondroitin sulfate E proteoglycans and heparin proteoglycans in the secretory granules of human lung mast cells, *Proc. Natl. Acad. Sci. USA*, *85*: 2284 (1988).
82. D. D. Metcalfe, C. E. Bland, and S. I. Wasserman, Biochemical and functional characterization of proteoglycans isolated from basophils of patients with chronic myelogenous leukemia, *J. Immunol.*, *132*: 1943 (1984).
83. S. O. Koiset, Oversulfated chondroitin sulfate proteoglycan in cultured human peritoneal macrophages, *Biochem. Biophys. Res. Commun.*, *139*: 377 (1986).
84. D. D. Metcalfe, J. Litvin, and S. I. Wasserman, The isolation, identification and characterization of sulfated glycosaminoglycans synthesized in vitro by human eosinophils, *Biochim. Biophys. Acta*, *715*: 196 (1982).
85. R. P. MacDermott, R. E. Schmidt, J. P. Caufield, A. Hein, G. T. Bartley, J. Ritz, S. F. Schlossman, K. F. Austen, and R. L. Stevens, Proteoglycans in cell-mediated cytotoxicity. Identification, localization and exocytosis of a chondroitin sulfate proteoglycan from human cloned natural killer cells during target cell lysis, *J. Exp. Med.*, *162*: 1771 (1985).

86. D. C. Seldin, N. Seno, K. F. Austen, and R. L. Stevens, Analysis of polysulfated chondroitin disaccharides by high-performance liquid chromatography, *Anal. Biochem.*, *141*: 291 (1984).
87. D. D. Metcalfe, H. L. Thompson, S. J. Klebanoff, and W. R. Henderson, Jr., Oxidative degradation of rat mast-cell heparin proteoglycan, *Biochem. J.*, *272*: 51 (1990).
88. R. L. Stevens, S. Avraham, M. C. Gartner, G. A. Bruns, K. F. Austen, and J. H. Weis, Isolation and characterization of a cDNA that encodes the peptide core of the secretory granule proteoglycan of human promyelocytic leukemia HL-60 cells, *J. Biol. Chem.*, *263*: 7287 (1988).
89. R. W. Yurt, R. W. Leid, K. F. Austen, and J. E. Silbert, Native heparin from rat peritoneal mast cells, *J. Biol. Chem.*, *252*: 518 (1977).
90. D. D. Metcalfe, N. A. Soter, S. I. Wasserman, and K. F. Austen, Identification of sulfated mucopolysaccharides including heparin in the lesional skin of a patient with mastocytosis, *J. Invest. Dermatol.*, *74*: 210 (1980).
91. D. D. Metcalfe, R. A. Lewis, J. E. Silbert, R. D. Rosenberg, S. I. Wasserman, and K. F. Austen, Isolation and characterization of heparin from human lung, *J. Clin. Invest.*, *64*: 1537 (1979).
92. R. L. Stevens, E. Razin, K. F. Austen, A. Hein, J. P. Caufield, N. Seno, K. Schmid, and F. Akiyama, Synthesis of chondroitin SO_4 E glycosaminoglycan onto *p*-nitrophenyl-β-D-xyloside and its localization to secretory granules of rat serosal mast cells and mucosal bone marrow-derived mast cells, *J. Biol. Chem.*, *258*: 5977 (1983).
93. D. D. Metcalfe, S. I. Wasserman, and K. F. Austen, Isolation and characterization of sulphated mucopolysaccharides from rat leukaemic (RBL-1) basophils, *Biochem. J.*, *185*: 367 (1980).
94. D. C. Seldin, K. F. Austen, and R. L. Stevens, Purification and characterization of protease-resistant secretory granule proteoglycans containing chondroitin sulfate di-B and heparin-like glycosaminoglycans from rat basophilic leukemia cells, *J. Biol. Chem.*, *260*: 11131 (1985).
95. H. R. Katz, K. F. Austen, B. Caterson, and R. L. Stevens, Secretory granules of heparin-containing rat serosal mast cells also possess highly sulfated chondroitin sulfate proteoglycans, *J. Biol. Chem.*, *261*: 13393 (1986).
96. L. Enerbäck, Mucosal mast cells in the rat and in man, *Int. Arch. Allergy Appl. Immunol.*, *82*: 249 (1987).
97. E. Ruoslahti, and Y. Yamaguchi, Proteoglycans as modulators of growth factor activities, *Cell*, *64*: 867 (1991).
98. L. Kjellen, and U. Lindahl, Proteoglycans: Structures and interactions, *Annu. Rev. Biochem.*, *60*: 443 (1991).
99. D. D. Metcalfe, M. Kaliner, and M. A. Donlon, The mast cell, *CRC Crit. Rev. Immunol.*, *3*: 23 (1981).
100. L. G. Oscarsson, G. Pejler, and U. Lindahl, Location of the antithrombin-binding sequence in the heparin chain, *J Biol. Chem.*, *264*: 296 (1989).
101. S. P. Peters, D. W. MacGlashan, E. S. Schulman, R. P. Schleimer, E. C. Hayes, J. Rokach, N. F. Adkinson, and L. M. Lichtenstein, Arachidonic acid metabolism in purified human lung mast cells, *J. Immunol.*, *132*: 1972 (1984).

102. A. M. Dvorak, H. F. Dvorak, S. P. Peters, E. S. Schulman, D. W. MacGlashan, K. Pyne, V. S. Harvey, S. J. Galli, and L. M. Lichtenstein, Lipid bodies: Cytoplasmic organelles important to arachidonate metabolism in macrophages and basophils, *J. Immunol.*, *131*: 2965 (1983).
103. R. A. Lewis, N. A. Soter, P. T. Diamond, K. F. Austen, J. A. Oates, and L. J. Roberts, Prostaglandin D_2 generation after activation of rat and human mast cells with anti-IgE, *J. Immunol.*, *129*: 1627 (1982).
104. I. D. Lawrence, J. A. Warner, V. L. Cohan, W. C. Hubbard, A. Kagey-Sobotka, and L. M. Lichtenstein, Purification and characterization of human skin mast cells, *J. Immunol.*, *139*: 3062 (1987).
105. R. C. Benyon, C. Robinson, and M. K. Church, Differential release of histamine and eicosanoids from human skin mast cells activated by IgE-dependent and non-immunological stimuli, *Br. J. Pharmacol.*, *97*: 898 (1989).
106. C. C. Fox, A. M. Dvorak, S. P. Peters, A. Kagey-Sobotka, and L. M. Lichtenstein, Isolation and characterization of human intestinal mucosal mast cells, *J. Immunol.*, *135*: 483 (1985).
107. C. Robinson, Mast cells and newly-generated lipid mediators, *Immunology and Medicine: Mast Cells, Mediators and Disease* (S. T. Holgate, ed.), Kluwer Academic Publishers, London, p. 149 (1988).
108. D. W. MacGlashan, Jr., S. P. Peters, J. Warner, and L. M. Lichtenstein, Characteristics of human basophil sulfidopeptide leukotriene release: Releasability defined as the ability of basophils to respond to dimeric cross-links, *J. Immunol.*, *136*: 2231 (1986).
109. M. E. Rothenberg, J. P. Caulfield, K. F. Austen, A. Hein, K. Edmiston, P. E. Newburger, and R. L. Stevens, Biochemical and morphological characterization of basophilic leukocytes from two patients with myelogenous leukemia, *J. Immunol.*, *138*: 2616 (1987).
110. J. MacDermot, C. R. Kelsey, and K. A. Waddell, Synthesis of leukotriene B_4, and prostanoids by human alveolar macrophages: Analysis by gas chromatography/mass spectrometry, *Prostaglandins*, *27*: 163 (1984).
111. O. Oelz, R. Oelz, H. R. Knapp, B. J. Sweetman, and J. A. Oates, Biosynthesis of prostaglandin D_2. I. Formation of prostaglandin D_2 by human platelets, *Prostaglandins*, *13*: 225 (1977).
112. O. Hayaishi, Molecular mechanisms of sleep–wake regulation: Roles of prostaglandins D_2 and E_2, *FASEB J.*, *5*: 2575 (1991).
113. Y. Urade, M. Ujihara, Y. Horiguchi, K. Ikai, and O. Hayaishi. The major source of endogenous prostaglandin D_2 production is likely antigen-presenting cells: Localization of glutathione-requiring prostaglandin D synthetase in histiocytes, dendritic, and Kupffer cells in various rat tissues, *J. Immunol.*, *143*: 2982 (1989).
114. D. J. Heavey, P. Lumley, S. E. Barrow, M. B. Murphy, P. P. A. Humphrey, and C. T. Dollery, Effects of intravenous infusions of prostaglandin D_2 in man, *Prostaglandins*, *28*: 755 (1984).
115. D. J. Heavey, A. Kobza-Black, S. E. Barrow, G. C. Chappell, M. W. Greaves, and C. T. Dollery, Prostaglandin D_2 and histamine release in cold urticaria, *J. Allergy Clin. Immunol.*, *78*: 458 (1986).
116. U. Pipkorn, D. Proud, L. M. Lichtenstein, A. Kagey-Sobotka, P. Norman, and

R. M. Naclerio, Inhibition of mediator release in allergic rhinitis by pretreatment with topical glucocorticosteroids, *N. Engl. J. Med.*, *316*: 1506 (1987).

117. J. J. Murray, A. B. Tonel, A. R. Brash, L. J. Roberts, P. Gosset, R. Workman, A. Capron, and J. A. Oates, Release of prostaglandin D_2 into human airways during acute antigen challenge, *N. Engl. J. Med.*, *315*: 800 (1986).

118. C. R. W. Beasley, C. Robinson, R. L. Featherstone, J. G. Varley, C. C. Hardy, M. K. Church, and S. T. Holgate, 9-alpha, 11-beta-Prostaglandin F_2, a novel metabolite of prostaglandin D_2, is a potent contractile agonist of human and guinea pig airways, *J. Clin. Invest.*, *79*: 978 (1987).

119. L. J. Roberts II, B. J. Sweetman, R. A. Lewis, K. F. Austen, and J. A. Oates, Increased production of prostaglandin D_2 in patients with systemic mastocytosis, *N. Engl. J. Med.*, *303*: 1400 (1980).

120. R. W. Fuller, C. M. Dixon, C. T. Dollery, and P. J. Barnes, Prostaglandin D_2 potentiates airway responsiveness to histamine and methacholine, *Am. Rev. Respir. Dis.*, *133*: 252 (1986).

121. C. C. Hardy, C. Robinson, A. E. Tattersfield, and S. T. Holgate, The bronchoconstrictor effect of inhaled prostaglandin D_2 in normal and asthmatic men, *N. Engl. J. Med.*, *311*: 209 (1984).

122. E. J. Goetzl, and W. C. Pickett, Novel structural determinants of the human neutrophil chemotactic activity of leukotriene B, *J. Exp. Med.*, *153*: 482 (1981).

123. N. A. Soter, R. A. Lewis, E. J. Corey, and K. F. Austen, Local effects of synthetic leukotrienes (LTC_4, LTD_4, LTE_4 and LTB_4) in human skin, *J. Invest. Dermatol.*, *80*: 115 (1983).

124. D. C. Mills, and D. E. MacFarlane, Stimulation of human platelet adenylate cyclase by prostaglandin D_2, *Thromb. Res.*, *5*: 401 (1974).

125. S. C. Bischoff, T. Brunner, A. L. De Weck, and C. A. Dahinden, Interleukin 5 modifies histamine release and leukotriene generation by human basophils in response to diverse agonists, *J. Exp. Med.*, *172*: 1577 (1990).

126. S. C. Bischoff, A. L. De Weck, and C. A. Dahinden, Interleukin 3 and granulocyte/macrophage-colony-stimulating factor render human basophils responsive to low concentrations of complement component C3a, *Proc. Natl. Acad. Sci. USA*, *87*: 6813 (1990).

127. F. Levi-Schaffer, K. F. Austen, J. P. Caulfield, A. Hein, P. M. Gravallese, and R. L. Stevens, Co-culture of human lung-derived mast cells with mouse 3T3 fibroblasts: Morphology and IgE-mediated release of histamine, prostaglandin D_2, and leukotrienes, *J. Immunol.*, *139*: 494 (1987).

128. A. W. Ford-Hutchinson, M. A. Bray, M. V. Doig, M. E. Shipley, and M. J. H. Smith, Leukotriene B_4, a potent chemotactic and aggregating substance released from polymorphonuclear leukocytes, *Nature*, *286*: 264 (1980).

129. S. Dahlen, J. Bjork, P. Hedqvist, K. E. Arfors, S. Hammarstrom, J. A. Lindgren, and B. Samuelsson, Leukotrienes promote plasma leakage and leukocyte adhesion in postcapillary venules: In vivo effects with relevance to the acute inflammatory response, *Proc. Natl. Acad. Sci. USA*, *78*: 3887 (1981).

130. R. L. Hoover, M. J. Karnovsky, K. F. Austen, E. J. Corey, and R. A. Lewis, Leukotriene B_4 action on endothelium mediates augmented neutrophil/endothelial adhesion, *Proc. Natl. Acad. Sci. USA*, *81*: 2191 (1984).

131. K. F. Austen, E. J. Corey, J. M. Drazen, and A. G. Leitch, The effect of indomethacin on the contractile response of the guinea-pig lung parenchymal strip to leukotrienes B_4, C_4, D_4 and E_4, *Br. J. Pharmacol.*, *80*: 47 (1982).
132. R. A. Lewis, K. F. Austen, and R. J. Soberman, Leukotrienes and other products of the 5-lipoxygenase pathway—biochemistry and relation to pathobiology in human diseases, *N. Engl. J. Med.*, *323*: 645 (1990).
133. G. D. Phillips, and S. T. Holgate, Interaction of inhaled LTC_4 with histamine and PGD_2 on airway caliber in asthma, *J. Appl. Physiol.*, *66*: 304 (1989).
134. M. A. Brown, J. H. Pierce, C. J. Watson, J. Falco, J. N. Ihle, and W. E. Paul, B cell stimulatory factor-1/interleukin-4 mRNA is expressed by normal and transformed mast cells, *Cell*, *50*: 809 (1987).
135. J. D.-E. Young, C.-C. Liu, G. Butler, Z. A. Cohn, and S. J. Galli, Identification, purification, and characterization of a mast cell-associated cytolytic factor related to tumor necrosis factor, *Proc. Natl. Acad. Sci. USA*, *84*: 9175 (1987).
136. M. Steffen, M. Abboud, G. K. Potter, Y. P. Yung, and M. A. S. Moore, Presence of tumour necrosis factor or a related factor in human basophil/mast cells, *Immunology*, *66*: 445 (1989).
137. L. M. Klein, R. M. Lavker, W. L. Matis, and G. F. Murphy, Degranulation of human mast cells induces an endothelial antigen central to leukocyte adhesion, *Proc. Natl. Acad. Sci. USA*, *86*: 8972 (1989).
138. C. Heusser, and M. K. Church, "IL-4 in human mast cells," *XIVth International Congress of Allergology and Clinical Immunology* (1991).
139. L. J. Walsh, G. Trinchieri, H. A. Waldorf, D. Whitaker, and G. F. Murphy, Human dermal mast cells contain and release tumor necrosis factor α, which induces endothelial leukocyte adhesion molecule 1, *Proc. Natl. Acad. Sci. USA*, *88*: 4220 (1991).
140. S. J. Galli, and K. F. Austen, *Mast Cell and Basophil Differentiation in Health and Disease*, Raven Press, New York (1989).
141. R. L. Stevens, T. D. Lee, D. C. Seldin, K. F. Austen, A. D. Befus, and J. Bienenstock, Intestinal mucosal mast cells from rats infected with *Nippostrongylus brasiliensis* contain protease-resistant chondroitin sulfate di-B proteoglycans, *J. Immunol.*, *137*: 291 (1986).
142. W. E. Serafin, H. R. Katz, K. F. Austen, and R. L. Stevens, Complexes of heparin proteoglycans, chondroitin sulfate E proteoglycans, and [^{3}H]diisopropyl fluorophosphate-binding proteins are exocytosed from activated mouse bone marrow-derived mast cells, *J. Biol. Chem.*, *261*: 15017 (1986).
143. E. Razin, R. L. Stevens, K. F. Austen, J. P. Caulfield, A. Hein, F. T. Liu, M. Clabby, G. Nabel, H. Cantor, and S. Friedman, Cloned mouse mast cells derived from immunized lymph node cells and from foetal liver cells exhibit characteristics of bone marrow-derived mast cells containing chondroitin sulphate E proteoglycan, *Immunology*, *52*: 563 (1984).
144. R. V. Tantravahi, R. L. Stevens, K. F. Austen, and J. H. Weis, A single gene in mast cells encodes the core peptides of heparin and chondroitin sulfate proteoglycans, *Proc. Natl. Acad. Sci. USA*, *83*: 9207 (1986).
145. S. Gibson, and H. R. Miller, Mast cell subsets in the rat distinguished im-

munohistochemically by their content of serine proteinases, *Immunology*, *58*: 101 (1986).

146. H. R. P. Miller, J. F. Huntley, G. F. J. Newlands, A Mackellar, D. A. Lammas, and D. Wakelin, Granule proteinases define mast cell heterogeneity in the serosa and the gastrointestinal mucosa of the mouse, *Immunology*, *65*: 559 (1988).
147. G. F. Newlands, S. Gibson, D. P. Knox, R. Grencis, D. Wakelin, and H. R. Miller, Characterization and mast cell origin of a chymotrypsin-like proteinase isolated from intestines of mice infected with *Trichinella spiralis*, *Immunology*, *62*: 629 (1987).
148. D. J. Heavey, P. B. Ernst, R. L. Stevens, A. D. Befus, J. Bienenstock, and K. F. Austen, Generation of leukotriene C_4, leukotriene B_4, and prostaglandin D_2 by immunologically activated rat intestinal mucosa mast cells, *J. Immunol.*, *140*: 1953 (1988).
149. J. M. Mencia-Huerta, E. Razin, E. W. Ringel, E. J. Corey, D. Hoover, K. F. Austen, and R. A. Lewis, Immunologic and ionophore-induced generation of leukotriene B_4 from mouse bone marrow-derived mast cells, *J. Immunol.*, *120*: 1885 (1983).
150. H. R. Katz, R. P. LeBlanc, and S. W. Russell, Two classes of mouse mast cells delineated by monoclonal antibodies, *Proc. Natl. Acad. Sci. USA*, *80*: 5916 (1983).
151. H. R. Katz, G. A. Schwarting, P. A. LeBlanc, K. F. Austen, and R. L. Stevens, Identification of the neutral glycosphingolipids of murine mast cells: Expression of Forssman glycolipid by the serosal by not the bone marrow-derived subclass, *J. Immunol.*, *134*: 2617 (1985).
152. H. R. Katz, E. T. Dayton, F. Levi-Schaffer, A. C. Benson, K. F. Austen, and R. L. Stevens, Coculture of mouse IL-3-dependent mast cells with 3T3 fibroblasts stimulates synthesis of globopentaosylceramide (Forssman glycolipid) by fibroblasts and surface expression on both populations, *J. Immunol.*, *140*: 3090 (1988).
153. A. A. Irani, and L. B. Schwartz, Neutral proteases as indicators of human mast cell heterogeneity, *Neutral Proteases of Mast Cells* (L. B. Schwartz, ed.) S. Karger, Basel, p.146 (1990).
154. R. Eliakim, L. Gilead, M. Ligumsky, E. Okon, D. Rachmilewitz, and E. Razin, Histamine and chondroitin sulfate E proteoglycan released by cultured human colonic mucosa: Indication for possible presence of E mast cells, *Proc. Natl. Acad. Sci. USA*, *83*: 461 (1986).
155. M. Swieter, R. Midura, C. Oliver, E. Berenstein, V. Hascall, S. E. Mergenhagen, and R. P. Siraganian, Rat basophilic leukemia (RBL-2H3) cells acquire responsiveness to compound 48/80 by coculture with 3T3 fibroblasts [abstract], *FASEB J.*, *5*: A1085 (1991).
156. M. K. Church, G. J. K. Pao, and S. T. Holgate. Characterization of histamine secretion from mechanically dispersed human lung mast cells: Effects of anti-IgE, calcium ionophore A 23187, compound 48/80, and basic polypeptides, *J. Immunol.*, *129*: 2116 (1982).
157. K. C. Flint, K. B. P. Leung, F. L. Pearce, B. N. Hudspith, J. Brostoff, and N. M. Johnson, Human mast cells recovered by bronchoalveolar lavage: Their

morphology, histamine release and the effects of sodium cromoglycate, *Clin. Sci.*, *68*: 427 (1985).

158. W. Schmutzler, K. Delmich, D. Eichelberg, S. Gluck, T. Greven, H. Jurgensen, K. P. Riesener, G. Risse, and P. Pult, The human adenoidal mast cell. Susceptibility to different secretagogues and secretion inhibitors, *Int. Arch. Allergy Appl. Immunol.*, *77*: 177 (1985).
159. M. K. Church, and J. Hiroi, Inhibition of IgE-dependent histamine release from human dispersed lung mast cells by anti-allergic drugs and salbutamol, *Br. J. Pharmacol.*, *90*: 421 (1987).
160. L. S. Cleggs, M. K. Church, and S. T. Holgate, Histamine secretion from human skin slices induced by anti-IgE and artificial secretagogues and the effects of sodium cromoglycate and salbutamol, *Clin. Allergy*, *15*: 321 (1985).
161. T. Sonoda, Y. Kanayama, H. Hara, C. Hayashi, M. Tadokoro, T. Yonezawa, and Y. Kitamura, Proliferation of peritoneal mast cells in the skin of W/W^v mice that genetically lack mast cells, *J. Exp. Med.*, *160*: 138 (1984).
162. Y. Kitamura, H. Matsuda, and K. Hatanaka, Clonal nature of mast cells clusters formed in W/W^v mice after bone marrow transplantation, *Nature*, *281*: 154 (1979).
163. Y. Kitamura, M. Shimada, S. Go, H. Matsuda, K. Hatanaka, and M. Seki, Distribution of mast-cell precursors in hematopoeitic and lymphopoietic tissues of mice, *J. Exp. Med.*, *150*: 482 (1979).
164. K. Hatanaka, Y. Kitamura, and Y. Nishimune, Local development of mast cells from bone marrow-derived precursors in the skin of mice, *Blood*, *53*: 142 (1979).
165. Y. Kitamura, M. Shimada, K. Hatanaka, and Y. Miyano, Development of mast cells from grafted bone marrow cells in irradiated mice, *Nature*, *268*: 442 (1977).
166. Y. Kitamura, and S. Go, Decreased production of mast cells in SI/SI^d anemic mice, *Blood*, *53*: 492 (1979).
167. E. S. Russell, Hereditary anemias of the mouse: A review for geneticists, *Adv. Genet.*, *20*: 357 (1979).
168. Y. Kitamura, and K. Hatanaka, Decrease of mast cells in W/W^v mice and their increase by bone marrow transplantation, *Blood*, *52*: 447 (1978).
169. S. J. Galli, and Y. Kitamura, Genetically mast-cell-deficient W/W^v and Sl/Sl^d mice. Their value for the analysis of the roles of mast cells in biologic responses in vivo, *Am. J. Pathol.*, *127*: 191 (1987).
170. E. N. Geissler, M. A. Ryan, and D. E. Housman, The dominant-white spotting *(W)* locus of the mouse encodes the c-*kit* proto-oncogene, *Cell*, *55*: 549 (1988).
171. B. Chabot, D. A. Stephenson, V. M. Chapman, P. Besmer, and A. Bernstein, The proto-oncogene c-*kit* encoding a transmembrane tyrosine kinase receptor maps to the mouse *W* locus, *Nature*, *335*: 88 (1988).
172. Y. Yarden, W. J. Kuang, T. Yang Feng, L. Coussens, S. Munemitsu, T. J. Dull, E. Chen, J. Schlessinger, U. Francke, and A. Ullrich, Human proto-oncogene c-*kit*: A new cell surface receptor tyrosine kinase for an unidentified ligand, *EMBO J.*, *6*: 3341 (1987).
173. K. M. Zsebo, D. A. Williams, E. N. Geissler, et al., Stem cell factor is encoded at the *Sl* locus of the mouse and is the ligand for the c-*kit* tyrosine kinase receptor, *Cell*, *63*: 213 (1990).
174. K. M. Zsebo, J. Wypych, I. K. McNiece, et al., Identification, purification, and

biological characterization of hematopoietic stem cell factor from buffalo rat liver conditioned medium, *Cell*, *63*: 195 (1990).

175. D. M. Anderson, S. D. Lyman, A. Baird, J. M. Wignall, J. Eisenman, C. Rauch, C. J. March, H. S. Boswell, S. D. Gimpel, D. Cosman, and D. E. Williams, Molecular cloning of mast cell growth factor, a hematopoietin that is active in both membrane bound and soluble forms, *Cell*, *63*: 235 (1990).
176. J. G. Flanagan, and P. Leder, The *kit* ligand: A cell surface molecule altered in steel mutant fibroblasts, *Cell*, *63*: 185 (1990).
177. E. Huang, E. Nocka, D. R. Beir, T.-Y. Chu, J. Buck, H.-W. Lahm, D. Wellner, P. Leder, and P. Besmer, The hematopoietic growth factor *kl* is encoded by the *sl* locus and is the ligand of the c-*kit* receptor, the gene product of the *W* locus, *Cell*, *63*: 225 (1990).
178. O. N. Witte, Steel locus defines new multipotent growth factor, *Cell*, *63*: 5 (1990).
179. L. Pevny, M. C. Simon, E. Robertson, W. H. Klein, S. F. Tsai, V. Dagati, S. H. Orkin, and F. Costantini, Erythroid differentiation in chimaeric mice blocked by a targeted mutation in the gene for transcription factor GATA-1, *Nature*, *349*: 257 (1991).
180. L. Zon, M. F. Gurish, D. S. Reynolds, K. F. Austen, R. L. Stevens, and S. H. Orkin, Control of transcription of protease genes in mast cells by the DNA-binding factor GATA-1 [abstract], *FASEB J.*, *5*: A1085 (1991).
181. J. R. Gordon, P. R. Burd, and S. J. Galli, Mast cells as a source of multifunctional cytokines, *Immunol. Today*, *11*: 458 (1990).
182. P. Valent, O. Majdic, D. Maurer, M. Bodger, M. Muhm, and P. Bettelheim, Further characterization of surface membrane structures expressed on human basophils and mast cells, *Int. Arch. Allergy Appl. Immunol.*, *91*: 198 (1990).
183. T. F. Huff, and D. E. Justus. Mast cell differentiation in cultures of T cell-depleted mesenteric lymph node cells from *Nippostrongylus brasiliensis*-infected mice, *Int. Arch. Allergy Appl. Immunol.*, *85*: 137 (1988).
184. R. I. Ashman, D. L. Jarboe, D. H. Conrad, and T. F. Huff, The mast cell-committed progenitor: In vitro generation of committed progenitors from bone marrow, *J. Immunol.*, *146*: 211 (1991).
185. R. I. Ashman, and T. F. Huff, Induction and trafficing of mast cell-committed progenitors from bone marrow [abstract], *FASEB J.*, *5*: A1005 (1991).
186. H. L. Thompson, P. D. Burbelo, B. Segui-Real, Y. Yamada, and D. D. Metcalfe, Laminin promotes mast cell attachment, *J. Immunol.*, *143*: 2323 (1989).
187. J. A. Leftwich, E. H. Westin, and T. F. Huff, Expression of c-*kit* by mesenteric lymph node cells from *Nippostrongylus brasiliensis* mice and colonies developing from these cells in response to 3T3-conditioned medium [abstract], *FASEB J.*, *5*: A1085 (1991).
188. D. L. Marquardt, and L. L. Walker, Regulation of expression of protein kinase C isotypes in mouse bone marrow mast cells (MBMMC) [abstract], *FASEB J.*, *5*: A1008 (1991).
189. M. Takagi, T. Nakahata, K. Koike, T. Kobayashi, K. Tsuji, S. Kojima, T. Hirano, A. Miyajima, K.-I. Arai, and T. Akabane, Stimulation of connective tissue-type mast cell proliferation by crosslinking of cell-bound IgE, *J. Exp. Med.*, *170*: 233 (1989).

190. T. Nakahata, T. Kobayashi, T. Ishiguro, K. Tsusi, K. Naganuma, O. Ando, Y. Yagi, K. Tadokoro, and T. Akabane, Extensive proliferation of mature connective tissue type mast cells in vitro, *Nature*, *324*: 65 (1986).
191. K. Otsu, T. Nakano, Y. Kanakura, H. Asai, H. R. Katz, K. F. Austen, R. L. Stevens, S. J. Galli, and Y. Kitamura, Phenotypic changes of bone marrow-derived mast cells after intraperitoneal transfer into W/W^v mice that are genetically deficient in mast cells, *J. Exp. Med.*, *165*: 615 (1987).
192. Y. Kanakura, H. Thompson, T. Nakano, T. Yamamura, H. Asai, Y. Kitamura, D. D. Metcalfe, and S. J. Galli, Multiple bidirectional alterations of phenotype and changes in proliferative potential during the in vitro and in vivo passage of clonal mast cell populations derived from mouse peritoneal mast cells, *Blood*, *72*: 877 (1988).
193. Y. Kitamura, T. Sonoda, T. Nokano, and Y. Kanayama, Probable dedifferentiation of mast cells in moust connective tissues, *Curr. Top. Dev. Biol.*, *20*: 325 (1986).
194. H. Ginsburg, and D. Lagunoff, The in vitro differentiation of mast cells. Cultures of cells from immunized mouse lymph nodes and thoracic duct lymph on fibroblast monolayers, *J. Cell Biol.*, *35*: 685 (1967).
195. H. Ginsburg, The in vitro differentiation and culture of normal mast cells from mouse thymus, *Ann. N. Y. Acad. Sci.*, *113*: 612 (1963).
196. H. Ginsburg, Growth and differentiation of cells of lymphoid origin on embryo cell monolayers, *Wistar Inst. Symp. Monogr.*, *4*: 21 (1965).
197. T. Ishizaka, W. König, M. Kurata, L. E. Mauser, and K. Ishizaka, Immunologic properties of mast cells from rats infected with *Nippostrongylus brasiliensis*, *J. Immunol.*, *115*: 1078 (1975).
198. T. Ishizaka, H. Okudaira, L. E. Mauser, and K. Ishizaka, Development of rat mast cells in vitro. I. Differentiation of mast cells from thymus cells, *J. Immunol.*, *116*: 747 (1976).
199. J. W. Schrader, S. J. Lewis, I. Clark-Lewis, and J. G. Culvenor, The persisting (P) cell: Histamine content, regulation by a T cell-derived factor, origin from a bone marrow precursor, and relationship to mast cells, *Proc. Natl. Acad. Sci. USA*, *78*: 323 (1981).
200. S. Hasthorpe, A hemopoietic cell line dependent upon a factor in pokeweed mitogen-stimulated spleen cell conditioning medium, *J. Cell. Physiol.*, *105*: 379 (1980).
201. G. J. Nabel, S. J. Galli, A. M. Dvorak, H. F. Dvorak, and H. Cantor, Inducer T lymphocytes synthesize a factor that stimulates proliferation of cloned mast cells, *Nature*, *291*: 332 (1981).
202. Y. P. Yung, R. Eger, G. Tertian, and M. A. Moore, Long-term in vitro culture of murine mast cells. II. Purification of a mast cell growth factor and its dissociation from TCGF, *J. Immunol.*, *127*: 794 (1981).
203. G. Tertian, Y. P. Yung, D. Guy-Grand, and M. A. Moore, Long-term in vitro culture of murine mast cells. I. Description of a growth factor-dependent culture technique, *J. Immunol.*, *127*: 788 (1981).
204. K. Nagao, K. Yokoro, and S. A. Aaronson, Continuous lines of basophil mast cells derived from normal mouse bone marrow, *Science*, *212*: 333 (1981).
205. E. Razin, C. Cordon-Cardo, and R. A. Good, Growth of a pure population of

mast cells in vitro with conditioned medium derived from concanavalin A-stimulated splenocytes, *Proc. Natl. Acad. Sci. USA*, *78*: 2559 (1981).

206. J. N. Ihle, J. R. Keller, S. Oroszlan, L. E. Henderson, T. D. Copeland, F. W. Fitch, M. B. Prystowsky, E. Goldwasser, J. W. Schrader, E. W. Palaszynski, M. Dy, and B. Lebel, Biologic properties of homogeneous interleukin 3. I. Demonstration of WEHI-3 growth factor activity, mast cell growth factor activity, P cell-stimulating factor activity, colony-stimulating factor activity, and histamine-producing cell-stimulating factor activity, *J. Immunol.*, *131*: 282 (1983).

207. F. Levi-Schaffer, K. F. Austen, P. M. Gravallese, and R. L. Stevens, Coculture of interleukin 3-dependent mouse mast cells with fibroblasts results in a phenotypic change of the mast cells [published erratum appears in *Proc. Natl. Acad. Sci. USA*, *83*(20):7805 (1986)], *Proc. Natl. Acad. Sci. USA*, *83*: 6485 (1986).

208. H. Ginsburg, D. Ben-Shazar, and E. Ben-David, Mast cell growth on fibroblast monolayers: Two-cell entities, *Immunology*, *45*: 371 (1982).

209. J. Fujita, H. Nakayama, H. Onoue, Y. Kanakura, T. Nakano, H. Asai, S. I. Takeda, T. Honjo, and Y. Kitamura. Fibroblast-dependent growth of mouse mast cells in vitro: Duplication of mast cell depletion in mutant mice of W/W^v genotype, *J. Cell. Physiol.*, *134*: 78 (1988).

210. T. Nakano, T. Sonoda, C. Hayashi, A. Yamatodani, Y. Kanayama, T. Yamamura, H. Asai, T. Yonezawa, Y. Kitamura, and S. J. Galli, Fate of bone marrow-derived cultured mast cells after intracutaneous, intraperitoneal and intracenous transfer into genetically mast cell deficient W/W mice. Evidence that cultured mast cells can give rise to both connective tissue type and mucosal mast cells, *J. Exp. Med.*, *162*: 1025 (1985).

211. T. Kobayashi, T. Nakano, T. Nakahata, H. Asai, Y. Yagi, K. Tsuji, A. Komiyama, P. Akabane, S. Kojima, and Y. Kitamura, Formation of mast cell colonies in methylcellulose by mouse peritoneal cells and differentiation of these cloned cells in both the skin and the gastric mucosa of W/W mice: Evidence that a common precursor can give rise to both "connective" tissue-type and "mucosal" mast cells, *J. Immunol.*, *136*: 1378 (1986).

212. D. L. Jarboe, J. S. Marshall, T. R. Randolph, A. Kukolja, and T. F. Huff, The mast cell-committed progenitor. I. Description of a cell capable of IL-3 independent proliferation and differentiation without contact with fibroblasts, *J. Immunol.*, *142: 2405 (1989).*

213. D. L. Jarboe, and T. F. Huff, The mast cell-committed progenitor. II. W/W^v mice do not make mast cell-committed progenitors and Sl/Sl^d fibroblasts do not support development of normal mast cell-committed progenitors, *J. Immunol.*, *142*: 2418 (1989).

214. K. Jozaki, A. Kuriu, S. Hirota, H. Onoue, Y. Ebi, S. Adachi, J. Y. Ma, S. Tarui, and Y. Kitamura, Bone marrow-derived cultured mast cells and peritoneal mast cells as targets of a growth activity secreted by BALB/3T3 fibroblasts, *Exp. Hematol.*, *19*: 185 (1991).

215. H. Matsuda, Y. Kannan, H. Ushio, Y. Kiso, T. Kanemoto, H. Suzuki, and Y. Kitamura, Nerve growth factor induces development of connective tissue-type mast cells in vitro from murine bone marrow cells, *J. Exp. Med.*, *174*: 7 (1991).

216. J. Nafziger, M. Arock, J.-J. Guillosson, and J. Wietzerbin, Specific high-affinity

receptors for interferon-gamma on mouse bone marrow-derived mast cells: Inhibitory effect of interferon-gamma on mast cell precursors, *Eur. J. Immunol.*, *20*: 113 (1990).
217. P. Valent, G. Schmidt, J. Besemer, P. Mayer, G. Zenke, E. Liehl, W. Hinterberger, K. Lechner, D. Maurer, and P. Bettelheim, Interleukin-3 is a differentiation factor for human basophils, *Blood*, *73*: 1763 (1989).
218. A. M. Dvorak, H. Saito, P. Estrella, S. Kissell, N. Arai, and T. Ishizaka, Ultrastructure of eosinophils and basophils stimulated to develop in human cord blood mononuclear cell cultures containing recombinant human interleukin-5 or interleukin-3, *Lab. Invest.*, *61*: 116 (1989).
219. B. Volc-Platzer, P. Valent, T. Radaszkiewicz, P. Mayer, P. Bettelheim, and K. Wolff, Recombinant human interleukin 3 induces proliferation of inflammatory cells and keratinocytes in vivo, *Lab. Invest.*, *64*: 557 (1991).
220. P. Valent, J. Besemer, M. Muhm, O. Majdic, K. Lechner, and P. Bettelheim, Interleukin 3 activates human blood basophils via high-affinity binding sites, *Proc. Natl. Acad. Sci. USA*, *86*: 5542 (1989).
221. P. Valent, J. Besemer, C. Sillaber, J. H. Butterfield, R. Eher, O. Majdic, K. Kishi, W. Klepetko, F. Eckersberger, K. Lechner, and P. Bettelheim, Failure to detect IL-3-binding sites on human mast cells, *J. Immunol.*, *145*: 3432 (1990).
222. T. Furitsu, H. Saito, A. Dvorak, L. B. Schwartz, A. A. Irani, J. F. Burdick, K. Ishizaka, and T. Ishizaka, Development of human mast cells in vitro, *Proc. Natl. Acad. Sci. USA*, *86*: 10039 (1989).
223. A. A. Irani, and L. B. Schwartz, In vitro development of MC_T cells from human fetal liver [abstract], *J. Allergy Clin. Immunol.*, *85*: 172 (1990).
224. A. M. Irani, S. S. Craig, G. DeBlois, C. O. Elson, N. M. Schechter, and L. B. Schwartz, Deficiency of the tryptase-positive, chymase-negative mast cell type in gastrointestinal mucosa of patients with defective T lymphocyte function, *J. Immunol.*, *138*: 4381 (1987).
225. S. S. Craig, N. M. Schechter, and L. B. Schwartz, Ultrastructural analysis of maturing human T and TC mast cells in situ, *Lab. Invest.*, *60*: 147 (1989).
226. H. L. Thompson, P. D. Burbelo, and D. D. Metcalfe, Regulation of adhesion of mouse bone marrow-derived mast cells to laminin, *J. Immunol.*, *145*: 3425 (1990).
227. H. L. Thompson, P. D. Burbelo, Y. Yamada, H. K. Kleinman, and D. D. Metcalfe, Identification of an amino acid sequence in the laminin A chain mediating mast cell attachment and spreading, *Immunology*, *72*: 144 (1991).
228. S. I. Wasserman, The mast cell and synovial inflammation. Or, what's a nice cell like you doing in a joint like this, *Arthritis Rheum.*, *27*: 841 (1984).
229. A. A. Irani, N. Golzar, G. DeBlois, B. Gruber, and L. B. Schwartz, Distribution of mast cell subsets in rheumatoid arthritis and osteoarthritis synovial [abstract], *Arthritis Rheum.*, *30*: 66 (1987).
230. M. C. Castells, A. M. Irani, and L. B. Schwartz, Evaluation of human peripheral blood leukocytes for mast cell tryptase, *J. Immunol.*, *138*: 2184 (1987).
231. L. B. Schwartz, J. W. Yunginger, J. S. Miller, R. Bokhari, and D. Dull, The time course of appearance and disappearance of human mast cell tryptase in the circulation after anaphylaxis, *J. Clin. Invest.*, *83*: 1551 (1989).
232. C. Oh, S. Suzuki, I. Nakashima, K. Yamashita, and K. Nakano, Histamine

synthesis by non-mast cells through mitogen-dependent induction of histidine decarboxylase, *Immunology.*, *65*: 143 (1988).

233. M. Shalit, L. B. Schwartz, N. Golzar, C. von Allman, M. Valenzano, P. Fleekop, P. C. Atkins, and B. Zweiman, Release of histamine and tryptase in vivo after prolonged cutaneous challenge with allergen in humans, *J. Immunol.*, *141*: 821 (1988).
234. E. N. Charlesworth, A. F. Hood, N. A. Soter, A. Kagey-Sobotka, P. S. Norman, and L. M. Lichtenstein, Cutaneous late-phase response to allergen. Mediator release and inflammatory cell infiltration, *J. Clin. Invest.*, *83*: 1519 (1989).
235. S. M. Walden, D. Proud, R. Bascom, L. M. Lichtenstein, A. Kagey-Sobotka, N. F. Adkinson, Jr., and R. M. Naclerio, Experimentally induced nasal allergic responses, *J. Allergy Clin. Immunol.*, *81*: 940 (1988).
236. L. B. Schwartz, D. D. Metcalfe, J. S. Miller, H. Earl, and T. Sullivan, Tryptase levels as an indicator of mast-cell activation in systemic anaphylaxis and mastocytosis, *N. Engl. J. Med.*, *316*: 1622 (1987).
237. J. W. Yunginger, D. R. Nelson, D. L. Squillace, R. T. Jones, K. E. Holley, B. A. Hyma, L. Biedrzycki, K. G. Sweeney, W. Q. Sturner, and L. B. Schwartz, Laboratory investigation of deaths due to anaphylaxis, *J. Forensic Sci.*, *36*: 857 (1991).
238. S. E. Wenzel, A. A. Fowler 3d, and L. B. Schwartz, Activation of pulmonary mast cells by bronchoalveolar allergen challenge. In vivo release of histamine and tryptase in atopic subjects with and without asthma, *Am. Rev. Respir. Dis.*, *137*: 1002 (1988).
239. M. C. Liu, E. R. Bleecker, L. M. Lichtenstein, A. Kagey-Sobotka, Y. Niv, T. L. McLemore, S. Permutt, D. Proud, and W. C. Hubbard, Evidence for elevated levels of histamine, prostaglandin D_2, and other bronchoconstricting prostaglandins in the airways of subjects with mild asthma, *Am. Rev. Respir. Dis.*, *142*: 126 (1990).
240. P. H. Howarth, J. Wilson, R. Djukanovic, S. Wilson, K. Britten, A. Walls, W. R. Roche, and S. T. Holgate, Airway inflammation and atopic asthma: A comparative bronchoscopic investigation, *Int. Arch. Allergy Appl. Immunol.*, *94*: 266 (1991).
241. N. N. Jarjour, W. J. Calhoun, L. B. Schwartz, and W. W. Busse, Elevated bronchoalveolar lavage fluid histamine levels in allergic asthmatics are associated with increased airway obstruction, *Am. Rev. Respir. Dis.*, *144*: 83 (1991).
242. J. B. Sedgwick, W. J. Calhoun, G. J. Gleich, H. Kita, J. S. Abrams, L. B. Schwartz, B. Volovitz, M. Ben-Yaakov, and W. W. Busse, Immediate and late airway response of allergic rhinitis patients to segmental antigen challenge: Characterization of eosinophil and mast cell mediators, *Am. Rev. Respir. Dis.*, *144*: 1274 (1991).
243. J. V. Bosso, L. B. Schwartz, and D. D. Stevenson, Tryptase and histamine release during aspirin-induced respiratory reactions, *J. Allergy Clin. Immunol.*, *88*: 830 (1991).
244. D. H. Broide, S. Eisman, J. W. Ramsdell, P. Ferguson, L. B. Schwartz, and S. I. Wasserman, Airway levels of mast cell-derived mediators in exercise-induced asthma, *Am. Rev. Respir. Dis.*, *141*: 563 (1990).

245. G. Mayrhofer, and H. Bazin, Nature of the thymus dependency of mucosal mast cells. III. Mucosal mast cells in nude mice and nude rats, in B rats and in a child with the Di George syndrome, *Int. Arch. Allergy Appl. Immunol.*, *64*: 320 (1981).
246. G. Mayrhofer, The nature of the thymus dependency of mucosal mast cells. II. The effect of thymectomy and of depleting recirculating lymphocytes on the response to *Nippostrongylus brasiliensis*, *Cell. Immunol.*, *47*: 312 (1979).
247. G. Mayrhofer, The nature of the thymus dependency of mucosal mast cells. I. An adaptive secondary response to challenge with *Nippostrongylus brasiliensis*, *Cell. Immunol.*, *47*: 304 (1979).
248. E. J. Ruitenberg, and A. Elgersma, Absence of intestinal mast cell response in congenitally athymic mice during *Trichinella spiralis* infection, *Nature*, *264*: 258 (1976).
249. R. G. Woodbury, H. R. Miller, J. F. Huntley, G. F. Newlands, A. C. Palliser, and D. Wakelin, Mucosal mast cells are functionally active during spontaneous expulsion of intestinal nematode infections in rat, *Nature*, *312*: 450 (1984).
250. K. B. Madden, Jr., J. F. Urban, H. J. Zittener, J. W. Schrader, F. D. Finkelman, and I. M. Katona, Antibodies to IL-3 and IL-4 suppress helminth-induced intestinal mastocytosis [abstract], *FASEB J.*, *5*: A1012 (1991).

12

Cytokines in Atopic Allergic Inflammation

Douglas S. Robinson
Royal Free Hospital and Royal Brompton National Heart and Lung Institute, London, England

A. Barry Kay
Royal Brompton National Heart and Lung Institute, London, England

I. ATOPIC ALLERGIC INFLAMMATION

Atopic allergic disease affects about 30% of the population in the United Kingdom. The clinical spectrum is wide and includes rhinitis, asthma, conjunctivitis, and dermatitis. *Atopy* literally means "out of place" and was originally defined to describe allergic reactions in which "the individuals as a group possess a peculiar capacity to become sensitive to certain proteins to which their environment and habits of life frequently expose them" (1). It was clear that atopy was an hereditary predisposition for forming allergen-specific tissue-sensitizing antibody. This was originally referred to as "reagin" (2), but in the 1960s was characterized as a novel immunoglobulin class: immunoglobulin E (IgE) (3,4). The aim of the present chapter is to examine the role of cytokines in controlling the inflammatory response in atopic allergic disease and to discuss possible cellular and molecular targets for therapeutic intervention.

Although associations between IgE and both asthma symptoms and bronchial hyperresponsiveness have been described (5,6), asthma occurs in nonatopic individuals, and not all atopics are asthmatic: the relation between findings in atopic allergic inflammation and the pathogenesis of asthma remains to be fully clarified.

Experimental allergen challenge protocols have been used widely to study human atopic inflammation. Allergen-induced responses in the skin, nose, and lung have been separated into an early response, occurring within a few minutes of challenge, which may be followed by a late response, which develops some 3–6 h later and may persist for up to 24 h (7). This separation may reflect the artificial nature of the model, and atopic disease probably results from a mixture of the mediator-dependent changes in early responses together with the cellular inflammatory changes that characterize late responses (8). Despite these reservations, the use of allergen challenge models of atopic allergic inflammation has yielded considerable information on the cellular basis of atopic disease.

II. IN VITRO ACTIONS OF CYTOKINES POTENTIALLY INVOLVED IN ATOPIC ALLERGIC INFLAMMATION

Cytokines are molecules, other than antibodies, produced by lymphocytes and other cell types that are involved in signaling between cells of the immune system. The group includes the interleukins (ILs), interferon-gamma (IFN-γ), granulocyte–macrophage colony-stimulating factor (GM-CSF), tumor necrosis factor (TNF), and some that still remain uncharacterized. In general, cytokines act as hemopoietic factors in the growth, differentiation, and maturation of white cells, red cells, and platelets, and also act as inflammatory mediators. They have a special role in the growth and differentiation of T and B lymphocytes.

A. Immunoglobulin E Regulation

The regulation of human IgE synthesis is discussed in detail elsewhere in this volume. In summary, IgE synthesis appears to be dependent on interleukin-4 (IL-4)(9) and IL-6 (10), and is enhanced by IL-5 (11), whereas IFN-γ (9), and IL-12 (13) are inhibitory. These cytokines act in concert with an array of cell surface molecules and other signals to activate or repress different B-cell immunoglobulin gene segments toward the procedure IgE mRNA transcript (14).

B. Differentiation of Inflammatory Cells

Eosinophil differentiation (reviewed in detail in Chap. 13) from human cord blood cells can be induced by IL-3, IL-5, and GM-CSF (15–17). Murine eosinophil differentiation appears to be T-cell-dependent (18), and eosinophilia was produced in transgenic mice constitutively expressing T-cell IL-5 genes (19). Furthermore, antibodies to IL-5 ablated the eosinophil response to helminthic infection in mice (20).

Interleukin-5 (unlike IL-3 and GM-CSF) is not active in neutrophil differentiation (17); thus, it may be an important determinant of the eosinophil preponderance of the inflammatory infiltrate in atopic allergic inflammation.

Human basophil growth and differentiation appear to be dependent on principally IL-3, although IL-4, IL-5, and GM-CSF are also active (21). Although mast cells and basophils have previously been thought to be related cell types, current evidence supports differences in stem cell origin and cytokine control of differentiation (22). Furthermore, there appear to be profound differences between rodent and human mast cells. Whereas IL-3, IL-4, IL-9, and IL-10 promote murine mast cell growth, IL-3, IL-4, and IL-5 did not induce mast cell differentiation from human cord blood cultures (15), and human mast cells do not appear to express IL-3 receptors (23). Whether stem cell factor (SCF) is the mast cell growth factor described in human cord blood/3T3 fibroblast coculture is not yet known (21,24).

The major T-cell growth factor is IL-2 (25), although IL-1 (26), IL-4 (27), IL-7 (28), and IL-12 (29) may also promote the growth of different functional subsets of T cells.

C. Recruitment of Inflammatory Cells

The recruitment of inflammatory cells to sites of allergic inflammation involves chemoattraction, adhesion to vascular endothelium, diapedesis through the endothelial cell wall, interaction with intracellular cell matrix, and maintenance of cell viability within tissues.

1. Chemoattractants

Interleukin 2, IL-3, IL-5, and GM-CSF are reported to be eosinophil chemoattractants (30–32). Also, GM-CSF had some neutrophil chemoattractant activity, but this was not seen for IL-3 and IL-5 (32). Interleukin 5 enhanced locomotor responses of eosinophils, but not neutrophils to platelet-activating factor (PAF) and leukotriene B_4 (LTB_4), and this effect was less in normal subjects than in those with circulating eosinophilia, suggesting a possible prior exposure in vivo in subjects with eosinophilia (32). Several members of the IL-8 superfamily appear to exert selective chemoattractant activity on different cell types. Thus, IL-8 is chemotactic for neutrophils and T lymphocytes, but not for monocytes or eosinophils. Macrophages inflammatory protein-1α (MIP-1α) is chemotactic for cytotoxic and naive T cells, whereas RANTES appears to selectively attract memory T cells, monocytes, and eosinophils, but not neutrophils (33–35). Production of a specific pattern of cytokines might thus act to attract specific cell types to sites of inflammation.

2. Adhesion

Once attracted to a site of inflammation, blood leukocytes must adhere to and traverse the vascular endothelium and intracellular matrix (36). Initial arrest of the leukocyte within the vascular lumen and attachment to the endothelium is dependent on the local flow characteristics and the interaction of specific adhesion molecules on both the leukocyte and endothelial cell. Several different adhesion pathways have been described. The interactions involved in the "normal" trafficking of T-cell subsets to different anatomical sites are not described here, rather this chapter will focus on the actions of cytokines to modulate adhesion of leukocytes during inflammation.

Selectins. The selectin family of membrane glycoproteins are important determinants of initial leukocyte–endothelial interaction (Table 1). L-selectin is constitutively expressed at the leukocyte surface and interacts with ligands (including E-selectin) on the endothelial surface to establish initial contact (37,38). Activation of leukocytes by cytokines leads to rapidly increased avidity of L-selectin–ligand interaction (39), followed by shedding of L-selectin from the leukocyte cell surface by proteolytic cleavage (40). This initial adhesion phase allows rolling of granulocytes or transient attachment of lymphocytes, permitting further adhesion pathways to come into play (41,42).

Endothelial cells activated by IL-1 or TNF-α also express E-selectin (43). E-selectin mRNA is rapidly induced, and surface expression is detectable at 1 h, maximal at 4–8 h, and is over by 24–48 h (41). Anti-E-selectin antibodies inhibit neutrophil and eosinophil adhesion to endothelial cells in vitro (44), and E-selectin may also activate subsequent CD11b/CD18-dependent binding and transmigration (45). E-selectin also mediates adhesion of a subpopulation of resting memory T lymphocytes to endothelium (46).

P-selectin is very rapidly expressed on activated endothelial cells, being externalized from secretory granules within seconds in response to stimuli, such as histamine or thrombin (47). It is rapidly reinternalized and may cooperate with surface-bound platelet-activating factor (PAF) in initial attachment, as in the rolling of neutrophils along the endothelial surface (48). P-selectin expression does not appear to be cytokine-dependent (41). Both P-selectin and E-selectin recognize sialyl-Lewis X on leukocytes, but the counterreceptors active in inflammation are not yet known (49).

Integrins. The integrin superfamily of adhesion receptors appear to be crucial to enhancing leukocyte–endothelial contact, and to margination and transmigration of activated cells at sites of inflammation (36).

The β_2-integrin CD11a/CD18 (LFA-1) is constitutively expressed by granulocytes and lymphocytes, and interaction with the ligand intercellular adhesion molecule 1 (ICAM-1) is potentially increased at sites of inflammation by increased expression of ICAM-1 (in response to TNF-α and IL-1) and by a cytokine-induced increase in avidity of LFA-1 for its ligand (50,51). CD11a/CD18 also interacts with ICAM-2, which is constitutively expressed by endothelial cells (36). CD11b/CD18 (Mac-1, CR3) is not expressed by lymphocytes, but is rapidly externalized from intracytoplasmic stores by neutrophils in response to GM-CSF and IL-3, and rather more slowly up-regulated by

Table 1 Adhesion Molecules in Atopic Allergic Inflammation: Effects of Cytokines

Adhesion molecule	Cell types expressing[a]	Effect of cytokines	Ligand	Effects of cytokines
Selectins				
L-selectin	Leukocytes	Constitutive	Sialyl Lewis X ? E-selectin	?
E-selectin	Endothelial cells	IL-1 and TNF-α[b]	Sialyl Lewis X	?
P-selectin	Leukocytes	None known	Sialyl Lewis X	?
Integrins				
β_1				
CD11a/CD18 (LFA-1)	T, No, Mo	GM-CSF[c]	ICAM-1	IL-1, TNF-α[b]
	Eo, Ba	IL-3, Il-5, GM-CSF[c]		
CD11b/CD18 (MAC-1, CR3)	No, Mo	GM-CSF[c]	ICAM-1	IL-1, TNF-α[b]
	Eo, Ba	IL-3, IL-5, GM-CSF[c]		
β_2 VLA_4 (α_4/β_2)	T, Eo, Ba	Constitutive	VCAM-1	IL-1, TNF-α, IL-4[b]

[a]T, T cells; No, neutrophils; Mo, monocytes; Eo, eosinophils; Ba, basophils.
[b]Increased expression.
[c]Increased avidity of adhesion.

eosinophils in response to IL-3, IL-5, and GM-CSF (52,53). This molecule also interacts with ICAM-1, and avidity is increased following leukocyte activation (41). Monoclonal antibodies directed against CD11a/CD18 and CD11b/CD18 blocked granulocyte adhesion to endothelial cells in vitro, but antibodies directed against the third β_2 integrin CD11c/CD18 had little effect (54). The LFA-1–ICAM-1 interaction is a costimulatory pathway in T-cell activation (36), although the role of intracellular signaling by adhesion receptors in modulation of granulocyte function is unknown.

The β_1-integrin family of proteins share a common β chain (CD49) and are variably expressed by T-cell subsets (55). These molecules were first detected as very late activation molecules in T cells (56); therefore, they are termed VLA-1–6. They interact with intracellular matrix proteins, such as fibronectin, collagen, and laminin, and thus may be important in transmigration of lymphocytes into tissues (55). VLA-4 also has a specific counterreceptor on endothelial cells activated by IL-1, TNF-α, and IL-4, termed vascular cell adhesion molecule 1 (VCAM-1) (57,58). VLA-4 is also constitutively expressed by eosinophils and basophils, but not by neutrophils, and monoclonal antibodies that block interaction with VCAM-1 inhibit adhesion of both eosinophils and basophils to endothelium in vitro (59,60).

Several mediators, including cytokines, may act in a juxtacrine manner to modulate adhesion processes. An example is platelet-activating factor (PAF), which is rapidly coexpressed at the endothelial cell surface with P-selectin and may then act on neutrophils to up-regulate CD11/CD18 adhesion (48). Interleukin-8 is released from endothelial cells (61), but may also associate at the cell surface (41). It activates CD11/CD18-dependent neutrophil adhesion (62), but decreases E-selectin-dependent adhesion (63) and, thereby, may act in the transition from rolling to margination and subsequent diapedesis. A similar juxtacrine role has been described for MIP-1β in T-cell adhesion to high endothelial venules (64).

Cytokine actions on adhesion systems relevant to allergic inflammation are summarized in Table 1. In brief, IL-1 and TNF-α activate endothelial cells, leading to increased expression of E-selectin, ICAM-1, and VCAM-1 (36,57). Interleukin-4 induces endothelial VCAM-1 expression (58). These cytokines also lead to endothelial cell production of mediators, such as PAF and nitric oxide, and cytokines, such as IL-8 and GM-CSF (41). Granulocyte–macrophage CSF and IL-8 promote neutrophil adhesion, whereas IL-3, IL-5, and GM-CSF promote adhesion of eosinophils and basophils. T-cell activation and acquisition of memory status lead to a variety of changes in adhesion receptor expression (65), some of which may be dependent on cytokine actions.

Thus, IL-3, IL-4, and IL-5 have a potential role in adhesion of eosinophils, basophils, and memory $CD4^+$ T cells, rather than neutrophils, at sites of allergic inflammation both through selective up-regulation of VCAM-1 on endothelium and through activation of eosinophil and basophil, but not neutrophil, integrin-dependent adhesion mechanisms.

3. Leukocyte Survival and Activation in Tissues

Once present at sites of allergic inflammation, it is likely that cytokines play an important role in maintaining eosinophil (and probably basophil) viability, and prime for functional activation, such as generation of lipid mediators and degranulation. Interleukin-3 (66), IL-5 (67), and GM-CSF (68) will maintain eosinophil viability in vitro, whereas only GM-CSF has this effect on neutrophils. This action may be caused by inhibition of apoptosis (69,70). Interestingly, eosinophil interaction with fibronectin will prolong survival in vitro, and this appears to be due to IL-3 induction from eosinophils in an autocrine fashion (71). Consequently, interaction of leukocytes with the endothelial cell membrane after migration into the tissues may exert an important effect on their survival and function. Interleukin-3 (66), IL-5 (72), and GM-CSF (68) also act to increase eosinophil LTC_4 production, superoxide generation, cytotoxicity, and phagocytosis, and also enhance basophil functional responses, such as histamine release (73). Although some direct effect of IL-5 and GM-CSF on eosinophil degranulation was detectable, the major effect of cytokines seems to be to prime and enhance eosinophil and basophil function triggered by pathways, such as ligation of IgG receptors (74). Although IL-3, IL-5, and GM-CSF appear to be pivotal in eosinophil- and basophil-mediated inflammation, other cytokines have been implicated in functional regulation of these cells. Tumor necrosis factor-α activates eosinophil cytotoxicity for schistosomal larvae (75), and IFN-γ activates eosinophils over a longer time course than IL-3, IL-5, and GM-CSF, leading to phenotypic changes (76) and stimulating expression of cytokine mRNA (77), at least in vitro.

The role of neutrophils in atopic allergic inflammation is uncertain (78), but cytokines of the IL-8 superfamily may be released by several cell types (including epithelial cells, T cells, eosinophils, and endothelial cells) and, together with GM-CSF, tend to enhance neutrophil persistence and functional activation (79).

T-lymphocyte proliferation is dependent on IL-2 (25), and local production of IL-4 (80), IFN-γ (81,82), and IL-10 (83) may determine the expansion of T-helper (Th)1 or Th2 effector populations during inflammatory responses (discussed later).

III. THE CELLULAR SOURCE OF CYTOKINES IN ATOPIC ALLERGIC INFLAMMATION

Many studies have described a cellular inflammatory infiltrate rich in mononuclear cells and eosinophils at sites of atopic allergic inflammation (84). Allergen challenge studies have demonstrated an influx of $CD4^+$ T cells, eosinophils, basophils, and neutrophils at differing time points after allergen challenge in atopic subjects (7). A complex array of cytokines are likely involved in coordinating this cellular inflammatory response. Which cells are the important players in orchestrating responses to allergen? The answers to this question and the unraveling of mechanisms that control cytokine production in inflammation are vital to progress in therapeutic intervention in atopic and other forms of inflammatory disease.

A. T Lymphocytes

The T lymphocytes recognize specific peptide sequences of antigen (epitopes) when these are presented by major histocompatibility complex (MHC) products on antigen-presenting cells. T cells have the capacity to produce a wide array of cytokines, particularly following repeated antigenic stimulation. Animal experiments suggest T-cell-dependence of eosinophil differentiation (18,85), and several lymphoid malignancies are associated with eosinophilia (86), which also accompanies IL-2 therapy (and, hence, expansion of activated T cells) in humans (87). The critical role of T cells in IgE synthesis is reviewed elsewhere in the volume. An important advance in the understanding of T-cell control of inflammatory response was the description of subsets of murine $CD4^+$ T-helper clones based on expression of mRNA for and production of differing patterns of cytokines (Fig. 1) (88). The functional consequences of this dichotomy were

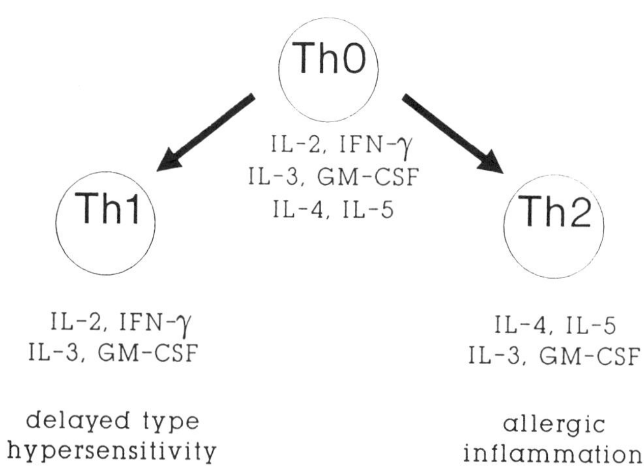

Figure 1 Th1 and Th2 cytokine profiles: Th1 and Th2 clones are hypothesized to originate from a Th0 precursor that produces cytokines of both Th1 and Th2 patterns. (From Ref. 95.)

immediately apparent and were rapidly confirmed in experimental animals (89,90). The Th1 clones producing IL-2 and IFN-γ, but little or no IL-4 and IL-5, were poor helpers for immunoglobulin production, tending to induce IgG2a synthesis (91). When transferred to the footpad of recipient mice, these T cells produced a delayed-type hypersensitivity (DTH) lesion (92) that was dependent on production of IFN-γ (93). In contrast, Th2 clones that produced IL-4 and IL-5, but little or no IFN-γ and IL-2, provided help for IgE and IgG1 synthesis (91) and did not induce DTH (92). It was suggested that the Th2-type T cells might be important determinants of IgE and eosinophil and basophil responses in atopic allergy and helminth infections (89), and this hypothesis was supported by increased isolation of Th2 clones from animals parasitized with the helminth parasite *Nippostrongylus brasiliensis* (94). Although clones with many different cytokines patterns are described in both mice and humans, it is now clear that Th1 and Th2 patterns of cytokines represent polar extremes of T-cell responses. Since these two patterns are mutually inhibitory—IFN-γ inhibits expansion of Th2 clones (81), and IL-10 inhibits Th1 cytokine production (95)—either one of these patterns might be expected to predominate during prolonged local antigenic stimulation.

B. Mast Cells

In 1989, several different groups described production of cytokines from murine mast cell lines in response to IgE cross-linking or ionophore response (96–98). It appears that human mast cells also have the capacity to synthesize a wide variety of cytokines. Mast cell-derived cytokines include IL-3, IL-4, IL-5, GM-CSF, and TNF-α. Mast cells also appear to have the capacity to store cytokines (TNF-α) and, thus, may have the capacity to release cytokines early on during response to allergen (99). This may be an important determinant of the ensuing inflammatory cell response.

C. Eosinophils

Human eosinophils have the capacity to produce a variety of cytokines in vitro as described in more detail in Chapter 13. Human peripheral blood eosinophils, stimulated with ionophore or IFN-γ, expressed mRNA and immunoreactivity for GM-CSF (77; Fig. 2). Eosinophil production of transforming growth factor-α (TGF-α; 100), IL-3 (101), IL-6 (102), and IL-8 (103) has also been described in vitro. In situ hybridization studies demonstrate expression of IL-5 mRNA by human eosinophils in the gut mucosa in celiac disease (104) and in lung eosinophils in allergen-induced asthmatic responses (105). These cytokines have clear potential to amplify the atopic allergic response.

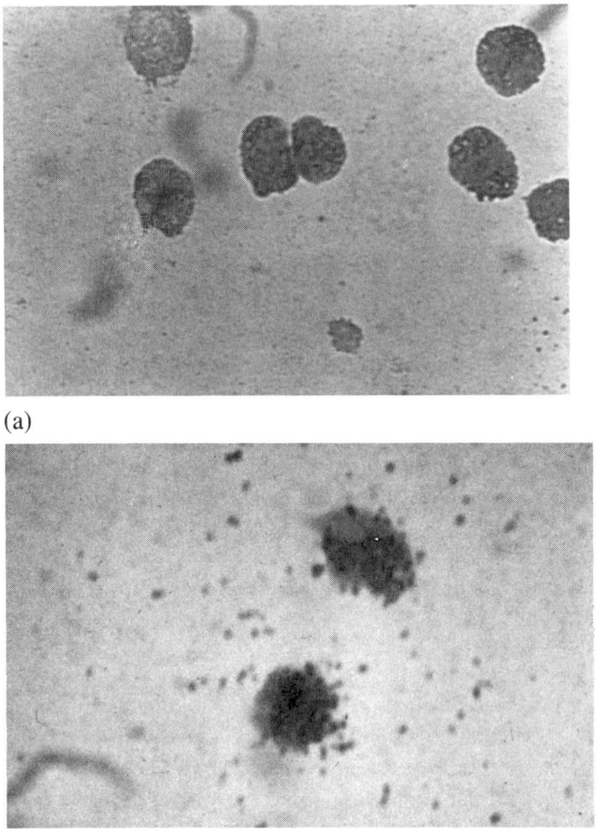

(a)

(b)

Figure 2 GM-CSF production by human peripheral blood eosinophils: (a) Immunoreactivity of GM-CSF is localized to peripheral blood eosinophils stimulated with calcium ionophore. Positive cells appear red using alkaline phosphatase–antialkaline phosphatase and Fast Red to localize antibody staining. (b) GM-CSF mRNA is localized to eosinophils. In situ hybridization was performed using radiolabeled cRNA probes, and mRNA is visualized as silver grains in the photographic emulsion. (Courtesy of Drs. Redwan Moqbel and Qutayba Hamid.)

D. Basophils

Production of IL-3 and GM-CSF by murine splenic non-B, non-T cells is described, and human splenic non-B, non-T cells produce IL-4 (21,106). These findings may represent basophil cytokine elaboration, and basophils are thus potential players in cytokine production in allergic inflammation.

E. Other Cell Types

Epithelial cells, far from being the passive victims of attack by eosinophil granule proteins (107), may also contribute to their own demise through their capacity to synthesize IL-6, IL-8, and GM-CSF (108,109). Endothelial cell IL-8 and GM-CSF may play an important role in leukocyte adhesion and, hence, influx to sites of allergic inflammation. Alveolar macrophages from atopic individuals have the capacity to produce GM-CSF (110), and monocyte–macrophages may also be an important source of IL-1 and TNF-α (111).

Thus, a wide variety of cell types have the capacity to produce cytokines. In IgE-mediated atopic allergic inflammation, it might be expected that Th2-type T cells, mast cells, and eosinophils are the major contributors to the cytokine network. What evidence is there for production of cytokines in human subjects in vivo, and which cell types are implicated?

IV. CYTOKINES IN VIVO IN ATOPIC ALLERGIC INFLAMMATION

A. T-Cell Clones

Initial examination of human T-cell clones did not support a Th1–Th2 dichotomy of cytokine production (112,113), and it is clear that a range of cytokine profiles may occur. However, in 1990 Wierenga et al. reported Th2-type cytokine profiles (IL-4 but no IFN-γ) from allergen-specific T-cell clones derived from atopic donors, whereas clones derived from nonatopics showed a Th1-type pattern (114). This was not a function of antigen-presenting cells (APC), since APC from histocompatible nonatopics would support the Th2-type clones without changing the cytokine profile. These findings were confirmed by the isolation of Th2-type clones specific for allergen and helminth parasites, and Th1-type clones specific for purified protein derivative of *Mycobacterium tuberculosis* (115). Furthermore, Th2-type T cells were also cloned from sites of allergic inflammation, such as the conjunctiva in vernal conjunctivitis (116) and the skin in atopic dermatitis (117).

B. Peripheral Blood Cells

Flow cytometric analysis of peripheral blood lymphocytes from subjects with acute asthma supported the concept of activation of $CD4^+$ T cells in asthma (118,119). Peripheral blood T cells from atopic asthmatic subjects produced eosinophil survival factors in vitro, and similar activity was detected in the sera of these subjects. Use of neutralizing antibodies suggested that GM-CSF was the preponderant activity in T-cell supernatants, and IL-5 and GM-CSF were present in the serum (119). Interleukin-5 has also been detected by enzyme-linked immunosorbent assay (ELISA) in the serum from symptomatic asthmatics (120). In acute, severe asthma IFN-γ was detected in serum (121).

C. Atopic Allergic Disease

Immunocytochemical studies suggest activation of $CD4^+$ T cells in the bronchial mucosa in atopic asthma (122–124). Cytokines have been studied by detection of mRNA or by detection of the cytokine protein by specific antibodies. Messenger RNA may not always be translated into protein product, and detection of cytokine protein is not conclusive proof of origin at the site of detection. Ideally, one would combine the two approaches, although this is not always feasible. By use of in situ hybridization, increased proportions of bronchoalveolar lavage (BAL) cells from atopic asthmatic subjects were shown to express mRNA for IL-2, IL-3, IL-4, IL-5, and GM-CSF when compared with those of nonatopic control subjects, but there was no difference between the two in numbers of IFN-γ mRNA-positive BAL cells (125). Immunomagnetic separation of BAL cells was used to isolate the BAL T-cell population, and in situ hybridization showed that IL-4 and IL-5 mRNAs were associated with BAL T cells from atopic asthmatic subjects (125). The TNF-α mRNA-positive cells were also increased in frequency in BAL from asthmatics compared with that from nonatopic control subjects (126). Walker et al. detected IL-4 and IL-5 in concentrated BAL fluid from subjects with atopic asthma, but not from nonatopic controls (127). The TNF-α, GM-CSF, and IL-6 immunoreactivity (but not IL-4) was increased in concentrated BAL fluid from subjects with acute asthma when compared with that from those without symptoms (128). Interleukin-5 mRNA was detected in bronchial biopsies from subjects with symptomatic asthma, but not from those with asymptomatic disease or control subjects (129). More recently, we have extended our studies of asthmatic subjects and shown significant correlations between the proportion of BAL cells expressing mRNA for both IL-4 and IL-5 with measures of airflow obstruction and bronchial responsiveness (130).

Immunoreactivity for IL-4 was localized to mast cells in bronchial biopsies from atopic asthmatic subjects and nasal biopsies from subjects with perennial rhinitis. Purified skin mast cells rapidly released IL-4 after anti-IgE stimulation, and it was suggested that IL-4 may be stored in mast cell granules (131). No immunoreactivity for IL-4 was detected in T lymphocytes in these studies. One possible explanation for the apparent discrepancy between mRNA and immunocytochemical findings is that T-cell production of IL-4 in vivo is localized to areas of cell–cell contact, as has been demonstrated in murine T-cell–B-cell hybridomas in vitro (132). It is possible that mast cell-derived IL-4 directs the expansion of Th2 effector cells during the development of atopic allergic inflammation, although the precise cellular interactions remain to be clarified (133).

Eosinophil cytokine production in vivo is suggested by the demonstration of GM-CSF mRNA and protein in nasal polyps, with localization of mRNA to eosinophils (134), and by the demonstration of IL-5 mRNA in eosinophils in the gut mucosa of patients with celiac disease (104). The contribution of cytokines from eosinophils and basophils in atopic inflammation remains to be established.

Expression of mRNA and cytokine protein for IL-6, IL-8, and GM-CSF by bronchial epithelial cells separated from bronchial biopsy specimens from atopic asthmatic subjects has also been described (108).

D. Allergen Challenge Studies

The allergen-induced late inflammatory response in skin, nose, and lung has been used to study allergen-induced events in atopic allergic inflammation. Skin biopsies from atopic subjects obtained 24 h after intracutaneous injection of allergen are characterized by infiltration with activated eosinophils and $CD4^+$ T lymphocytes (135,136). In situ hybridization demonstrated increased numbers of cells expressing mRNA for IL-3, IL-4, IL-5, and GM-CSF, when compared with biopsies from skin injected with the diluent control, but no difference was found between the two in numbers of cells expressing mRNA for IL-2 and IFN-γ (137). The same pattern of cytokine mRNA expression was detected in nasal biopsies obtained 24 h after local allergen challenge of atopic subjects (138). By combination of immunocytochemical analysis and in situ hybridization, the phenotypes of cells expressing cytokine mRNA have recently been studied in late nasal responses (139). Most IL-5 $mRNA^+$ cells were also $CD3^+$ (i.e., T cells), although a few (16.4%) were IL-5^+/$tryptase^+$ (mast cells).

Bronchoalveolar lavage and bronchial biopsy after airway allergen challenge have been used to study allergen-induced late asthmatic responses. Several investigators have used local allergen challenge protocols where allergen extract is instilled directly into a segmental airway. Broide and co-workers studied GM-CSF mRNA and protein expression after local allergen challenge of subjects with mild asthma (140). Combined immunocytochemical tests and in situ hybridization suggested a preponderant T-lymphocyte source for GM-CSF after local challenge, with a contribution from alveolar macrophages. More recently, the same authors have also demonstrated expression of IL-5 and GM-CSF mRNA by BAL eosinophils after endobronchial allergen challenge (105). Sedgwick and co-workers detected IL-5 protein in BAL fluid obtained 48 h after endobronchial challenge (141). We have examined mRNA expression in BAL and bronchial biopsies obtained 24 h after allergen inhalation challenge in subjects with atopic asthma (142, 143). Results were compared with those obtained 24 h after diluent control challenge of the same subjects: each acting as his or her own control. When allergen and diluent challenge were compared there was a significant increase in BAL eosinophil and $CD4^+$ T-cell activation, and in activated eosinophils ($EG2^+$) and

CD25 expression in bronchial mucosal biopsies after allergen challenge. In bronchoalveolar lavage cells, there were significant increases in IL-4, IL-5, and GM-CSF $mRNA^+$ cells after allergen challenge, compared with diluent control, and in biopsies there were significant increases in IL-5 and GM-CSF $mRNA^+$ cells. There were no differences between allergen and diluent challenge in numbers of BAL or biopsy cells expressing mRNA for IL-2 and IFN-γ. Immunomagnetic separation of BAL T cells showed that most BAL cells positive for IL-4 and IL-5 mRNA after allergen challenge were T lymphocytes. Examples of signals for IL-5

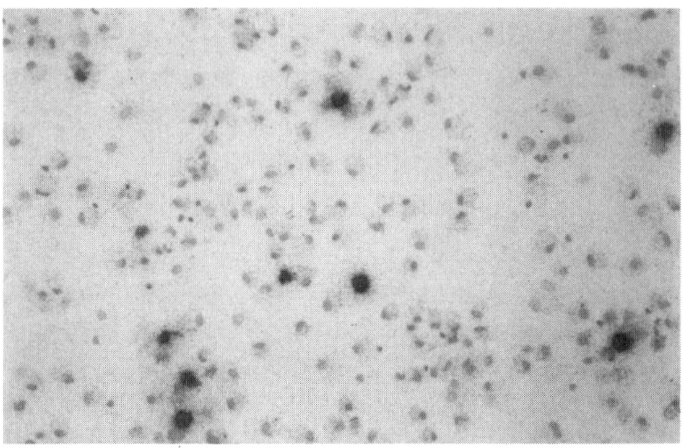

(a)

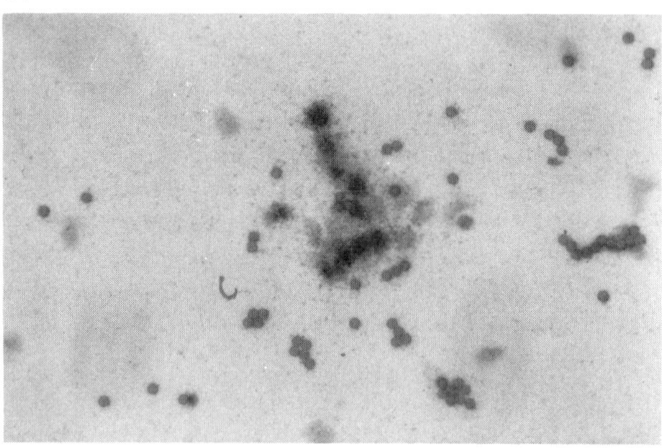

(b)

Figure 3 Interleukin-5 mRNA expression by bronchoalveolar lavage cells obtained after allergen inhalation challenge of atopic asthmatics: (a) In situ hybridization with radiolabeled cRNA probe. Positive BAL cells are overlain by dense deposits of silver grain in the photographic emulsion. (b) After immunomagnetic selection of CD2-positive cells, using Dynabeads, IL-5 mRNA signals are seen to be closely associated with CD2-positive cells.

mRNA in BAL cells and immunomagnetically separated BAL T cells from atopic asthmatics after allergen challenge are shown in Figure 3.

Thus, in summary, allergen challenge studies suggest allergen-induced production of IL-3, IL-4, IL-5, and GM-CSF from T lymphocytes, together with IL-5 and GM-CSF from eosinophils, and IL-5 from mast cells. Endobronchial challenge does not accurately reproduce natural allergen exposure, and the time course of cytokine production in atopic allergic inflammation remains to be established. It has yet to be shown conclusively that cytokines are important initiators of allergic inflammation in humans: it is possible that other mediators may act in early inflammatory responses, which are then amplified and modulated by cytokine release. The effects of cytokine antagonists or blocking monoclonal antibodies have not yet been studied in human allergic disease. However, animal studies have shown inhibition of antigen-induced eosinophil influx to guinea pig lung and mouse trachea by anti-IL-5 (144,145).

V. THERAPEUTIC POTENTIAL OF CYTOKINE BLOCKADE IN ATOPIC ALLERGIC INFLAMMATION

What is the effect of pharmacological intervention in atopic allergic disease on cytokine production? It is widely recognized that the most effective intervention in asthma is corticosteroids (146). Since corticosteroids inhibit cytokine production in vitro, we have recently explored the hypothesis that part of the beneficial effect of corticosteroids in asthma may result from similar actions in vivo (147). In a double-blind, placebo-controlled study of prednisolone treatment in patients with moderate symptomatic asthma, there was a significant reduction in bronchial hyperresponsiveness in those subjects randomized to receive prednisolone when compared with those receiving placebo (Fig. 4). Bronchoalveolar lavage was performed before and after 2 weeks of either prednisolone or placebo therapy in each subject. There was a significant reduction in BAL eosinophils and the proportions of BAL cells expressing mRNA for IL-4 and IL-5 after prednisolone treatment when compared with placebo (see Fig. 4). Thus, clinical response to corticosteroid therapy in atopic asthma was associated with reduction in IL-4 and IL-5 mRNA expression. How else might one modulate the cytokine response in inflammatory responses?

In vitro studies of T-cell cytokine production have shown that IFN-γ favors the expansion of Th1 clones, and IL-4 favors expansion of Th2 clones, both in mice and in humans (80–82). Exogenous administration of these cytokines or of blocking antibodies to them had profound effects on the response of different mouse strains to infection with *Leishmania major* (148). BALB/c mice tend to produce a rapidly fatal response, with a predominant expansion of Th2 T cells, whereas resistant strains, such a C57BL/6, produce an IFN-γ-predominant Th1

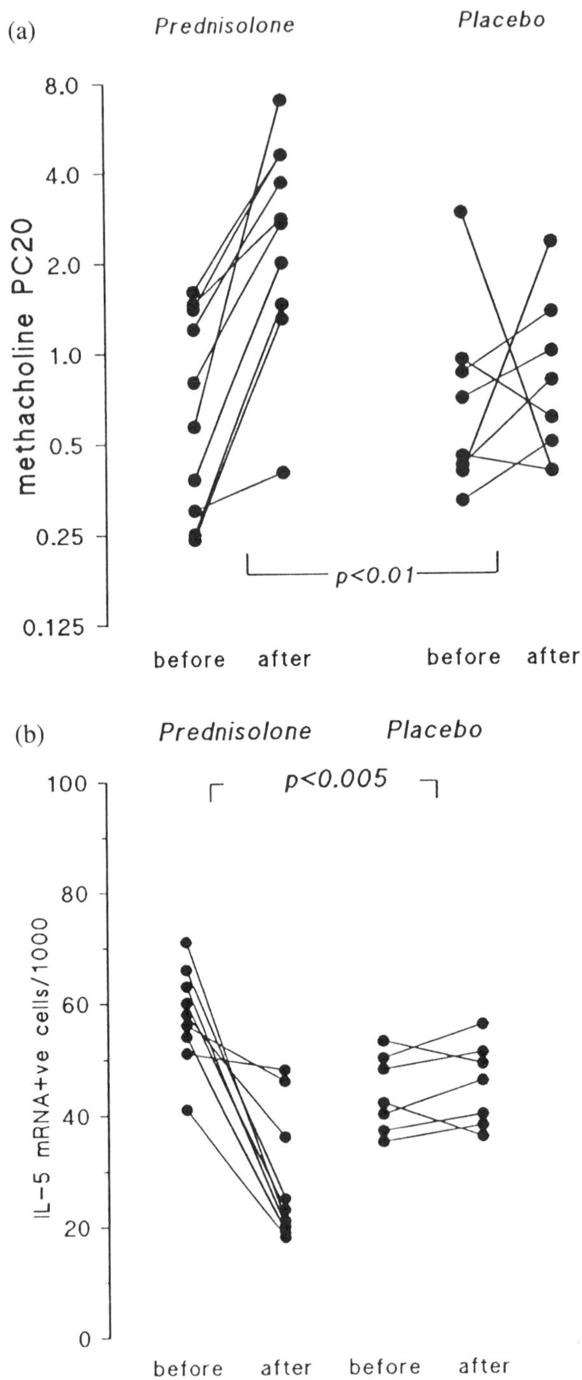
(a)
Prednisolone
Placebo
8.0
4.0
2.0
1.0
0.5
0.25
0.125
methacholine PC20
p<0.01
before
after
before
after
(b)
Prednisolone
Placebo
p<0.005
100
80
60
40
20
0
IL-5 mRNA+ve cells/1000
before
after
before
after

response to infection. Antibodies to IL-4 converted the BALB/c to a healing response (148), and anti-IFN-γ caused a rapidly fatal response to infection in the C57BL/6 strain (149). Thus, Th1 or Th2 patterns of cytokines are associated with marked differences in inflammatory response and outcome to infection, and these can be manipulated by blocking antibodies (150). Exogenous IFN-γ has improved human responses to drug therapy in leishmaniasis (151). The T-cell cytokine response may also directly influence the pathological response in leprosy in which different patterns of cytokine production were found in lepromatous and tuberculoid skin lesion (152). Thus manipulation of T-cell cytokine response to allergen may allow clinical modulation of atopic allergic inflammation. However, exogenous cytokine therapy may be associated with profound immunological effects, and IFN-γ had no effect on allergic rhinitis symptoms or IgE synthesis (153).

The Th1 and Th2 murine T-cell clones differ in preferred antigen-presenting cell (154), intracellular signaling (155), response to tolerogenic signal (156), and response to prostaglandin E_2 (157). If similar differences in activation requirements or differential induction of anergy are possible in human T cells, this could provide powerful tools for manipulation of T-cell-dependent inflammatory processes.

Allergen-injection immunotherapy was described in 1911 as a treatment for atopic allergic rhinitis (158). In a placebo-controlled study of allergen immunotherapy, successful immunotherapy was associated with a change in the cytokine profile in allergen-induced late reactions, with an increase in IL-2 and IFN-γ mRNA expression (159). In a recent study, reported in abstract form, immunotherapy decreased production of IL-3, IL-5, and GM-CSF during the cutaneous late-phase response (160). Thus immunotherapy may modulate cytokine production in response to allergen.

The genes for many cytokines and cytokine receptors have been cloned. The genes for IL-3, IL-4, IL-5, and GM-CSF are closely associated on human chromosome 5q31 (161). Since these genes are often (although not always; 162) coexpressed, both common and individual transcriptional activators or repressors must exist. Definition of these factors may allow specific regulation of gene expression for these cytokines. Antisense oligonucleotides are sequences of RNA that have complementary sequence to, and thus bind specifically to, a target

Figure 4 Effect of prednisolone treatment on bronchial hyperreactivity and IL-5 gene expression in asthma: (a) Methacholine airway reactivity before and after 2 weeks of treatment of symptomatic asthmatic subjects with either prednisolone (0.6 mg/kg per day) or placebo in a double-blind study. (PC_{20}, concentration of methacholine producing a 20% fall in forced expiratory volume in 1 s from baseline value.) (b) Numbers of BAL cells per 1000 total on cytocentrifuge preparations giving positive signals for IL-5 mRNA by in situ hybridization after either prednisolone or placebo treatment. (Comparison of treatments is by Mann Whitney U test on the differences before and after prednisolone or placebo treatment.)

RNA sequence. These agents have been used in vitro to inhibit cytokine expression (163), and they may have future therapeutic potential. The receptors for IL-3, IL-5, and GM-CSF share a common subunit (164,165). It should be possible to develop specific cytokine–receptor antagonists: indeed anti-IL-1R (166) and anti-IL-4R (167) delay allograft rejection in experimental animals. Perhaps blockade of IL-4R and IL-5R would be logical therapeutic targets in atopic allergic inflammation.

VI. CONCLUSION

There is now considerable evidence that atopic allergic inflammation represents a response to a particular "Th2-type" pattern of cytokine production. Whether this results from activation of human Th2-equivalent T cells in vivo, and what the relative contributions of mast cells, eosinophils, basophils, and other cell types is to the modulation and amplification of allergic inflammation, remains to be clarified. Specific intervention at the level of individual cytokines or cell types may hold promise for future advance in the therapy of allergic disease.

REFERENCES

1. A. F. Coca, and R. A. On the classification of the phenomena of hypersensitiveness, *J. Immunol.*, *8*: 163 (1923).
2. C. von Pirquet, Allergie, *Munch. Med. Wochenscher.*, *30*: 1457 (1906).
3. K. Ishizaka, and T. Ishizaka, Identification of cE-antibodies as a carrier of reaginic activity, *J. Immunol.*, *99*: 1187 (1967).
4. S. G. O. Johansson, and H. Bennich, Immunological studies of an atypical (myeloma) immunoglobulin, *Immunology*, *13*: 381 (1967).
5. B. Burrows, F. D. Martinez, M. Halonen, R. A. Barbee, and M. G. Cline, Association of asthma with serum IgE levels and skin-test reactivity to allergens, *N. Engl. J. Med.*, *320*: 271 (1989).
6. M. R. Sears, B. Burrows, E. M. Flannery, G. P. Herbison, C. J. Hewitt, and M. D. Holdaway, Relation between airway responsiveness and serum IgE in children with asthma and in apparently normal children, *N. Engl. J. Med.*, *325*: 1067 (1991).
7. S. R. Durham, Late asthmatic responses, *Respir. Med.*, *84*: 263 (1990).
8. R. P. Schleimer, Effects of glucocorticoids on inflammatory cells relevant to their therapeutic applications in asthma, *Am. Rev. Respir. Dis.*, *141*: S59 (1990).
9. G. Del Prete, E. Maggi, P. Parronchi, I. Chretien, A. Tiri, D. Macchia, M. Ricci, J. Banchereau, J. de Vries, and S. Romagnani, IL-4 is an essential factor for the IgE synthesis induced in vitro by human T cell clones and their supernatants, *J. Immunol.*, *140*: 4193 (1988).
10. D. Vercelli, H. H. Jabara, K. Arai, T. Yokota, and R. S. Geha, Endogenous IL-6 plays an obligatory role in IL-4-induced human IgE synthesis, *Eur. J. Immunol.*, *19*: 1419 (1989).

11. J. Pene, F. Rousset, F. Briere, I. Chretien, J. Wademan, J.-Y. Bonnefoy, and J. E. De Vries, Interleukin-5 enhances interleukin-4 induced IgE production by normal human B cells. The role of soluble CD23 antigen, *Eur. J. Immunol.*, *18*: 929 (1988).
12. H. Kimata, A. Yoshida, C. Ishioka, I. Lindley, and H. Mikawa, Interleukin 8 (IL-8) selectively inhibits immunoglobulin E production induced by IL-4 in human B cells, *J. Exp. Med.*, *176*: 1227 (1992).
13. M. Kiniwa, M. Gateley, R. Chizzonite, C. Fargeas, and G. Delespesse, Recombinant interleukin-12 suppresses the synthesis of immunoglobulin E by interleukin-4 stimulated human lymphocytes, *J. Clin. Invest.*, *90*: 262 (1992).
14. R. S. Geha, Regulation of IgE synthesis in humans, *J. Allergy Clin. Immunol.* *90*: 143 (1992).
15. H. Saito, K. Hatake, A. M. Dvorak, K. M. Leiferman, A. D. Donnenberg, N. Arai, K. Ishizaka, and K. Ishizaka, Selective differentiation and proliferation of haematopoietic cells induced by recombinant human interleukins, *Proc. Natl. Acad. Sci. USA*, *85*: 2288 (1988).
16. H. D. Campbell, W. Q. J. Tucker, Y. Hort, M. E. Martinson, G. Mayo, E. J. Clutterbuck, C. J. Sanderson, and I. G. Young, Molecular cloning, nucleotide sequence, and expression of the gene encoding human eosinophil differentiation factor (interleukin 5), *Proc. Natl. Acad. Sci. USA*, *84*: 6629 (1987).
17. Y. Sonoda, N. Arai, and M. Ogawa, Humoral regulation of eosinophilpoiesis in vitro: Analysis of the targets of interleukin-3, granulocyte/macrophage colony-stimulating factor (GM-CSF) and interleukin-5, *Leukemia*, *3*: 14 (1989).
18. A. Basten, and P. B. Beeson, Mechanism of eosinophilia. II. Role of the lymphocyte, *J. Exp. Med.*, *131*: 1288 (1970).
19. L. A. Dent, M. Strath, A. L. Mellor, and C. J. Sanderson, Eosinophilia in transgenic mice expressing interleukin-5, *J. Exp. Med.*, *172*: 1425 (1990).
20. R. L. Coffman, B. W. P. Seymour, S. Hudak, J. Jackson, and D. Rennick, Antibody to interleukin-5 inhibits helminth-induced eosinophilia in mice, *Science*, *245*: 308 (1989).
21. J. A. Denburg, Basophil and mast cell lineages in vitro and in vivo, *Blood*, *79*: 846 (1992).
22. P. Valent, L. K. Ashman, W. Hintenberg, F. Eckersberger, O. Majdic, K. Lechner, and P. Bettelheim, Mast cell typing: Demonstration of a distinct hemopoietic cell type and evidence for immunophenotypic relationship to mononuclear phagocytes, *Blood*, *73*: 1778 (1989).
23. P. Valent, J. Besemer, C. H. Sillake, J. H. Butterfield, R. Eher, O. Majdic, K. Kishi, W. Klepetko, F. Eckersberger, K. Lechner, and P. Bettelheim, Failure to detect IL-3 binding sites on human mast cells, *J. Immunol.*, *145*: 3422 (1990).
24. A. S. Kirshenbaum, S. W. Kessler, J. P. Goff, and D. D. Metcalfe, Demonstration of the origin of human mast cells from $CD34^+$ bone marrow progenitor cells, *J. Immunol.*, *146*: 1410 (1991).
25. D. A. Cantrell, and K. A. Smith, Transient expression of interleukin-2 receptors, *J. Exp. Med.*, *158*: 1895 (1983).
26. L. A. Greenbaum, J. B. Horowitz, A. Woods, T. Pasqaulini, E.-P. Reich, and K. Bottomly, Autocrine growth of $CD4^+$ T cells. Differential effects of IL-1 on helper and inflammatory T cells, *J. Immunol.*, *140*: 1555 (1988).

27. H. Spits, H. Yssel, Y. Takebe, N. Arai, T. Yokota, F. Lee, K.-I. Arai, J. Banchereau, and J. de Vries, Recombinant interleukin-4 promotes the growth of human T cells, *J. Immunol.*, *139*: 1142 (1987).
28. M. Londei, A. Verhoef, C. Hawrylowicz, J. Groves, P. de Bardinis, and M. Feldmann, Interleukin 7 is a growth factor for mature human T cells, *Eur. J. Immunol.*, *20*: 425 (1990).
29. M. K. Gately, B. B. Desai, A. G. Wolitzky, P. M. Quinn, C. M. Dwyer, F. J. Podlaski, P. C. Familletti, F. Sinigaglia, R. Chizonnite, U. Gubler, and A. S. Stern. Regulation of human lymphocyte proliferation by a heterodimeric cytokine, IL-12 (cytotoxic lymphocyte maturation factor), *J. Immunol.*, *147*: 874 (1991).
30. T. H. Rand, D. S. Silberstein, H. Kornfeld, and P. F. Weller, Human eosinophils express functional interleukin-2 receptors, *J. Clin. Invest.*, *88*: 825 (1991).
31. J. M. Wang, A. Rambaldi, A. Biondi, Z. G. Chen, C. J. Sanderson, and A. Mantovani, Recombinant human interleukin-5 is a selective eosinophil chemoattractant, *Eur. J. Immunol.*, *19*: 701 (1989).
32. R. Sehmi, A. J. Wardlaw, O. Cromwell, K. Kurihara, P. Waltmann, and A. B. Kay, Interleukin-5 selectively enhances the chemotactic response of eosinophils obtained from normal but not eosinophilic subjects, *Blood*, *79*: 2952 (1992).
33. S. D. Wolpe, and A. Cerami, Macrophage inflammatory proteins 1 and 2: Members of a novel superfamily of cytokines, *FASEB J.*, *3*: 2565 (1989).
34. T. J. Schall, K. Bacon, K. J. Toy, and D. V. Goeddel, Selective attraction of monocytes and T lymphocytes of the memory phenotype by cytokine RANTES, *Nature*, *347*: 669 (1990).
35. Y. Kameyoshi, A. Dorschner, A. I. Mallet, E. Christophers, and J.-M. Schroder, Cytokine RANTES released from thrombin-stimulated platelets is a potent attractant for human eosinophils, *J. Exp. Med.*, *176*: 587 (1992).
36. T. A. Springer, Adhesion receptors of the immune system, *Nature*, *346*: 425 (1990).
37. M. Bevilacqua, E. Butcher, B. Furie, M. Gallatin, M. Gimbrone, J. Harlan, K. Kishimoto, L. Lasky, R. McEver, J. Paulson, S. Rosen, B. Seed, M. Siegelman, T. Springer, L. Stoolman, T. Tedder, A. Varki, D. Wagner, I. Weissman, and G. Zimmerman, Selectins: A family of adhesion receptors, *Cell*, *67*: 233 (1991).
38. T. F. Tedder, G. M. Isaacs, T. J. Ernst, G. D. Demetri, D. A. Adler, and C. M. Disteche, Isolation and chromosomal location of cDNAs encoding a novel human lymphocyte cell surface molecule, LAM-1, *J. Exp. Med.*, *170*: 123 (1989).
39. O. Spertini, G. S. Kansas, J. M. Munro, J. D. Griffin, and T. F. Tedder, Regulation of leukocyte migration by activation of the leukocyte adhesion molecule-1 (LAM-1) selectin, *Nature*, *349*: 691 (1991).
40. T. K. Kishimoto, M. A. Jutila, E. L. Berg, and E. C. Butcher, Neutrophil Mac-1 and MEL-14 adhesion proteins inversely regulated by chemotactic factors, *Science*, *245*: 1239 (1989).
41. G. A. Zimmerman, S. M. Prescott, and T. M. McIntyre, Endothelial cell interactions with granulocytes: Tethering and signalling molecules, *Immunol. Today*, *13*: 93 (1992).

42. Y. Shimizu, W. Newman, Y. Tanaka, and S. Shaw, Lymphocyte interactions with endothelial cells, *Immunol. Today*, *13*: 106 (1992).
43. M. P. Bevilacqua, S. Stengelin, M. A. Gimbrone, and B. Seed, Endothelial leukocyte adhesion molecule-1: An inducible receptor for neutrophils related to complementary regulatory proteins and lectins, *Science*, *243*: 1160 (1989).
44. T. K. Kishomoto, R. A. Warnock, M. A. Jutila, E. C. Butcher, C. Lane, D. C. Anderson, and C. W. Smith, Antibodies against human neutrophil LECAM-1 (LAM-1/Leu-8/DREG-56 antigen) and endothelial cell ELAM-1 inhibit a common CD18-independent adhesion pathway in vitro, *Blood*, *78*: 805 (1991).
45. S. K. Lo, S. Lee, R. A. Ramos, R. Lobb, M. Rosa, G. Chi-Rosso, and S. D. Wright, Endothelial–leukocyte adhesion molecule-1 stimulates the adhesive activity of leukocyte integrin CR3 (CD11b/CD18, Mac-1, $\alpha_m\beta_2$) on human neutrophils, *J. Exp. Med.*, *173*: 1493 (1991).
46. L. J. Picker, R. A. Warnock, A. R. Burns, C. M. Doerschuk, E. L. Berg, and E. C. Butcher, The neutrophil selectin LECAM-1 presents carbohydrate ligands to the vascular selectins ELAM-1 and GMP-140, *Cell*, *66*: 921 (1991).
47. R. Hattori, K. K. Hamilton, R. D. Fugate, R. P. McEver, and P. J. Sims, Stimulated secretion of endothelial von Willebrand factor is accompanied by rapid redistribution to the cell surface of the intracellular granule membrane protein GMP-140, *J. Biol. Chem.*, *264*: 7768 (1989).
48. D. E. Lorant, K. D. Patel, T. M. McIntyre, R. P. McEver, S. M. Prescott, and G. A. Zimmerman, Coexpression of GMP-140 and PAF by endothelium stimulated by histamine or thrombin: A juxtacrine system for adhesion and activation of neutrophils, *J. Cell Biol.*, *115*: 223 (1991).
49. M. J. Polley, M. L. Phillips, E. Wayner, E. Nudelman, A. K. Singhal, S.-I. Hakomori, and J. C. Paulson, CD62 and endothelial cell–leukocyte adhesion molecule-1 (ELAM-1) recognize the same carbohydrate ligand, sialyl-Lewis X, *Proc. Natl. Acad. Sci. USA*, *88*: 6224 (1991).
50. M. S. Diamond, D. E. Staunton, S. D. Marlin, and T. A. Springer, Binding of the integrin Mac-1 (CD11b/CD18) to the third immunoglobulin-like domain of ICAM-1 (CD54) and its regulation by glycosylation, *Cell*, *65*: 961 (1991).
51. M. L. Hibbs, H. Xu, S. A. Stacker, and T. A. Springer, Regulation of adhesion to ICAM-1 by the cytoplasmic domain of LFA-1 integrin β subunit, *Science*, *251*: 1611 (1991).
52. D. H. Jones, D. C. Anderson, B. L. Burr, H. E. Rudloff, C. W. Smith, S. S. Krater, and F. C. Schlamstieg, Quantitation of intracellular Mac-1 (CD11b/CD18) pools in human neutrophils, *J. Leukocyte Biol*, *44*: 535 (1988).
53. A. Hartnell, A. B. Kay, and A. J. Wardlaw, IL-3 induced upregulation of CR3 expression in human eosinophils is inhibited by dexamethasone, *Immunology*, *77*: 488 (1992).
54. G. M. Walsh, A. Hartnell, A. J. Wardlaw, K. Kurihara, C. J. Sanderson, and A. B. Kay, IL-5 enhances the in vitro adhesion of human eosinophils, but not neutrophils, in a leukocyte integrin (CD11/18)-dependent manner, *Immunology*, *71*: 258 (1990).
55. M. E. Hemler, VLA proteins in the integrin family: Structures, functions, and their role on leukocytes, *Annu. Rev. Immunol.*, *8*: 365 (1990).

56. M. E. Hemler, J. G. Jacobson, M. B. Brenner, D. Mann, and J. L. Strominger, VLA-1: A T cell surface antigen which defines a novel late stage of human T cell activation, *Eur. J. Immunol.*, *15*: 502 (1985).
57. L. Osborn, C. Hession, P. Tizard, C. Vassalo, S. Luhowskyj, G. Chi-Rosso, and R. Lobb, Direct expression cloning of vascular cell adhesion molecule 1, a cytokine-induced endothelial protein that binds to lymphocytes, *Cell*, *59*: 1203 (1989).
58. M. H. Thornhill, and D. O. Haskard, IL-4 regulates endothelial cell activation by IL-1, tumor necrosis factor, or IFN-gamma, *J. Immunol.*, *145*: 865 (1990).
59. G. M. Walsh, A. Hartnell, J.-J., Mermod, A. B. Kay, and A. J. Wardlaw, Human eosinophil, but not neutrophil, adherence to IL-1 stimulated HUVEC is $\alpha4\beta1$ (VLA-4) dependent, *J. Immunol.*, *146*: 3419 (1991).
60. R. P. Schleimer, S. A. Sterbinsky, J. Kaiser, C. A. Bickel, D. A. Klunk, K. Tomioka, W. Newman, F. W. Luscinskas, M. A. Gimbrone, B. W. McIntyre, and B. S. Bochner, IL-4 induces adherence of human eosinophils and basophils but not neutrophils to endothelium, *J. Immunol.*, *148*: 1086 (1992).
61. J.-M. Schroder, and E. Christophers, Secretion of novel and homologous neutrophil activating peptides by LPS-stimulated human endothelial cells, *J. Immunol.*, *142*: 244 (1989).
62. P. A. Detmers, S. K. Lo, E. Olsen-Egbert, A. Walz, M. Bagglioni, and Z. A. Cohn, Neutrophil-activating protein 1/interleukin 8 stimulates the binding activity of the leukocyte adhesion receptors CD11b/CD18 on human neutrophils, *J. Exp. Med.*, *171*: 1155 (1990).
63. M. A. Gimbrone, M. S. Obin, A. F. Brock, E. A. Luis, P. E. Hass, C. A. Herbert, Y. K. Yip, D. W. Leung, D. G. Lowe, and W. J. Kohr, Endothelial interleukin 8: A novel inhibitor of leukocyte–endothelial interactions, *Science*, *246*: 1601 (1989).
64. Y. Tanaka, D. H. Adams, and S. Shaw, Proteoglycans on endothelial cells present adhesion-inducing cytokines to leukocytes, *Immunol. Today*, *14*: 111 (1993).
65. D. L. Wallace, and P. C. L. Beverley, Phenotypic changes associated with activation of $CD45RA^+$ and $CD45RO^+$ T cells, *Immunology*, *69*: 460 (1990).
66. M. E. Rothenberg, W. F. Owen, Jr., D. S. Silberstein, J. Woods, R. J. Soberman, K. F. Austen, and R. L. Stevens, Human eosinophils have prolonged survival, enhanced functional properties, and become hypodense when exposed to interleukin 3, *J. Clin. Invest.*, *81*: 1986 (1988).
67. A. F. Lopez, C. J. Sanderson, J. R. Gamble, H. D. Campbell, I. G. Young, and M. A. Vadas, Recombinant human interleukin 5 is a selective activator of human eosinophil function, *J. Exp. Med.*, *167*: 219 (1988).
68. W. F. Owen, Jr., M. E. Rothenberg, D. S. Silberstein, J. Gasson, R. L. Stevens, K. F. Austen, and R. J. Soberman, Regulation of human eosinophil viability, density, and function by granulocyte/macrophage colony-stimulating factor in the presence of 3T3 fibroblasts, *J. Exp. Med.*, *166*: 129 (1987).
69. E. Her, J. Frazer, K. F. Austen, and W. F. Owen, Jr., Eosinophil hematopoietins antagonize the programmed cell death of eosinophils, *J. Clin. Invest.*, *88*: 1982 (1991).
70. M. Stern, L. Meagher, J. Savill, and C. Haslett, Apoptosis in human eosinophils.

Programmed cell death in the eosinophil leads to phagocytosis by macrophages and is modulated by IL-5, *J. Immunol.*, *148*: 3543 (1992).

71. A. R. E. Anwar, R. Moqbel, G. M. Walsh, A. B. Kay, and A. J. Wardlaw, Adhesion to fibronectin prolongs eosinophil survival, *J. Exp. Med.*, *177*: 819 (1993).
72. M. E. Rothenberg, J. Petersen, R. L. Stevens, D. S. Silberstein, D. T. McKenzie, K. F. Austen, and W. F. Owen, Jr., IL-5 dependent conversion of normodense human eosinophils to the hypodense phenotype uses 3T3 fibroblasts for enhanced viability, accelerated hypodensity, and sustained antibody-dependent cytotoxicity, *J. Immunol.*, *143*: 2311 (1989).
73. K. Hirai, M. Yamaguchi, Y. Misaki, T. Takaishi, K. Ohta, Y. Morita, K. Ito, and T. Miyamoto, Enhancement of human basophil histamine release by interleukin 5, *J. Exp. Med.*, *172*: 1525 (1990).
74. T. Fujisawa, R. Abu-Ghazaleh, H. Kita, C. J. Sanderson, and G. J. Gleich, Regulatory effect of cytokines on eosinophil degranulation, *J. Immunol.*, *144*: 642 (1990).
75. D. S. Silberstein, and J. R. David, TNF enhances eosinophil toxicity to *Schistosoma mansoni* larvae, *Proc. Natl. Acad. Sci. USA*, *83*: 1055 (1986).
76. T. Valerius, R. Repp, J. R. Kalden, and E. Platzer, Effects of IFN on human eosinophils in comparison with other cytokines, *J. Immunol.*, *145*: 2950 (1990).
77. R. Moqbel, Q. Hamid, S. Ying, J. Barkans, A. Hartnell, A. Tsicopoulos, A. J. Wardlaw, and A. B. Kay, Expression of mRNA for the granulocyte/macrophage colony-stimulating factor (GM-CSF) in activated human eosinophils, *J. Exp. Med.*, *174*: 749 (1991).
78. T. Fujisawa, G. M. Kephart, B. H. Gray, and G. J. Gleich, The neutrophil and chronic allergic inflammation. Immunochemical localization of neutrophil elastase, *Am. Rev. Respir. Dis.*, *141*: 689 (1990).
79. E. J. Leonard, A. Skeel, T. Yoshimura, K. Noer, S. Kutvirt, and K. van Epps, Leukocyte specificity and binding of human neutrophil attractant/activation protein 1, *J. Immunol.*, *144*: 1323 (1990).
80. S. L. Swain, A. D. Weinberg, M. English, and G. Huston, IL-4 directs the development of Th2-like helper effectors, *J. Immunol.*, *145*: 3796 (1990).
81. T. F. Gajewski, and F. W. Fitch, Anti-proliferative effect of IFN-gamma in immune regulation. III Differential selection of Th1 and Th2 murine helper T lymphocyte clones using recombinant IL-2 and IFN-gamma, *J. Immunol.*, *143*: 15 (1989).
82. E. Maggi, P. Parronchi, R. Manetti, C. Simonelli, M.-P. Piccini, F. S. Rugiu, M. De Carli, M. Ricci, and S. Romagnani, Reciprocal regulatory effects of IFN-gamma and IL-4 on the in vitro development of human Th1 and Th2 clones, *J. Immunol.*, *148*: 2142 (1992).
83. D. F. Fiorentino, M. W. Bond, and T. R. Mosmann, Two types of mouse T helper cells. IV. Th2 clones secrete a factor that inhibits cytokine production by Th1 clones, *J. Exp. Med.*, *170*: 2081 (1989).
84. R. Djukanovic, W. R. Roche, J. W. Wilson, C. R. W. Beasley, O. P. Twentyman, P. H. Howarth, and S. T. Holgate, Mucosal inflammation in asthma. *Am. Rev. Respir. Dis.*, *142*: 434 (1990).

85. C. J. Sanderson, Interleukin-5, eosinophils, and disease, *Blood*, *79*: 3101 (1992).
86. D. Catovsky, C. Bernasconi, P. J. Verdonck, A. Postma, J. Hows, J. van der Does, A. van den Berg, J. K. H. Rees, G. Castelli, E. Morra, and D. A. G. Galton, The association of eosinophilia with lymphoblastic leukaemia or lymphoma, *Br. J. Haematol.*, *45*: 523 (1980).
87. Y. Yamaguchi, T. Suda, H. Shiozaki, Y. Miura, Y. Hitoshi, A. Tominaga, K. Takatsu, and T. Kasahara, Role of IL-5 in IL-2-induced eosinophilia. In vivo and in vitro expression of IL-5mRNA by IL-2, *J. Immunol.*, *145*: 873 (1990).
88. T. R. Mosmann, H. Cherwinski, M. W. Bond, M. A. Giedlin, and R. L. Coffman, Two types of murine helper T cell clones, *J. Immunol.*, *136*: 2348 (1986).
89. T. R. Mosmann, and R. L. Coffman, Two types of mouse helper T cell clone. Implications for immune regulation, *Immunol. Today*, 8: 223 (1987).
90. T. R. Mosmann, and R. L. Coffman, TH1 and TH2 cells: Different patterns of lymphokine secretion lead to different functional properties, *Annu. Rev. Immunol.*, *7*: 145 (1989).
91. T. L. Stevens, A. Bossie, V. M. Sanders, R. Fernandez-Botran, R. L. Coffman, T. R. Mosmann, and E. S. Vitetta, Regulation of antibody isotype secretion by subsets of antigen-specific helper T cells, *Nature*, *334*: 255 (1988).
92. D. J. Cher, and T. R. Mosmann, Two types of murine helper T cell clone. II. Delayed-type hypersensitivity is mediated by Th1 clones, *J. Immunol.*, *138*: 3688 (1987).
93. T. A. T. Fong, and T. R. Mosmann, The role of IFN-gamma in delayed-type hypersensitivity mediated by Th1 clones, *J. Immunol.*, *143*: 2887 (1989).
94. N. E. Street, J. H. Schumacher, T. A. T. Fong, H. Bass, D. F. Fiorentino, J. A. Leverah, and T. R. Mosmann, Heterogeneity of mouse helper T cells: Evidence from bulk cultures and limiting dilution cloning for precursors of Th1 and Th2 cells, *J. Immunol.*, *144*: 1629 (1990).
95. D. F. Fiorentino, A. Zlotnick, T. R. Mosmann, M. Howard, K. W. Moore, and A. O'Garra, IL-10 acts on the antigen presenting cell to inhibit cytokine production by Th1 cells, *J. Immunol.*, *146*: 3444 (1991).
96. M. Plaut, J. H. Pierce, C. J. Watson, J. Hanley-Hyde, R. P. Nordan, and W. E. Paul, Mast cell lines produce lymphokines in response to cross-linkage of FcϵR1 or to calcium ionophores, *Nature*, *329*: 64 (1989).
97. P. R. Burd, H. W. Rogers, C. A. Gordon, C. A. Martin, S. Jayaraman, S. Wilson, A. M. Dvorak, S. Galli, and M. E. Dorf, Interleukin 3-dependent and independent mast cells stimulated with IgE and antigen express multiple cytokines, *J. Exp. Med.*, *170*: 245 (1989).
98. A. Wodnar-Filipowicz, C. H. Heusser, and C. Moroni, Production of the haemopoietic growth factors GM-CSF and interleukin-3 by mast cells in response to IgE receptor-mediated activation, *Nature*, *339*: 150 (1989).
99. J. R. Gordon, and S. J. Galli, Release of both preformed and newly synthesized tumor necrosis factorα (TNF-α)/cachectin by mouse mast cells stimulated via the FCeR1. A mechanism for the sustained action of mast cell-derived TNF-α during IgE dependent biological responses, *J. Exp. Med.*, *174*: 103 (1991).
100. D. T. W. Wong, P. F. Weller, S. J. Galli, A. Elovic, T. H. Rand, G. T. Gallagher,

T. Chiang, M. Y. Chou, K. Matossian, J. McBride, and R. Todd, Human eosinophils express transforming growth factor α, *J. Exp. Med.*, *172*: 673 (1990).

101. H. Kita, T. Ohnishi, Y. Okubo, D. Weiler, J. S. Abrams, and G. J. Gleich, GM-CSF and interleukin 3 release from human peripheral blood eosinophils and neutrophils, *J. Exp. Med.*, *174*: 745 (1991).
102. Q. Hamid, J. Barkans, Q. Meng, S. Ying, J. S. Abrams, A. B. Kay, and R. Moqbel. Human eosinophils synthesize and secrete interleukin-6 in vitro, *Blood*, *80*: 1496 (1992).
103. R. K. Braun, T. T. Hansel, I. J. M. de Vries, S. Rihs, K. Blaser, F. Erard, and C. Walker, Human peripheral blood eosinophils have the capacity to produce IL-8 [abstract], *FASEB J.*, *6*: A1612 (1992).
104. P. Desreumaux, A. Janin, J. F. Colombel, L. Prin, J. Plumas, D. Emilie, G. Torpier, A. Capron, and M. Capron, Interleukin 5 messenger RNA expression by eosinophils in the intestinal mucosa of patients with coeliac disease, *J. Exp. Med.*, *175*: 293 (1992).
105. D. H. Broide, M. M. Paine, and G. S. Firestein, Eosinophils express interleukin 5 and granulocyte macrophage-colony-stimulating factor mRNA at sites of allergic inflammation in asthmatics, *J. Clin. Invest.*, *90*: 1414 (1992).
106. M.-P. Piccinni, D. Macchia, P. Parronchi, M.-G. Giudizi, D. Bani, R. Alterini, A. Grossi, M. Ricci, E. Maggi, and S. Romagnani, Human bone marrow non-B, non-T cells produce IL-4 in response to cross-linkage of Fc_e and Fc_c receptors, *Proc. Natl. Acad. Sci. USA*, *88*: 8656 (1991).
107. G. J. Gleich, The eosinophil and bronchial asthma: Current understanding, *J. Allergy Clin. Immunol.*, *85*: 422 (1990).
108. M. Marini, E. Vittori, J. Hollemborg, and S. Mattoli, Expression of the potent inflammatory cytokines, granulocyte–macrophage-colony-stimulating factor and interleukin 6 and interleukin 8, in bronchial epithelial cells of patients with asthma, *J. Allergy Clin. Immunol.*, *89*: 1001 (1992).
109. O. Cromwell, Q. Hamid, C. J. Corrigan, J. Barkans, Q. Meng, P. D. Collins, and A. B. Kay, Expression and generation of IL-6, IL-8, and GM-CSF by human bronchial epithelial cells and enhancement by IL-1β and TNF-α, *Immunology*, *77*: 330 (1992).
110. C. J. Howell, J.-L. Pujol, A. E. G. Crea, R. Davidson, A. J. H. Gearing, P. H. Godard, and T. H. Lee. Identification of an alveolar macrophage-derived activity in bronchial asthma that enhances leukotriene C_4 generation by human eosinophils stimulated by ionophore A23187 as a granulocyte-macrophage colony-stimulating factor, *Am. Rev. Respir. Dis.*, *140*: 1340 (1989).
111. L. Borish, J. J. Mascali, and L. J. Rosenwasser, IgE-dependent cytokine production by human peripheral blood mononuclear phagocytes, *J. Immunol.*, *146*: 63 (1991).
112. X. Paliard, R. de Waal Malefijt, H. Yssel, D. Blanchard, I. Chretien, J. Abrams, J. de Vries, and H. Spits, Simultaneous production of IL-2, Il-4, and IFN-gamma by activated human $CD4^+$ and $CD8^+$ T cell clones, *J. Immunol.*, *141*: 849 (1988).
113. D. T. Umetsu, H. H. Jabara, R. H. DeKruyff, A. K. Abbas, J. S. Abrams, and R. S. Geha, Functional heterogeneity among human inducer T cell clones, *J. Immunol.*, *140*: 4211 (1988).

114. E. A. Wierenga, M. Snoek, C. de Groot, I. Chretien, J. de Bos, H. M. Jansen, and M. L. Kapsenberg, Evidence for compartmentalization of functional subsets of $CD4^+$ T lymphocytes in atopic patients, *J. Immunol.*, *144*: 4651 (1990).
115. P. Parronchi, D. Macchia, M.-P. Piccini, C. Biswas, C. Simonelli, E. Maggi, M. Ricci, A. A. Ansari, and S. Romagnani, Allergen and bacterial antigen-specific T cell clones established from atopic donors show a different profile of cytokine production, *Proc. Natl. Acad. Sci. USA*, *88*: 4538 (1991).
116. E. Maggi, P. Biswas, G. Del Prete, P. Parronchi, D. Macchia, C. Simonelli, L. Emmi, M. De Carli, A. Tiri, M. Ricci, and S. Romagnani, Accumulation of Th2-like helper T cells in the conjunctiva of patients with vernal conjunctivitis, *J. Immunol.*, *146*: 1169 (1991).
117. F. C. van Reijsen, C. A. F. M. Bruijnzeel-Koomen, F. S. Kalthoff, E. Maggi, S. Romagnani, J. K. T. Westland, and G. C. Mudde, Skin-derived aeroallergen-specific T-cell clones of Th2 phenotype in patients with atopic dermatitis, *J. Allergy Clin. Immunol.*, *90*: 184 (1992).
118. C. J. Corrigan, A. Hartnell, and A. B. Kay, T lymphocyte activation in acute severe asthma, *Lancet 1*: 1129 (1988).
119. C. Walker, J.-C. Virchow, P. L. B. Bruijnzeel, and K. Blaser, T cell subsets and their soluble products regulate eosinophilia in allergic and nonallergic asthma, *J. Immunol.*, *146*: 1829 (1991).
120. C. J. Corrigan, A. Haczku, V. Gemou-Engesaeth, S. Doi, Y. Kikuchi, K. Takatsu, S. R. Durham, and A. B. Kay, CD4 T-lymphocyte activation in asthma is accompanied by increased serum concentrations of interleukin-5: Effect of glucocorticoid therapy, *Am. Rev. Respir. Dis.*, *147*: 540 (1993).
121. C. J. Corrigan, and A. B. Kay, CD4 T-lymphocyte activation in acute severe asthma. Relationship to disease severity and atopic status, *Am. Rev. Respir. Dis.*, *141*: 970 (1990).
122. C. Walker, M. K. Kaegi, P. Braun, and K. Blaser, Activated T cells and eosinophilia in bronchoalveolar lavages from subjects with asthma correlated with disease severity, *J. Allergy Clin. Immunol.*, *88*: 935 (1991).
123. D. S. Robinson, A. M. Bentley, A. Hartnell, A. B. Kay, and S. R. Durham, Activated memory T helper cells in bronchoalveolar lavage fluid from patients with atopic asthma: Relation to asthma symptoms, lung function and bronchial responsiveness, *Thorax*, *48*: 26 (1993).
124. M. Azzawi, B. Bradley, P. K. Jeffery, A. J. Frew, A. J. Wardlaw, G. Knowles, B. Assoufi, J. V. Collins, S. Durham, and A. B. Kay, Identification of activated T lymphocytes and eosinophils in bronchial biopsies in stable atopic asthma, *Am. Rev. Respir. Dis.*, *142*: 1407 (1990).
125. D. S. Robinson, Q. Hamid, S. Ying, A. Tsicopoulos, J. Barkans, A. M. Bentley, C. J. Corrigan, S. R. Durham, and A. B. Kay, Predominant Th2-like bronchoalveolar T-lymphocyte population in atopic asthma, *N. Engl. J. Med.*, *326*: 298 (1992).
126. S. Ying, D. S. Robinson, V. Varney, Q. Meng, A. Tsicopoulos, R. Moqbel, S. R. Durham, and A. B. Kay, TNFα mRNA expression in allergic inflammation, *Clin. Exp. Allergy*, *21*: 745 (1991).

127. C. Walker, E. Bode, L. Boer, T. T. Hansel, K. Blaser, and J. C. Virchow, Jr., Allergic and nonallergic asthmatics have distinct patterns of T-cell activation and cytokine production in peripheral blood and bronchoalveolar lavage, *Am. Rev. Respir. Dis.*, *146*: 109 (1992).
128. D. H. Broide, M. Lotz, A. J. Cuomo, D. A. Coburn, E. C. Federman, and S. I. Wasserman, Cytokines in symptomatic asthma airways, *J. Allergy Clin. Immunol.*, *89*: 958 (1992).
129. Q. Hamid, M. Azzawi, S. Ying, R. Moqbel, A. J. Wardlaw, C. J. Corrigan, B. Bradley, S. R. Durham, J. V. Collins, P. K. Jeffrey, D. J. Quint, and A. B. Kay, Expression of mRNA for interleukin-5 in mucosal bronchial biopsies from asthma, *J. Clin. Invest.*, *87*: 1541 (1991).
130. D. S. Robinson, S. Ying, A. M. Bentley, Q. Meng, J. North, S. R. Durham, A. B. Kay, and Q. Hamid, Relationships among numbers of bronchoalveolar lavage cells expressing mRNA for cytokines, asthma symptoms, and airway methacholine responsiveness in atopic asthma, *J. Allergy Clin. Immunol.*, *92*: 397 (1993).
131. P. Bradding, I. H. Feather, P. H. Howarth, R. Mueller, J. A. Roberts, K. Britten, J. P. A. Bews, T. C. Hunt, Y. Okayama, C. H. Heusser, G. R. Bullock, M. K. Church, and S. T. Holgate. Interleukin 4 is localized to and released by human mast cells, *J. Exp. Med.*, *176*: 1381 (1992).
132. A. Kupfer, T. R. Mosmann, and H. Kupfer, Polarized expression of cytokines in cell conjugates of helper T cells and splenic B cells, *Proc. Natl. Acad. Sci. USA*, *88*: 775 (1991).
133. S. Romagnani, Induction of Th1 and Th2 responses: A key role for the "natural" immune response? *Immunol. Today*, *13*: 379 (1992).
134. I. Ohno, R. Lea, S. Finotto, J. Marshall, J. Denburg, J. Dolovich, J. Gauldie, and M. Jordana, Granulocyte/macrophage colony-stimulating factor (GM-CSF) gene expression by eosinophils in nasal polyposis, *Am. J. Respir. Cell. Mol. Biol.*, *5*: 505 (1991).
135. A. J. Frew, and A. B. Kay, The relationship between infiltrating $CD4^+$ lymphocytes, activated eosinophils, and the magnitude of the allergen-induced late-phase cutaneous reaction in man, *J. Immunol.*, *141*: 4158 (1988).
136. Q. Hamid, J. Barkans, D. S. Robinson, S. R. Durham, and A. B. Kay, Coexpression of CD25 and CD3 in atopic allergy and asthma, *Immunology*, *75*: 659 (1992).
137. A. B. Kay, S. Ying, V. Varney, M. Gaga, S. R. Durham, R. Moqbel, A. J. Wardlaw, and Q. Hamid, Messenger RNA expression of the cytokine gene cluster interleukin 3 (IL-3), IL-4, IL-5, and granulocyte/macrophage colony-stimulating factor, in allergen-induced late-phase cutaneous reactions in atopic subjects, *J. Exp. Med.*, *173*: 775 (1991).
138. S. R. Durham, S. Ying, V. A. Varney, M. R. Jacobson, R. M. Sudderick, I. S. Mackay, A. B. Kay, and Q. A. Hamid, Cytokine messenger RNA expression for IL-3, IL-4, IL-5 and granulocyte/macrophage-colony stimulating factor in the nasal mucosa after local allergen provocation: Relationship to tissue eosinophilia, *J. Immunol.*, *148*: 2390 (1992).
139. S. Ying, Q. Hamid, J. Barkans, R. Moqbel, S. R. Durham, and A. B. Kay,

Phenotype of cells expressing interleukin 5 mRNA in the nasal mucosa after local allergen provocation [abstract], *J. Allergy Clin. Immunol.*, *91*: 252 (1993).
140. D. H. Broide, and G. S. Firestein, Endobronchial allergen challenge in asthma: Demonstration of cellular source of granulocyte macrophage colony-stimulating factor by in situ hybridization, *J. Clin. Invest.*, *88*: 1048 (1991).
141. J. B. Sedgwick, W. J. Calhoun, G. J. Gleich, H. Kita, J. S. Abrams, L. B. Schwartz, B. Volovitz, M. Ben-Yaakov, and W. W. Busse, Immediate and late airway response of allergic rhinitis patients to segmental antigen challenge, *Am. Rev. Respir. Dis.*, *144*: 1274 (1991).
142. A. M. Bentley, Q. Meng, D. S. Robinson, Q. Hamid, A. B. Kay, and S. R. Durham, Increases in activated T lymphocytes, eosinophils, and cytokine mRNA expression for interleukin 5 and granulocyte/macrophage colony-stimulating factor in bronchial biopsies after allergen inhalation challenge in atopic asthmatics, *Am. J. Respir. Cell. Mol. Biol.*, *8*: 35 (1993).
143. D. S. Robinson, Q. Hamid, A. M. Bentley, S. Ying, A. B. Kay, and S. R. Durham, Activation of $CD4^+$ T cells, increased Th2-type cytokine mRNA expression, and eosinophil recruitment in bronchoalveolar lavage after allergen inhalation challenge in atopic asthmatics, *J. Allergy Clin. Immunol.*, *92*: 313 (1993).
144. A. R. Gulbenkian, R. W. Egan, X. Fernandez, H. Jones, W. Kreutner, T. Kung, F. Payvandi, L. Sullivan, J. A. Zurcher, and A. S. Watnick, Interleukin 5 modulates eosinophil accumulation in allergic guinea pig lung, *Am. Rev. Respir. Dis.*, *146*: 263 (1992).
145. H. Nakajima, I. Iwamoto, S. Tomoe, R. Matsumura, H. Tomioka, K. Takatsu, and S. Yoshida, $CD4^+$ T-lymphocytes and interleukin 5 mediate antigen-induced eosin- ophil infiltration into the mouse trachea, *Am. Rev. Respir. Dis.*, *146*: 374 (1992).
146. National Heart, Lung, and Blood Institute, Guidelines for the diagnosis and management of asthma, *J. Allergy Clin. Immunol.*, *88*: 425 (1991).
147. D. S. Robinson, Q. Hamid, S. Ying, A. M. Bentley, B. Assoufi, S. R. Durham, and A. B. Kay, Prednisolone treatment in bronchial asthma. Decreased bronchial responsiveness is accompanied by reduction in bronchoalveolar lavage eosinophilia and modulation of IL-4, IL-5 and IFN-gamma cytokine gene expression, *Am. Rev. Respir. Dis.*, *148*: 401 (1993).
148. M. D. Sadick, F. P. Heinzel, B. J. Holaday, R. T. Pu, R. S. Dawkins, and R. M. Locksley, Cure of murine leishmaniasis with anti-interleukin 4 monoclonal antibody. Evidence for a T cell-dependent, interferon gamma-independent mechanism, *J. Exp. Med.*, *171*: 115 (1990).
149. M. Belosevic, D. S. Finbloom, P. H. Van Der Meide, M. V. Slayter, and C. A. Nacy, Administration of monoclonal anti-IFN gamma antibodies in vivo abrogates natural resistance of C3H/HeN mice to infection with *Leishmania major*, *J. Immunol.*, *143*: 266 (1989).
150. R. L. Coffman, K. Varkila, and P. Scott, Role of cytokines in the differentiation of $CD4^+$ T-cell subsets in vivo, *Immunol. Rev.*, *123*: 189 (1991).
151. R. Badaro, E. Falcoff, F. S. Badaro, E. M. Carvalho, D. Pedral-Sampaio, A. Barral, J. S. Carvalho, M. Barral-Netto, M. Brandely, L. Silva, J. C. Bina, R. Teixeira, R. Falcoff, H. Rocha, J. L. Ho, and W. D. Johnson, Jr., Treatment of

visceral leishmaniasis with pentavalent antimony and interferon gamma, *N. Engl. J. Med.*, *322*: 16 (1990).
152. M. Yamamura, K. Uyemura, R. J. Deans, K. Weinberg, T. H. Rea, B. R. Bloom, and R. L. Modlin, Defining protective responses to pathogens: Cytokine profiles in leprosy lesions, *Science*, *254*: 277 (1991).
153. J. T. C. Li, J. W. Yunginger, C. E. Reed, H. S. Jaffe, D. R. Nelson, and G. J. Gleich, Lack of suppression of IgE production by recombinant interferon gamma: A controlled trial in patients with allergic rhinitis, *J. Allergy Clin. Immunol.*, *85*: 934 (1990).
154. T. F. Gajewski, M. Pinnas, T. Wong, and F. W. Fitch, Murine Th1 and Th2 clones proliferate optimally in response to distinct antigen-presenting cell populations, *J. Immunol.*, *146*: 1750 (1991).
155. T. F. Gajewski, S. R. Schell, and F. W. Fitch, Evidence implicating utilization of different T cell receptor-associated signalling pathways by Th1 and Th2 clones, *J. Immunol.*, *144*: 4110 (1990).
156. M. E. Williams, A. H. Lichtman, and A. K. Abbas, Anti-CD3 antibody induces unresponsiveness to IL-2 in Th1 clones but not in Th2 clones, *J. Immunol.*, *144*: 1208 (1990).
157. M. Betz, and B. S. Fox, Prostaglandin E_2 inhibits production of Th1 lymphokines but not of Th2 lymphokines, *J. Immunol.*, *146*: 108 (1991).
158. L. Noon, Prophylactic inoculation against hay fever, *Lancet*, *1*: 1572 (1911).
159. V. A. Varney, Q. Hamid, M. Gaga, S. Ying, M. Jacobson, A. J. Frew, A. B. Kay, S. R. Durham, Influence of grass pollen immunotherapy on cellular infiltration and cytokine mRNA expression during allergen-induced late-phase cutaneous response, *J. Clin. Invest.*, *92*: 644 (1993).
160. E. N. Charlesworth, W. A. Nish, M. G. Charlesworth, S. Valtier, T. George, and K. M. Leiferman, Standard high-dose immunotherapy decreases the production of IL-3, IL-5 and GM-CSF during the cutaneous late-phase response (LPR) to antigen [abstract], *J. Allergy Clin. Immunol.*, *91*: 252 (1993).
161. B. H. Van Leuwen, M. E. Martinson, C. G. Webb, and I. G. Young, Molecular organization of the cytokine gene cluster, involving the human IL-3, IL-4, IL-5 and GM-CSF genes on chromosome 5, *Blood 73*: 1142 (1989).
162. R. Bacchetta, R. De Waal Malefijt, H. Yssel, J. Abrams, J. E. De Vries, H. Spits, and M. G. Roncarolo, Host-reactive $CD4^+$ and $CD8^+$ T cell clones from a human chimera produce IL-5, IL-2, IFN-gamma, and granulocyte/macrophage-colony stimulating factor but not IL-4, *J. Immunol.*, *144*: 902 (1990).
163. G. M. Segal, T. D. Smith, M. C. Heinrich, F. S. Ey, and G. C. Bagby, Specific repression of granulocyte-macrophage colony-stimulating factor gene expression in interleukin-1 stimulated endothelial cells with antisense oligonucleotides,
164. T. Kitamura, N. Sato, K.-I. Arai, and A. Miyajima, Expression cloning of the human IL-3 receptor cDNA reveals a shared β subunit for the human IL-3 and GM-CSF receptors, *Cell*, *66*: 1165 (1991).
165. J. Tavernier, R. Devos, S. Cornelis, T. Tuypens, J. Van der Heyden, W. Fiers, and G. Plaetinck, A human high affinity interleukin-5 receptor (IL-5R) is composed of an IL-5-specific α chain and β chain shared with the receptor for GM-CSF, *Cell*, *66*: 1175 (1991).

166. W. C. Fanslow, J. E. Sims, H. Sassenfeld, P. J. Morrissey, S. Gillis, S. Dower, and M. B. Widmer, Regulation of alloreactivity in vivo by a soluble form of the interleukin-1 receptor, *Science*, *248*: 739 (1991).
167. W. C. Fanslow, K. N. Clifford, L. S. Park, A. S. Rubin, R. F. Voice, M. P. Beckmann, and M. B. Widmer, Regulation of alloreactivity in vivo by IL-4 and the soluble IL-4 receptor, *J. Immunol.*, *147*: 535 (1991).

13

Immunoglobulin E-Dependent Inflammation: Eosinophils and Other Inflammatory Cells

Peter F. Weller
Harvard Medical School and Beth Israel Hospital, Boston, Massachusetts

I. INTRODUCTION

The inflammatory responses that develop following initiation by IgE-dependent mechanisms involve the participation of many types of recruited leukocytes. The initial activation and subsequent evolution of IgE-mediated inflammatory processes are important in reactions that follow single, discrete exposures to antigens that trigger IgE-dependent activation as well as in reactions that occur in ongoing, subacute, or chronic diseases in which IgE-dependent mechanisms may participate. Such chronic diseases include asthma, and the inflammatory components are determinants of the clinical manifestations and pathological sequelae of such diseases. In this chapter, we will consider recruitment and activation of cells that participate in IgE-dependent inflammation. The delineation of the cells involved in such inflammation has been based on clinical and pathological observations of patients and on studies of experimentally elicited IgE-dependent inflammation in sites that include the skin, the nose, and the lungs. A major cellular component characteristically recruited to sites of IgE-dependent responses is the eosinophil, and this chapter will focus on the eosinophilic leukocyte: the mechanisms that may lead to its recruitment, and the products that it may elaborate that contribute to the evolution of inflammatory responses. Although eosinophils are characteristically preponderant, both clinically and experimentally, in IgE-initiated cellular responses (1–3), these leukocytes are not alone, and other cells, including $CD4^+$T lymphocytes, are important. The collaborative interactions that undoubtedly occur among these cells, locally recruited and activated within tissue sites, further contribute to the evolution and resolution of IgE-dependent inflammation.

II. EOSINOPHILS

A. Mechanisms of Eosinophil Recruitment

Eosinophils are recruited into sites of early- and late-phase IgE-mediated reactions, as well as into tissue sites of some other immunologically mediated diseases. In IgE-mediated reactions, eosinophils will be present to varying extents with other cell types, including neutrophils, basophils, lymphocytes, and monocytes, as discussed later. Since eosinophils are a minority of circulating blood leukocytes, even in atopic individuals, specific mechanisms exist that are responsible for the recruitment of eosinophils in large numbers into tissue sites. The release of chemotactic factors alone cannot account for a preponderant infiltration by eosinophils, since eosinophil chemoattractants, such as C5a and platelet-activating factor (PAF), are equally active in stimulating neutrophil migration (4). The mobilization of eosinophils into an inflammatory site, as for any other type of blood-borne leukocyte, involves adhesion to, and transmigration through, the vascular endothelium. The processes underlying adhesion and transmigration can be regulated by differential expression of adhesion molecules on both eosinophils and endothelial cells and act in concert with the local production of cell-selective chemotactic factors that promote tissue migration of eosinophils.

1. Eosinophil Mobilization from the Vasculature

For eosinophils to exit the blood stream to enter tissues, they must initially adhere to, and then transmigrate across, vascular endothelial cells. The differential expression of surface adhesion molecules by various types of leukocytes, and the specific regulation of adherence ligand expression on endothelial cells, likely contribute to preferential eosinophil accumulation in specific inflammatory states. The process of leukocyte–endothelial cell recognition involves three or more steps, beginning with a transient, reversible adhesion, and culminating with transmigration through the endothelial monolayer (5). The initial stimulus for eosinophil migration probably involves enhanced expression of endothelial adhesion molecules induced by inflammatory cytokines, such as interleukin (IL)-1 and tumor necrosis factor-α (TNF-α) (6) and IL-4. Eosinophil adherence to the activated endothelium may employ eosinophil surface integrins expressed constitutively as well as some whose increased expression is induced by inflammatory mediators (7–9).

Eosinophil–endothelial interactions may be mediated by several adherence pathways (Fig. 1) (10). Like neutrophils, monocytes, and lymphocytes, eosinophils adhere to intercellular adhesion molecule 1 (ICAM-1; CD54) by means of lymphocyte function-associated antigen-1 (LFA-1; CD11a–CD18 complex) (11–14). Eosinophil expression of CD11a is constitutive, whereas eosinophil CD11b expression is up-regulated by various cytokines and inflammatory mediators (8,15). Consequently, the adherence of stimulated eosinophils is heavily dependent on the CD11–CD18 adhesion complex (12). A role for other adhesion pathways is indicated by histopathological observations in the leukocyte-deficiency syndrome, in which there are genetic deficiencies in CD18 and consequent impairments in the abilities of neutrophils to emigrate from the vasculature into sites of inflammation. Despite the deficits in CD18-mediated adherence in this syndrome, lymphocytes, monocytes, and eosinophils are recruited into inflammatory lesions (16), providing in vivo evidence that eosinophils, similar to lymphocytes and monocytes, possess mechanisms of adherence independent of CD18.

Eosinophils, like neutrophils and mononuclear leukocytes, express L-selectin, which is involved in initial adherence to endothelial cells during shear flow conditions, but then is shed from the cell surface (17,18). Eosinophils also bind to P-selectin, induced to be expressed on endothelial cells, which may mediate shear-resistant adhesion and functions to inhibit cell spreading and superoxide anion generation (19). Eosinophils, similarly to neutrophils, also use a non-CD18-dependent pathway to bind to E-selectin (endothelial adhesion molecule-1; ELAM-1), which, in contrast with ICAM, is found only on activated endothelial cells (11–14, 18).

Another pathway for eosinophil adherence is based on the expression of very

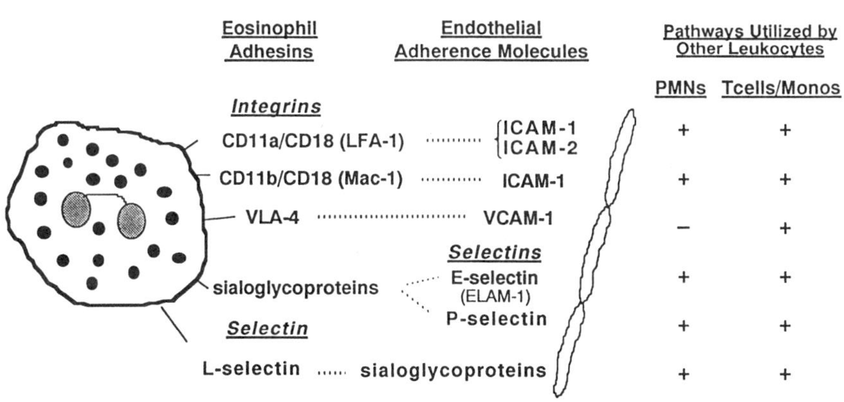

Figure 1 Pathways of adherence of human eosinophils to vascular endothelial cells. (From Ref. 10.)

late antigen-4 (VLA-4; CD49d/CD29) by eosinophils. VLA-4 is constitutively expressed on eosinophils from both normal and eosinophilic donors and on basophils, but is not expressed on neutrophils (11–13, 20, 21). VLA-4, the $\alpha_4\beta_1$ integrin, is also expressed on lymphocytes and mononuclear cells and binds to vascular cell adhesion molecule-1 (VCAM-1) as well as fibronectin (22).

The involvement of the three adherence pathways in mediating eosinophil adherence to activated endothelial cells and recombinant adherence molecules has been demonstrated in vitro (11–13, 20, 21). With TNF-α-activated human umbilical vein endothelial cells that express ICAM-1, E-selectin, and VCAM-1, combinations of monoclonal antibodies, blocking two or more of these adherence pathways, yielded greater inhibition of eosinophil adherence than did blockade of any one adhesion pathway (11,13). Blockade of both CD18- and VLA-4-mediated pathways produced complete inhibition of eosinophil adherence to the TNF-α-treated endothelial cells (11). Since VLA-4–VCAM interactions mediate adherence of eosinophils, basophils, lymphocytes, and monocytes, the concomitant recruitment of eosinophils, basophils, and mononuclear leukocytes, without neutrophils, into specific sites of IgE-mediated or other immunological reactions might be promoted by enhanced expression of VCAM on local vascular endothelium.

The possibility that VCAM expression might be preferentially elicited in sites of IgE-mediated reactions is suggested by the capacity of IL-4, which is produced by the T-helper type 2 (Th2) subset of lymphocytes, to elicit VCAM expression on endothelial cells. Interleukin-4, either alone (21) or together with TNF-α or IL-1α (23,24), stimulated endothelial cell VCAM expression in vitro and promoted VLA-4–VCAM-dependent adherence of eosinophils and basophils, but not neutrophils, to endothelium (21). However, eosinophil transendothelial migration, does not inevitably follow adhesion (25,26). Eosinophils from patients with allergic asthma, but not eosinophils isolated from nonallergic donors, were able to adhere to and transmigrate through endothelial cells grown on a human fibroblast-derived extracellular matrix if the endothelial cells had been treated with IL-1, TNF, or IL-4 (25,26). The in vivo "priming" of eosinophils from allergic patients could be elicited in vitro by preincubation of normal eosinophils with IL-3, IL-5, or granulocyte–macrophage colony-stimulating factor (GM-CSF) (25,26). In contrast with the other two cytokines studied for their effects on endothelial cells, only IL-4 treatment of endothelium stimulated the transmigration of eosinophils and not neutrophils. Eosinophil transmigration was both VLA-4–VCAM- and CD11–CD18-dependent. Thus, IL-4 production, by promoting VCAM expression on endothelial cells, would enable adherence and transmigration of eosinophils, but not neutrophils, as well as basophils, lymphocytes, and monocytes.

In limited studies to date, augmented expression of endothelial E-selectin and ICAM-1 and, less uniformly, VCAM has been found in sites of cutaneous late-phase reactions (27,28). Furthermore, IL-4 expression in vivo, in both IL-4-expressing, transplanted tumors and in IL-4 transgenic mice, has resulted in marked tissue accumulation of eosinophils, as well as the induction of allergic-type reactions (29,30).

Investigations in animal models of allergic airway disease extend these in vitro studies of eosinophil–endothelial cell interactions. Antagonism of cell adherence to ICAM-1 by intravenous administration of an anti-ICAM-1 monoclonal antibody led to decreased eosinophil infiltration and attenuated airway hyperresponsiveness in a primate model of chronic airway inflammation (31). Administration of an anti-E-selectin monoclonal antibody blocked neutrophil accumulation and late-phase airway obstruction in acute antigen-elicited airway inflammation in primates (32).

The roles of the $\alpha_4\beta_1$ integrin, VLA-4, as expressed on eosinophils, lymphocytes, and monocytes, or of other α_4-containing integrins, may not be limited to activities involved in leukocyte recruitment. In addition, engagement of α_4 integrins may also participate in cellular activation pertinent to IgE-mediated inflammation, as suggested by recent studies in a sheep model of asthma (33). In control sheep, ascarid antigen challenge produced early and late increases in specific lung resistance, as well as airway hyperresponsiveness, continuing for 14 days after antigen challenge. Systemic treatment with a blocking anti-α_4 monoclonal antibody, 30 min before antigen challenge, did not affect the early increase in specific lung resistance, but significantly inhibited the late-phase increase at 6–8 h by 75% and the postantigen challenge airway hyperresponsiveness. Airway administration of anti-α_4 antibody likewise was effective. Inhibition of leukocyte recruitment did not account for the activity of anti-α_4 antibody, since the antibody treatment neither diminished the eosinopenia or lymphopenia that followed antigen challenge, nor altered the composition of leukocytes recovered by bronchoalveolar lavage. Thus, inhibition of α_4 integrin with anti-α_4 antibody may have inhibited activation of specific leukocytes within airway tissues. These findings indicated a role for α_4 integrins, probably not based on their activities as adherence proteins, in processes that lead to airway late-phase responses and persisting airway hyperresponsiveness following antigen challenge. Thus, integrins and adherence molecules may be involved both in recruitment and activation of cells involved in the evolution and expression of IgE-mediated inflammation.

2. *Eosinophil Chemoattraction Within Tissues*

In addition to the processes involved in eliciting the migration of eosinophils (and other leukocytes) through the endothelium, another mechanism likely to be involved in the recruitment and activation of eosinophils within tissues is their response to certain chemotactic agents. The early recognition that diffusates of guinea pigs and human lungs, following IgE-mediated antigenic challenge, contained an activity [eosinophilic chemotactic activity of anaphylaxis (ECF-A)] that elicited chemotaxis of eosinophils (34,35), but not neutrophils, prompted the search for chemoattractants that would solely account for the selective recruitment of eosinophils to sites of allergic reactions. No specific eosinophil chemoattractant has yet been identified. The two tetrapeptides (Val-Gly-Ser-Glu and Ala-Gly-Ser-Glu), initially believed to mediate the eosinophil chemoattractant activity of ECF-A, have not been potent eosinophil chemoattractants (4). A reinvestigation of ECF-A activity obtained from sensitized and challenged guinea pig lung has recently identified leukotriene B_4 (LTB_4) and 8(*S*),15(*S*)-dihydroxy-5,9,11,13(*Z,E,Z,E*)-eicosatetraenoic acid as the ECF-A eosinophil chemoattractants, at least in guinea pigs (36). Neither of these chemoattractant lipids was specific for eosinophils in comparison with neutrophils (36). Thus, no single agent has been demonstrated that acts as a chemoattractant only for eosinophils. All chemoattractants for eosinophils have activity for at least some other types of leukocytes (Table 1). The final preferential recruitment of any cell type into tissue sites of inflammation results from an interplay of multiple mechanisms and pathways (5). In this interplay, despite their lack of absolute cellular specificity for eosinophils, eosinophil chemoattractants are likely to have roles in eosinophilic recruitment and activation.

Investigations of the relative potencies of several eosinophil chemoattractants (see Table 1) have demonstrated that PAF and the complement factor C5a are among the most potent, in comparison with other agents, such as LTB_4, histamine, and formylmethionylleucyl-phenylalanine (fMLP) (4,37). All of these mediators are active on neutrophils and eosinophils, although PAF, in comparative in vitro assays, can exhibit greater potency for eosinophils (37). Platelet-activating factor can also stimulate eosinophil degranulation and activation in vitro (38,39). It induces a local eosinophil accumulation after intracutaneous administration into atopic subjects (40) and promotes infiltration of eosinophils into bronchoalveolar lavage in a guinea pig model of asthma (41). Some of the effects of PAF of eosinophil accumulation may also be attributable to the abilities of PAF to enhance binding of eosinophils to endothelial cells (7) and promote the transendothelial migration of eosinophils (42,43).

In addition to the humoral mediators, specific polypeptide cytokines act as eosinophil chemoattractants. GM-CSF, IL-3, and IL-5, the hematopoietic cytokines that promote eosinophilopoiesis in vivo and enhance eosinophil survival and function in vitro (1), will stimulate eosinophil migration (see Table 1). Interleukin-5, like IL-4, is produced by the Th2-like subset of $CD4^+$ T cells, whereas IL-3 and GM-CSF are elaborated by both Th1 and Th2 subsets. Eosinophils themselves can elaborate GM-CSF, IL-3, and IL-5 (44–47). Interleukin-3 and GM-CSF on a molar basis are more potent as eosinophil chemoattractants than C5a or PAF, but similar to these humoral mediators also stimulate migration of neutrophils and basophils (48,49). Interleukin-5 stimulates

Table 1 Human Eosinophil Chemoattractants

Eosinophil chemoattractant	Molar potency (~ED_{50})	Other responding cells[a]	Notes	Ref.
Humoral mediators				
PAF	10^{-7}–10^{-6} M	N, M, B		4,55,56
C5a	10^{-7}–10^{-6} M	N, M, B		4,55,56
LTB_4	10^{-6} M	N, M		4,37
fMLP	10^{-6} M	N, M	Eos requires priming	4,37,48
8(*S*),15(*S*)-diHETE	10^{-6} M	N		36,308
5,15-diHETE	10^{-6} M	N		308
5-oxo-15-HETE	10^{-7}	N		308,309
Cytokines				
LCF	10^{-12}–10^{-11} M	$CD4^+$ L and M, ?B		55
IL-2	10^{-12} M	IL-$2R^+$ L and M, ?B		56
IL-3 and GM-CSF	10^{-9}–10^{-8} M	N, B		48,51
IL-5	10^{-11} M	?B		51,310
RANTES	10^{-8} M	Memory/helper L, M, B		70,71
IL-8	10^{-8} M	N, L, B	Eos requires priming	48
Other				
Cyclophilin	10^{-8} M	N		311

[a]Cell types are abbreviated: N, neutrophils; M, monocytes; L, lymphocytes; B, basophils; Eos, eosinophils.
Source: Modified from Ref. 10.

eosinophil, but not neutrophil, migration (50,51). Although not yet reported, IL-5 might stimulate migration of basophils, which are otherwise responsive to IL-5 (52). It also increases eosinophil adherence to endothelial cells (9). Preincubation of eosinophils from normal donors with picomolar concentrations of IL-5, IL-3, and GM-CSF markedly increases eosinophil chemotaxis toward PAF, LTB_4, fMLP, and IL-8 (48,51,53). The release of these cytokines in vivo may result in an increased responsiveness, or priming, of eosinophils to migratory stimuli. Eosinophil stimulation promoter (ESP), a lymphokine activity initially described in the mouse, which stimulates eosinophil migration and functional responses, has recently been shown to be attributable to the combined activities of IL-5 and GM-CSF (54).

Two lymphocyte-derived cytokines, lymphocyte chemoattractant factor (LCF) and IL-2, are active as eosinophil chemoattractants with ED_{50}s of 10^{-12}–10^{-11} M, making them the most potent currently known eosinophil chemoattractants (55,56). Lymphocyte chemoattractant factor is a 56,000-Da homotetrameric basic glycoprotein that is elaborated by $CD8^+$ human T lymphocytes (57,58). It uses CD4 as its receptor and stimulates the migration of $CD4^+$ lymphocytes and monocytes (57). CD4, at lower levels than on $CD4^+$ T lymphocytes, is expressed on eosinophils obtained from normal and eosinophilic donors (55,59). With eosinophils from both normal and eosinophilic donors, LCF was 100- to 1000-fold more potent as an eosinophil chemoattractant than either C5a or PAF (55). It did not stimulate migration of neutrophils, which lack CD4 expression; hence, LCF differs from C5a and PAF in being appreciably more potent as an eosinophil chemoattractant and in being more specific for eosinophils than for neutrophils (55). Since histamine elicits LCF release from $CD8^+$ lymphocytes in vitro (60,61), the histamine-stimulated elaboration of LCF at sites of IgE-mediated allergic responses could contribute to the recruitment and activation of both eosinophils and $CD4^+$ mononuclear cells.

Eosinophils, which express detectable p55 (CD25) IL-2 receptor subunits (56,62,63), migrate in response to IL-2, with responses elicited at 10^{-12} M (56). This chemokinetic migration of eosinophils is mediated by high-affinity IL-2 receptors, as evidenced by the molar potency of IL-2 as an eosinophil chemoattractant and the capacities of blocking monoclonal antibodies to either the p55 or p75 IL-2 receptor subunits to competitively inhibit IL-2-induced eosinophil migration (56). In chronic asthma, airway tissues are infiltrated with eosinophils and $CD4^+$ mononuclear cells, with many IL-2 receptor-bearing mononuclear cells (64). In IgE-mediated cutaneous (65) and pulmonary (66,67) late-phase reactions, eosinophils and $CD4^+$ lymphocytes are prominent in the resulting cellular infiltrates. Thus, the cytokines LCF and IL-2 would be candidate mediators to contribute to the preferential tissue influx of eosinophils, lymphocytes, and mononuclear cells.

Interleukin-8 and RANTES are members of a structurally related chemokine superfamily of lower relative molecular mass (M_r) cytokines that have cell-selective chemotactic activities (68). Interleukin-8, which is chemotactic for neutrophils and T lymphocytes, induces eosinophil migration only after in vitro priming with GM-CSF or IL-3 (48). Since bronchial epithelial cells of asthmatic patients elaborate both GM-CSF and IL-8 (69), even epithelial cells might have a role in local eosinophil recruitment. RANTES, which is chemotactic for monocytes and "memory" T lymphocytes, but not for neutrophils (68), is a potent eosinophil chemoattractant (70,71). Local production of RANTES by activated T cells (68) also would represent a candidate mechanism to contribute to the accumulation of eosinophils and $CD4^+$ T lymphocytes that develops during IgE-initiated late-phase reactions.

B. Eosinophils as Inflammatory Cells and Their Products

Eosinophils are a source of several types of products that may contribute to inflammatory responses. These eosinophil-derived products include specific lipid mediators, peptides, cytokines, and preformed granule proteins.

1. Lipid Mediators

Platelet Activating Factor. Eosinophils synthesize the phospholipid mediator PAF, 1-alkyl-2-acetyl-*sn*-glycero-3-phosphocholine. Human eosinophils contain 1-alkyl-2-lyso-*sn*-glycero-3-phosphocholine:acetyl-CoA acetyltransferase, the PAF-synthesizing enzyme that acetylates lysophospholipid to form PAF (72). The activity of this PAF-forming enzyme in eosinophils increases in a dose-, time-, and Ca^{2+}/Mg^{2+}-dependent fashion following treatment of eosinophils with the calcium ionophore A23187 or the eosinophil chemoattractants: C5a, fMLP, or eosinophil chemotactic factor of anaphylaxis peptide (ECF-A) (72). The enzyme that catabolizes PAF, 1-alkyl-2-lyso-*sn*-glycero-3-phosphocholine:acetyl-CoA acetylhydrolase, is also present in eosinophils (73); but the activity of this enzyme was not affected by stimulation of eosinophils with C5a, fMLP, or ionophore A23187 (72). In these early studies, 10^6 eosinophils released about 1–2 fmol (~1–2 pg) of PAF in response to 10^{-6}M A23187 (72), and the activity of the PAF-forming acetyltransferase was greater with eosinophils from eosinophilic donors than with either normal donor-derived eosinophils or neutrophils (73).

More recent studies have documented the ability of human eosinophils to generate even greater quantities of PAF. In response to 5×10^{-6} M A23187, maximal intracellular amounts of PAF of about 35 ng PAF per 10^6 eosinophils developed after 15 min of stimulation, whereas maximal extracellular amounts of about 25 ng/10^6 eosinophils were detected later at 45–60 min, by which time intracellular PAF levels had declined (74). Following stimulation with IgG-coated

Sepharose particles, normal-density eosinophils formed PAF (about 2 ng PAF per 10^6 eosinophils) that initially remained largely cell-associated; levels of released PAF increased to about 2 ng of extracellular PAF after 45–60 min (74). The PAF formation was greater with eosinophils of normal density than with either low-density eosinophils or neutrophils; and for both groups of eosinophils and for neutrophils, the amounts of newly formed PAF were usually greater intracellularly than those released extracellularly (74). Both PAF turnover and incorporation of [^{3}H]PAF into phosphatidylcholine were greater with low-density eosinophils; accordingly, net levels of PAF assayed in these eosinophils may have underestimated rates of synthesis owing to heightened catabolism of some newly formed PAF (74). Another study has also quantitated the production of PAF by normodense eosinophils stimulated with ionophore A23187 or unopsonized zymosan (75). With 10 μM A23187, eosinophils formed 4.2 ng PAF per 10^6 eosinophils at 15 min, with 90% of the generated PAF remaining cell-associated; and with 5×10^8 zymosan particles, a plateau level of 0.11 ng PAF per 10^6 eosinophils was attained at 30 min, with 91% of the PAF remaining cell-associated (75). The formation of PAF, as well as of LTC_4, stimulated with either A23187 or unopsonized zymosan, was inhibited by quinacrine (mepacrine), but not nedocromil sodium (75).

Platelet-activating-factor has multiple effects pertinent to its potential roles as a mediator involved in the pathogenesis of IgE-mediated inflammatory reactions (76–79). It stimulates platelet aggregation and degranulation, enhances vascular permeability, induces bronchoconstriction, and stimulates migration and activation of neutrophils. It has several effects on eosinophils, including stimulating eosinophil migration (4,39,80), superoxide production and degranulation (38,81), eicosanoid formation through the cyclooxygenase (82) and lipoxygenase (83) pathways, and conversion to an activated hypodense phenotype (84). Endogenous synthesis and release of PAF by eosinophils may be involved in stimulating the respiratory burst of eosinophils elicited by opsonized zymosan, since WEB 2086, a PAF antagonist, blocks superoxide production by eosinophils incubated with opsonized zymosan (85). Platelet-activating factor is formed by many cell types involved in inflammation, including neutrophils, monocytes, alveolar macrophages, and platelets, as well as eosinophils.

Eicosanoids. Eicosanoids are formed from the oxidative metabolism of arachidonic acid or other polyunsaturated fatty acids, which may proceed by three potential enzymatic pathways: the cyclooxygenase, lipoxygenase, and monooxygenase pathways. As in other cells, the cyclooxygenase pathway was the first to be recognized in eosinophils (86). Subsequently, the 5- and 15-lipoxygenase pathways of eicosanoid formation have been well studied in eosinophils; but little is known in eosinophils about the activity of products of the monooxygenase pathway, which is mediated by cytochrome P-450 enzymes.

Human eosinophils and neutrophils do exhibit one difference in cytochrome P-450 function: eosinophils do not form ω-oxidation derivatives of LTB_4 (87,88), whereas neutrophils, by a cytochrome P-450 enzyme with LTB_4 ω-hydrolase activity (89), catabolically transform LTB_4 into 20-hydroxyl- and 20-carboxyl-LTB_4 (87,88). Thus, the cyclooxygenase and lipoxygenase pathways represent the major identified routes of eicosanoid formation in eosinophils (Fig. 2).

Prostaglandin endoperoxide synthase, also termed prostaglandin H synthase (PGH) or cyclooxygenase, is the first enzyme involved in the pathway for formation of various prostaglandins (PG), thromboxanes (TX), and prostacyclin (PEI_2). Prostaglandin endoperoxide synthase catalyzes two reactions, a cyclooxygenase reaction that inserts two molecules of oxygen into substrate arachidonic acid to form PGG_2 and an endoperoxidase reaction that reduces PGG_2 to its 15-hydroxy analogue, PGH_2. Other distinct enzymes subsequently act on PGG_2 to form the specific derivatives of the cyclooxygenase pathway. The capacity of eosinophils to form PGE_2 was first indicated in 1975 (86,90), and recent studies have confirmed the ability of eosinophils to elaborate PGE_2. Eosinophils stimulated in vitro with the calcium ionophore A23187 and opsonized zymosan released 18–77 pg PGE_2 per 10^6 eosinophils (91). Parsons and Roberts also found that eosinophils stimulated with 2 μM A23187 produced 483 pg PGE_2 per 10^6 eosinophils (92). These investigators, who demonstrated that human eosinophils transform PGD_2 to isomeric PGF_2 compounds, found that the preponderant cyclooxygenase pathway product of human eosinophils, as quan-

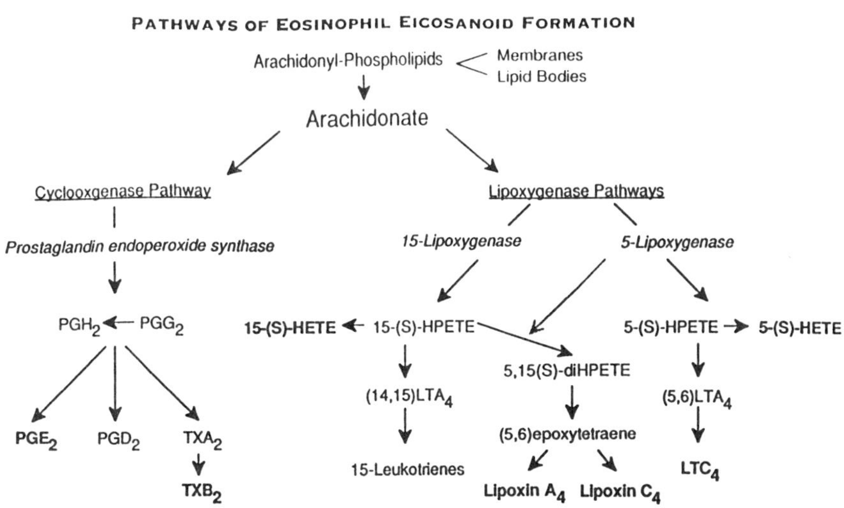

Figure 2 Oxidative metabolism of arachidonate by eosinophils. Principal products are shown in bold face. (From Ref. 241.)

titated by negative-ion chemical ionization mass spectrometry, was thromboxane B_2 (TXB_2); 2247 pg/10^6 eosinophils) with smaller amounts of $PGF_{2\alpha}$, $PGD_{2\alpha}$, and 6-keto-$PGF_{1\alpha}$ (265, 50, and 6 pg/10^6 eosinophils, respectively) (92). Another study, that specifically sought and excluded the presence of contaminating platelets (which with their preponderant production of TXA_2 might have contributed to the measured levels of TXB_2), used human peritoneal eosinophils from peritoneal dialysis bags, and also found TXB_2 to be the preponderant eosinophil prostanoid, with smaller amounts of PGE_2 and 6-keto-$PGF_{1\alpha}$, as measured by radioimmunoassays (93). When 5×10^6 eosinophils were stimulated with 1 μM A23187, about 9.5, 3.4, and 0.5 ng of immunoreactive TXB_2, PGE_2, and 6-keto-$PGF_{1\alpha}$, respectively, were released in 1 h (93). Increased levels of PGE_2, which were correlated with levels of eosinophils, in bronchoalveolar lavage fluids from two patients with eosinophilic pneumonia have been noted (94). The roles of the eosinophil-derived prostanoids as mediators of inflammation or as paracrine mediators of activities of other cell types remain uncertain. Exogenous PGE_2 can inhibit eosinophil degranulation (95), but autocrine effects of prostanoids on the functioning of eosinophils themselves have not been evaluated.

In leukocytes, 5-lipoxygenase catalyzes two enzymatic activities. 5-Lipoxygenase inserts one oxygen molecule into the 5 position of arachidonic acid (or other polyunsaturated fatty acids) to form 5*S*-hydroperoxyeicosatetraenoic acid (5-HPTE) and then transforms 5-HPETE into an unstable allylic epoxide, LTA_4. In most species, except guinea pigs, the subsequent metabolism of LTA_4 differs between eosinophils and neutrophils. In neutrophils, LTA hydrolase, a specific epoxide hydrolase, enzymatically hydrolyses LTA_4 to LTB_4. In contrast, in eosinophils, a specific glutathione *S*-transferase, LTC_4 synthase, catalyzes the adduction of glutathione to LTA_4 to form LTC_4. Although LTC_4 was initially described as a product of calcium ionophore-stimulated human granulocytes, early studies with human (87) and horse (96) eosinophils documented that LTC_4 was the major 5-lipoxygenase pathway derivative in eosinophils, whereas neutrophils elaborated LTB_4. This preferential generation of LTC_4 by human eosinophils has been evaluated by many research groups (88,97–106). Problems in isolating eosinophil preparations fully devoid of neutrophils, potential differences in the amounts of LTC_4 formed by normodense eosinophils in comparison with hypodense eosinophils (those more contaminated with neutrophils when isolated on metrizamide or Percoll-density gradients), potential transcellular exchange of LTA_4 between neutrophils and eosinophils or between neutrophils and platelets (which cannot synthesize LTA_4, but can use leukocyte-derived LTA_4 to form LTC_4 (107–109), contribute to the difficulty in ascertaining whether neutrophils themselves can form any LTC_4. Nevertheless, eosinophils are the primary (if not exclusive) source of LTC_4 made by granulocytes, although other cell types, including basophils, macrophages, monocytes, and mast cell populations, can also form LTC_4 (110). Therefore, LTB_4 is the principal product of the 5-lipoxygenase pathway in human neutrophils, whereas LTC_4 is the preponderant 5-lipoxygenase product of human eosinophils (87). Leukotriene C_4 and its peptidolytic derivatives, LTD_4 and LTE_4, the three sulfidopeptide leukotrienes, have potent smooth-muscle contractile, vasoactive, and mucous secretory activities; whereas LTB_4 exhibits contrastingly different activities and stimulates leukocyte migration, aggregation, oxidative burst activity, and degranulation (111,112).

Human eosinophils, unlike neutrophils, contain 15-lipoxygenase (113). This enzyme catalyzes both the hydroperoxidation of arachidonic acid, with the insertion of one molecule of oxygen at position 15 to form 15-hydroperoxyeicosatetraenoic acid (15-HPETE), and the insertion of molecular oxygen into other polyunsaturated fatty acids and phospholipids (114–116). Eicosanoids formed by the 15-lipoxygenase pathway of arachidonate metabolism include the monohydroxy acid, 15-hydroxyeicosatetraenoic acid (15-HETE), and the dihydroxy acids, 8,15-diHETE, and 14,15-diHETE, as well as compounds requiring the additional activities of other lipoxygenase enzymes, the trihydroxy acids, lipoxins A and B.

Studies with porcine and, later, human leukocytes established that eosinophils were the predominant granulocyte expressing 15-lipoxygenase (117–119). The principal product of purified human eosinophils, incubated with 100 μM arachidonate and 10 μM A23187, was 15-HETE derived from 15-HPETE, along with quantities of 8,15-diHETEs (119). Although human neutrophils have been reported to express 15-lipoxygenase activity (120–123), it is likely that contaminating eosinophils were the source of 15-lipoxygenase activity detected among neutrophils. Because 15-HETE formation can be as high as 1000 ng/10^6 eosinophils (113,119,124), even minor eosinophil contamination could account for 15-HETE previously noted in neutrophils. 15-Lipoxygenase activity can be correlated with the numbers of eosinophils present among human granulocyte preparations (106,125). In addition, 15-lipoxygenase protein has not been detected by immunofluorescent localization studies within human peripheral blood neutrophils, monocytes, or lymphocytes (125), and RNA blots using 15-lipoxygenase cDNA have not detected 15-lipoxygenase mRNA in leukocyte preparations, unless they contain eosinophils (125). Nevertheless, 15-lipoxygenase may be inducible in cells under some conditions, as suggested by observations of 15-lipoxygenase in lipid-laden, monocyte-derived macrophages of atherosclerotic lesions (126). Hence, 15-lipoxygenase gene expression may be regulated in eosinophils and conceivably in other clases of leukocytes not normally expressing appreciable levels of this enzyme.

The 15-lipoxygenase of human eosinophils has been purified to homogeneity (113,127), and 15-lipoxygenase from human reticulocytes has been molecularly cloned (128) and expressed in eukaryotic and prokaryotic cells (129,130). It is a 70,000-M_r protein distinct from, but sharing homologies with, human leukocyte 5-lipoxygenase and other lipoxygenases (127). Eosinophil 15-lipoxygenase is activated by phosphatidylcholine and calcium and is competitively inhibited by ATP (113), in contrast with leukocyte 5-lipoxygenase, which is stimulated by ATP (131,132). On the basis of subcellular fractionation studies, 15-lipoxygenase is primarily a cytososlic enzyme in the eosinophil, with 75–90% of activity recovered from cytosolic fractions and little activity in granule or microsomal fractions (127). Correspondingly, immunofluorescent localization studies have indicated a primarily cytosolic localization of the enzyme in eosinophils as well as in airway epithelial cells and mature reticulocytes (125). The 15-lipoxygenase activity of purified eosinophil enzyme is most active when assayed with relatively high (80–100 μM) concentrations of substrate arachidonic acid (113); likewise, 15-lipoxygenase activity of intact eosinophils is most prominently expressed when cells are incubated with high, albeit cytotoxic, concentrations (50–250 μM) of exogenous arachidonate (97,113,133,134).

The activities of 15-lipoxygenase pathway products include stimulation of mucus release from cultured human airway (15-HETE) (135), chemotaxis of human and canine neutrophils (8,15-diHETE) (136,137), inhibition of natural killer cell activity (138), and contraction of guinea pig and human bronchi (139,140). In addition to its presence in eosinophils, 15-lipoxygenase pathway is present in human lung homogenates (141) and isolated human airway epithelial cells (142) and in reticulocytes. Expression of 15-lipoxygenase activity may down-regulate 5-lipoxygenase activity, since 15-HPETE inhibits human leukocyte 5-lipoxygenase (143), and 15-HETE inhibits eosinophil LTC_4 formation (133). Given the active expression of 15-lipoxygenase in mature reticulocytes, in which dissolution of mitochondria takes place during this stage of erythrocyte development (144), 15-lipoxygenase has been suggested to play a major role in lipid and membrane remodeling (125,129) as well as a role in forming paracrine mediators. The activities and consequences of 15-lipoxygenase actions on substrate unsaturated fatty acids esterified in phospholipids have yet to be fully evaluated (116).

Lipoxins are acyclic eicosanoids that contain a conjugated tetraene structure and three alcohol groups. The two principal compounds in this series of eicosanoids are positional isomers, lipoxin A_4 (LXA_4; 5*S*,6*R*,15*S*-trihydroxy-7,9,13-*trans*-11-*cis*-eicosatetraenoic acid) and lipoxin B_4 (LXB_4; 5*S*,14*R*,15*S*-trihydroxy-6,10,12-*trans*-8-*cis*-eicosatetraenoic acid) (145,146). Lipoxins are formed by the sequential actions of two lipoxygenase enzymes, including various potential combinations of the 5-, 12-, and 15-lipoxygenases. Eosinophil-enriched human granulocytes form specific lipoxin products of double lipoxygenation (147,148). Serhan et al. (147) demonstrated that eosinophil-enriched human leukocytes released LSA_4, and not LXB_4, when stimulated with ionophore A23187, and that the quantity of LSA_4 formed correlated with the numbers of eosinophils present. Lipoxin A_4 formation was detectable in these A23187-stimulated leukocytes with and without exogenous arachidonate. Whether eosinophils form LXA_4 solely as a result of the intracellular actions of their endogenous 5- and 15-lipoxygenase enzymes, or whether transcellular interactions among eosinophils or eosinophils and neutrophils (or even contaminating platelets) are involved, has not been ascertained. Although the eosinophil with its content of both lipoxygenases would be capable of endogenous lipoxin biosynthesis, the role of transcellular interactions in lipoxin formation is also well documented. Lipoxin formation during neutrophil–platelet interactions can result from the transformation of neutrophil LTA_4 by platelet 12-lipoxygenase (107,109,149). Analogously, eosinophil-derived 15-HETE might be acted on by neutrophil 5-lipoxygenase or platelet 12-lipoxygenase.

The biological activities of lipoxins have yet to be fully defined (145). Lipoxin A_4 has been measured in the bronchoalveolar lavage fluid of patients with pulmonary diseases (150), although the cell sources of this lipoxin remain unknown. It can block LTB_4-induced inflammation in vivo, including plasma leakage and leukocyte migration (151). At 10^{-10} M, LXA_4 inhibits neutrophil chemotaxis to either LTB_4 or fMLP and blocks LTB_4- and fMLP-induced phosphatidylinositol hydrolysis and calcium mobilization in neutrophils (152,153). LXA_4 has inhibited the cytotoxic actions of human natural killer cells (154). Other actions of lipoxins, including induction of vasodilation and antagonism of LTD_4-induced responses on renal microcirculation, have been described (145). In addition to these potential paracrine activities of LXA_4, it may have intracellular messenger functions (145). It is a potent activator of protein kinase C, more potent than even diacylglyceride (155), and selectively stimulates the γ-subspecies of protein kinase C (156).

Another series of lipoxins, potentially resulting from interactions between the 5- and 15-lipoxygenase pathways in eosinophils and involving a glutathione-*S*-transferase, has been detected by Steinhilber and Roth (148), who demonstrated compounds named lipoxins C_4, D_4, and E_4 in extracts of eosinophil-enriched human granulocytes, incubated with either arachidonic acid or 15-HETE, and stimulated with A23187. LXC_4, like LTC_4, contains glutathione and is generated by conjugation of a 5(6)-epoxytetraene intermediate. Both LXD_4 and LXE_4 are formed by sequential peptidolytic cleavage of the glutathione moiety with γ-glutamyltransferase and dipeptidase, analogous to the enzymatic steps in the formation of LTD_4 and LTE_4. With ionophore A23187-stimulated eosinophils, in the absence of exogenous arachidonate or 15-HETE, no sulfidopeptide lipoxins were formed, but in the presence of exogenous arachidonate and, especially, of exogenous 15-HETE, increasing quantities of LXC_4 were formed (148). Leukotriene formation predominated in ionophore-stimulated eosinophils without added 15-HETE, whereas sulfidopeptide lipoxin formation was predominant when eosinophils were incubated with 20 μM 15-HETE (148). The biological activities of these novel lipoxins have not been defined.

2. *Peptide Mediators*

In addition to lipid mediators, eosinophils can elaborate two specific neuropeptides, substance P and vasoactive intestinal peptide (VIP). Eosinophil-enriched granulomas, formed in mouse livers around *Schistosoma mansoni* ova, contained immunoreactive substance P, which coeluted on high-performance liquid chromatography (HPLC) with authentic substance P peptide (157). Weinstock et al. (157) demonstrated that murine eosinophils isolated from these granulomas contained immunoreactive substance P localized by immunohistochemistry to eosinophil cytoplasmic granules. Antisera to substance P, as well as antisera reactive with other tachykinins, including substance K, neuropeptide K, and neurokinin B, but minimally with substance P, both specifically stained eosinophils (158). The staining with the latter antisera raised the possibility that other neuropeptides were present in murine eosinophils; but in situ hybridization studies with murine granuloma-derived eosinophils using separate oligonucleotide antisense probes encoding substance P and substance K/neuropeptide K detected only preprotachykinin mRNAs encoding substance P and not those encoding substance K (157,158). Secretagogues, both A23187 and histamine, released substance P from murine eosinophils detectable in a reverse hemolytic plaque assay (159). Histamine was active at concentrations between 10^{-9} and 10^{-5} M, and its releasing activity was antagonized by the H_1 receptor antagonist diphenhydramine, and not by the H_2 receptor antagonist cimetidine (159).

Immunoreactive neuropeptides have been found in extracts of human eosinophils. Aliakbari et al. (160) detected 72 fmol of immunoreactive vasoactive intestinal peptide (VIP) and 21 fmol of substance P per 10^7 eosinophils. Likewise, murine eosinophils have been implicated as a source of VIP. Extracts of eosinophil-rich hepatic granulomas from *S. mansoni*-infected mice contained VIP that was assayable by radioimmunoassay and coeluted on HPLC with synthetic VIP (161). In situ hybridization with an antisense, but not a sense, oligonucleotide probe for VIP or preprovasoactive intestinal peptide mRNA was reactive with eosinophils isolated from the granulomas and from an elicited peritoneal exudate (161). Substance P can enhance proliferation of smooth-muscle cells, fibroblasts, and endothelial cells; can induce vasodilatation, vascular permeability, and smooth-muscle contractility; and can have immunomodulatory effects, including stimulation of lymphocyte proliferation, enhancement of immunoglobulin secretion, and elicitation of mast cell degranulation and histamine release (162–164). In addition to these multiple potential activities of eosinophil-derived substance P, VIP, at least with murine cells, can stimulate IL-5 release from splenic and granuloma T cells isolated from mice infected with *Schistosoma mansoni* (165).

3. *Eosinophil Cytokines*

The capacity of eosinophils from both humans and mice to elaborate cytokines has recently been demonstrated. The recent recognition that eosinophils, similar to many other leukocytes and cells, can synthesize and release cytokines identifies a new mechanism whereby eosinophils may interact with and influence other cell types within their surroundings.

Transforming Growth Factor-α. The first cytokine to be demonstrated as being formed by human eosinophils, both in vivo and in vitro, was transforming growth factor-α (TGF-α) (166). In in situ hybridization studies with a ^{35}S-riboprobe for human TGF-α, most eosinophils infiltrating interstitial tissues adjacent to two colonic adenocarcinomas and two oral squamous cell carcinomas labeled specifically for hTGF-α mRNA. Evaluation of peripheral blood leukocytes from a patient with the idiopathic hypereosinophilic syndrome demonstrated that 80% of the eosinophils were positive for hTGF-α mRNA, whereas neutrophils and monocytes were negative for TGF-α transcripts. Northern blot analyses, with a ^{32}P-labeled human TGF-α cDNA probe, confirmed the specificity of hybridization to TGF-α mRNA. In addition, polymerase chain reaction (PCR) product restriction enzyme analysis, followed by Southern blot hybridization, confirmed the identification of the eosinophil-associated transcript as hTGF-α. The formation of TGF-α protein by human eosinophils was demonstrated both by radioimmunoassay of TGF-α in culture supernatants of unstimulated eosinophils from two hypereosinophilic donors and by immunohistochemical detection of TGF-α protein in almost half of the peripheral blood eosinophils from a hypereosinophilic syndrome patient. In contrast, peripheral blood eosinophils from only two of seven normal donors had detectable TGF-α mRNA, and none of these eosinophils contained immunohistochemically detectable TGF-α protein. Additional studies of other lesions have provided further evidence that eosinophils are a cellular source of TGF-α. In human oral cancers, infiltrating eosinophils were the preponderant source of TGF-α (167). Likewise, eosinophils in human nasal polyps were shown to be a source of TGF-α (168). Eosinophils in these lesions not only contained TGF-α mRNA transcripts detectable by in situ hybridization, but also contained immunohistochemically detectable TGF-α protein. The expression of TGF-α mRNA within eosinophils in various neoplastic and inflammatory colonic lesions has also been documented (169). Thus, human eosinophils can form and elaborate the multifunctional cytokine TGF-α.

Observations in other species also have demonstrated that eosinophils are capable of TGF-α production. Hamster eosinophils, both normal marrow eosinophils and eosinophils in connective tissues adjacent to chemically elicited cheek pouch carcinomas, exhibited TGF-α mRNA and protein detectable by in situ hybridization and immunohistochemical analyses, respectively (170). In healing cutaneous wounds in rabbits, eosinophils are prominent infiltrating cells. Most rabbit eosinophils in healing wounds contain TGF-α mRNA and protein as detected by in situ hybridization and immunohistochemistry (171). Likewise,

TGF-α is detectable during cutaneous wound healing in hamsters (172). The observations with rabbit, hamster, and human eosinophils all document the eosinophil as a source of TGF-α and further suggest that the expression of TGF-α within eosinophils may be subject to regulation by other cytokines or stimuli.

Transforming growth factor-α activates the same receptor employed by epidermal growth factor (EGF) and is elaborated by neoplastic cells as well as some normal cells, including activated macrophages (173,174). TGF-α is mitogenic for fibroblasts and stimulates epithelial cell proliferation and angiogenesis (175,176). The production of TGF-α by eosinophils would constitute a means whereby eosinophils might stimulate fibroblast proliferation and induce epithelial cell proliferation, as occur in asthmatic airways (177,178).

Transforming Growth Factor-β. Human eosinophils also synthesize transforming growth factor-β_1 (TGF-β_1) (179). Northern blot analyses using RNA isolated from eosinophils from a patient with the idiopathic hypereosinophilic syndrome detected the 2.5-kb TGF-β_1 mRNA transcript. In situ hybridization and immunohistochemistry of peripheral blood leukocytes from two patients with the idiopathic hypereosinophilic syndrome and two other patients with blood eosinophilia detected both TGF-β_1 mRNA and protein in eosinophils. In these patients, no other blood leukocyte contained TGF-β_1 mRNA by in situ hybridization, whereas other leukocytes contained detectable TGF-β_1 protein by immunohistochemistry. Eosinophils from two of four normal donors labeled weakly for TGF-β_1 mRNA by in situ hybridization, but eosinophils from none of these four normal donors contained TGF-β_1 protein detectable by immunohistochemistry (179). Eosinophils in nasal polyps also contain TGF-β_1 mRNA, as demonstrated by in situ hybridization and TGF-β_1 protein by immunohistochemical staining (168,180).

The TGF-β_1 is one of three distinct molecular forms of TGF-β, a multifunctional regulator of cell growth produced by many types of normal and transformed cells (181). One of the prominent effects of TGF-β is its stimulation of extracellular matrix formation (181). Hence, elaboration of this cytokine by eosinophils might contribute to the development of tissue fibrosis.

Interleukin-1. Both murine and human eosinophils elaborate IL-1α (182,183). Eosinophils from mice infected with the parasite *Mesocestoides corti* contained IL-1α mRNA detectable by in situ hybridization following exposure of eosinophils to lipopolysaccharide (LPS) or to indomethacin (183). Interleukin-1 bioactivity was present in supernatants of murine eosinophils stimulated with LPS or ionophore A23187. Human eosinophils contained IL-1α mRNA transcripts and immunohistochemically detectable IL-1α protein after stimulation with phorbol myristate acetate (PMA) (182). Northern blot analyses of PMA-stimulated eosinophils corroborated the presence of IL-1α mRNA (182).

Interleukin-1 can have many biological effects (184,185). One potential role of IL-1 released by eosinophils would be to serve as a costimulatory factor for lymphocytes during eosinophil antigen presentation. Human eosinophils can express the class II major histocompatibility complex (MHC) protein HLA-DR (186). Enhanced expression of HLA-DR on eosinophils has been demonstrated with eosinophils recovered from sputum (187) and from segmental lung lavages following airway antigen challenge (188). Eosinophils can function as MHC-restricted, HLA-DR-dependent antigen-presenting cells in stimulating lymphocyte proliferation (182). Interleukin-1 elaborated by other antigen-presenting cells plays a significant role in facilitating antigen-dependent lymphocyte responses (189). Thus, in addition to its proinflammatory functions, eosinophil-derived IL-1α could serve as a costimulator of eosinophil-induced antigen presentation and lymphocyte stimulation.

Granulocyte–Macrophage Colony-Stimulating Factor and Interleukin-3. Both GM-CSF and IL-3 are active in inducing eosinophilopoiesis and in activating mature eosinophils (1), which retain receptors for these cytokines (190–192). Human eosinophils elaborate each of these cytokines. Kita et al. (45) have shown that human eosinophils, following stimulation with the calcium ionophore, ionomycin, release eosinophil survival-enhancing activity, the production of which was inhibited with cyclosporine and of which activity was abolished with anti-IL-3 and anti-GM–CSF-neutralizing monoclonal antibodies. Both IL-3 and GM-CSF were measured by immunoassays in supernatants of ionomycin-stimulated eosinophil supernatants, in amounts less than produced by ionomycin-stimulated mononuclear cells, but greater than by similarly stimulated human neutrophils (45). Moreover, Moqbel et al. (46) demonstrated, by in situ hybridization, that stimulated eosinophils contained mRNA for GM-CSF. Eosinophils stimulated with calcium ionophore A23187 or interferon-γ yielded detectable in situ hybridization for GM-CSF mRNA in 15–27% of eosinophils, with 4–6% of eosinophils stimulated with IL-3 or IL-5, yielding positive GM-CSF mRNA hybridization. Stimulated eosinophils were immunoreactive with polyclonal anti-GM-CSF antisera providing evidence for translation of GM-CSF mRNA transcripts (46). Broide et al. demonstrated, by in situ hybridization, that eosinophils recovered by bronchoalveolar lavage following airway antigen challenge were positive for GM-CSF mRNA (as for IL-5 mRNA, see next section) (193). Together, these studies establish that eosinophils can release two cytokines that may exert autocrine activities by activating functional responses of mature eosinophils as well as potentially stimulating the activities of other cell types, including basophils, neutrophils, monocytes, macrophages, and endothelial cells (194–199).

Interleukin-5. The initial detection of IL-5 mRNA within human eosinophils was made by Desreumaux and colleagues, who employed in situ hybridization to detect mRNA for IL-5 in duodenal mucosal infiltrating eosinophils and blood eosinophils from patients with celiac disease (47). In studies by Broide et al., eosinophils recovered by bronchoalveolar lavage following airway antigen challenge were positive for IL-5 mRNA, with some, but not all, eosinophils also containing GM-CSF mRNA transcripts, as detected by dual label in situ hybridization (193). In contrast with the mouse, IL-5 does not stimulate functions of B lymphocytes in humans (200). The activities of IL-5 on mature human eosinophils (200) and basophils (52) and the capability of eosinophils, in addition to T lymphocytes, to elaborate this cytokine provide a mechanism whereby eosinophils may generate a cytokine with either autocrine eosinophil activities or paracrine actions on selected leukocytes.

Interleukin-6. Human eosinophils contain mRNA for IL-6, detectable by in situ hybridization, and interferon-γ increased the intensity and frequency of IL-6 $mRNA^+$ eosinophils (201). Synthesis of IL-6 by human eosinophils was demonstrable by immunohistochemical detection with a polyclonal anti-human IL-6 antibody and by immunoassay of released IL-6 from cultured unstimulated and, to a greater extent, interferon-γ-stimulated eosinophils. Thus, another potentially proinflammatory cytokine may be elaborated by eosinophils.

Tumor Necrosis Factor-α and Macrophage Inflammatory Protein-1α. Recent work was established that human eosinophils can be sources of both tumor necrosis factor-α (TNF-α) and macrophage inflammatory protein (MIP)-1α (202). By in situ hybridization, 44–100% of blood eosinophils from five patients with hypereosinophilia and four normal subjects exhibited intense hybridization signals for TNF-α mRNA (202). The TNF-α protein was detectable by immunohistochemical techniques in blood eosinophils of hypereosinophilic subjects, and purified blood eosinophils from three atopic donors exhibited cycloheximide-inhibitable spontaneous release of TNF-α in vitro. Many blood eosinophils from hypereosinophilic donors exhibited intense labeling for MIP-1α mRNA, whereas eosinophils of normal donors demonstrated only weak or undetectable hybridization signals for MIP-1α mRNA (202). Most tissue eosinophils infiltrating nasal polyps were strongly positive for both TNF-α and MIP-1α mRNA. By Northern blot analysis, highly enriched blood eosinophils from a patient with the idiopathic hypereosinophilic syndrome exhibited differential expression of TNF-α and MIP-1α mRNA. Thus, human eosinophils represent a potential source of two additional proinflammatory cytokines, TNF-α and MIP-1α (202).

4. Eosinophil Granule Constituents

Specific granules of eosinophils contain lysosomal hydrolases as well as most proteins unique to eosinophils (203). The characteristic crystalloid core present in eosinophil-specific granules is composed of major basic protein (MBP); the non–core granule matrix contains eosinophil cationic protein (ECP), eosinophil-derived neurotoxin (EDN), and eosinophil peroxidase (EPO) (204,205). Each of these proteins is cationic and their biochemical and functional properties have been extensively studied (206).

Major basic protein, named because it is one of the most abundant cationic eosinophil granule proteins, is a 14,000 M_r protein, rich in arginine residues. Eosinophil MBP has been sequenced and cloned (207–210). It has no recognized enzymatic activity, but is toxic for helminthic parasites, tumor cells, and host cells (206,211).

Eosinophil cationic protein and EDN are similar, but structurally, distinct proteins (212,213), which have been sequenced and cloned (214–217). ECP is an 18,000–21,000 M_r, markedly cationic polypeptide, which has bacteriocidal (218) and helminthotoxic activities (219) and, like MBP, is toxic to host cells. Both ECP and EDN can elicit the Gordon phenomenon, which constituted the basis for naming EDN. The Gordon phenomenon, first described as a cerebrocerebellar dysfunction in test rabbits following intracerebral injection of Hodgkin's disease patients' lymph nodes, and subsequently shown to be attributable to eosinophils infiltrating these lymph nodes, is evoked when either eosinophil granule constituent, ECP or EDN, is injected into the cerebrospinal fluid or brains of rabbits or guinea pigs (206,213). Both ECP and EDN display partial sequence identity with pancreatic ribonuclease and express ribonuclease activity, with EDN about 100-fold more potent as a ribonuclease than ECP (213,217).

Eosinophil peroxidase, an enzyme distinct from myeloperoxidase of neutrophils and monocytes, consists of two polypeptides of about 15,000 and 55,000 M_r (220,221). EPO has been cloned (222,223). By using hydrogen peroxide and halide ions, including chloride or preferentially bromide (224,225), it catalyzes the formation of hypohalous (e.g., hypochlorous or hypobromous) acid. In the presence of hydrogen peroxide and halide ions, EPO is toxic for helminthic and protozoan parasites, bacteria, and tumor and host cells (206,226–229).

The cationic granule constituents of eosinophils have activities on diverse cells (206), including damaging airway epithelial cells, promoting bronchial hyperresponsiveness, causing mast cell and basophil secretion, and contracting smooth muscle (3,230,231). Release of these granule proteins, therefore, has the capacity to cause dysfunction and damage to other cell types at sites of IgE-mediated reactions.

5. Activation of Eosinophils

Eosinophils circulating in the blood of eosinophilic patients often include a proportion of cells that appear "activated" and are metabolically more active (232) and less dense ("hypodense"), when isolated on density gradients (1), than normal donor-derived eosinophils. The ability of eosinophils from patients with asthma or related allergic diseases to release LTC_4 in response to ionophore A23187 stimulation has been evaluated in many studies (233–240), often with the finding that eosinophils from allergic patients released more LTC_4 than did normal eosinophils (reviewed in 241). Less dense eosinophils are present in greater numbers in the blood of asthmatic patients than in that of normal donors (240,242,243). Differences in the formation of LTC_4 related to eosinophil density are exemplified in the study by Hodges et al., who evaluated eosinophils stimulated with 1 μg/mL A23187 in the presence of the hypochlorous acid scavenger L-serine (240). Eosinophils with densities >1.093 g/mL from asthmatic and normal subjects released 19.1±4.2 and 23.9±5.0 ng LTC_4 per 10^6 eosinophils in 15 min. In contrast, lower-density eosinophils (≤1.093 g/mL) from asthmatic subjects released significantly less LTC_4 than did eosinophils of similar densities from normal subjects (41.6±3.0 vs 79.0±6.7 ng LTC_4 per 10^6 eosinophils, $p<0.05$). There were no differences between the two study groups in LTC_4 metabolism, time course of extracellular LTC_4 release, or dose response to A23187. Thus, hypodense eosinophils elaborated greater quantities of LTC_4 than did eosinophils of normal density, whether obtained from normal or asthmatic subjects (240). However, since eosinophils from asthmatic subjects had decreased capacity for LTC_4 synthesis, compared with eosinophils of similar densities from normal donors, the ability of eosinophils to produce LTC_4 was regulated by factors not necessarily reflected in their cell density (240).

The activation of functional responses of eosinophils by in vivo exposures to cytokines and other mediators is likely to account for some of the enhanced responses observed with eosinophils derived from patients. Cocultivation of eosinophils with endothelial cells (244), or exposures of eosinophils to various cytokines in vitro, including GM-CSF (197,245,246), IL-3 (196), IL-5 (247), TNF-α (248), and monocyte-derived factors (249–252) can enhance LTC_4 release elicitable with low concentrations of A23187. Furthermore, cytokines, such as GM-CSF, IL-3, and IL-5, can prolong the survival of eosinophils, convert eosinophils to having a "hypodense" phenotype, and enhance degranulation and superoxide generation (253).

In addition to the activating effects on functional responses of eosinophils provided by specific cytokines, it is highly likely that additional activating stimuli for eosinophils are provided during the processes of cellular recruitment and migration within tissues. The studies of Sedgwick et al. have established that recruited eosinophils, recovered from bronchoalveolar lavage following antigen challenge, exhibit further evidence of cellular activation not displayed even by circulating "hypodense" eosinophils (188). Thus, the functional responses of eosinophils (and probably other leukocytes) will be governed by exposures to soluble compounds, such as cytokines and humoral inflammatory mediators, and by stimulation that may be derived through eosinophil surface proteins, such as integrins, interacting with endothelial or epithelial cells and cellular and matrix components within tissue sites.

III. IMMUNOGLOBULIN E-DEPENDENT INFLAMMATION

Following the triggering of mast cells, an early response attributable to the release of mast cell-derived components and products subsides within 1–2 h and is followed by the development of a later response. Such late-phase responses occur in the skin, nose, and lungs. Although there are some differences in timing between these different organs in the onset of late-phase responses, studies of these late-phase reactions have provided data on the recruited cells and their products participating in IgE-dependent reactions.

A. Cutaneous Late-Phase Reactions

The occurrence of late-phase cutaneous reactions has been recognized since the initial description in 1921 by Prausnitz and Küstner (254). Following intradermal introduction of antigen, the immediate wheal-and-flare reaction subsides within about 90 min, to be followed by a late-phase reaction that peaks between the 6–12 h and resolves by 24 h. The reaction is characterized by edema, erythema, pruritis, tenderness, hyperalgesia, and hyperthermia. Both the immediate and the late-phase reactions to antigen are dependent on IgE (255,256); and other mast cell-degranulating agents, such as compound 48/80, also induce early and late responses (257). In the early period, there is the appearance of multiple mediators released by mast cells, including tryptase (258,259), histamine (258–260), eosinophil chemoattractant activity (261), LTB_4 (262), LTC_4 (263), PGD_2 (259,260), and mast cell granule contents (264).

Early biopsy studies of evolving cutaneous late-phase reactions demonstrated that neutrophils were early cellular constituents increasing by 30–60 min after initiation, followed after several hours by increasing numbers of eosinophils and mononuclear cells (255,256,265). Studies of neutrophils recovered from skin windows during early-reaction periods demonstrated that the numbers of recruited neutrophils correlated with the intensity of mast cell mediator release and that the recruited neutrophils exhibited evidence, such as enhanced oxidative metabolism and CR3 expression, of being activated (266–268). Evidence of neutrophil degranulation following IgE-mediated cutaneous late-phase reactions has been obtained by staining for neutrophil elastase, an azurophilic or primary granule constituent. As early as 15 min after antigen instillation small focal areas of extracellular elastase deposition were detectable by indirect immunofluorescence, and the extracellular elastase deposition was extensive by 1–3 h and persisted for up to 56 h (269). Likewise, neutrophil-derived myeloperoxidase (270) and lactoferrin (271) were detected in these lesions. In skin chamber studies, neutrophils were the preponderant cellular-infiltrating cell, with numbers increasing from 5 h peaking about 10 h, and persisting through 12 h (260). Others, evaluating lesions biopsied at 6, 24, and 48 h, have found that elastase-positive neutrophils peak at 6 h and have declined extensively by 24 and 48 h (65). Skin chamber fluids obtained 1 h after antigen challenge contain as yet uncharacterized moieties capable of enhancing neutrophil superoxide anion generation and lactoferrin release (272).

Eosinophil infiltration in cutaneous reactions, as assessed in biopsies and skin window techniques, increases after 3 h, continues for 12 h or more in skin blister fluids, and for 24 h or more in biopsy tissues (260). As early as 15 min after antigen instillation small focal areas of extracellular eosinophil granule MBP and EDN deposition are detectable by indirect immunofluorescence, and the extracellular deposition of these two eosinophil granule proteins is extensive by 1–3 h and persists for up to 56 h (269). Electron microscopy showed degenerating eosinophils and free eosinophil granules in the dermis (269). With a monoclonal antibody, EG2, that recognizes an activated form of eosinophil granule ECP (273), "activated" eosinophils have been found infiltrating skin sites 6, 24, and 48 h after intradermal allergen challenge, with the magnitude of the number of activated eosinophils correlating at 6 h with the diameter of the late-phase reaction (65). Moreover, there was a strong correlation between the numbers of infiltrating $CD4^+$ lymphocytes and activated eosinophils at 24 h (65). Studies of blister fluids overlying IgE-mediated allergic skin reactions have detected increasing levels of eosinophil granule MBP and EDN from the first to the fifth hours (274). Although imprints of skin bases demonstrated only variable numbers of eosinophils at antigen challenged sites, with neutrophils being the preponderant cells, immunofluorescent staining for eosinophil MBP demonstrated extensive extracellular deposition of this protein (274). Hence, it appears that extensive degranulation of eosinophils had occurred. This extensive eosinophil degranulation contributes to underestimation of the quantitative roles of eosinophils because of an inability to detect and enumerate intact, histologically recognizable eosinophils, a difficulty also noted in atopic dermatitis (275).

Basophils have been detected as later cellular infiltrates in skin biopsies 8 h after challenge and in skin blister chamber fluids, with numbers increasing between 6 and about 12 h (260). At the latter times, up to 12 h, when mast cell-derived PGD_2 was declining, a second rise in histamine occurred, possibly attributable to release from the infiltrating basophils (260). Basophils do not elaborate PGD_2, but do release LTC_4.

Although mononuclear leukocytes have been recognized from the earliest histological studies (255,256) as major later-appearing components of intracutaneous late-phase reactions, the nature of the lymphocytes recruited into dermal lesions has been evaluated only more recently. Frew and Kay demonstrated increased numbers of $CD3^+$ lymphocytes at 6, 24, and 48 h, with increases in $CD4^+$, but not $CD8^+$, lymphocytes at these times (65). Only a few cells staining with Leu M3, which stains most blood monocytes, were found (65). Kay and colleagues have also compared the dermal infiltrates in cutaneous late-phase reactions and classic delayed hypersensitivity reactions (276). Studies in mice, depleted of $CD4^+$ or $CD8^+$ T cells or given anti-IL-5 antisera, have shown that $CD4^+$, but not $CD8^+$ T cells, contribute to the eosinophilia found in cutaneous late-phase reactions, and that IL-5 also is involved in this eosinophil recruitment (277).

The foregoing studies, therefore, have delineated several leukocyte types that are recruited into sites of IgE-mediated cutaneous reactions. Since these leukocytes must be attracted from the bloodstream, adhesion proteins are likely elicited on local vascular endothelial cells that participate in the initial adherence and attraction processes for intravascular leukocytes. Although further studies will be needed to fully define the temporal sequences of expression of endothelial adherence molecules in these reactions, skin biopsy studies have already provided some evidence for the enhanced expression of specific adherence molecules. At skin biopsy sites taken 6 h after antigen or saline control injections, Kyan-Aung et al. demonstrated increased expression of ICAM-1 and E-selectin (28). Although some lesions exhibited enhanced VCAM-1 staining, this was not a consistent occurrence (28). Leung and colleagues demonstrated enhanced E-selectin expression in biopsied skin test sites as early as 3–4 h after ragweed or mite antigen, but not saline, challenge, and the enhanced expression was greatest at 6 h and still present at 24 h (278). In skin organ cultures, the presence of combined rabbit antihuman TNF-α and IL-1α and β inhibited allergen-induced expression of vascular E-selectin (278).

B. Nasal Late-Phase Reactions

Late-phase nasal responses occur between 4 and 12 h after intranasal exposure to antigen and are characterized by nasal congestion and rhinitis (279,280). During the first 3 h following antigen challenge, eosinophils, but not neutrophils or mononuclear cells, are increased (281). During the late-phase response, from 4 to 12 h, neutrophils, eosinophils, and mononuclear cells increase (281). Levels of eosinophil MBP increased in nasal washings from early and late responses, whereas EDN was increased in late-phase responses (282). Whereas mast cell-derived tryptase, PGD_2, and histamine are found following the early-phase response (283), basophils increase during late-phase reactions and are a source of later-appearing histamine (284–286). Immunohistological examination of nasal mucosa, obtained 24 h after grass pollen application, demonstrated increases in specific cells. There were no increases in $CD45^+$ leukocytes or $CD3^+$ or $CD8^+$ lymphocytes; but in comparison with control sites, at allergen sites there were significant increases of eosinophils (anti-MBP staining), neutrophils (anti-elastase staining), $CD4^+$ T lymphocytes and IL-2 receptor ($CD25^+$)-bearing cells (287). The recruitment of eosinophils in allergen-induced rhinitis correlates with enhanced numbers of cells expressing mRNA for IL-3, IL-4,

IL-5, and GM-CSF (288). Thus, the inflammatory cells elicited in nasal late-phase reactions include eosinophils, neutrophils, basophils, $CD4^+$ T lymphocytes, and other mononuclear cells.

C. Pulmonary Late-Phase Reactions

Early studies of pulmonary late-phase reactions in mild asthmatics were conducted by Metzer and colleagues (289,290). Following bronchopulmonary challenge and total lung lavage, eosinophils and neutrophils were elevated 4 h later, with eosinophil, but not neutrophil, numbers remaining elevated at 24 h. Forth-eight hours following local antigen challenge, significant increases in macrophages, lymphocytes, neutrophils, and eosinophils were found, and eosinophil and lymphocyte numbers remained elevated at 96 h after challenge (289). In de Monchy's early work, the appearance of eosinophils in airways was demonstrated with late-phase reactions (291). At 6 h after antigen challenge eosinophils and ECP were elevated in bronchoalveolar lavages from only those who developed late-phase reactions (291). Diaz et al. found that the cells that were increased 6 h after antigen challenge in those developing late-phase reactions, included lymphocytes, neutrophils, and eosinophils (67). Rossi et al., evaluating recruited cells at baseline and 4, 24, and 72 h after allergen inhalation challenge, found that increases in eosinophils at 4 h were associated with the later development of late-phase reactions, whereas eosinophils numbers were significantly increased at 24 h in those with only early-phase airway reactions (292). Aalbers et al. evaluated the bronchial and bronchoalveolar lavage cell profiles in patients experiencing single early or dual, early and late, phase reactions to inhaled house dust mite allergen. Increased numbers of bronchial eosinophils and eosinophils, straining with the EG2 monoclonal antibody for activated ECP, and increases in total blood IgE were associated with the occurrence of late-phase reactions (293). The sputum of asthmatics, obtained 32 h after allergen challenge, contained increased numbers of eosinophils and metachromatic cells (basophils and mast cells) (294). The magnitude of the late-phase response correlated with the increased metachromatic cells (294). Dupuis et al. reported a dose-dependent recruitment of inflammatory cells after segmental antigen challenge. Atopic volunteers challenged with ragweed antigen E (Amb a I) experienced a twofold increase in macrophages, a fourfold increase in lymphocytes, a 90-fold increase in neutrophils and an 800-fold increase in eosinophils 24 h after antigen challenge (295).

The recruitment of eosinophils in late-phase pulmonary responses leads to an increase not only in their numbers, but also in products derived from them. Sedgwick et al. studied early (12 min) and late (48 h) changes in cells and mediators following segmental antigen challenge of allergic rhinitis patients (296). In the early samples, mediators derived from mast cells, histamine and tryptase, were increased, but there was no increase in inflammatory cells. In contrast, at 48 h there were increases in low- and normal-density eosinophils and in concentrations of eosinophil granule proteins, MBP, ECP, EDN, and EPO, free in lavaged fluids (296). Leukotriene C_4, but not tryptase, concentrations were elevated in the late-phase samples. The functional characteristics of eosinophils recruited into the airways 48 h after segmental antigen challenge, in comparison with blood eosinophils, have also been evaluated by Sedgwick and colleagues (188). Bronchoalveolar lavage eosinophils could be resolved into three different densities (1.100, 1.095, and 1.090 g/mL) whereas blood eosinophils exhibited normal density of 1.100 g/mL. The lavage eosinophils of all three densities had greater superoxide generation and adherence with fMLP activation than did blood eosinophils. Both blood and lavage eosinophils, however, responded similarly to PMA. Lavage eosinophils had increased expression of CD11b/CD18 and HLA-DR. Lavage eosinophils had a greater and more sustained increase in extracellular Ca^{2+} than did blood eosinophils (188). Cytokines, including IL-5 (296) and GM-CSF (297), are elevated in late-phase lavage fluids. Eosinophils have been shown by in situ hybridization to be one source of these two cytokines in late phase reactions (193).

The responses of lymphocyte populations as assessed in the peripheral blood, bronchoalveolar lavage fluids, and biopsied airway tissues following antigen challenge are dynamic during the evolution of pulmonary IgE-mediated inflammation. In lavaged fluids from unchallenged atopic asthmatics, some have found no increases in $CD4^+$ and $CD8^+$ lymphocytes, but have noted mild increases in bronchoalveolar lavage $CD3^+$ lymphocytes expressing activation markers, CD25 and HLA-DR (298). As assessed 6 h after antigen challenge, both $CD4^+$ and $CD8^+$ lymphocytes increased in lavage fluids in those with early- and with dual early- and late-phase reactions (299). In those with only early responses, there was an increase in relative numbers of $CD4^+$ lymphocytes in the blood and in $CD8^+$ lymphocytes in the lavages fluids (299). Evidence for recruitment of $CD4^+$ T lymphocytes from the blood into the lungs following bronchial antigen challenge in asthmatics has come from the studies of Gerblich et al., who have shown that 48 h after antigen challenge, blood $CD4^+$ T-cell numbers are decreased, whereas $CD4^+$ lymphocytes were recovered in greater percentages from the lungs (300,301). Acutely, after antigen challenge, there is a decreased recovery of lymphocytes from the lavage fluids of antigen challenged asthmatics. Levels of $CD3^+$, $CD4^+$, and to a lesser extent $CD8^+$ lymphocytes declined at 10 min in asthmatic, but not normal subjects, suggesting mobilization of these cells into or onto the airway tissues (302).

Although analyses of cells and mediators recovered by lavage of airways following antigen challenge can provide information reflecting the evolution of IgE-initiated inflammation within airways, analyses of the airway tissues will provide more direct information for events occurring at the sites of such inflammation. In bronchial biopsies from atopic asthmatics obtained 24 h following inhalational challenge with allergen or control diluent, Bentley et al. found increases in $EG2^+$ eosinophils and IL-2 receptor-expressing ($CD25^+$) cells but no differences in numbers of total leukocytes ($CD45^+$), T lymphocytes ($CD3^+$, $CD4^+$, or $CD8^+$), elastase-positive neutrophils, macrophages ($CD68^+$), or mast cell subtypes (MCT^+ or $MCTC^+$) (303). By in situ hybridization studies, there were increased numbers of cells expressing mRNA for IL-5 and GM-CSF after allergen challenge, compared with diluent control (303). There was an inverse correlation between the numbers of cells with mRNA detectable for IL-4 and IFN-γ (303). The authors suggested that cytokines possibly from activated T lymphocytes may contribute to local accumulation of eosinophils during allergen-induced asthma. In stable asthmatics, analyses have indicated the presence of Th-2-like $CD4^+$ lymphocytes in asthma, with their preferential elaboration of specific cytokines, including IL-5 and GM-CSF (304). Others have detected GM-CSF in BAL 19 h after antigen in atopic subjects with allergen-induced late-phase reactions (297) and have found significant increases in IL-5 concentrations that correlated with the presence of eosinophils and eosinophil granule proteins in late 48-h lavage samples after segmental antigen challenge (296). However, eosinophils (193) recovered from lavage fluids are also sources of these cytokines, in addition to any potential contributions from lymphocytes (303).

IV. CONCLUSIONS

Investigations of IgE-mediated inflammatory responses have benefited from understandings of the regulation and activities of a variety of leukocytes. Studies of the evolution of IgE-initiated late-phase reactions in the skin, nose, and lungs have demonstrated the sequential recruitment of leukocytes and a progressively expansive involvement of eosinophils, T lymphocytes, and other leukocytes durng the late-phase reactions. The applications of experimental approaches to tissue sites of IgE-mediated reactions, such as in situ hybridization to detect cytokine mRNA transcripts and immunohistochemistry to detect proteins released from involved leukocytes, are enabling a further definition of the events involved in the sequential development of late-phase reactions. Although isolated late-phase responses are amenable to study, the roles of cumulative late-phase responses, as may occur in allergic asthma or rhinitis, will have to further consider the development of even later immunopathological sequelae, such as fibrosis. Hence, whereas mucosal inflammation in the airways can be evaluated with acute antigen challenge studies (305), the pathology of chronic asthma in notable in that increased numbers of eosinophils and mast cells and heightened subepithelial collagen deposition are found in atopic asthmatics compared with nonatopic asthmatics or normal subjects on endobronchial biopsies (306,307). Thus, the events that begin with IgE-triggered mast cell activation are followed by the progressive recruitment and activation of eosinophils, lymphocytes, and other leukocytes. Delineation of the interactions that occur among these recruited leukocytes that modulate their activation and functioning within tissue sites will further elucidate the immunopathogenesis of IgE-mediated inflammation.

ACKNOWLEDGMENTS

The author's studies have been supported in part by NIH grants AI20241, AI22571, AI31982, and HL46563.

REFERENCES

1. P. F. Weller, The immunobiology of eosinophils, *N. Engl. J. Med.*, *324*: 1110 (1991).
2. T. B. Nutman, E. A. Ottesen, and S. G. Cohen, The eosinophil, eosinophilia, and eosinophil-related disorders. III. Clinical assessments and eosinophil related disorders, *Allergy Proc.*, *10*: 33 (1989).
3. G. J. Gleich, The eosinophil and broncial asthma: Current understanding, *J. Allergy Clin. Immunol.*, *85*: 422 (1990).
4. A. J. Waldlaw, R. Moqbel, O. Cromwell, and A. B. Kay, Platelet activating factor. A potent chemotactic and chemokinetic factor for human eosinophils, *J. Clin. Invest.*, *78*: 1701 (1986).
5. E. C. Butcher, Leukocyte–endothelial cell recognition: Three (or more) steps to specificity and diversity, *Cell*, *67*: 1033 (1991).
6. J. S. Pober, and R. S. Cotran, Cytokines and endothelial cell biology, *Physiol. Rev.*, *70*: 427 (1990).
7. A. M. Lamas, C. M. Mulroney, and R. P. Schleimer, Studies on the adhesive interaction between purified human eosinophils and cultured vascular endothelial cells, *J. Immunol.*, *140*: 1500 (1988).
8. K. J. I. Thorne, B. A. Richardson, G. Mazza, and A. E. Butterworth, A new method for measuring eosinophil activating factors, based on the expression of CR3 alpha-chain (CD11b) on the surface of the activated eosinophils, *J. Immunol. Methods*, *133*: 47 (1990).
9. G. M. Walsh, A. J. Wardlaw, A. Hartnell, C. J. Sanderson, and A. B. Kay, Interleukin-5 enhances the in vitro adhesion of human eosinophils, but not neutrophils, in a leukocyte integrin (CD11/18)-dependent manner, *Int. Arch. Allergy Appl. Immunol.*, *94*: 174 (1991).
10. M. B. Resnick, and P. F. Weller, Mechanisms of eosinophil recruitment, *Am. J. Respir. Cell. Mol. Biol.*, *8*: 349 (1993).
11. P. F. Weller, T. H. Rand, S. E. Goelz, G. Chi-Rosso, and R. J. Lobb, Human eosinophil adherence to vascular endothelium mediated by binding to VCAM-1 and ELAM-1, *Proc. Natl. Acad. Sci. USA*, *88*: 7430 (1991).
12. A. Dobrina, R. Menegazzi, T. M. Carlos, E. Nardon, R. Cramer, T. Zacchi, J. M. Harlan, and P. Patriarc, Mechanisms of eosinophil adherence to cultured

vascular endothelial cells. Eosinophils bind to the cytokine-induced ligand vascular cell adhesion molecule-1 via the very late activation antigen-4 integrin receptor, *J. Clin. Invest.*, *88*: 20 (1991).

13. B. S. Bochner, F. W. Luscinskas, M. A. J. Gimbrone, W. Newman, S. A. Sterbinsky, C. P. Derse-Anthony, D. Klunk, and R. P. Schleimer, Adhesion of human basophils, eosinophils, and neutrophils to interleukin 1-activated human vascular endothelial cells: Contributions of endothelial cell adhesion molecules, *J. Exp. Med.*, *173*: 1553 (1991).
14. U. Kyan-Aung, D. O. Haskard, and T. H. Lee, Vascular cell adhesion molecule-1 in eosinophil adhesion to cultured human umbilical vein endothelial cells in vitro, *Am. J. Respir. Cell. Mol. Biol.*, *5*: 445 (1991).
15. A. Hartnell, R. Moqbel, G. M. Walsh, B. Bradley, and A. B. Kay, Fc gamma and CD11/CD18 receptor expression on normal density and low density eosinophils, *Immunology 69*: 264 (1990).
16. D. C. Anderson, F. C. Schmalstieg, M. J. Finegold, B. J. Hughes, R. Rothlein, L. J. Miller, S. Kohl, M. F. Tosi, R. L. Jacobs, T. C. Waldrop, A. S. Goldman, W. T. Shearer, and T. A. Springer, The severe and moderate phenotypes of heritable Mac-1, LFA-1 deficiency: Their quantitative definition and relation to leukocyte dysfunction and clinical features, *J. Infect. Dis.*, *152*: 668 (1985).
17. J. B. Smith, R. D. Kunjummen, T. K. Kishimoto, and D. C. Anderson, Expression and regulation of L-selectin on eosinophils from human adults and neonates, *Pediatr. Res.*, *32*: 465 (1992).
18. M. P. Bevilacqua, and R. M. Nelson, Selectins, *J. Clin. Invest.*, *91*: 379 (1993).
19. M. A. Vadas, C. M. Lucas, J. R. Gamble, A. F. Lopez, M. P. Skinner, and M. C. Berndt, Regulation of eosinophil function by P selectin, *Eosinophils in Allergy and Inflammation* (G. J. Gleich and A. B. Kay, eds.), Marcel Dekker, New York, 1994, p. 69.
20. G. M. Walsh, J. J. Mermod, A. Hartnell, A. B. Kay, and A. J. Wardlaw, Human eosinophil, but not neutrophil, adherence to IL-1-stimulated human umbilical vascular endothelial cells in $alpha_4$ $beta_1$ (very late antigen-4) dependent, *J. Immunol.*, *146*: 3419 (1991).
21. R. P. Schleimer, S. A. Sterbinsky, J. Kaiser, C. A. Bickel, D. A. Klunk, K. Tomoika, W. Newman, F. W. Luscinskas, M. A. Gimbrone, Jr., B. W. McIntyre, and B. S. Bochner, Interleukin-4 induces adherence of human eosinophils and basophils but not neutrophils to endothelium: Association with expression of VCAM-1, *J. Immunol.*, *148*: 1086 (1992).
22. M. J. Elices, L. Osborn, Y. Takada, C. Crouse, S. Luhowskyj, M. E. Hemler, and R. R. Lobb, VCAM-1 on activated endothelium interacts with the leukocyte integrin VLA-4 at a site distinct from the VLA-4/fibronectin binding site, *Cell*, *60*: 577 (1990).
23. B. Masinovsky, D. Urdal, and W. M. Gallatin, IL-4 acts synergistically with IL-1β to promote lymphocyte adhesion to microvascular endothelium by induction of vascular cell adhesion molecule-1, *J. Immunol.*, *145*: 2886 (1990).
24. M. H. Thornhill, S. M. Wellicome, D. L. Mahiouz, J. S. S. Lanchbury, U. Kyan-Aung, and D. O. Haskard, Tumor necrosis factor combines with IL-4 or IFN-γ to selectively enhance endothelial cell adhesiveness for T cells. The

contribution of vascular cell adhesion molecule-1-dependent and -independent binding mechanisms, *J. Immunol.*, *146*: 592 (1991).
25. R. Moser, J. Fehr, L. Olgiati, and P. L. Bruijnzeel, Migration of primed human eosinophils across cytokine-activated endothelial cell monolayers, *Blood*, *79*: 2937 (1992).
26. R. Moser, J. Fehr, and P. L. Bruijnzeel, IL-4 controls the selective endothelium-driven transmigration of eosinophils from allergic individuals, *J. Immunol.*, *149*: 1432 (1992).
27. R. P. Schleimer, and B. S. Bochner, Endothelial leukocyte adhesion molecule-1 and intercellular adhesion molecule-1 mediate the adhesion of eosinophils to endothelial cells in vitro and are expressed by endothelium in allergic cutaneous inflammation in vivo [letter], *J. Immunol.*, *147*: 380 (1991).
28. U. Kyan-Aung, D. O. Haskard, R. N. Poston, M. H. Thornhill, and T. H. Lee, Endothelial leukocyte adhesion molecule-1 and intercellular adhesion molecule-1 mediate the adhesion of eosinophils to endothelial cells in vitro and are expressed by endothelium in allergic cutaneous inflammation in vivo, *J. Immunol.*, *146*: 521 (1991).
29. R. I. Tepper, D. A. Levinson, B. Z. Stanger, J. Campos-Torres, A. K. Abbas, and P. Leder, IL-4 induces allergic-like inflammatory disease and alters T cell development in transgenic mice, *Cell*, *62*: 457 (1990).
30. R. I. Tepper, R. L. Coffman, and P. Leder, An eosinophil-dependent mechanism for the antitumor effect of interleukin-4, *Science*, *257*: 548 (1992).
31. C. D. Wegner, R. H. Gundel, P. Reilly, N. Haynes, G. Letts, and G. Rothlein, Intercellular adhesion molecule-1 (ICAM-1) in the pathogenesis of asthma, *Science*, *247*: 456 (1990).
32. R. H. Gundel, C. D. Wegner, C. A. Torcellini, C. C. Clarke, N. Haynes, R. Rothlein, C. W. Smith, and L. G. Letts, Endothelial leukocyte adhesion molecule-1 mediates antigen-induced acute airway inflammation and late-phase airway obstruction in monkeys, *J. Clin. Invest.*, *88*: 1407 (1991).
33. W. M. Abraham, M. W. Sielczak, A. Ahmed, A. Cortes, I. T. Lauredo, J. Kim, B. Pepinsky, C. D. Benjamin, D. R. Leone, R. R. Lobb, and P. F. Weller, Anti-α_4 intergrin mediates antigen-induced late bronchial responses and prolonged airway hyperresponsiveness in sheep, *J. Clin. Invest.* (in press).
34. A. B. Kay, and K. F. Austen, The IgE mediated release of an eosinophil leukocyte chemotactic factor from human lung, *J. Immunol.*, *107*: 899 (1971).
35. A. B. Kay, D. J. Stechschulte, and K. F. Austen, An eosinophil leukocyte chemotactic factor of anaphylaxis, *J. Exp. Med.*, *133*: 602 (1971).
36. R. Sehmi, O. Cromwell, G. W. Taylor, and A. B. Kay, Identification of guinea pig eosinophil chemotactic factor of anaphylaxis as leukotriene-B_4 and 8(*S*),15(S)-dihydroxy-5,9,11,13(*Z*,*E*,*Z*,*E*)-eicosatetranoic acid, *J. Immunol.*, *147*: 2276 (1991).
37. E. Morita, J. M. Schroder, and E. Christophers, Differential sensitivities of purified human eosinophils and neutrophils to defend chemotaxins, *Scand. J. Immunol.*, *29*: 709 (1989).
38. C. Kroegel, T. Yukawa, G. Dent, P. Venge, K. F. Chung, and P. J. Barnes, Stimulation of degranulation from human eosinophils by platelet-activating factor, *J. Immunol.*, *142*: 3518 (1989).

39. N. Tamura, D. K. Agrawal, F. A. Suliaman, and R. G. Townley, Effects of platelet activating factor on the chemotaxis of normodense eosinophils from normal subjects. *Biochem. Biophys. Res. Commun.*, *142*: 638 (1987).
40. E. Henocq, and B. B. Vargaftig, Skin eosinophilia in atopic patients, *J. Allergy Clin. Immunol.*, *81*: 691 (1988).
41. A. Lellouch-Tubiana, J. Lefort, M. T. Simon, A. Pfister, and B. B. Vargaftig, Eosinophil recruitment into guinea pig lungs after PAF-acether and allergen administration, *Am. Rev. Respir. Dis.*, *137*: 948 (1988).
42. C. M. Morland, S. J. Wilson, S. T. Holgate, and W. R. Roche, Selective eosinophil leukocyte recruitment by transendothelial migration and not by leukocyte-endothelial adhesion, *Am. J. Respir. Cell. Mol. Biol.*, *6*: 557 (1992).
43. T. B. Casale, R. A. Erger, and M. M. Little, Platelet-activating factor-induced human eosinophil transendothelial migration: Evidence for a dynamic role of the endothelium, *Am. J. Respir. Cell. Mol. Biol.*, *8*: 77 (1993).
44. I. Ohno, R. Lea, S. Finotto, J. Marshall, J. Denburg, J. Dolovich, J. Gauldie, and M. Jordana, Granulocyte/macrophage colony-stimulating factor (GM-CSF) gene expression by eosinophils in nasal polyposis, *Am. J. Respir. Cell. Mol. Biol.*, *5*: 505 (1991).
45. H. Kita, T. Ohnishi, Y. Okubo, D. Weiler, J. S. Abrams, and G. J. Gleich, Granulocyte/macrophage colony-stimulating factor and interleukin 3 release from human peripheral blood eosinophils and neutrophils, *J. Exp. Med.*, *174*: 745 (1991).
46. R. Moqbel, Q. Hamid, S. Ying, J. Barkans, A. Hartnell, A. Tsicopoulos, A. J. Wardlaw, and A. B. Kay, Expression of mRNA and immunoreactivity for the granulocyte/macrophage colony-stimulating factor in activated human eosinophils, *J. Exp. Med.*, *174*: 749 (1991).
47. P. Desreumaux, A. Janin, J. F. Colombel, L. Prin, J. Plumas, D. Emiliee, G. Torpier, A. Capron, and M. Capron, Interleukin-5 messenger RNA expression by eosinophils in the intestinal mucosa of patients with coeliac disease, *J. Exp. Med.*, *175*: 293 (1992).
48. R. A. Warringa, L. Koenderman, P. T. Kok, J. Kreukniet, and P. L. Bruijnzeel, Modulation and induction of eosinophil chemotaxis by granulocyte-macrophage colony-stimulating factor and interleukin-3, *Blood*, *77*: 2694 (1991).
49. M. Yamaguchi, K. Hirai, S. Shoji, T. Takaishi, K. Ohta, Y. Morita, S. Suzuki, and K. Ito, Haemopoietic growth factors induce human basophil migration in vitro, *Clin. Exp. Allergy*, *22*: 379 (1992).
50. C. J. Sanderson, H. D. Campbell, and I. G. Young, Molecular and cellular biology of eosinophil differentiation factor (interleukin 5) and its effects on human and mouse B cells, *Immunol. Rev.*, *102*: 29 (1988).
51. R. Sehmi, A. J. Wardlaw, O. Cromwell, K. Kurihara, P. Waltmann, and A. B. Kay, Interleukin-5 selectively enhances the chemotactic response of eosinophils obtained from normal but not eosinophilic subjects, *Blood*, *79*: 2952 (1992).
52. S. C. Bischoff, T. Brunner, A. L. De Weck, and C. A. Dahinden, Interleukin 5 modifies histamine release and leukotriene generation by human basophils in response to diverse agonists, *J. Exp. Med.*, *172*: 1577 (1990).
53. R. A. Warringa, H. J. J. Mengelers, P. H. M. Kuijper, J. A. M. Raaijmakers, P.

L. B. Bruijnzeel, and L. Koenderman, In vivo priming of platelet-activating factor-induced eosinophil chemotaxis in allergic asthmatic individuals, *Blood*, *79*: 1836 (1992).

54. W. E. Secor, S. J. Stewart, and D. G. Colley, Eosinophils and immune mechanisms. VI. The synergistic combination of granulocyte–macrophage colony-stimulating factor and IL-5 accounts for eosinophil-stimulation promoter activity in *Schistosoma mansoni*-infected mice, *J. Immunol.*, *144*: 1484 (1990).
55. T. H. Rand, W. W. Cruikshank, D. M. Center, and P. F. Weller, CD4-mediated stimulation of human eosinophils: Lymphocyte chemoattractant factor and other CD4-binding ligands elicit eosinophil migration, *J. Exp. Med.*, *173*: 1521 (1991).
56. T. H. Rand, D. S. Silberstein, H. Kornfield, and P. F. Weller, Human eosinophils express functional interleukin 2 receptors, *J. Clin. Invest.*, *88*: 825 (1991).
57. W. W. Cruikshank, J. S. Berman, A. C. Theodore, J. Bernardo, and D. M. Center, Lymphokine activation of $T4^+$ lymphocytes and monocytes, *J. Immunol.*, *138*: 3817 (1987).
58. W. W. Cruikshank, J. S. Berman, A. C. Theodore, J. Bernardo, and D. M. Center, Lymphocyte chemoattractant factor (LCF) induces CD4-dependent intracytoplasmic signalling in lymphocytes, *J. Immunol.*, *146*: 2928 (1991).
59. D. R. Lucey, D. I. Dorsky, A. Nicholson-Weller, and P. F. Weller, Human eosinophils express CD4 protein and bind HIV-1 GP120, *J. Exp. Med.*, *169*: 327 (1989).
60. D. M. Center, W. W. Cruikshank, J. S. Berman, and D. J. Beer, Functional characteristics of histamine receptor-bearing mononuclear cells. I. Selective production of lymphocyte chemoattractant lymphokine utilizing histamine as a ligand, *J. Immunol.*, *131*: 1854 (1983).
61. J. S. Berman, D. J. Beer, W. W. Cruikshank, and D. M. Center, Chemoattractant lymphokines specific for the helper/inducer T-lymphocyte subset, *Cell. Immunol.*, *95*: 105 (1985).
62. J. Plumas, V. Gruart, D. Aldebert, M. J. Truong, M. Capron, A. Capron, and L. Prin, Human eosinophils from hypereosinophilic patients spontaneously express the p55 but not the p75 interleukin 2 receptor subunit, *Eur. J. Immunol.*, *21*: 1265 (1991).
63. D. Riedel, A. Lindemann, M. Brach, R. Mertelsmann, and F. Hermann, Granulocyte–macrophage colony-stimulating factor and interleukin-3 induce surface expression of interleukin-2 receptor p55-chain and CD4 by human eosinophils, *Immunology*, *70*: 258 (1990).
64. M. Azzawi, B. Bradley, P. K. Jeffrey, A. J. Frew, A. J. Wardlaw, G. Knowles, B. Assoufi, J. V. Collins, S. Durham, and A. B. Kay, Identification of activated T lymphocytes and eosinophils in bronchial biopsies in stable atopic asthma, *Am. Rev. Respir. Dis.*, *142*: 1407 (1990).
65. A. J. Frew, and A. B. Kay, The relationship between infiltrating $CD4^+$ lymphocytes, activated eosinophils, and the magnitude of the allergen-induced late phase cutaneous reaction in man, *J. Immunol.*, *141*: 4158 (1988).
66. A. J. Frew, C. J. Corrigan, P. Maestrelli, J. J. Tsai, K. Kurihara, R. E. O'Hehir, A. Hartnell, O. Cromwell, and A. B. Kay, T lymphocytes in allergen-induced late-phase reactions and asthma, *Int. Arch. Allergy Appl. Immunol.*, *88*: 63 (1989).

67. P. Diaz, M. C. Gonzalez, F. R. Galleguillos, P. Ancic, P. Cromwell, D. Shepherd, S. R. Durham, G. J. Gleich, and A. B. Kay, Leukocytes and mediators in bronchoalveolar lavage during allergen-induced late-phase asthmatic reactions, *Am. Rev. Respir. Dis.*, *139*: 1383 (1989).
68. T. J. Schall, Biology of the RANTES/SIS cytokine family, *Cytokine*, *3*: 165 (1991).
69. M. Marini, E. Vittori, J. Hollemborg, and S. Mattoli, Expression of the potent inflammatory cytokines, granulocyte–macrophage colony-stimulating factor, interleukin-6 and interleukin-8, in bronchial epithelial cells of patients with asthma, *J. Allergy Clin. Immunol.*, *89*: 1001 (1992).
70. Y. Kameyoshi, A. Dorschner, A. I. Mallet, E. Christophers, and J. M. Schroder, Cytokine RANTES released by thrombin-stimulated platelets is a potent attractant for human eosinophils, *J. Exp. Med.*, *176*: 587 (1992).
71. A. Rot, M. Krieger, T. Brunner, S. C. Bischoff, T. J. Schall, and C. A. Dahinden, RANTES and macrophage inflammatory protein 1α induce the migration and activation of normal human eosinophil granulocytes, *J. Exp. Med.*, *176*: 1489 (1992).
72. T.-C. Lee, D. J. Lenihan, B. Malone, L. L. Roddy, and S. I. Wasserman, Increased biosynthesis of platelet-activating factor in activated human eosinophils, *J. Biol. Chem.*, *259*: 5526 (1984).
73. T.-C. Lee, B. Malone, S. I. Wasserman, V. Fitzgerald, and F. Snyder, Activities of enzymes that metabolize platelet-activating factor (1-alkyl-2-acetyl-*sn*-glycero-3-phosphocholine) in neutrophils and eosinophils from humans and the effect of a calcium ionophore, *Biochem. Biophys. Res. Commun.*, *105*: 1303 (1982).
74. O. Cromwell, A. J. Wardlaw, A. Champion, R. Moqbel, D. Osei, and A. B. Kay, IgG-dependent generation of platelet-activating factor by normal and low density human eosinophils, *J. Immunol.*, *145*: 3862 (1990).
75. L. A. Burke, A. E. Crea, J. R. Wilkinson, J. P. Arm, B. W. Spur, and T. H. Lee, Comparison of the generation of platelet-activating factor and leukotriene C_4 in human eosinophils stimulated by unopsonized zymosan and by the calcium ionophore A23187: The effects of nedocromil sodium, *J. Allergy Clin. Immunol.*, *85*: 26 (1990).
76. P. J. Barnes, K. F. Chung, and C. P. Page, Platelet-activating factor as a mediator of allergic disease, *J. Allergy Clin. Immunol.*, *81*: 919 (1988).
77. S. M. Prescott, G. A. Zimmerman, and T. M. McIntyre, Platelet-activating factor, *J. Biol. Chem.*, *265*: 17381 (1990).
78. P. M. Henson, PAF—a perspective, *Am. J. Respir. Cell. Mol. Biol.*, *1*: 263 (1989).
79. W. R. J. Henderson, Eicosanoids and platelet-activating factor in allergic respiratory diseases, *Am. Rev. Respir. Dis.*, *143*: S86 (1991).
80. C. E. Sigal, F. H. Valone, M. J. Holtzman, and E. J. Goetzl, Preferential human eosinophil chemotactic activity of the platelet-activating factor (PAF) 1-*O*-hexadecyl-2-acetyl-*sn*-3-phosphocholine (AGEPC), *J. Clin. Immunol.*, *7*: 179 (1987).
81. C. Kroegel, T. Yukawa, J. Westwick, and P. J. Barnes, Evidence for two platelet activating factor receptors on eosinophils: Dissociation between PAF-induced intracellular calcium mobilization degranulation and superoxide anion generation in eosinophils, *Biochem. Biophys. Res. Commun.*, *162*: 511 (1989).

82. M. A. Giembycz, C. Kroegel, and P. J. Barnes, Platelet activating factor stimulates cyclooxygenase activity in guinea pig eosinophils. Concerted biosynthesis of thromboxane A_2 and E-series prostaglandins, *J. Immunol.*, *144*: 3489 (1990).
83. P. L. Bruynzeel, L. Koenderman, P. T. Kok, M. L. Hameling, and J. Verhagen, Platelet-activating factor (PAF-acether) induced leukotriene C_4 formation and luminol dependent chemiluminescence by human eosinophils, *Pharmacol. Res. Commun.*, *18* (Suppl): 61 (1986).
84. T. Yukawa, C. Kroegel, P. Evans, T. Fukuda, K. F. Chung, and P. J. Barnes, Density heterogeneity of eosinophil leucocytes: Induction of hypodense eosinophils by platelet-activating factor, *Immunology*, *68*: 140 (1989).
85. A. T. Tool, L. Koenderman, P. T. Kok, M. Blom, D. Roos, and A. J. Verhoeven, Release of platelet-activating factor is important for the respiratory burst induced in human eosinophils by opsonized particles, *Blood*, *79*: 2729 (1992).
86. T. Hubscher, Role of the eosinophil in the allergic reactions. II. Release of prostaglandins from human eosinophilic leukocytes, *J. Immunol.*, *114*: 1389 (1975).
87. P. F. Weller, C. W. Lee, D. W. Foster, E. J. Corey, K. F. Austen, and R. A. Lewis, Generation and metabolism of 5-lipoxygenase pathway leukotrienes by human eosinophils: Predominant production of leukotriene C_4, *Proc. Natl. Acad. Sci. USA*, *80*: 7626 (1983).
88. J. Verhagen, P. L. B. Bruynzeel, J. A. Koedam, G. A. Wassink, M. de Boer, G. K. Terpstra, J. Kreukniet, G. A. Veldink, and J. F. G. Vliegenthart, Specific leukotriene formation by purified human eosinophils and neutrophils, *FEBS Lett.*, *168*: 23 (1984).
89. S. Shak, and I. M. Goldstein, Leukotriene B_4 ω-hydroxylase in human polymorphonuclear leukocytes—partial purification and identification as a cytochrome P-450, *J. Clin. Invest.*, *76*: 1218 (1985).
90. T. Hubscher, Role of the eosinophil in the allergic reactions. I. EDI—a eosinophil-derived inhibitor of histamine release, *J. Immunol.*, *114*: 1379 (1975).
91. P. L. B. Bruijnzeel, and J. Verhagen, Lipid metabolism by eosinophils, *Eosinophils in Asthma* (J. Morley, and I. Colditz, eds.), Academic Press, San Diego, p. 69 (1989).
92. W. G. Parsons III, and L. J. Roberts II, Transformation of prostaglandin D_2 to isomeric prostaglandin F_2 compounds by human eosinophils. A potential mast cell-eosinophil interaction, *J. Immunol.*, *141*: 2413 (1988).
93. M. L. Foegh, Y. T. Maddox, and P. W. Ramwell, Human peritoneal eosinophils and formation of arachidonate cylooxygenase products, *Scand. J. Immunol.*, *23*: 599 (1986).
94. F. Ogushi, T. Ozaki, T. Kawano, and S. Yasuoka, PGE_2 and PGF_2 alpha content in bronchoalveolar lavage fluid obtained from patients with eosinophilic pneumonia, *Chest*, *91*: 204 (1987).
95. H. Kita, R. I. Abu-Ghazaleh, G. J. Gleich, and R. T. Abraham, Regulation of Ig-induced eosinophil degranulation by adenosine 3′,5′-cyclic monophosphate, *J. Immunol.*, *146*: 2712 (1991).
96. A. Jörg, W. R. Henderson, R. C. Murphy, and S. J. Klebanoff, Leukotriene generation by eosinophils, *J. Exp. Med.*, *155*: 390 (1982).

97. W. R. Henderson, J. B. Harley, and A. S. Fauci, Arachidonic acid metabolism in normal and hypereosinophilic syndrome eosinophils: Generation of leukotrienes B_4, C_4, D_4 and 15-lipoxygenase products, *Immunology*, *51*: 679 (1984).
98. R. J. Shaw, O. Cromwell, and A. B. Kay, Preferential generation of leukotriene C_4 by human eosinophils, *Clin. Exp. Immunol.*, *56*: 716 (1984).
99. R. J. Shaw, G. M. Walsh, O. Cromwell, R. Moqbel, C. J. Spry, and A. B. Kay, Activated human eosinophils generate SRS-A leukotrienes following IgG-dependent stimulation, *Nature*, *316*: 150 (1985).
100. P. L. Bruynzeel, P. T. Kok, M. L. Hamelink, A. M. Kijne, and J. Verhagen, Exclusive leukotriene C_4 synthesis by purified human eosinophils induced by opsonized zymosan, *FEBS Lett.*, *189*: 350 (1985).
101. T. Kajita, Y. Yui, H. Mita, N. Taniguchi, H. Saito, T. Mishima, and T. Shida, Release of leukotriene C_4 from human eosinophils and its relation to the cell density, *Int. Arch. Allergy Appl. Immunol.*, *78*: 406 (1985).
102. N. Tamura, D. K. Agrawal, and R. G. Townley, Leukotriene C_4 production from human eosinophils in vitro. Role of eosinophil chemotactic factors on eosinophil activation, *J. Immunol.*, *141*: 4291 (1988).
103. H. J. Ziltener, P.-A. Chavaillez, and A. Jörg, Leukotriene formation by eosinophil leukocytes. Analysis with ion-pair high pressure liquid chromatography and effect of the respiratory burst, *Hoppe-Seylers Z. Physiol. Chem.*, *364*: 1029 (1983).
104. P. Borgeat, B. Fruteau de Laclos, H. Rabinovitch, S. Picard, P. Braquet, J. Hébert, and M. Laviolette, Eosinophil-rich human polymorphonuclear leukocyte preparations characteristically release leukotriene C_4 on ionophore A23187 challenge, *J. Allergy Clin. Immunol.*, 310 (1984).
105. W. F. Owen, Jr., R. J. Soberman, T. Yoshimoto, A. L. Sheffer, R. A. Lewis, and K. F. Austen, Synthesis and release of leukotriene C_4 by human eosinophils, *J. Immunol.*, *138*: 532 (1987).
106. M. Laviolette, S. Picard, P. Braquet, and P. Borgeat, Comparison of 5- and 15-lipoxygenase activities in blood and alveolar leukocyte preparations from normal subjects and patients with eosinophilia, *Prostaglandins Leukotriene Med.*, *23*: 191 (1986).
107. C. Edenius, J. Haeggström, and J. A. Lindgren, Transcellular conversion of endogenous arachidonic acid to lipoxins in mixed human platelet-granulocyte suspensions, *Biochem. Biophys. Res. Commun.*, *157*: 801 (1988).
108. C. Edenius, K. Heidvall, and J. A. Lindgren, Novel transcellular interaction: Conversion of granulocyte-derived leukotriene A_4 to cysteinyl-containing leukotrienes by human platelets, *Eur. J. Biochem.*, *178*: 81 (1988).
109. S. Fiore, and C. N. Serhan, Formation of lipoxins and leukotrienes during receptor-mediated interactions of human platelets and recombinant human granulocyte/macrophage colony-stimulating factor-primed neutrophils, *J. Exp. Med.*, *172*: 1451 (1990).
110. M. J. Holtzman, Arachidonic acid metabolism. Implications of biological chemistry for lung function and disease, *Am. Rev. Respir. Dis.*, *143*: 188 (1991).
111. T. H. Lee, and K. F. Austen, Arachidonic acid metabolism by the 5-lipoxygenase pathway, and the effects of alternative dietary fatty acids, *Adv. Immunol.*, *39*: 145 (1986).

112. R. A. Lewis, K. F. Austen, and R. J. Soberman, Leukotrienes and other products of the 5-lipoxygenase pathway. Biochemistry and relation to pathobiology in human diseases, *N. Engl. J. Med.*, *323*: 645 (1990).
113. E. Sigal, D. Grunberger, J. R. Cashman, C. S. Craik, G. H. Caughey, and J. A. Nadel, Arachidonate 15-lipoxygenase from human eosinophil-enriched leukocytes: Partial purification and properties, *Biochem. Biophys. Res. Commun.*, *150*: 376 (1988).
114. J. J. Murray, and A. R. Brash, Rabbit reticulocyte lipoxygenase catalyzes specific 12(S) and 15(S) oxygenation of arachidonoyl-phosphatidylcholine, *Arch. Biochem. Biophys.*, *263*: 514 (1988).
115. G. Jung, D. C. Yang, and A. Nakao, Oxygenation of phosphatidylcholine by human polymorphonuclear leukocyte 15-lipoxygenase, *Biochem. Biophys. Res. Commun.*, *130*: 559 (1985).
116. A. W. Ford-Hutchinson, Arachidonate 15-lipoxygenase; characteristics and potential biological significance, *Eicosanoids*, *4*: 65 (1991).
117. R. L. Maas, A. R. Brash, and J. A. Oates, A second pathway of leukotriene biosynthesis in porcine leukocytes, *Proc. Natl. Acad. Sci. USA*, *78*: 5523 (1981).
118. R. L. Maas, J. Turk, J. A. Oates, and A. R. Brash, Formation of a novel dihydroxy acid from arachidonic acid by lipoxygenase-catalyzed double oxygenation in rat mononuclear cells and human leukocytes, *J. Biol. Chem.*, *257*: 7056 (1982).
119. J. Turk, R. L. Maas, A. R. Brash, L. J. Roberts 2nd, and J. A. Oates, Arachidonic acid 15-lipoxygenase products from human eosinophils, *J. Biol. Chem.*, *257*: 7068 (1982).
120. R. J. Soberman, 5- And 15(omega-6)-lipoxygenases from human polymorphonuclear leukocytes, *Methods Enzymol.*, *163*: 344 (1988).
121. B. Fruteau de Laclos, P. Braquet, and P. Borgeat, Characteristics of leukotriene (LT) and hydroxyeicosatetraenoic acid (HETE) synthesis in human leukocytes in vitro: Effect of arachidonic acid concentration, *Prostaglandins Leukotrienes Med.*, *13*: 47 (1984).
122. J. McGuire, J. McGee, N. Crittenden, and F. Fitzpatrick, Cell damage unmasks 15-lipoxygenase activity in human neutrophils, *J. Biol. Chem.*, *260*: 8316 (1985).
123. R. C. Nichols, and J. Y. Vanderhoek, 5-Hydroxyeicosanoids selectively stimulate the human neutrophil 15-lipoxygenase to use endogenous substrate, *J. Exp. Med.*, *171*: 367 (1990).
124. M. J. Holtzman, A. Pentland, N. L. Baenziger, and J. R. Hansbrough, Heterogeneity of cellular expression of arachidonate 15-lipoxygenase: Implications for biological activity, *Biochim. Biophys. Acta*, *1003*: 204 (1989).
125. J. A. Nadel, D. J. Conrad, I. F. Ueki, and E. Sigal, Immunocytochemical localization of arachidonate 15-lipoxygenase in erythrocytes, leukocytes, and airway cells, *J. Clin. Invest.*, *87*: 1139 (1991).
126. S. Ylä-Herttuala, M. E. Rosenfeld, S. Parthasarathy, E. Sigal, T. Särkioja, J. L. Witztum, and D. Steinberg, Gene expression in macrophage-rich human atherosclerotic lesions. 15-Lipoxygenase and acetyl low density lipoprotein receptor messenger RNA colocalize with oxidation specific lipid-protein adducts, *J. Clin. Invest.*, *87*: 1146 (1991).
127. E. Sigal, D. Grunberger, C. S. Craik, G. H. Caughey, and J. A. Nadel,

Arachidonate 15-lipoxygenase (ω-6 lipoxygenase) from human leukocytes. Purification and structural homology to other mammalian lipoxygenases, *J. Biol. Chem.*, *263*: 5328 (1988).
128. E. Sigal, C. S. Craik, E. Highland, D. Grunberger, L. L. Costello, R. A. F. Dixon, and J. A. Nadel, Molecular cloning and primary structure of human 15-lipoxygenase, *Biochem. Biophys. Res. Commun.*, *157*: 457 (1988).
129. E. Sigal, D. Grunberger, E. Highland, C. Gross, R. A. F. Dixon, and C. S. Craik, Expression of cloned human reticulocyte 15-lipoxygenase and immunological evidence that 15-lipoxygenases of different cell types are related, *J. Biol. Chem.*, *265*: 5113 (1990).
130. D. L. Sloane, C. S. Craik, and E. Sigal, The expression of active human reticulocyte 15-lipoxygenase in bacteria, *Biomed. Biochim. Acta*, *49*: S11 (1990).
131. C. A. Rouzer, and B. Samuelsson, On the nature of the 5-lipoxygenase reaction in human leukocytes: Enzyme purification and requirement for multiple stimulatory factors, *Proc. Natl. Acad. Sci. USA*, *82*: 6040 (1985).
132. K. Ochi, T. Yoshimoto, S. Yamamoto, K. Taniguchi, and T. Miyamoto, Arachidonate 5-lipoxygenase of guinea pig peritoneal polymorphonuclear leukocytes. Activation by adenosine 5′-triphosphate, *J. Biol. Chem.*, *258*: 5754 (1983).
133. P. L. Bruynzeel, P. T. Kok, R. J. Vietor, and J. Verhagen, On the optimal conditions of LTC_4 formation by human eosinophils in vitro, *Prostaglandins Leukotriene Med.*, *20*: 11 (1985).
134. E. Morita, J. M. Schroder, and E. Christophers, Production of 15-hydroxyeicosatetraenoic acid by purified human eosinophils and neutrophils, *Scand. J. Immunol.*, *32*: 497 (1990).
135. Z. Marom, J. H. Shelhamer, F. Sun, and M. Kaliner, Human airway monohydroxyeicosatetraenoic acid generation and mucus release, *J. Clin. Invest.*, *72*: 122 (1983).
136. S. Shak, D. Perez, and I. M. Goldstein, A novel dioxygenation product of arachidonic acid possesses potent chemotactic activity for human polymorphonuclear leukocytes, *J. Biol. Chem.*, *258*: 14948 (1983).
137. C. M. Kirsch, E. Sigal, T. D. Djokiic, P. D. Graf, and J. A. Nadel, An in vivo chemotaxis assay in the dog trachea: Evidence for chemotactic activity of 8, 15-diHETE, *J. Appl. Physiol.*, *64*: 1792 (1988).
138. U. Ramstedt, C. N. Serhan, U. Lundberg, H. Wigzell, and B. Samuelsson, Inhibition of human natural killer cell activity by (14*R*, 15*S*)-14,15-dihydroxy-5*Z*,8*Z*,10*E*,12*E*-icosatetraenoic acid, *Proc. Natl. Acad. Sci. USA*, *81*: 6914 (1984).
139. C. N. Serhan, K. C. Nicolaou, S. E. Webber, C. A. Veale, S.-E. Dahlem, T. J. Puustinen, and B. Samuelsson, Lipoxin A. Stereochemistry and biosynthesis, *J. Biol. Chem.*, *261*: 16340 (1986).
140. S.-E. Dahlen, L. Franzen, J. Raud, C. N. Serhan, P. Westlund, E. Wikstrom, T. Bjorck, H. Matsuda, S. E. Webber, C. A. Veale, T. Puustinen, J. Haeggstrom, K. C. Nicolaou, and B. Samuelsson, Actions of lipoxin A_4 and related compounds in smooth muscle preparations and on the microcirculation in vivo, *Lipoxins: Chemistry, Biosynthesis and Biological Activities* (P.-Y. K. Wong, and C. N. Serhan, eds.), Plenum Publishing, New York, p.107 (1988).
141. M. Hamberg, P. Hedqvist, and K. Radegran, Identification of 15-hydroxy-

5,8,11,13-eicosatetraenoic acid (15-HETE) as a major metabolite of arachidonic acid in human lung, *Acta Physiol. Scand.*, *110*: 219 (1980).

142. J. A. Hunter, W. E. Finkbeiner, J. A. Nadel, E. J. Goetzl, and M. J. Holtzman, Predominant generation of 15-lipoxygenase metabolites of arachidonic acid by epithelial cells from human trachea, *Proc. Natl. Acad. Sci. USA*, *82*: 4633 (1985).
143. J. R. Cashman, C. Lambert, and E. Sigal, Inhibition of human leukocyte 5-lipoxygenase by 15-HPETE and related eicosanoids, *Biochem. Biophys. Res. Commun.*, *155*: 38 (1988).
144. S. M. Rapoport, T. Schewe, R. Wiesner, W. Halangk, P. Ludwig, M. Janicke-Hohne, C. Tannert, C. Hiebsch, and D. Klat, The lipoxygenase of reticulocytes. Purification, characterization and biological dynamics of the lipoxygenase: Its identity with the respiratory inhibitors of the reticulocyte, *Eur. J. Biochem.*, *96*: 545 (1979).
145. C. N. Serhan, Lipoxins: Eicosanoids carrying intra- and intercellular messages, *J. Bioenerg. Biomembr.*, *23*: 105 (1991).
146. C. N. Serhan, and B. Samuelsson, Lipoxins: A new series of eicosanoids (biosynthesis, stereochemistry, and biological activities), *Adv. Exp. Med. Biol.*, *229*: 1 (1988).
147. C. N. Serhan, U. Hirsch, J. Palmblad, and B. Samuelsson, Formation of lipoxin A by granulocytes from eosinophilic donors, *FEBS Lett.*, *217*: 242 (1987).
148. D. Steinhilber, and H. J. Roth, New series of lipoxins isolated from human eosinophils, *FEBS Lett.*, *255*: 143 (1989).
149. C. N. Serhan, and K.-A. Sheppard, Lipoxin formation during human neutrophil–platelet interactions. Evidence for the transformation of leukotriene A_4 by platelet 12-lipoxygenase in vitro, *J. Clin. Invest.*, *85*: 772 (1990).
150. T. H. Lee, A. E. G. Crea, V. Gant, B. W. Spur, B. E. Marron, K. C. Nicolaou, E. Reardon, M. Brezinski, and C. N. Serhan, Identification of lipoxin A_4 and its relationship to the sulfidopeptide leukotrienes C_4, D_4, and E_4 in the bronchoalveolar lavage fluids obtained from patients with selected pulmonary diseases, *Am. Rev. Respir. Dis.*, *141*: 1453 (1990).
151. P. Hedqvist, J. Raud, U. Palmertz, J. Haeggstrom, K. C. Nicolaou, and S. E. Dahlen, Lipoxin A_4 inhibits leukotriene B_4-induced inflammation in the hamster cheek pouch, *Acta Physiol. Scand.*, *137*: 571 (1989).
152. T. H. Lee, C. E. Horton, A. U. Kyan, D. Haskard, A. E. Crea, and B. W. Spur, Lipoxin A_4 and lipoxin B_4 inhibit chemotactic responses of human neutrophils stimulated by leukotriene B_4 and *N*-formyl-L-methionyl-L-leucyl-L-phenylalanine, *Clin. Sci.*, *77*: 195 (1989).
153. B. M. Grandordy, H. Lacroix, E. Mavoungou, S. Krilis, A. E. Crea, B. W. Spur, and T. H. Lee, Lipoxin A_4 inhibits phosphoinositide hydrolysis in human neutrophils, *Biochem. Biophys. Res. Commun.*, *167*: 1022 (1990).
154. U. Ramstedt, C. N. Serhan, K. C. Nicolaou, S. E. Webber, H. Wigzell, and B. Samuelsson, Lipoxin A-induced inhibition of human natural killer cell cytotoxicity: Studies on stereospecificity of inhibition and mode of action, *J. Immunol.*, *138*: 266 (1987).
155. A. Hansson, C. N. Serhan, J. Haeggström, M. Ingelman-Sundberg, B. Samuelsson, and J. Morris, Activation of protein kinase C by lipoxin A and other

eicosanoids. Intracellular action of oxygenation products of arachidonic acid., *Biochem. Biophys. Res. Commun.*, *134*: 1215 (1986).

156. M. S. Shearman, Z. Naor, K. Sekiguchi, A. Kishimoto, and Y. Nishizuka, Selective activation of the gamma-subspecies of protein kinase C from bovine cerebellum by arachidonic acid and lipoxygenase products, *FEBS Lett.*, *243*: 177 (1989).
157. J. V. Weinstock, A. Blum, J. Walder, and R. Walder, Eosinophils from granulomas in murine schistosomiasis mansoni produce substance P, *J. Immunol.*, *141*: 961 (1988).
158. J. V. Weinstock, and A. M. Blum, Tachykinin production in granulomas of murine schistosomiasis mansoni, *J. Immunol.*, *142*: 3256 (1989).
159. J. V. Weinstock, and A. M. Blum, Release of substance P by granuloma eosinophils in response to secretagogues in murine schistosomiasis mansoni, *Cell. Immunol.*, *125*: 380 (1990).
160. J. Aliakbari, S. P. Sreedharan, C. W. Turck, and E. J. Goetzl, Selective localization of vasoactive intestinal peptide and substance P in human eosinophils, *Biochem. Biophys. Res. Commun.*, *148*: 1440 (1987).
161. J. V. Weinstock, and A. M. Blum, Detection of vasoactive intestinal peptide and localization of its mRNA within granulomas of murine schistosomiasis, *Cell Immunol.*, *125*: 291 (1990).
162. J. N. Baraniuk, and M. A. Kaliner, Neuropeptides and nasal secretion, *J. Allergy Clin. Immunol.*, *86*: 620 (1990).
163. J. P. McGillis, M. Mitsuhashi, and D. G. Payan, Immunomodulation by tachykinin neuropeptides, *Ann. N. Y. Acad. Sci.*, *594*: 85 (1990).
164. D. G. Payan, Neuropeptides and inflammation: The role of substance P, *Annu. Rev. Med.*, *40*: 341 (1989).
165. R. Mathew, J. Weinstock, A. Blum, G. Cook, A. Metwali, and R. Felman, Vasoactive intestinal peptide (VIP) regulates interleukin 5 (IL5) production in murine schistosomiasis mansoni, *FASEB J.*, *5*: A1093 (1991).
166. D. T. W. Wong, P. F. Weller, S. J. Galli, T. H. Rand, A. Elovic, T. Chiang, M. Y. Chou, G. T. Gallagher, K. Matossian, J. McBride, and R. Todd, Human eosinophils express transforming growth factor α, *J. Exp. Med.*, *172*: 673 (1990).
167. R. Todd, M. Y. Chou, K. Matossian, G. T. Gallagher, R. B. Donoff, and D. T. Wong, Cellular sources of transforming growth factor-alpha in human oral cancer, *J. Dent. Res.*, *70*: 917 (1991).
168. A. Elovic, D. T. W. Wong, P. F. Weller, K. Matossian, and S. J. Galli, Co-expression of TGF-α and TGF-β_1 mRNA and product by eosinophils in nasal polyps, *J. Allergy Clin. Immunol.* (in press).
169. S. Emery, A. Elovic, D. Wong, P. F. Weller, and S. J. Galli, Expression of transforming growth factor-alpha (TGF-α) mRNA by human eosinophils in diverse pathological processes, *FASEB J.*, *5*: A1695 (1991).
170. A. Elovic, S. J. Galli, P. F. Weller, A. L. C. Chang, T. Chiang, M. Y. Chou, B. Donoff, G. T. Gallagher, K. Matossian, J. McBride, M. Tsai, R. Todd, and D. T. W. Wong, Production of transforming growth factor alpha by hamster eosinophils, *Am. J. Pathol.*, *137*: 1425 (1990).
171. R. Todd, B. R. Donoff, T. Chiang, M. Y. Chou, A. Elovic, G. T. Gallagher,

and D. T. Wong, The eosinophil as a cellular source of transforming growth factor alpha in healing cutaneous wounds, *Am. J. Pathol.*, *138*: 1307 (1991).

172. D. T. W. Wong, R. B. Donoff, J. Yang, B.-Z. Song, K. Matossian, N. Nagura, A. Elovic, J. McBride, G. Gallagher, R. Todd, T. Chiang, C. M. Yung, S. J. Galli, and P. F. Weller, Sequential expression of TGF-α and TGF-β_1 during cutaneous wound healing in the hamster, *Am. J. Pathol.*, *143*: 130 (1993).
173. D. K. Madtes, E. W. Raines, K. S. Sakariassen, R. K. Assoian, M. B. Sporn, G. I. Bell, and R. Ross, Induction of transforming growth factor-α in activated human alveolar macrophages, *Cell*, *53*: 285 (1988).
174. D. A. Rappolee, D. Mark, M. J. Brenda, and Z. Werb, Wound macrophages express TGF-α and other growth factors in vivo: Analysis by mRNA phenotyping, *Science*, *241*: 708 (1988).
175. J. Massagué, Transforming growth factor-α, *J. Biol. Chem.*, *265*: 21393 (1990).
176. R. M. Lyons, and H. L. Moses, Transforming growth factors and the regulation of cell proliferation, *Eur. J. Biochem.*, *187*: 467 (1990).
177. C. E. P. Brewster, P. H. Howarth, R. Djukanovic, J. Wilson, S. T. Holgate, and W. R. Roche, Myofibroblasts and subepithelial fibrosis in bronchial asthma, *Am. J. Respir. Cell. Mol. Biol.*, *3*: 507 (1990).
178. W. R. Roche, R. Beasley, J. H. Williams, and S. T. Holgate, Subepithelial fibrosis in the bronchi of asthmatics, *Lancet*, *1*: 520 (1989).
179. D. T. W. Wong, A. Elovic, K. Matossian, N. Nagura, J. McBride, J. R. Gordon, T. H. Rand, S. J. Galli, and P. F. Weller, Eosinophils from patients with blood eosinophilia express transforming growth factor β_1, *Blood*, *78*: 2702 (1991).
180. I. Ohno, R. G. Lea, K. C. Flanders, D. A. Clark, D. Banwatt, J. Dolovich, J. Denburg, C. B. Harley, J. Gauldie, and M. Jordana, Eosinophils in chronically inflamed human upper airway tissues express transforming growth factor beta 1 gene (TGF beta 1), *J. Clin. Invest.*, *89*: 1662 (1992).
181. J. A. Barnard, R. M. Lyons, and H. L. Moses, The cell biology of transforming growth factor beta, *Biochim. Biophys. Acta*, *1032*: 79 (1990).
182. P. F. Weller, T. H. Rand, T. Barrett, A. Elovic, D. T. Wong, and R. W. Finberg, Accessory cell function of human eosinophils: HLA-DR dependent, MHC-restricted antigen-presentation and interleukin-1α formation, *J. Immunol.*, *150*: 2554 (1993).
183. V. del Pozo, B. de Andres, E. Martin, N. Maruri, J. M. Zubeldia, P. Palomino, and C. Lahoz, Murine eosinophils and IL-1: αIL-1 mRNA detection by in situ hybridization. Production and release of IL-1 from peritoneal eosinophils, *J. Immunol.*, *144*: 3117 (1990).
184. S. Akira, T. Hirano, T. Taga, and T. Kishimoto, Biology of multifunctional cytokines: IL 6 and related molecules (IL 1 and TNF), *FASEB J.*, *4*: 2860 (1990).
185. C. A. Dinarello, Biology of interleukin 1, *FASEB J.*, *2*: 108 (1988).
186. D. R. Lucey, A. Nicholson-Weller, and P. F. Weller, Mature human eosinophils have the capacity to express HLA-DR, *Proc. Natl. Acad. Sci. USA*, *86*: 1348 (1989).
187. T. T. Hansel, J. B. Braunstein, C. Walker, K. Blaser, P. L. B. Bruijnzeel, J. C. Virchow, Jr., and C. Virchow, Sputum eosinophils from asthmatics express ICAM-1 and HLA-DR, *Clin. Exp. Immunol.*, *86*: 271 (1991).

188. J. B. Sedgwick, W. J. Calhoun, R. F. Vrtis, M. E. Bates, P. K. McAllister, and W. W. Busse, Comparison of airway and blood eosinophil function after in vivo antigen challenge, *J. Immunol.*, *149*: 3710 (1992).

189. C. T. Weaver, and E. R. Unanue, The costimulatory function of antigen-presenting cells, *Immunol. Today*, *11*: 49 (1990).

190. A. F. Lopez, J. M. Eglinton, D. Gillis, L. S. Park, S. Clark, and M. A. Vadas, Reciprocal inhibition of binding between interleukin 3 and granulocyte-macrophage colony-stimulating factor to human eosinophils, *Proc. Natl. Acad. Sci. USA*, *86*: 7022 (1989).

191. E. Ingley, and I. G. Young, Characterization of a receptor for interleukin-5 on human eosinophils and the myeloid leukemia line HL-60, *Blood*, *78*: 339 (1991).

192. M. Migita, N. Yamaguchi, S. Mita, S. Higuchi, Y. Hitoshi, Y. Yoshida, M. Tomonaga, I. Matsuda, A. Tominaga, and K. Takatsu, Characterization of the human IL-5 receptors on eosinophils, *Cell. Immunol.*, *133*: 484 (1991).

193. D. H. Broide, M. M. Paine, and G. S. Firestein, Eosinophils express interleukin 5 and granulocyte–macrophage colony-stimulating factor mRNA at sites of allergic inflammation in asthmatics, *J. Clin. Invest.*, *90*: 1414 (1992).

194. A. F. Lopez, J. M. Eglinton, A. B. Lyons, P. M. Tapley, L. B. To, L. S. Park, S. C. Clark, and M. A. Vadas, Human interleukin-3 inhibits the binding of granulocyte–macrophage colony-stimulating factor and interleukin-5 to basophils and strongly enhances their functional activity, *J. Cell. Physiol.*, *145*: 69 (1990).

195. D. S. Silberstein, and J. R. David, The regulation of human eosinophil function by cytokines, *Immunol. Today*, *8*: 380 (1987).

196. M. E. Rothenberg, W. F. Owen, Jr., D. S. Silberstein, J. Woods, R. J. Soberman, K. F. Austen, and R. L. Stevens, Human eosinophils have prolonged survival, enhanced functional properties, and become hypodense when exposed to human interleukin 3, *J. Clin. Invest.*, *81*: 1986 (1988).

197. W. F. Owen, Jr., M. F. Rothenberg, D. S. Silberstein, J. C. Gasson, R. L. Stevens, K. F. Austen, and R. J. Soberman, Regulation of human eosinophil viability, density, and function by granulocyte/macrophage colony-stimulating factor in the presence of 3T3 fibroblasts, *J. Exp. Med.*, *166*: 129 (1987).

198. P. F. Weller, Eosinophils and fibroblasts: The medium in the mesenchyme, *Am. J. Respir. Cell. Mol. Biol.*, *1*: 267 (1989).

199. C. Ruef, and D. L. Coleman, Granulocyte–macrophage colony-stimulating factor: Pleiotropic cytokine with potential clinical usefulness, *Rev. Infect. Dis.*, *12*: 41 (1990).

200. C. J. Sanderson, Interleukin-5, eosinophils, and disease, *Blood*, *79*: 3101 (1992).

201. Q. Hamid, J. Barkans, Q. Meng, S. Ying, J. S. Abrams, A. B. Kay, and R. Moqbel, Human eosinophils synthesize and secrete interleukin-6, in vitro, *Blood*, *80*: 1496 (1992).

202. J. J. Costa, K. Matossian, W. J. Beil, D. T. W. Wong, J. R. Gordon, A. M. Dvorak, P. F. Weller, and S. J. Galli, Human eosinophils can express the cytokines TNF-α and MIP-1α, *J. Clin. Invest.*, *91*: 2673 (1993).

203. A. M. Dvorak, S. J. Ackerman, and P. F. Weller, Subcellular morphology and biochemistry of eosinophils, *Blood Cell Biochemistry*: *Megakaryocytes, Platelets,*

Macrophages and Eosinophils (J. R. Harris, ed.), Plenum Publishing, London, p.237. (1990).
204. M. S. Peters, M. Rodriquez, and G. J. Gleich, Localization of human eosinophil granule major basic protein, eosinophil cationic protein, and eosinophil-derived neurotoxin by immunoelectron microscopy, *Lab. Invest.*, *54*: 656 (1986).
205. A. Egeston, J. Aluments, C. V. Meclenburg, M. Palmegren, and I. Olsson, Localization of eosinophil cationic protein, major basic protein, and eosinophil peroxidase in human eosinophils by immunoelectron microscopic technique, *J. Histochem. Cytochem.*, *34*: 1399 (1986).
206. G. J. Gleich, and C. R. Adolphson, The eosinophilic leukocyte: Structure and function, *Adv. Immunol.*, *39*: 177 (1986).
207. R. L. Barker, D. A. Loegering, K. C. Arakawa, L. A. Pease, and G. J. Gleich, Cloning and sequence analysis of the human gene encoding eosinophil major basic protein, *Gene*, *86*: 285 (1990).
208. T. L. Wasmoen, M. P. Bell, D. A. Loegering, G. J. Gleich, F. G. Prendergast, and D. J. McKean, Biochemical and amino acid sequence analysis of human eosinophil granule major basic protein, *J. Biol. Chem.*, *263*: 12559 (1988).
209. R. L. Barker, G. J. Gleich, and L. R. Pease, Acidic precursor revealed in human eosinophil granule major basic protein cDNA, *J. Exp. Med.*, *168*: 1493 (1988).
210. M. McGrogan, C. Simonsen, R. Scott, J. Griffith, N. Ellis, J. Kennedy, D. Campanelli, C. Nathan, and J. Gabay, Isolation of a complementary DNA clone encoding a precursor to human eosinophil major basic protein, *J. Exp. Med.*, *168*: 2295 (1988).
211. A. E. Butterworth, D. L. Wassom, G. J. Gleich, D. A. Loegering, and J. R. David, Damage to schistosomula of *Schistosoma mansoni* induced directly by eosinophil major basic protein, *J. Immunol.*, *122*: 221 (1979).
212. I. Olsson, A.-M. Persson, and I. Winqvist, Biochemical properties of the eosinophil cationic protein (ECP) and studies of its biosynthesis in vitro in marrow cells from patients with eosinophilia, *Blood*, *67*: 498 (1986).
213. G. J. Gleich, D. A. Loegering, M. P. Bell, J. L. Checkel, S. J. Ackerman, and D. J. McKean, Biochemical and functional similarities between eosinophil-derived neurotoxin and eosinophil cationic protein: Homology with ribonuclease, *Proc. Natl. Acad. Sci. USA*, *83*: 3146 (1986).
214. H. F. Rosenberg, S. J. Ackerman, and D. G. Tenen, Human eosinophil cationic protein. Molecular cloning of a cytotoxin and helminthotoxin with ribonuclease activity, *J. Exp. Med.*, *170*: 163 (1989).
215. H. F. Rosenberg, D. G. Tenen, and S. J. Ackerman, Molecular cloning of the human eosinophil-derived neurotoxin: a member of the ribonuclease gene family, *Proc. Natl. Acad. Sci. USA*, *86*: 4460 (1989).
216. K. J. Hamann, R. L. Barker, D. A. Loegering, L. A. Pease, and G. J. Gleich, Sequence of human eosinophil-derived neurotoxin cDNA: Identity of deduced amino acid sequence with nonsecretory ribonucleases, *Gene*, *83*: 161 (1989).
217. R. L. Barker, D. A. Loegering, R. M. Ten, K. J. Hamann, L. R. Pease, and G. J. Gleich, Eosinophil cationic protein cDNA. Comparison with other toxic cationic proteins and ribonucleases, *J. Immunol.*, *143*: 952 (1989).
218. R. I. Lehrer, D. Szklarek, A. Barton, T. Ganz, K. J. Hamann, and G. J. Gleich,

Antibacterial properties of eosinophil major basic protein and eosinophil cationic protein, *J. Immunol.*, *142*: 4428 (1989).

219. S. J. Ackerman, G. J. Gleich, D. A. Loegering, B. A. Richardson, and A. E. Butterworth, Comparative toxicity of purified human eosinophil granule cationic proteins for schistosomula of *Schistosoma mansoni*, *Am. J. Trop. Med. Hyg.*, *34*: 735 (1985).
220. R. L. Olsen, and C. Little, Purification and some properties of myeloperoxidase and eosinophil peroxidase from human blood, *Biochem. J.*, *209*: 781 (1983).
221. P. F. Weller, S. J. Ackerman, and J. A. Smith, Eosinophil granule cationic proteins: Major basic protein is distinct from the smaller subunit of eosinophil peroxidase, *J. Leukocyte Biol.*, *43*: 1 (1988).
222. K. Sakamaki, M. Tomonaga, K. Tsukui, and S. Nagata, Molecular cloning and characterization of a chromosomal gene for human eosinophil peroxidase, *J. Biol. Chem.*, *264*: 16828 (1989).
223. R. M. Ten, L. R. Pease, D. J. McKean, M. P. Bell, and G. J. Gleich, Molecular cloning of the human eosinophil peroxidase. Evidence for the existence of a peroxidase multigene family, *J. Exp. Med.*, *169*: 1757 (1989).
224. S. J. Weiss, S. T. Test, C. M. Eckmann, D. Roos, and S. Regiani, Brominating oxidants generated by human eosinophils, *Science*, *234*: 200 (1986).
225. A. N. Mayeno, A. J. Curran, R. L. Roberts, and C. S. Foote, Eosinophils preferentially use bromide to generate halogenating agents, *J. Biol. Chem.*, *264*: 5660 (1989).
226. E. C. Jong, W. R. Henderson, and S. J. Klebanoff, Bactericidal activity of eosinophil peroxidase, *J. Immunol.*, *124*: 1378 (1980).
227. E. C. Jong, A. A. F. Mahmoud, and S. J. Klebanoff, Peroxidase-mediated toxicity to schistosomula of *Schistosoma mansoni*, *J. Immunol.*, *126*: 468 (1981).
228. N. M. Nogueira, S. J. Klebanoff, and Z. A. Cohn, *Trypanosoma cruzi*: Sensitization to macrophage killing by eosinophil peroxidase, *J. Immunol.*, *128*: 1705 (1982).
229. E. C. Jong, and S. J. Klebanoff, Eosinophil-mediated mammalian tumor cell cytotoxicity: Role of the peroxidase system. *J. Immunol.*, *124*: 1949 (1980).
230. G. J. Gleich, N. A. Flavahan, T. Fujisawa, and P. M. Vanhoutte, The eosinophil as a mediator of damage to respiratory epithelium: A model for bronchial hyperreactivity, *J. Allergy Clin. Immunol.*, *81*: 776 (1988).
231. S. R. White, S. Ohno, N. M. Munoz, G. J. Gleich, C. Abrahams, J. Solway, and A. R. Leff, Epithelium-dependent contraction of airway smooth muscle by eosinophil MBP, *Am. J. Physiol.*, *259*: 294 (1990).
232. D. A. Bass, W. H. Grover, J. C. Lewis, P. Szejda, L. R. DeChatelet, and C. E. McCall, Comparison of human eosinophils from normals and patients with eosinophilia, *J. Clin. Invest.*, *66*: 1265 (1980).
233. S. R. Wang, L. Liang, S. S. Wang, C. L. Yu, B. N. Chiang, and S. H. Han, Enhanced production of LTC_4 in leukocytes from allergic asthmatic subjects, *Adv. Prostaglandin Thromboxane Leukotriene Res.*, *19*: 548 (1989).
234. T. Aizawa, G. Tamura, H. Ohtsu, and T. Takishima, Eosinophil and neutrophil production of leukotriene C_4 and B_4: Comparison of cells from asthmatic subjects and healthy donors, *Ann. Allergy*, *64*: 287 (1990).

235. F. Kohi, H. Miyagawa, D. K. Agrawal, A. K. Bewtra, and R. G. Townley, Generation of leukotriene B_4 and C_4 from granulocytes of normal controls, allergic rhinitis, and asthmatic subjects, *Ann. Allergy*, *65*: 228 (1990).
236. U. Schauer, A. Eckhart, R. Muller, D. Gemsa, and C. H. Rieger, Enhanced leukotriene C_4 production by peripheral eosinophilic granulocytes from children with asthma, *Int. Arch. Allergy Appl. Immunol.*, *90*: 201 (1989).
237. U. Schauer, U. Daume, R. Muller, F. Riedel, D. Gemsa, and C. H. Rieger, Relationship between LTC_4 generation of hypodense eosinophils and bronchial hyperreactivity in asthmatic children, *Int. Arch. Allergy Appl. Immunol.*, *92*: 82 (1990).
238. N. Taniguchi, H. Mita, H. Saito, Y. Yui, T. Kajita, and T. Shida, Increased generation of leukotriene C_4 from eosinophils in asthmatic patients, *Allergy*, *40*: 571 (1985).
239. H. F. Kauffman, B. van der Belt, J. G. R. de Monchy, H. Boelens, G. H. Koëter, and K. de Vries, Leukotriene C_4 production by normal-density and low-density eosinophils of atopic individuals and other patients with eosinophilia, *J. Allergy Clin. Immunol.*, *79*: 611 (1987).
240. M. K. Hodges, P. F. Weller, N. P. Gerard, S. J. Ackerman, and J. M. Drazen, Heterogeneity of leukotriene C_4 production by eosinophils from asthmatic and normal subjects, *Am. Rev. Respir. Dis.*, *138*: 799 (1988).
241. P. F. Weller, Eicosanoids, cytokines and other mediators elaborated by eosinophils, *Eosinophils: Biological and Clinical Aspects* (S. Makino, and T. Fukuda, eds.), CRC Press, Boca Raton, p.125 (1992).
242. T. Fukuda, S. L. Dunnette, C. E. Reed, S. J. Ackerman, M. S. Peters, and G. J. Gleich, Increased numbers of hypodense eosinophils in the blood of patients with bronchial asthma. *Am. Rev. Respir. Dis.*, *132*: 981 (1985).
243. T. Fukuda, and S. Makino, Eosinophil heterogeneity, *Eosinophils in Asthma* (J. Morley, and I. Colditz, eds.), Academic Press, San Diego, p.125 (1989).
244. M. E. Rothenberg, W. F. Owen, Jr., D. S. Silberstein, R. J. Soberman, K. F. Austen, and R. L. Stevens, Eosinophils cocultured with endothelial cells have increased survival and functional properties, *Science*, *237*: 645 (1987).
245. D. S. Silberstein, W. F. Owen, J. C. Gasson, J. F. DiPersio, D. W. Golde, J. C. Bina, R. Soberman, K. F. Austen, and J. R. David, Enhancement of human eosinophil cytotoxicity and leukotriene synthesis by biosynthetic (recombinant) granulocyte–macrophage colony-stimulating factor. *J. Immunol.*, *137*: 3290 (1986).
246. C. J. Howell, J. L. Pujol, A. E. Crea, R. Davidson, A. J. Gearing, P. Godard, and T. H. Lee, Identification of an alveolar macrophage-derived activity in bronchial asthma that enhances leukotriene C_4 generation by human eosinophils stimulated by ionophore A23187 as a granulocyte–macrophage colony-stimulating factor, *Am. Rev. Respir. Dis.*, *140*: 1340 (1989).
247. M. E. Rothenberg, J. Petersen, R. L. Stevens, D. S. Silberstein, D. T. McKenzie, K. F. Austen, and W. F. Owen, Jr., IL-5-dependent conversation of normodense human eosinophils to the hypodense phenotype uses 3T3 fibroblasts for enhanced viability, accelerated hypodensity, and sustained antibody-dependent cytotoxicity, *J. Immunol.*, *143*: 2311 (1989).

248. R. Roubin, P. P. Elsas, W. Fiers, and A. J. Dessein, Recombinant human tumour necrosis factor (rTNF)$_2$ enhances leukotriene biosynthesis in neutrophils and eosinophils stimulated with the Ca^{2+} ionophore A23187, *Clin. Exp. Immunol.*, *70*: 484 (1987).
249. P. Fitzharris, R. Moqbel, K. J. Thorne, B. A. Richardson, A. Hartnell, O. Cromwell, A. E. Butterworth, and A. B. Kay, The effects of eosinophil activating factor on IgG-dependent sulphidopeptide leukotriene generation by human eosinophils, *Clin. Exp. Immunol.*, *66*: 673 (1986).
250. P. X. Elsas, M. I. Elsas, and A. J. Dessein, Eosinophil cytotoxicity enhancing factor: Purification, characterization and immunocytochemical localization on the monocyte surface, *Eur. J. Immunol.*, *20*: 1143 (1990).
251. A. J. Dessein, T. H. Lee, P. Elsas, J. Ravalese III, D. Silberstein, J. R. David, K. F. Austen, and R. A. Lewis, Enhancement by monokines of leukotriene generation by human eosinophils and neutrophils stimulated with calcium ionophore A23187, *J. Immunol.*, *136*: 3829 (1986).
252. P. Elsas, T. H. Lee, H. L. Lenzi, and A. J. Dessein, Monocytes activate eosinophils for enhanced helminthotoxicity and increased generation of leukotriene C_4, *Ann. Inst. Pasteur Immunol.*, *138*: 97 (1987).
253. P. F. Weller, Cytokine regulation of eosinophil function, *Clin. Immunol. Immunopathol.*, *62*(1 Pt 2): S55 (1992).
254. C. Prausnitz, and H. Küstner, Studien über die überempfindlichkeit, *Zentralbl. Bakteriol.*, *86*: 160 (1921).
255. G. O. Solley, G. J. Gleich, R. E. Jordon, and A. L. Schroeter, Late phase of immediate wheal and flare skin reaction. Its dependence on IgE antibodies, *J. Clin. Invest.*, *58*: 408 (1976).
256. J. Dolovich, F. E. Hargreave, R. Chalmers, K. J. Shier, J. Gauldie, and J. Bienenstock, Late cutaneous allergic responses in isolated IgE-dependent reactions, *J. Allergy Clin. Immunol.*, *52*: 38 (1973).
257. P. Atkins, G. R. Green, and B. Zweiman, Histologic studies of human skin test exposure to ragweed, compound 48/80, and histamine, *J. Allergy Clin. Immunol.*, *51*: 263 (1973).
258. L. B. Schwartz, P. C. Atkins, T. R. Bradford, P. Fleekop, M. Shalit, and B. Zweiman, Release of tryptase together with histamine during immediate cutaneous response to allergen, *J. Allergy Clin. Immunol.*, *80*: 857 (1987).
259. P. C. Atkins, L. B. Schwartz, N. F. Adkinson, C. von Allmen, M. Valenzano, and B. Zweiman, In vivo antigen-induced cutaneous mediator release: Simultaneous comparisons of histamine, tryptase, and prostaglandin D_2 release and the effect of oral corticosteroid administration, *J. Allergy Clin. Immunol.*, *86*: 360 (1990).
260. E. N. Charlesworth, A. F. Hood, N. A. Soter, A. Kagey Sobotka, P. S. Norman, and L. M. Lichtenstein, Cutaneous late-phase response to allergen. Mediator release and inflammatory cell infiltration, *J. Clin. Invest.*, *83*: 1519 (1989).
261. S. Ting, B. Zweiman, R. M. Lavker, and E. H. Dunsky, In vivo release of eosinophil chemoattractant activity in human allergic skin reactions, *J. Immunol.*, *127*: 557 (1981).
262. M. Shalit, F. H. Valone, P. C. Atkins, W. D. Ratnoff, E. J. Goetzl, and B.

Zweiman, Late appearance of phsopholipid platelet-activating factor and leukotriene B_4 in human skin after repeated antigen challenge, *J. Allergy Clin. Immunol.*, *83*: 691 (1989).
263. S. F. Talbot, P. C. Atkins, E. J. Goetzl, and B. Zweiman, Accumulation of leukotriene C_4 and histamine in human allergic skin reactions, *J. Clin. Invest.*, *76*: 650 (1985).
264. M. S. Kaminer, R. M. Lavker, L. J. Walsh, D. Whitaker, B. Zweiman, and G. F. Murphy, Extracellular localization of human connective tissue mast cell granule contents, *J. Invest. Dermatol.*, *96*: 857 (1991).
265. R. I. Slott, and B. Zweiman, Histologic studies of human skin test responses to ragweed and compound 48/80. II. Effects of corticosteroid therapy, *J. Allergy Clin. Immunol.*, *55*: 232 (1975).
266. M. Shalit, C. von Allmen, P. C. Atkins, and B. Zweiman, Increased expression of CR3 (C3bi receptor) on neutrophils in human inflammatory skin reactions, *J. Clin. Immunol.*, *7*: 456 (1987).
267. M. Shalit, D. E. Campbell, C. von Allmen, P. C. Atkins, S. D. Douglas, and B. Zweiman, Neutrophil activation in human inflammatory skin reactions, *J. Allergy Clin. Immunol.*, *80*: 87 (1987).
268. P. D. Fleekop, P. C. Atkins, C. von Allmen, M. C. Valenzano, M. Chalit, and B. Zweiman, Cellular inflammatory responses in human allergic skin reactions, *J. Allergy Clin. Immunol.*, *80*: 140 (1987).
269. K. M. Leiferman, T. Fujisawa, B. H. Gray, and G. J. Gleich, Extracellular deposition of eosinophil and neutrophil granule proteins in the IgE-mediated cutaneous late phase reaction. *Lab. Invest.*, *62*: 579 (1990).
270. P. M. Bedard, B. Zweiman, and P. C. Atkins, Quantitation by myeloperoxidase assay of neutrophil accumulation at the site of in vivo allergic reactions, *J. Clin. Immunol.*, *3*: 84 (1983).
271. B. Zweiman, U. Kucich, M. Shalit, C. von Allmen, A. Moskovitz, G. Weinbaum, and P. C. Atkins, Release of lactoferrin and elastase in human allergic skin reactions, *J. Immunol.*, *144*: 3953 (1990).
272. K. Sato, B. Zweiman, A. R. Moskowitz, C. von Allman, A. Lane, and P. C. Atkins, Effects of skin-chamber fluids from human allergic reactions on neutrophil activation, *J. Allergy Clin. Immunol.*, *90*: 21 (1992).
273. P.-C. Tai, C. J. F. Spry, C. Peterson, P. Venge, and I. Olsson, Monoclonal antibodies distinguish between storage and secreted forms of eosinophil cationic protein, *Nature*, *309*: 182 (1984).
274. B. Zweiman, P. C. Atkins, C. von Allmen, and G. J. Gleich, Release of eosinophil granule proteins during IgE-mediated allergic skin reactions, *J. Allergy Clin. Immunol.*, *87*: 984 (1991).
275. K. M. Leiferman, S. J. Ackerman, H. A. Sampson, H. S. Haugen, P. Y. Venencie, and G. J. Gleich, Dermal deposition of eosinophil-granule major basic protein in atopic dermatitis. Comparison with onchocerciasis, *N. Engl. J. Med.*, *313*: 282 (1985).
276. M. Gaga, A. J. Frew, V. A. Varney, and A. B. Kay, Eosinophil activation and T lymphocyte infiltration in allergen-induced late phase skin reactions and classical delayed-type hypersensitivity, *J. Immunol.*, *147*: 816 (1991).

277. I. Iwamoto, S. Tomoe, H. Tomioka, K. Takatsu, and S. Yoshida, Role of CD4$^+$ T lymphocytes and interleukin-5 in antigen-induced eosinophil recruitment into the site of cutaneous late-phase reaction in mice, *J. Leukocyte Biol.*, *52*: 572 (1992).

278. D. Y. M. Leung, J. S. Pober, and R. S. Cotran, Expression of endothelial–leukocyte adhesion molecule-1 in elicited late phase allergic reactions, *J. Clin. Invest.*, *87*: 1805 (1991).

279. J. E. Dvoracek, J. W. Yunginger, E. B. Kern, R. E. Hyatt, and G. J. Gleich, Induction of nasal late-phase reactions by insufflation of ragweed-pollen extract, *J. Allergy Clin. Immunol.*, *73*: 363 (1984).

280. Z. Pelikan, Late nasal response (LNR)—its clinical characteristics, features, and possible mechanisms(s), *Late Phase Allergic Reactions* (W. Dorsch, ed.), CRC Press, Boca Raton, p.111 (1990).

281. R. Bascom, U. Pipkorn, L. M. Lichtenstein, and R. M. Naclerio, The influx of inflammatory cells into nasal washings during the late response to antigen challenge. Effect of systemic steroid pretreatment, *Am. Rev. Respir. Dis.*, *138*: 406 (1988).

282. R. Bascom, U. Pipkorn, D. Proud, S. Dunnette, G. J. Gleich, L. M. Lichtenstein, and R. M. Naclerio, Major basic protein and eosinophil-derived neurotoxin concentrations in nasal-lavage fluid after antigen challenge: Effect of systemic corticosteroids and relationship to eosinophil influx, *J. Allergy Clin. Immunol.*, *84*: 338 (1989).

283. D. Proud, G. S. Bailey, R. M. Naclerio, C. J. Reynolds, A. A. Cruz, P. A. Eggleston, L. M. Lichtenstein, and A. G. Togias, Tryptase and histamine as markers to evaluate mast cell activation during the responses to nasal challenge with allergen, cold, dry air, and hyperosmolar solutions, *J. Allergy Clin. Immunol.*, *89*: 1098 (1992).

284. R. Bascom, M. Wachs, R. M. Naclerio, U. Pipkorn, S. J. Galli, and L. M. Lichtenstein, Basophil influx occurs after nasal antigen challenge: effects of topical corticosteroid pretreatment, *J. Allergy Clin. Immunol.*, *81*: 580 (1988).

285. O. Iliopoulos, F. M. Baroody, R. M. Naclerio, B. S. Bochner, A. Kagey-Sobotka, and L. M. Lichtenstein, Histamine-containing cells obtained from the nose hours after antigen challenge have functional and phenotypic characteristics of basophils, *J. Immunol.*, *148*: 2223 (1992).

286. R. M. Naclerio, D. Proud, A. G. Togias, N. F. J. Adkinson, D. A. Meyers, A. Kagey-Sobotka, M. Plaut, P. S. Norman, and L. M. Lichtenstein, Inflammatory mediators in late antigen-induced rhinitis, *N. Engl. J. Med.*, *313*: 65 (1985).

287. V. A. Varney, M. R. Jacobson, R. M. Sudderick, D. S. Robinson, A. M. Irani, L. B. Schwartz, I. S. Mackay, A. B. Kay, and S. R. Durham, Immunohistology of the nasal mucosa following allergen-induced rhinitis. Identification of activated T lymphocytes, eosinophils, and neutrophils, *Am. Rev. Respir. Dis.*, *146*: 170 (1992).

288. S. R. Durham, S. Ying, V. A. Varney, M. R. Jacobson, R. M. Sudderick, I. S. Mackay, A. B. Kay, and Q. A. Hamid, Cytokine messenger RNA expression for IL-3, IL-4, IL-5, and granulocyte–macrophage colony-stimulating factor in the nasal mucosa after local allergen provocation—relationship to tissue eosinophilia, *J. Immunol.*, *148*: 2390 (1992).

289. W. J. Metzger, The use of bronchoalveolar lavage (BAL) to study late-phase allergic asthma (LAR), *Late Phase Allergic Reactions* (W. Dorsch, ed.), CRC Press, Boca Raton, p.197 (1990).
290. W. J. Metzger, H. B. Richerson, K. Worden, M. Monick, and G. W. Hunninghake, Bronchoalveolar lavage of allergic asthmatic patients following allergen bronchoprovocation, *Chest*, *89*: 477 (1986).
291. J. G. R. De Monchy, H. F. Kauffman, P. Venge, G. H. Koëter, H. M. Jansen, H. J. Sluiter, and K. De Vries, Bronchoalveolar eosinophilia during allergen-induced late asthmatic reactions, *Am. Rev. Respir. Dis.*, *131*: 373 (1985).
292. G. A. Rossi, E. Crimi, S. Lantero, P. Gianiorio, S. Oddera, P. Crimi, and V. Brusasco, Late-phase asthmatic reaction to inhaled allergen is associated with early recruitment of eosinophils in the airways, *Am. Rev. Respir. Dis.*, *144*: 379 (1991).
293. R. Aalbers, H. F. Kauffman, B. Vrugt, M. Smith, G. H. Koëter, W. Timens, and J. G. R. De Monchy, Bronchial lavage and bronchoalveolar lavage in allergen-induced single early and dual asthmatic responders, *Am. Rev. Respir. Dis.*, *147*: 76 (1993).
294. I. Pin, A. P. Freitag, P. M. O'Byrne, A. Girgis-Gabardo, R. M. Watson, J. Dolovich, J. A. Denburg, and F. E. Hargreave, Changes in the cellular profile of induced sputum after allergen-induced asthmatic responses, *Am. Rev. Respir. Dis.*, *145*: 1265 (1992).
295. R. Dupuis, D. S. Collins, Y. Y. Koh, M. Pollice, K. H. Albertine, J. E. Fish, and S. P. Peters, Effect of antigen dose on the recruitment of inflammatory cells to the lung by segmental antigen challenge, *J. Allergy Clin. Immunol.*, *89*: 850 (1992).
296. J. B. Sedgwick, W. J. Calhoun, G. J. Gleich, H. Kita, J. S. Abrams, L. B. Schwartz, B. Volovitz, M. Ben-Yaakov, and W. W. Busse, Immediate and late airway response of allergic rhinitis patients to segmental antigen challenge. Characterization of eosinophil and mast cell mediators, *Am. Rev. Respir. Dis.*, *144*: 1274 (1991).
297. M. Kato, M. C. Liu, B. A. Stealey, B. Friedman, L. M. Lichtenstein, S. Permutt, and R. P. Schleimer, Production of granulocyte/macrophage colony-stimulating factor in human airways during allergen-induced late-phase reactions in atopic subjects, *Lymphokine Cytokine Res.*, *11*: 287 (1992).
298. J. W. Wilson, R. Djukanovic, P. H. Howarth, and S. T. Holgate, Lymphocyte activation in bronchoalveolar lavage and peripheral blood in atopic asthma, *Am. Rev. Respir. Dis.*, *145*: 958 (1992).
299. M. C. Gonzalez, P. Diaz, F. R. Galleguillos, P. Ancic, O. Cromwell, and A. B. Kay, Allergen-induced recruitment of bronchoalveolar helper (OKT4) and suppressor (OKT8) T-cells in asthma. Relative increases in OKT8 cells in single early responders compared with those in late-phase responders, *Am. Rev. Respir. Dis.*, *136*: 600 (1987).
300. A. A. Gerblich, A. E. Campbell, and M. R. Schuyler, Changes in T-lymphocyte subpopulations after antigenic bronchial provocation in asthmatics, *N. Engl. J. Med.*, *310*: 1349 (1984).
301. A. A. Gerblich, H. Salik, and M. R. Schuyler, Dynamic T-cell changes in peripheral blood and bronchoalveolar lavage after antigen bronchoprocation in asthmatics, *Am. Rev. Respir. Dis.*, *143*: 533 (1991).

302. C. Gratziou, M. Carroll, A. Walls, P. H. Howarth, and S. T. Holgate, Early changes in T lymphocytes recovered by bronchoalveolar lavage after local allergen challenge of asthmatic airways, *Am. Rev. Respir. Dis.*, *145*: 1259 (1992).
303. A. M. Bentley, Q. Meng, D. S. Robinson, Q. Hamid, A. B. Kay, and S. R. Durham, Increases in activated T lymphocytes, eosinophils, and cytokine mRNA expression for interleukin-5 and granulocyte/macrophage colony-stimulating factor in bronchial biopsies after allergen challenge in atopic asthmatics, *Am. J. Respir. Cell. Mol. Biol.*, *8*: 35 (1993).
304. D. S. Robinson, Q. Hamid, S. Ying, A. Tsicopoulos, J. Barkans, A. M. Bentley, C. Corrigan, S. R. Durham, and A. B. Kay, Predominant TH-2 bronchoalveolar T-lymphocyte population in atopic asthma, *N. Engl. J. Med.*, *326*: 298 (1992).
305. R. Djukanovic, W. R. Roche, J. W. Wilson, C. R. W. Beasley, O. P. Twentyman, P. H. Howarth, and S. T. Holgate, Mucosal inflammation in asthma, *Am. Rev. Respir. Dis.*, *142*: 434 (1990).
306. R. Djukanovic, C. K. Lai, J. W. Wilson, K. M. Britten, S. J. Wilson, W. R. Roche, P. H. Howarth, and S. T. Holgate, Bronchial mucosal manifestations of atopy: A comparison of markers of inflammation between atopic asthmatics, atopic nonasthmatics and healthy controls, *Eur. Respir. J.*, *5*: 538 (1992).
307. R. Djukanovic, J. W. Wilson, K. M. Britten, S. J. Wilson, A. F. Walls, W. R. Roche, P. H. Howarth, and S. T. Holgate, Quantitation of mast cells and eosinophils in the bronchial mucosa of symptomatic atopic asthmatics and healthy control subjects using immunohistochemistry, *Am. Rev. Respir. Dis.*, *142*: 863 (1990).
308. E. Morita, J. M. Schroder, and E. Christophers, Identification of a novel and highly potent eosinophil chemotactic lipid in human eosinophils treated with arachidonic acid, *J. Immunol.*, *144*: 1893 (1990).
309. U. Schwenk, E. Morita, R. Engel, and J. M. Schroder, Identification of 5-oxo-15-hydroxy-6,8,11,13-eicosatetraenoic acid as a novel and potent human eosinophil chemotactic eicosanoid, *J. Biol. Chem.*, *267*: 12482 (1992).
310. J. M. Wang, A. Rambaldi, A. Biondi, Z. G. Chen, C. J. Sanderson, and A. Mantovani, Recombinant human interleukin 5 is a selective eosinophil chemoattractant, *Eur. J. Immunol.*, *19*: 701 (1989).
311. Q. Xu, M. C. Leiva, S. A. Fischkoff, R. E. Handschumacher, and C. R. Lyttle, Leukocyte chemotactic activity of cyclophilin, *J. Biol. Chem.*, *267*: 11968 (1992).

14

Molecular Characterization and Immune Recognition of Allergenic Peptides

Robyn E. O'Hehir, Gerard F. Hoyne, and Jonathan R. Lamb
Imperial College of Science, Technology, and Medicine, St. Mary's Hospital Medical School, London, England

I. INTRODUCTION

Over the last decade substantial progress has been made in cloning and sequencing genes that encode proteins from a variety of environmental allergens, and from information on nucleotide sequences it has been possible to predict the primary structure of different allergens. The practical application of this information is broad. It provides basic data on which the structure and biological activity of the different allergens may be defined. Perhaps more important, from an immunological standpoint, it allows the antigen specificity of the allergen-reactive immune repertoires of atopic and nonatopic individuals to be analyzed at the molecular level. This knowledge is required for development of experimental models of allergen-mediated immunomodulation as a forerunner to the design of therapeutics for clinical application.

The list of allergens being investigated as primary targets of immune recognition in specific atopic diseases is increasing and includes those from pollens, insect venom, house dust mites, and cat salivary gland proteins. It is beyond the scope of this chapter to address these allergens in detail and we will focus the discussion on aeroallergens derived from *Dermatophagoides* spp. (house dust mite; HDM). Apart from the clinical importance of HDM allergens, which induce pathogenic immune responses in approximately 10% of the population with symptoms ranging from extrinsic asthma to atopic dermatitis, they illustrate the complexity of inflammatory immune responses to allergens.

As for many of the allergens, the major advance was the cloning of the cDNAs for the major allergens of HDM (group I and II), which account for >95% of specific IgE in HDM-atopic individuals. This allowed the immunological analysis of these molecules to begin, and now the combination of recombinant DNA technology and conventional biochemical analysis will allow detailed characterization of the minor HDM allergens (groups III–VII).

Dissecting the response of human cells cultured in vitro has revealed extensive heterogeneity in the immune recognition of HDM-derived allergens that is paralleled by in vivo murine studies. Responsiveness to allergens is not a unique feature of the immune repertoire of atopic individuals, and it is now well documented that T cells from nonatopic subjects will proliferate to allergens and that their B cells produce IgG. However, the qualitative aspects of the immune response differ in the two groups of individuals. T cells expressing the "IL-4 dominant" (Th2) functional phenotype and driving Ig isotype switch to IgE are induced in atopic individuals, whereas T cells in nonatopic individuals that recognize HDM are characteristic of the "IFN-γ dominant" (Th1) subset of $CD4^+$ T cells (Fig. 1). It has been suggested that a component of the T-cell repertoire in atopic individuals is comprised of the Th1 (IFN-γ dominant) phenotype, and the situation is further complicated by the observation that both Th1- and Th2-type cells secrete IL-13 that appears to have similar effects to IL-4 on Ig isotype switch.

The differences between functionally distinct subpopulations of $CD4^+$ T cells have been investigated in detail. The kinetics of secretion of different cytokines and the potential of the majority of $CD4^+$ T cells to produce a range of these mediators influence the quality of the response. The mechanisms that select or preferentially expand Th2 cells as opposed to Th1 cells in atopic but not nonatopic individuals following exposure to allergens remain obscure. This question is of paramount importance to understanding the immunological basis of allergic diseases.

II. CHARACTERIZATION OF HDM ALLERGENS

The development of monoclonal antibody and molecular cloning techniques has allowed the characterization of HDM allergens. To date, seven allergenic proteins have been defined for *Dermatophagoides* spp. (1), which bind most of the IgE in sera from HDM-allergic patients. However, despite the large number of allergens, most patients tend to become sensitized to only a few mite proteins. Additionally, some of the HDM allergens may exist in more than one molecular weight form (1).

The predominant binding of IgE in sera from HDM-allergic patients is to the group I and II allergens (2). Cloning and sequencing of cDNAs encoding these proteins (3–6) have established that the group I allergens of *Dermatophagoides*

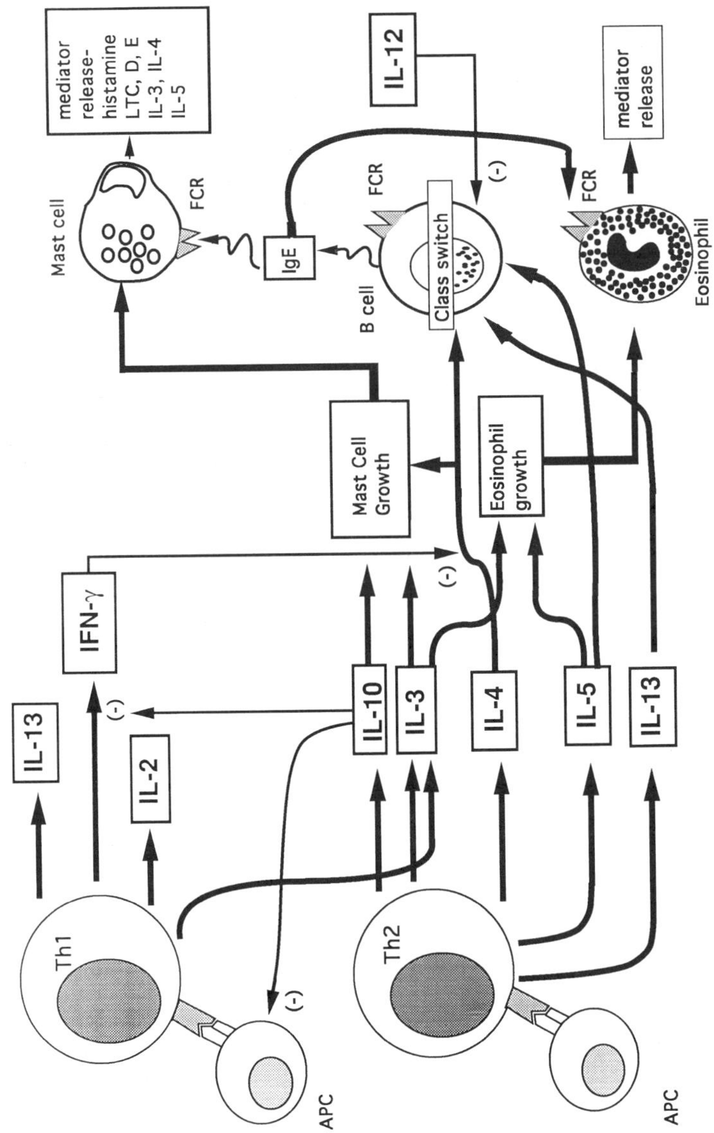

Figure 1 Cartoon model of the cellular and cytokine components of the allergic immune response.

pteronyssinus (*Der p* I) and *Dermatophagoides farinae* (*Der f* I) are 222- and 223-amino-acid polypeptides, respectively. They have a molecular weight of 25 kDa, share sequence homology with cysteine protease (3,4), and have 81% identity in their primary structure (4). The group II allergens, *Der p* II and *Der f* II, encode a 129-amino-acid-residue protein with a predicted M_r of 14 kDa (5,6). As with the group I allergens, there is a high degree of identity (88%) in the primary amino acid sequence structure (6). The group II allergens have enzymatic biological activities, in this case similar to that of lysozyme (7). The group III allergens exist as multiple-molecular-weight bands on SDS-PAGE gels of 30, 28, 21, and 17 kDa. Trypsin-like activity has been assigned by primary sequence derived from N-terminal amino acid analysis of purified *Der p* III (8), with in vivo autocatalysis as a likely explanation for the multiple-band forms.

The *Der p* IV allergen has been identified from spent mite medium as a 56- to 60-kDa protein (9) and is structurally homologous to α-amylase. Approximately 40% of HDM-sensitized individuals have serum IgE antbodies reactive with *Der p* IV. Molecular analysis of the cDNA sequence encoding *Der p* V reveals a 113-residue protein (10,11), but functional activity has yet to be assigned. Determination of the amino-terminal sequence data together with enzyme substrate specificity analysis has shown the 25-kDa group VI allergen to be equivalent to chymotrypsin (12). A cDNA clone encoding a 22-kDa polypeptide of 198 amino acid residues has recently been isolated (Shen et al., unpublished observations), which also binds IgE in 40% of sera from HDM-allergic subjects. Western blotting studies have shown that antibodies to this antigen react with components of 28, 26, and 22 kDa from mite extracts. This allergen has been assigned the name *Der p* VII.

III. IDENTIFICATION OF T-CELL EPITOPES

Investigations of T-cell proliferative responses to different species of *Dermatophagoides* spp., affinity-purified proteins, and fractionated HDM extracts insolubilized on nitrocellulose have all contributed to knowledge of the relative immunological importance of selected HDM proteins and the cross-reactivity of the T-cell repertoire reactive with HDM (13).

It has been established that the peripheral repertoires of atopic and nonatopic individuals include T cells that are reactive with HDM allergens (14–16); however, the quality of the T-cell help differs in the two groups. The most convincing evidence that the T-cell repertoires of both groups of subjects are comparable results from analysis of the fine antigen specificity of T-cell recognition using peptides (16). Variations observed in T-cell responses observed to individual epitopes may reflect primary differences in the HLA haplotype and antigen exposure, rather than specific regions of the molecules being associated with the induction of pathogens in atopic subjects.

Initial results of proliferative responses to affinity-purified group I and II allergens were extended by the use of electrophoretically separated HDM extracts insolubilized on nitrocellulose filters (17,18) and demonstrated the presence of T cells that recognize proteins of the group I–IV allergens (1,17). Subsequent clonal analysis has identified both species-specific and cross-reactive T-cell populations (19,20), whereas serological investigations suggested that the major component of the IgE response was cross-reactive for *D. pteronyssinus* and *D. farinae* (2).

Nucleotide sequences of cloned cDNAs have allowed the determination of the amino acid sequences of the major allergenic proteins of HDM. Therefore, as sequence data for each of the proteins become available, it is possible to construct T-cell epitope maps (Fig. 2). Analysis of the antigen specificity of T-cell clones reactive with *Der p* I, with truncated recombinant proteins and synthetic peptides, has identified several sites of T-cell recognition within *Der p* I (13,21,22). Yssel and colleagues mapped three distinct regions at residues 45–67, 94–104, and 117–143 by investigating T-cell clones from two atopic subjects (22). In a separate study, the polyclonal T-cell response of an individual with perennial rhinitis was limited to a single region of *Der p* I (residues 101–143) and was cross-reactive between *D. pteronyssinus* and *D. farinae*. However, to map the fine antigen specificity of the *Der p* I repertoire, T-cell clones were isolated, and unlike, the polyclonal response, they were HDM-species-specific. Comparison of the patterns of responsiveness to overlapping peptides demonstrated that a cluster of three overlapping T-cell epitopes, amino acids 107–119, 110–119, and 110–131 (Fig. 2) were recognized (23). Mapping of T-cell epitopes from *Der p* II was also carried out using both polyclonal and monoclonal T-cell populations. Peripheral blood T-cell analysis highlighted that T-cell epitopes are present in all regions of the protein, with the majority of HDM atopic and nonatopic individuals recognizing *Der p* II. However, T-cell epitopes located in regions 11–35 and 61–104 were immunodominant at the population level. Characterization of the specificity of a panel of T-cell clones revealed the presence of four distinct epitopes, 16–31, 22–40, 82–100, and 111–129 (20,24).

IV. IDENTIFICATION OF B-CELL EPITOPES OF HDM ALLERGENS

Atopic patients are susceptible to sensitization to a range of environmental allergens since they constitutively produce high levels of circulating IgE in the serum (25). It is widely appreciated that the development of IgE antibodies to HDM allergens is a major risk associated with the development of asthma (26,27). There has been considerable interest in recent years in the identification of B- and T-cell epitopes on the HDM allergens primarily for their potential use in immunotherapy, but clearly this information will also lead to a better

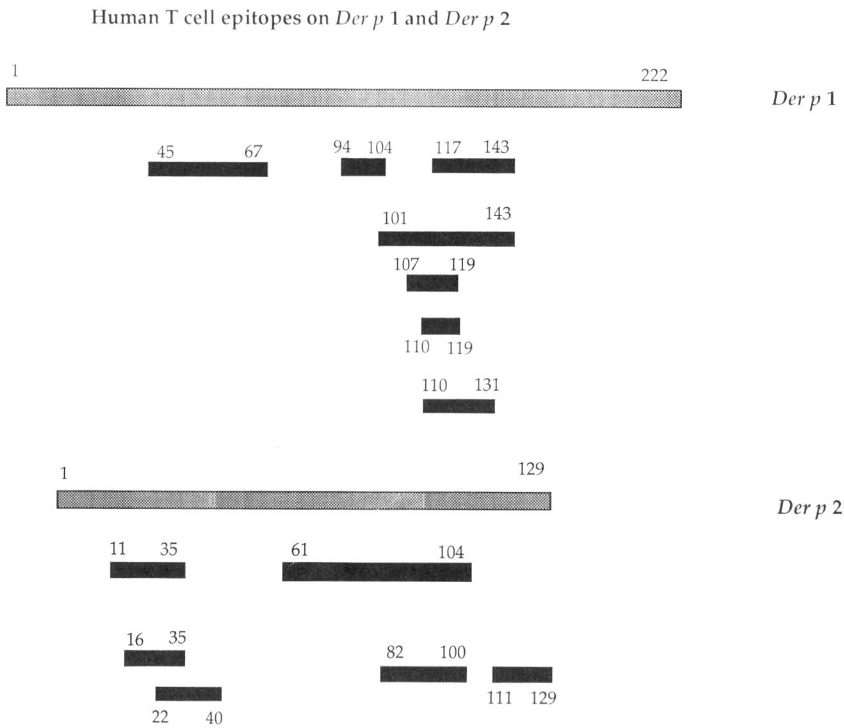

Figure 2 Human T-cell epitopes recognized on *Der p* I and *Der p* II. Epitopes were identified by lymphocyte proliferation in vitro using overlapping synthetic peptides derived from the primary sequence of each allergen.

understanding of immune recognition of this important group of allergens. B-cell epitopes, which bind specific antibodies, are conformationally dependent and usually located on the exposed surfaces of allergens. T-cell epitopes, however, are recognized in association with MHC gene products after processing of the proteins to peptidic form. The antigenic determinants recognized by antibodies are usually nonlinear and are determined by the tertiary structure of a protein antigen. However, recent studies have shown that some B-cell determinants can also be defined by short linear peptides (28–31).

IgE antibodies in the serum of HDM-reactive patients are usually directed against a number of different mite proteins when analyzed by Western blotting using whole HDM extract (32,33). However, as mentioned above, it is likely that only about seven proteins will account for most of the IgE binding in the sera of HDM-allergic patients (1). The first detailed characterization of antibody binding to the HDM allergens *Der p* I and *Der f* I was performed by Platts-Mills and colleagues (34–36). They produced a large panel of murine monoclonal antibodies (MAbs) against either *Der p* I or *Der f* I, and the binding specificity of individual MAbs was determined using cross-inhibition studies with radiolabeled antigen. The results of these studies revealed, first, that the majority of the hybridomas produced against one allergen (e.g., *Der p* I) showed minimal cross-reactivity to the other homologous antigen, *Der f* I. This result conflicted with the observation that polyclonal rabbit anti-*Der p* I serum contained both species-specific and cross-reactive antibodies with *Der f* I. Second, it was shown that the MAbs were restricted to a limited number of epitopes on the antigen and also that they were unable to inhibit binding of human allergic sera to radiolabeled *Der p* I (34,35). It was concluded from these studies that the murine B-cell epitopes were either distinct or lay very close to the human IgE antibody epitopes. In an independent study, the binding specificity of a panel of 19 anti-*Der p* I murine MAbs was compared to that of human allergic sera (37). It was found that most MAbs were not cross-reactive with the *Der f* I allergen, but five MAbs did show cross-reactivity with determinant(s) on the homologous antigen from the related species *D. microceras* (37). In contrast to the findings of Chapman and his colleagues, however, Lind et al. (37) found that one murine MAb could inhibit the binding of radiolabeled *Der p* I to immobilized human IgE from allergic serum by up to 65% and up to 24% of labeled *Der p* I binding to immobilized polyclonal rabbit antiserum. Hoyne (38) found that polyclonal mouse antisera raised against *Der p* I contained *Der f* I- cross-reactive antibodies, which is consistent with previous findings for both polyclonal rabbit and human allergic sera (29–35).

From the experimental design of the studies described above, it was not possible to identify the primary amino acid sequence on the allergen that was bound by either murine or human antibodies. With recombinant proteins of *Der p* I and *Der f* I, antibody binding to individual HDM allergens has been subjected to detailed analysis in vitro. One approach was to make a random fragment library to produce overlapping fragments of the *Der p* I protein by either sonication or restriction enzyme digestion of *Der p* I cDNA (29–31). The fragments were cloned into the expression vector pGEX, and recombinant peptides produced as fusion proteins were isolated by affinity chromatography. The fusion peptides produced in this way were then assessed for their ability to bind antibodies using radioimmune dot blot assays. Polyclonal rabbit anti-*Der p* I serum was found to bind a range of fragments of various sizes (29), and owing to the overlapping nature of the fragments used, it was possible to identify the linear epitopes recognized by antibodies. Rabbit anti-*Der p* I serum was found to have specificities for five different regions on the allergen. The dominant response was to 60–72, intermediate responses were made to 82–99 and 112–140, and weak responses to 34–47 and 166–194 (29). Rabbit polyclonal anti-*Der f* I serum displayed cross-reactivity to *Der p* I, and antibodies within the serum displayed a significant degree of overlap with the epitypic regions described above. The dominant region recognized on *Der p* I by anti-*Der f* I serum was 60–72, and intermediate binding was also reported to 34–47 and 166–194 (29).

Human sera from HDM-allergic patients contain antibodies of both IgG and IgE isotypes specific for epitopes on allergens to which the patients have become sensitized (31). In a recent study Greene et al. found that both *Der p* I-specific IgG and IgE antibodies bound a similar range of recombinant peptides, and thus there appeared to be an overlap in the IgE and IgG repertoires within individual HDM-allergic patients. Furthermore, these studies revealed that approximately five different regions on *Der p* I bound antibodies (1–56, 53–99, 98–140, 166–194, and 188–222) that spread along the length of the entire molecule (31). When results were compared from individual sera, however, no single dominant antibody-binding region was recognized by patients, but rather each patient's sera showed individual specificities for peptide binding.

The cDNA encoding *Der p* II was first isolated by Chua et al. (39) and a similar approach for studying antibody binding to this allergen was employed. The available data on B-cell epitopes for *Der p* II showed that it is extremely resistant to denaturation (e.g., boiling) and that IgE binding to *Der p* II was markedly increased after incubation of the native protein with detergents (40). Polyclonal human HDM-reactive allergic serum and polyclonal rabbit anti-*Der p* II serum contain antibodies that cross-react with the homologous allergen *Der f* II (39–41). Full-length recombinant *Der p* II expressed in *Escherichia coli* binds human IgE very well, but the determinants to which these antibodies bind are dependent on the conformational structure of the intact protein (41). Human IgE antibodies from allergic patients were found to bind very weakly or not at all to large recombinant fragments of *Der p* II in radioimmune dot blot assays (41). Thus the inability of the large peptides to form IgE-binding structures suggests that the antigenic determinants of *Der p* II are highly conformational and restricted. Polyclonal mouse antisera raised against *Der p* II can bind to native or recombinant *Der p* II and to various recombinant fragments in radioimmune dot blot assays (38). Binding of murine antisera was to all regions of the *Der p* II molecule and no single region was immunodominant in any of the haplotypes tested (38). The weak binding of human IgE to recombinant *Der p* II peptides has severely restricted further study of antibody binding to this molecule.

The information derived from the analysis of antibody binding to genetically engineered recombinant proteins of HDM group I and II allergens has begun to shed light on how the repertoires of allergen-specific IgG- and IgE-secreting B cells differ within individual allergic patients. The practical value of identifying the regions that contain antibody epitopes may be in the generation of allergoids of HDM allergens by site-directed mutagenesis. Such reagents could, presumably, be used in immunotherapy since they would maintain immunogenicity but would be nonallergenic and, thus, could be used to modulate allergic reactivity in HDM-reactive patients.

V. GENETIC INFLUENCES ON IMMUNE RESPONSIVENESS TO HDM ALLERGENS

Population studies investigating patients with IgE to specific allergens have demonstrated linkages between the expression of certain HLA class II molecules and immune responsiveness to a number of defined allergenic proteins. Studies on allergic immune reactivity to ragweed have illustrated a clear association between HLA-DR2Dw2 (DRB1*1501) and the presence of IgE antibodies specific for *Amb a* V (42,43), one of the smaller allergens. To date, population studies examining IgE responses to unfractionated HDM extracts have not identified an HLA association, perhaps owing to the structural complexity and multiplicity of T- and B-cell determinants in allergens from HDM. Recent epidemiological studies investigating IgE antibodies specific for distinct proteins of HDM initially suggested that HLA-DP class II molecules contribute to the IgE response of HDM atopic individuals, but now the association is less clear (44). However, studies of the MHC class II restriction specificity of HDM-reactive T cells isolated from atopic individuals demonstrate the marked heterogeneity of HLA class II molecules capable of presenting HDM-derived peptides (19,20,23,24,25). Using serological inhibition studies with MAbs directed at MHC class II framework antigens, panels of allogeneic-presenting cells, and murine fibroblasts transfected with HLA class II genes as APC, it was demonstrated by clonal analysis that a major component of the *D. farinae*–specific T-cell repertoire of a HDM atopic individual was restricted by products of the DRB3 genes (DRB3*0202) (19,45). Moreover, cloned T cells of the "IL-4 dominant" phenotype (Th2-like), were identified in T cells from the peripheral blood of this individual and also shown to be restricted by allelic forms of DRB1 (HLA-DR5 and -DR8) and DRB5 (HLA-DR2) genes (45). These studies were conducted before primary amino acid sequence was available for the group I and II allergens of *D. farinae*; therefore, the antigen specificity of these T-cell clones was not mapped at the epitope level. However, with the availability of the amino acid sequences of both the group I and II allergens of HDM, analysis of the antigen and restriction specificity of T-cell responses to these proteins is possible using synthetic peptides. Such analysis of a panel of *Der p* I–induced T-cell clones demonstrated that five regions of the protein, residues 45–67, 94–104, 107–119, 110–131, and 117–143, contain T-cell epitopes restricted by several different HLA class II molecules (21,23). T-cell clones specific for residues 45–67 and 117–143 were restricted by HLA-DR7 gene products, while those T cells reactive with amino acids 94–104 recognized antigen in association with HLA-DR2, -DRw11 (DR5), and -DR8 (23). The region 101–131 of *Der p* I contains three overlapping epitopes. The T-cell epitopes residues 110–131 and 110–119 are restricted by DRB1*0101 and DPB1*0402, respectively. Additionally, the T-cell epitope amino acids 107–119 was HLA-DP restricted and could be successfully presented to T cells bound to HLA-DPB1*0401, -DPB1*0402, and -DPB1*0501 bearing accessory cells (23).

Heterogeneity is also present in the HLA class II specificities that function as restriction elements in recognition of *Der p* II by T cells. Analysis of a panel of *Der p* II–specific T-cell clones all isolated from the same HDM-allergic individual showed that T-cell clones reactive with the amino-terminal region of *Der p* II residues 16–31 and 22–40 were present. From serological inhibition studies and the pattern of antigen-dependent proliferation observed in the presence of HLA-typed allogeneic APC, it was shown that T-cell recognition of peptide 16–31 was restricted by HLA-DQB1*0301, whereas 22–40 could be presented by both DQB1*0301 and DRB1*1101 [DR11(5)] class II molecules (20). An additional HLA-DQ (DQB1*0301) epitope was mapped to the carboxyl-terminus amino acids 111–129, and residues 82–105 contained a T-cell epitope restricted by DRB1*1101. Other groups have also identified *Der p* II–reactive T-cell clones restricted by HLA-DQ gene products (24). Together, these studies on the MHC class II restriction specificity of HDM-reactive T-cell clones illustrate that HLA-DP, -DQ, and -DR class II molecules of different specificity may function in the presentation of HDM peptides. Moreover, individual HDM-derived peptides may bind to HLA-DR and -DP or -DR and -DQ, but as yet no HDM peptide has been identified that is promiscuous in its binding to all three subsets of HLA class II molecules. Promiscuity of binding of immunodominant epitopes to more than a single HLA class II molecule would be a distinct advantage in the design of peptide-based immunotherapy.

VI. T-CELL SUBSETS

Analysis of clonal populations of murine $CD4^+$ T lymphocytes has allowed a division into functionally distinct subsets based on their in vitro cytokine profiles (46,47). Th1 cells produce predominantly IL-2 and IFN-γ but no IL-4 and are principally involved in delayed-type hypersensitivity responses, whereas Th2 cells secrete mainly IL-4, IL5, and IL-6 with no IL-2 or IFN-γ and provide help for IgE and IgG2a antibody production. In addition, Th1 and Th2 cells negatively regulate the opposing subset through the production of IFN-γ and IL-10, respectively (48). Another population termed Th0 cells secrete most cytokines, and it has been suggested that on antigen stimulation they differentiate into one of the other two subsets. In humans, however, a clear dichotomy of subsets is not observed although T cells that favor an "IL-4 dominant" (Th2-like) or "IFN-γ dominant" (Th1-like) functional phenotype have been identified (49–51). The resulting biological activity mediated by these subsets will depend on both temporal coordination and quantitative aspects of cytokine production.

The great diversity of human T cells with respect to their potential to secrete a range of cytokines offers the opportunity to selectively manipulate the synthesis of individual cytokines and thus induce changes in their functional profile and activity. The nature of the accessory cell, the concentration and physicochemical form of antigen, and the local cytokine milieu together act to determine the overall cytokine pattern and the quality of the ensuing immune response. Moreover, the selective induction and temporal coordination of cytokines may differ in activated versus anergized T cells (52), with differential signaling resulting in post-transcriptional regulation of selected cytokines, including IL-8 and TNF-α (53).

Despite the peripheral T-cell repertoires of both atopic and nonatopic individuals including recognition of HDM allergens, there is a difference in the quality of the T-cell help provided. T-cell clones derived from atopic individuals can support in vitro HDM-specific IgE synthesis from autologous B cells that is dependent on the cytokine IL-4, whereas cloned HDM-specific T cells from nonatopic individuals fail to do so, in either the presence or absence of exogenous IL-4 (15,54). Investigation of the cytokine potential of T cells from the two clinical groups confirmed that atopic individuals possess an "IL-4 dominant" lymphokine profile of HDM-specific T cells in contrast to the nonatopic individuals, who have an "IFN-γ dominant" functional phenotype. The synthesis of HDM-specific IgG can be supported by T cells from both groups.

VII. MODULATION OF HDM IMMUNE RESPONSIVENESS

Although reports of efficacy vary, allergen-specific immunotherapy with HDM extracts is used in many countries. A lack of detailed understanding of the basic principles of immunological tolerance induction has hampered understanding of the mechanisms of effective immunotherapy. The activation of $CD4^+$ T cells necessitates the cross-linking of specific T-cell-antigen receptors, by antigenic peptide fragments occupying the combining sites on MHC class II molecules exposed on the surface of APC (55, and see Chap. 6, this volume). Moreover, disruption of these interactions allows modulation of the allergic immune response. Incomplete activation in the absence of costimulatory signals may directly inhibit the function of selected populations of disease associated T cells (56) and induce nonresponsiveness. In contrast, nonstimulatory peptides may be used to compete with binding of dominant T-cell epitopes to the MHC class II molecules (57).

A. Functional Inactivation of HDM-Reactive T Cells

The clonal distribution of the TcR allows a high degree of specificity to be attained when specific peptides or proteins are used to delete or render T-cell populations nonresponsive (anergic). For clinical application of this approach in allergen-based immunotherapy, the identification of dominant T-cell epitopes at a population level is necessary. The modification of their structures may be required to remove IgE binding, minimizing the risk for anaphylaxis, which is currently a problem with allergen immunotherapy. It has been demonstrated in vitro that the HDM reactivity of a monoclonal population may be decreased using a low-dose, escalating protocol for hyposensitizing preactivated T cells (58). The HDM allergen was presented on irradiated autologous accessory cells to activate the cloned T cells before daily exposure to escalating concentrations of the unfractionated allergen mixture. Subsequent antigen rechallenge elicited a 42% decrease in the proliferative response. The decreased antigen responsiveness could not be attributed to cytolysis as enhanced nonspecific proliferation to exogenous IL-2 could be observed, as has been observed in other reports of human antigen specific nonresponsiveness (58).

Elucidation of immunodominant T-cell epitopes of the major HDM allergens, *Der p* I and *Der p* II, allowed models of in vitro anergy to be developed using supraoptimal concentrations of peptide (21,52). Antigen-specific anergy can be induced by the peptides either alone or in the presence of antigen-presenting cells. The peptides are complexed with MHC class II molecules expressed on human T cells and presented to other T cells in the absence of engagement of costimulatory molecules such as CD28. Despite the marked heterogeneity of the MHC class II restriction specificity of human T-cell recognition of HDM proteins, namely in the context of HLA-DR, -DP, or -DQ gene products (19–21,23,24,45), T cells using each of these restriction elements may be anergized (21,52). This model was developed by Lamb and colleagues in the investigation of tolerance induction of influenza-virus immune human T cells using a peptide derived from the carboxyl-terminus of influenza hemagglutinin (56). The state of antigen-specific nonresponsiveness was accompanied by phenotypic modulation (59), with down-regulation of the TcR (CD3) and up-regulation of the IL-2R (CD25). Subsequently, we observed that the receptor for costimulatory activity, CD28, is differentially expressed in activation versus anergy (59). The surface expression of CD28 is markedly enhanced on activated T cells, and this is paralleled by increased mRNA transcripts. However, in the anergic T cell, the induction of membrane CD28 is accompanied by the rapid decay of specific mRNA transcripts (60).

Functional changes were also detected in anergized HDM-reactive T cells. Secretion of the cytokines IL-2, IL-4, and IFN-γ was equivalent to that produced in activated T cells during the induction of anergy, but on subsequent rechallenge with the native ligand, IL-4 production was attenuated while the production IFN-γ was maintained (52). Isotype switching to IgE synthesis is largely regulated by the fine balance of these two cytokines, and hence the secretion pattern observed in the tolerant population would favor negative regulation of IgE production.

Similar phenotypic and lymphokine profile modulations were observed in in situ hybridization analysis of skin biopsy sections taken from patients treated with grass pollen extract desensitization (61). Moreover, recent studies investigating polyclonal cytokine changes in patients given immunotherapy for insect venom allergy suggest similar cytokine switches (McHugh et al., unpublished observations). Together these studies support the hypothesis that anergy induction, with a switch of allergen-specific T cells from the Th2-like to the Th1-like profile of cytokine secretion, may be the mechanism by which successful allergen immunotherapy occurs. The in vitro studies demonstrate that the major peptide epitopes in the absence of accessory cells are more effective as tolerogens than intact proteins. However, in vivo studies illustrate that both proteins and peptides may be highly effective when administered optimally.

The ability of bacterial superantigens at supraimmunogenic concentrations to tolerize T-cell families expressing the appropriate TcR-Vβ gene elements (62) offers an alternative method for functional inactivation of HDM-reactive T cells. Nonresponsiveness is accompanied by modulation of surface and functional phenotype with changes similar to those observed in peptide-mediated anergy. However, serological inhibition studies suggested that, unlike peptide-induced anergy, the toxins at concentrations that saturate MHC class II may inactivate T cells through direct interaction with TcR (63). Down-regulation of the TcR on a class II negative human T-cell line transfected with a human $V\beta3^+$ TcR, when it was incubated with the appropriate superantigen, supported this suggestion (64). Studies to establish whether there is restriction of TcR gene usage in HDM-specific T-cell responses using anchor PCR and sequencing have demonstrated long-lived clonal dominance in vivo with oligoclonality of the expressed HDM-reactive T-cell repertoire for Vβ3 and Vα8 gene products (63). A panel of HDM-reactive T-cell clones that had been independently isolated from the same individual over a period of 3 years was investigated and demonstrated that T cells from the same clonal precursors had been isolated four times in one case (Vβ21.5) and twice in another case (Vβ6.7a). Subsequently, the peripheral blood of the donor of the T cells was investigated some 6 years after the initial isolation of these clones. Specific oligonucleotide probes for these particular Vβ and Jβ regions revealed that the dominant clones could still be strongly detected using PCR on mononuclear cells from the peripheral blood of the donor. These studies, however, reflect the results of one individual and may be patient-specific and must be followed by wider studies on other patients. Nevertheless, the results enhance our understanding of the TcR recognition of environmental allergens and the life span of T-cell clones in vivo during a chronic immune response.

B. Inhibition of Responsiveness by Antigen Competition

The degeneracy of specific MHC class II alleles in their ability to bind different peptides is well established. Similarly, individual peptides usually demonstrate promiscuity in their binding to MHC class II proteins. But, despite this apparent permissivity (65–67), there is a wide range of different affinities of binding, which introduces an alternative approach for modulation of the allergic immune response due to the ability of nonstimulatory, non-allergen-derived peptides to compete for binding to MHC class II proteins. A peptide analogue of influenza virus hemagglutinin, residues 307–319, with a serine substituted for tyrosine at position 309, binds with high affinity to many of the MHC class II molecules capable of restricting T-cell recognition of HDM peptides (68). In coculture proliferation assays, the analogue, which is itself nonimmunogenic in individuals primed with hemagglutinin influenza virus, was able to inhibit both HDM-specific polyclonal and monoclonal T-cell populations from HDM-allergic subjects (68).

C. Inhibition of Responsiveness by TcR-Derived Peptides

The variable regions of both the alpha (Vα) and beta (Vβ) chains of the TcR have four hypervariable loops, known as complementarity-determining regions (CDR1, 2, 3, and 4). CDR3 appears the most diverse of these regions and is believed to be involved in antigen contact. The CDR1 and CDR2 regions are implicated in MHC binding (69), and CDR4 appears to participate in superantigen binding (70). Recent rodent experiments suggest that these regions of the TcR promote antigen and MHC binding and, in particular with reference to the Vβ-CDR2 region, serve as an antigenic determinant or recognition site for regulatory T cells (71–73). "Disease associated" T cells that preferentially use selected TcR gene products have been identified in animal models of experimental allergic encephalomyelitis (EAE). Encephalitogenic T cells in both rat and mouse EAE express predominantly TcVβ 8.2 genes, and immunization with a synthetic peptide corresponding to the CDR2 of TcR-Vβ8.2 prevents clinical signs of disease (71–76). Using this approach, we have tested the capacity of a CDR2 peptide derived from the human TcR-Vβ3 to modulate the polyclonal human T-cell response to HDM allergen. The TcR-Vβ3 was selected for these studies in light of our recent experiments demonstrating a bias for Vβ3 usage in a panel of clones from a HDM-allergic individual, with 50% of the clones studied by anchor PCR and sequence analysis expressing TcR-Vβ3 gene elements (67). The

Vβ3-CDR2 peptide could specifically inhibit the polyclonal response of HDM-primed, genetically unrelated individuals, both atopic and nonatopic, to unfractionated HDM extract (77). Depletion of $CD8^+$ T cells abolished the peptide-mediated inhibition of $CD4^+$ T-cell proliferation to HDM, suggesting that induction of a $CD8^+$ regulatory T-cell subset by the CDR2 peptide may modulate HDM-specific allergic T-cell responses.

D. Murine Models of Nonresponsiveness to HDM Allergens

Murine models have demonstrated that mucosally administered antigens, either oral or inhaled, are able to induce long-lasting immunological tolerance, affecting both the B- and T-cell compartments (78,79). The pathogenesis of peripheral tolerance induced in this manner is complex, with some evidence supporting a role for involvement of "suppressor" T (Ts) cells and other evidence suggesting clonal anergy (78–80). Resolution of the mechanisms underlying the development of oral and respiratory tolerance will have a practical application in determining both the regulation of allergic sensitization and the delivery of peptides and proteins via mucosal surfaces to induce nonresponsiveness in immunotherapy.

The T cells of naïve mice treated either orally or intranasally, on 3 consecutive days with supraoptimal doses of peptide containing T-cell epitopes of *Der p* I, are rendered unresponsive to a parenteral challenge with *Der p* I in CFA (81). The T cells from these tolerant mice have a decreased ability for lymphokine secretion (IL-2, IL-3/GM-CSF, and IFN-γ) after in vitro restimulation with either whole protein or peptide epitopes. The more immunogenic peptides were also more potent in their ability to induce tolerance, with peptides containing the immunodominant determinant (*Der p* I, 110–131) being more effective tolerogens than other peptides containing only the minor T-cell epitopes (p78–100 and p21–49) (81). Similarly, it has been shown that peptides derived from myelin basic protein are able to treat EAE in susceptible mice (80,82). Reproducibly, treating mice with a peptide from the immunodominant region of *Der p* I (111–139) is able to inhibit the response of T cells directed to subdominant epitopes in the antigen (p78–100 and 21–49). If *Der p* I–sensitized mice are treated intranasally with this immunodominant peptide 111–139 for 5 consecutive days, they are subsequently unresponsive to a further parenteral challenge with the allergen and the T cells fail to secrete IL-2 when restimulated in vitro with *Der p* I.

These experiments demonstrate that peptides may be administered to downregulate $CD4^+$ class II MHC-restricted T cells and also to mediate cross-regulation of T cells directed to other epitopes on the antigen (infectious tolerance) perhaps through the release of inhibitory cytokines such as IFN-γ and TGF-β.

Alternatively, a subset of regulatory T cells may be induced that possibly modulate the function of $CD4^+$ T cells.

VIII. CONCLUDING COMMENTS

There is compelling evidence that susceptibility to allergic diseases is multifactorial and polygenic, involving genes additional to those associated with regulation of immune responses. Nevertheless, the ability to manipulate specific components of the immune system that are selectively activated in response to allergen may offer an opportunity to reduce the severity of the clinical symptoms in some atopic individuals. Theoretically, intervention could be targeted at a number of different steps in the immunological pathways activated in the atopic response to allergens. The choice, however, is driven by the need to achieve specificity in any potential immunotherapeutic agent. The induction of $CD4^+$ T cells that overproduce IL-4 would appear to be a fundamental but unique difference in the way that atopics respond to allergen as compared to nonatopics. It appears that a proportion of allergen-reactive T cells are of the IFN-γ dominant phenotype. This adds to the complexity of the mechanisms that underlie selection of the Th2 phenotype in some individuals. The kinetics of secretion may be as important as quantitative differences in cytokine production in deciding whether an allergic inflammatory response is induced. It may be possible, therefore, to establish protective immunity prior to allergen exposure, and this strategy may have potential value in the treatment of seasonal allergy. However, the observation that mice that are deficient in IL-4 fail to make IgE suggests that regulating IL-4 production by manipulating Th2-like allergen-specific T cells or presenting antigen in a way to select Th1-like responses may dampen the symptoms of atopic disease.

The application of molecular genetic techniques, in particular, has been critical in facilitating the molecular characterization and immune recognition of allergenic peptides. This information is required, irrespective of the allergen being investigated, to develop allergen-mediated immunotherapy. The cloning of cDNAs encoding the genes of a number of allergens has provided probes for investigating the antigen specificity of allergen-dependent immune responses in both atopic and nonatopic individuals at the epitope level. They also allow in vivo experimental models to be established. With such reagents it is possible to determine the influence that HLA class II molecules have on allergen-specific IgE production. These points are well illustrated by reviewing the literature on the analysis of the molecular characterization of HDM-derived allergens and immune responses to these proteins. The heterogeneity and complexity of responses to HDM are representative of immune responses to allergens as a whole.

However, although progress on the characterization of allergens, and the analysis of immune responses to them, has been rapid, the mechanisms that direct the T-cell response toward the Th2 pathway and IgE in atopics, and IFN-γ and IL-2 synthesis in nonatopics, are still relatively obscure.

REFERENCES

1. W. R. Thomas, Mite allergens group I–VII. A catalogue of enzymes, *Clin. Exp. Allergy*, *23*: 350 (1993).
2. T. A. Platts-Mills and M. D. Chapman, Dust mites: immunology, allergic disease, and environmental control, *J. Allergy Clin. Immunol.*, *80*: 755 (1987).
3. K. Y. Chua, G. A. Stewart, W. R. Thomas, R. J. Simpson, R. J. Dilworth, T. M. Plozza, and K. J. Turner, Sequence analysis of cDNA coding for a major house dust mite allergen, *Der p* I. Homology with cysteine proteases, *J. Exp. Med.*, *167*: 175 (1988).
4. R. J. Dilworth, K. Y. Chua, and W. R. Thomas, Sequence analysis of cDNA coding for a major house dust mite allergen, *Der f* I, *Clin. Exp. Allergy 21*: 25 (1991).
5. K. Y. Chua, R. J. Dilworth, and W. R. Thomas, Expression of *Dermatophagoides pteronyssinus* allergen, *Der p* II, in *Escherichia coli* and the binding studies with human IgE, *Int. Arch. Allergy Appl. Immunol. 91*: 124 (1990).
6. M. Trudinger, K. Y. Chua, and W. R. Thomas, cDNA encoding the major mite allergen *Der f* II, *Clin. Exp. Allergy*, *21*: 33 (1991).
7. G. A. Stewart, C. H. Bird and P. J. Thompson, Do the group II allergens correspond to lysozyme? *J. Allergy Clin. Immunol.*, *90*: 141 (1992).
8. G. A. Stewart, L. A. Ward, R. J. Simpson, and P. J. Thompson, The group III allergen from the house dust mite *Dermatophagoides pteronyssinus* is a trypsin-like molecule, *Immunology*, *75*: 29 (1992).
9. F. R. Lake, L. D. Ward, P. J. Thompson, and G. A. Stewart, House dust mite–derived amylase: allergenicity and physiological characterization, *J. Allergy Clin. Immunol.*, *87*: 1035 (1991).
10. E. R. Tovey, M. C. Johnson, A. L. Roche, G. S. Gobon, and B. A. Baldo, Cloning and sequencing of cDNA expressing a recombinant house dust mite protein that binds human IgE and corresponds to an important low molecular weight allergen, *J. Exp. Med.*, *179*: 1457 (1989).
11. K-L. Lin, Y-C. Chua, W. R. Thomas, B-L. Chiang, and K-H Hsieh, Characterisation of *Der p* V allergen. cDNA analysis and IgE mediated reactivity to the recombinant protein, *J. Allergy Clin. Immunol.* (in press).
12. H. Yasueda, H. Mita, T. Shida, T. Ando, S. Sugiyama, and H. Yamakawa, Allergens from *Dermatophagoides* mites with chymotryptic activity, *Clin. Exp. Allergy*, *23*: 384 (1993).
13. R. E. O'Hehir, R. D. Garman, J. L. Greenstein, and J. R. Lamb, The specificity and regulation of T cell responsiveness to allergens, *Annu. Rev. Immunol.*, *9*: 76 (1991).
14. R. Halvorsen, V. Bosnes, and E. T. Thorsby, T cell responses to *Dermatophagoides farinae* allergen preparations in allergics and healthy controls, *Int. Arch. Allergy Appl. Immunol.*, *80*: 62 (1986).
15. R. E. O'Hehir, V. Bal, D. Quint, R. Moqbel, A. B. Kay, E. D. Zanders, and J. R. Lamb, An in vitro model of allergen-dependent IgE synthesis by human B

lymphocytes: comparison of the response of an atopic and a non-atopic individual to *Dermatophagoides* spp. (house dust mite), *Immunology*, *66*: 499 (1989).

16. R. E. O'Hehir, A. Verhoef, E, Panagiotopoulou, S. Keswani, J. D. Hayball, W. R. Thomas, and J. R. Lamb. Analysis of human T cell responses to the group II allergen of *Dermatophagoides* spp. (house dust mite): localisation of major antigenic sites, *J. Allergy Clin. Immunol.*, *92*: 105 (1993).
17. R. E. O'Hehir, D. B. Young, A. B. Kay, and J. R. Lamb, Cloned human T lymphocytes reactive with *Dermatophagoides farinae* (house dust mite): a comparison of T- and B-cell antigen recognition, *Immunology*, *62*: 635 (1987).
18. J. R. Lamb, R. E. O'Hehir, and D. B. Young, The use of nitrocellulose immunoblots for the analysis of antigen recognition by T lymphocytes, *J. Immunol. Meth.*, *110*: 1 (1988).
19. R. E. O'Hehir, D. D. Eckels, A. J. Frew, A. B. Kay, and J. R. Lamb, MHC class II restriction specificity of cloned human T lymphocytes reactive with *Dermatophagoides farinae* (house dust mite), *Immunology*, *64*: 627 (1988).
20. A. Verhoef, J. A. Higgins, C. J. Thorpe, S. G. E. Marsh, J. D. Hayball, J. R. Lamb, and R. E. O'Hehir, Clonal analysis of the atopic immune response to the group II allergen of *Dermatophagoides* spp: identification of HLA-DR and -DQ restricted T cell epitopes, *Int. Immunol.* (in press).
21. J. A. Higgins, J. R. Lamb, S. G. E. Marsh, S. Tonks, J. D. Hayball, S. Rosen-Bronson, J. J. Bodmer, and R. E. O'Hehir, Peptide-induced non-responsiveness of HLA-DP restricted human T cells reactive with *Dermatophagoides* spp. (house dust mite), *J. Allergy Clin. Immunol.*, *90*: 749 (1992).
22. H. Yssel, K. E. Johnson, P. V. Schneider, J. Wideman, A. Terr, R. Kasteein, and J. E. de Vries, T cell activation-inducing epitopes of the house dust mite allergen *Der p* I. Proliferation and lymphokine production patterns by *Der p* I–specific $CD4^+$ T cell clones, *J. Immunol.*, *148*: 738 (1992).
23. J. A. Higgins, C. J. Thorpe, J. D. Hayball, R. E. O'Hehir, and J. R. Lamb, Overlapping T cell epitopes in the group I allergen of *Dermatophagoides* spp. (house dust mite) restricted by HLA-DP and HLA-DR class II molecules, *J. Allergy Clin. Immunol.* (in press).
24. R. J. J. van Neerven, W. van'Hof, J. H. Ringrose, H. M. Jansen, R. C. Aalberse, E. A. Wierenga, and M. L. Kapsenberg, T cell epitopes of house dust mite major allergen *Der p* II, *J. Immunol.*, *151*: 2326 (1993).
25. B. Sibbald, Genetics of asthma and atopy: an overview, *Clin. Exp. Allergy*, *21*: S178 (1991).
26. T. A. Platts-Mills and A. L. deWeck, Dust mite allergens and asthma—a worldwide problem, *J. Allergy Clin. Immunol.*, *83*: 416 (1989).
27. S. M. Sears, G. P. Herbison, M. D. Holdaway, C. J. Hewitt, E. R. Flannery, and P. A. Silva, The relative risks of sensitivity to grass pollen, house dust mite and cat dander in the development of childhood asthma, *J. Allergy Clin. Immunol.*, *19*: 419 (1989).
28. H. Atassi and M. Z. Atassi, Antibody recognition of ragweed allergen Ra3: localization of the full profile of the continuous antigenic sites by synthetic overlapping peptides representing the entire protein chain, *Eur. J. Immunol.*, *16*: 229 (1986).

29. W. K. Greene, K. Y. Chua, G. A. Stewart, and W. R. Thomas, Antigenic analysis of group I house dust mite allergens using random fragments of *Der p* I expressed by recombinant DNA libraries. *Int. Arch. Allergy Appl. Immunol.*, *92*: 30 (1990).
30. W. K. Greene and W. R. Thomas, IgE binding structures of the major house dust mite allergen *Der p* I, *Mol. Immunol.*, *29*: 257 (1992).
31. W. K. Greene, J. G. Cyster, K. Y. Chua, R. M. O'Brien, and W. R. Thomas, IgE and IgG binding of peptides expressed from fragments of cDNA encoding the major house dust mite allergen *Der p* I, *J. Immunol.*, *147*: 3768 (1992).
32. E. R. Tovey and B. A. Baldo, Comparison by electroblotting of IgE-binding components in extracts from house dust mite bodies and spent mite culture, *J. Allergy Clin. Immunol.*, *79*: 93 (1987).
33. S. A. Ford and E. A. Tovey, The spectrum of low molecular weight house dust mite (*Dermatophagoides pteronyssinus*) allergens with emphasis on *Der p* II, *Clin. Exp. Allergy*, *20*: 27 (1989).
34. M. D. Chapman, W. M. Sutherland, and M. T. Platts, Recognition of two *Dermatophagoides pteronyssinus*–specific epitopes on antigen P1 by using monoclonal antibodies: binding to each epitope can be inhibited by serum from dust mite-allergic patients, *J. Immunol.*, *133*: 2488 (1984).
35. M. D. Chapman, P. W. Heymann, and M. T. Platts, Epitope mapping of two major inhalant allergens, *Der p* I and *Der f* I, from mites of the genus *Dermatophagoides*, *J. Immunol.*, *139*: 1479 (1987).
36. P. W. Heymann, M. D. Chapman, and M. T. Platts, Antigen *Der f* I from the dust mite *Dermatophagoides farinae*: structural comparison with *Der p* I from *Dermatophagoides pteronyssinus* and epitope specificity of murine IgG and human IgE antibodies. *J. Immunol.*, *137*: 2841 (1986).
37. P. Lind, O. C. Hansen, and N. Horn, The binding of mouse hybridoma and human IgE antibodies to the major fecal allergen, *Der p* I, of *Dermatophagoides pteronyssinus*. Relative binding site location and species specificity studied by solid-phase inhibition assays with radiolabelled antigen, *J. Immunol.*, *140*: 4256 (1988).
38. G. F. Hoyne, T cell recognition during mucosal and systemic responses, Ph.D. thesis, University of Western Australia (1992).
39. K. Y. Chua, C. R. Doyle, R. J. Simpson, K. J. Turner, G. A. Stewart, and W. R. Thomas, Isolation of cDNA coding for the major mite allergen *Der p* II by IgE plaque immunoassay, *Int. Arch. Allergy Appl. Immunol.*, *91*: 118 (1990).
40. E. R. Tovey, S. A. Ford, and B. A. Baldo, Enhanced immunodetection of blotted house dust mite protein allergens on nitrocellulose following blocking with Tween 20, *Electrophoresis*, *10*: 243 (1989).
41. K. Y. Chua, W. K. Greene, P. Kehal, and W. R. Thomas, IgE binding studies with large peptides expressed from *Der p* II cDNA constructs, *Clin. Exp. Allergy*, *21*: 161 (1991).
42. D. G. Marsh, S. H. Hsu, M. Roebber, E. Ehrlich-Kautzky, L. R. Freidhoff, D. A. Meyers, M. K. Pollard, and W. B. Bias. HLA-Dw2: a genetic marker for human immune response to short ragweed pollen allergen Ra5. I. Response resulting primarily from natural antigenic exposure, *J. Exp. Med.*, *155*: 1439 (1982).
43. D. G. Marsh, P. Zwollo, S. K. Huang, B. Ghosh, and A. A. Ansari, Molecular

studies of human immune response to allergens, *Cold Spring Harbor Symp. Quant. Biol.*, *54*: 459 (1990).
44. R. P. Young, B. P. Wordsworth, K. D. Pile, J. W. Dekker, J. I. Bell, W. O. Cookson, and J. M. Hopkin, HLA-DPB1*0401 increased in atopic disease, *Eur. J. Allergy Clin. Immunol.*, *47*: 8 (1992).
45. R. E. O'Hehir, B. Mach, C. Berte, R. Greenlaw, J-M. Tiercy, V. Bal, R. I. Lechler, J. Trowsdale, and J. R. Lamb, Direct evidence for a functional role of HLA-DRB1 and -DRB3 gene products in the recognition of *Dermatophagoides* spp. (house dust mite) by helper T lymphocytes, *Int. Immunol.*, *2*: 885 (1990).
46. T. R. Mosmann, H. Cherwinski, M. W. Bond, M. A. Giedlin, and R. L. Coffman, Two types of murine helper T cell clone. I. Definition according to profiles of lymphokine activities and secreted proteins, *J. Immunol.*, *136*: 2348 (1986).
47. T. R. Mosmann and R. L. Coffman, Th1 and Th2 cells: different patterns of lymphokine secretion lead to different functional properties, *Annu. Rev. Immunol.*, *7*: 145 (1989).
48. L. A. Thompson-Snipes, V. Dnar, M. W. Bond, T. R. Mosmann, K. W. Moore, and D. M. Rennick, Interleukin 10: a novel stimulatory factor for mast cells and their progenitors, *J. Exp. Med.*, *173*: 507 (1991).
49. E. A. Wierenga, M. Snoek, C. de Groot, I. Chretien, J. D. Bos, H. M. Jansen, and M. L. Kapsenberg. Evidence for compartmentalisation of functional subsets of $CD4^+$ T lymphocytes in atopic patients, *J. Immunol.*, *144*: 4651 (1990).
50. X. Paliard, R. de Waal-Malefijt, H. Yssel, D. Blanchard, I. Chretien, J. S. Abrams, J. E. de Vries, and H. Spits, Simultaneous production of interleukin 2, interleukin 4 and interferon gamma by activated human $CD4^+$ and $CD8^+$ T cell clones, *J. Immunol.*, *141*: 849 (1988).
51. E. Maggi, G. F. Del Prete, D. Macchia, P. Parronchi, A. Tiri, I. Chretien, M. Ricci, and S. Romagnani, Profiles of lymphokine activities and helper function for IgE in human T cell clones, *Eur. J. Immunol.*, *18*: 1045 (1988).
52. R. E. O'Hehir, H. Yssel, S. Verma, J. E. de Vries, H. Spits, and J. R. Lamb, Clonal analysis of differential lymphokine production in peptide and superantigen induced T cell anergy, *Int. Immunol.*, *8*: 819 (1991).
53. T. J. Schall, R. E. O'Hehir, D. V. Goeddel, and J. R. Lamb, Uncoupling of cytokine mRNA expression and protein secretion during the induction phase of T cell anergy, *J. Immunol.*, *148*: 381 (1992).
54. E. A. Wierenga, M. Snoek, J. D. Bos, H. M. Jansen, and M. L. Kapsenberg, Comparison of diversity and function of house dust mite–specific T lymphocyte clones from atopic and nonatopic donors, *Eur. J. Immunol.*, *20*: 1519 (1990).
55. R. H. Schwartz, T lymphocyte recognition of antigen in association with gene products of the major histocompatibility complex, *Annu. Rev. Immunol.*, *3*: 237 (1985).
56. J. R. Lamb, B. J. Skidmore, N. Green, J. M. Chiller, and M. Feldmann, Induction of tolerance in influenza-virus immune T lymphocytes clones with synthetic peptides of influenza haemagglutinin, *J. Exp. Med.*, *157*: 1434 (1983).
57. L. Adorini and Z. A. Nagy, Peptide competition for antigen presentation, *Immunol. Today*, *11*: 21 (1990).
58. J. R. Lamb and R. E. O'Hehir, Cellular and molecular mechanisms of T lymphocyte

unresponsiveness induced in vitro. In: *Thymus Update*, Vol. III (Kendall and Ritter, eds.), Harwood Academic Publishers (1989).
59. R. E. O'Hehir and J. R. Lamb, Induction of specific clonal anergy in human T lymphocytes by *Staphylococcus aureus* enterotoxins, *Proc. Natl. Acad. Sci. USA*, *87*: 8884 (1990).
60. R. A. Lake, R. E. O'Hehir, A. Verhoef, and J. R. Lamb, CD28 mRNA decays when activated T cells are functionally anergized with specific peptide, *Int. Immunol.*, *5*: 461 (1993).
61. V. A. Varney, Q. A. Hamid, M. Gaga, S. Ying, M. Jacobson, A. J. Frew, A. B. Kay, and S. R. Durham, Influence of grass pollen immunotherapy on cellular inflitration and cytokine mRNA expression during allergen-induced late-phase cutaneous responses, *J. Clin. Invest.*, *92*: 644 (1993).
62. R. E. O'Hehir, B. A. Aguilar, T. J. Schmidt, S. O. Gollnick, and J. R. Lamb, Functional inactivation of *Dermatophagoides* spp. (house dust mite) reactive human T cell clones, *Clin. Exp. Allergy*, *21*: 209 (1991).
63. C. R. A. Hewitt, J. R. Lamb, J. Hayball, M. Hill, M. J. Owen, and R. E. O'Hehir, MHC independent clonal T cell anergy by direct interaction of *Staphylococcus aureus* enterotoxin B with the T cell antigen receptor, *J. Exp. Med.*, *175*: 1493 (1992).
64. L. R. Wedderburn, R. E. O'Hehir, C. R. A. Hewitt, J. R. Lamb, and M. J. Owen, In vivo clonal dominance and limited T cell receptor usage in human $CD4^+$ T cell recognition of house dust mite allergens, *Proc. Natl. Acad. Sci. USA*, *90*: 8214 (1993).
65. R. Ceppellini, G. Frumento, G. B. Ferrara, R. Tosi, A. Chersi, and B. Pernis, Binding of labelled influenza matrix peptide to HLA-DR in living B-lymphoid cells, *Nature*, *339*: 392 (1989).
66. A. Sette, S. Buus, S. Colon, C. Miles, and H. M. Grey, Structural analysis of peptides capable of binding to more than one Ia antigen, *J. Immunol.*, *142*: 35 (1989).
67. R. Busch, G. Strang, K. Howland, and J. R. Rothbard, Degenerate binding of immunogenic peptides to HLA-DR proteins on B cell surfaces, *Int. Immunol.*, *2*: 443 (1990).
68. R. E. O'Hehir, R. Busch, J. R. Rothbard, and J. R. Lamb, An in vitro model of peptide-mediated immunomodulation of the human T cell response to *Dermatophagoides* spp. (house dust mite), *J. Allergy Clin. Immunol.*, *87*: 1120 (1991).
69. M. M. Davis and P. J. Bjorkman, T-cell antigen receptor genes and T-cell recognition, *Nature*, *334*: 395 (1988).
70. J. White, A. Pullen, K. Choi, P. Marrack, and J. W. Kappler, Antigen recognition of mutant Vβ3+ T cell receptors are consistent with an immunoglobulin-like structure for the receptor, *J. Exp. Med.*, *177*: 119 (1993).
71. A. A. Vandenbark, G. A. Hashim, and H. Offner, Immunization with a synthetic T-cell receptor V-region peptide protects against experimental autoimmune encephalomyelitis, *Nature*, *341*: 541 (1989).
72. A. Gaur, R. Haspel, J. P. Mayer, and C. G. Fathman, Requirement for $CD8^+$ cells in T cell receptor peptide-induced clonal unresponsiveness, *Science*, *259*: 91 (1993).
73. H. Acha-Orbea, D. J. Mitchell, L. Timmerman, et al., Limited heterogeneity of T

cell receptors from lymphocytes mediating autoimmune encephalomyelitis allows specific immune intervention, *Cell*, *54*: 263 (1988).

74. D. P. Gold, M. Vainiene, B. Celnik, et al., Characterization of the immune response to a secondary encephalitogenic epitope of basic protein in Lewis rats. II. Biased T cell receptor Vβ expression predominates in spinal cord infiltrating T cells, *J. Immunol.*, *148*: 1712 (1992).
75. M. D. Howell, S. T. Winters, T. Olee, H. C. Powell, D. J. Carlo, and S. W. Brostoff, Vaccination against experimental allergic encephalomyelitis with T cell receptor peptides, *Science*, *246*: 668 (1989).
76. H. Offner, G. A. Hashim, and A. A. Vandenbark, T cell receptor peptide therapy triggers autoregulation of experimental encephalomyelitis, *Science*, *251*: 430 (1991).
77. E. R. Jarman, C. M. Hawrylowicz, E. Panagiotopoulou, R. E. O'Hehir, and J. R. Lamb, Inhibition of human T cell responses to house dust mite allergens by a T cell receptor peptide (submitted).
78. P. G. Holt and C. McMenamin, Defence against allergic sensitization in the healthy lung: the role of inhalation tolerance, *Clin. Exp. Allergy*, *19*: 255 (1989).
79. A. M. Mowat, The regulation of immune responses to dietary antigens, *Immunol. Today*, *8*: 93 (1987).
80. A. Gaur, B. Wiers, A. Liu, J. B. Rothbard, and C. G. Fathman, Amelioration of autoimmune encephalomyelitis by myelin basic protein synthetic peptide-induced anergy, *Science*, *258*: 1491 (1992).
81. G. F. Hoyne, R. E. O'Hehir, D. C. Wraith, W. R. Thomas, and J. R. Lamb, *J. Exp. Med.*, *178*: 1783 (1993).
82. B. Metzler and D. C. Wraith, Inhibition of experimental autoimmune encephalomyelitis by inhalation but not oral administration of the encephalitogenic peptide: influence of MHC binding affinity, *Int. Immunol.*, *9*: 1159 (1993).

Index

About the Editors

ARNOLD I. LEVINSON is an Associate Professor of Medicine, an Associate Professor of Neurology in Medicine, and a member of both the Immunology Graduate Group and the University of Pennsylvania Cancer Center at the University of Pennsylvania School of Medicine, Philadelphia. An editorial board member of the *Journal of Clinical Immunology,* Dr. Levinson is the author or coauthor of over 100 professional papers, book chapters, and review articles. He is a Director of the American Board of Allergy and Immunology and a member of the American Society for Clinical Investigation, the American Association of Immunologists, the Clinical Immunology Society, and the American Academy of Allergy and Clinical Immunology, in which he serves as Chairman of its Basic and Clinical Immunology intersection. Dr. Levinson received the B.S. degree (1965) in psychology from the University of Maryland, College Park, and the M.D. degree (1969) from the University of Maryland, Baltimore.

YVONNE PATERSON is a Professor of Microbiology and a member of the University of Pennsylvania Cancer Center at the University of Pennsylvania School of Medicine, Philadelphia. A member of the National Institutes of Health Study Section on Allergy and Immunology, Dr. Paterson is the author or coauthor of over 60 professional papers. She is a member of the American Association of Immunologists, the American Society for Biochemistry and Molecular Biology, and the American Peptide Society, among others. Dr. Paterson received the B.Sc. Tech. (1963) and the M.Sc. (1966) degrees in biochemistry from the University of Manchester Institute of Science and Technology, United Kingdom, the B.A. degree (1968) in mathematics and philosophy from the Australian National University, Canberra, and the Ph.D. degree (1978) in biochemistry from the University of Melbourne, Australia.